Atlas of Animal Anatomy and Histology

Péter Lőw • Kinga Molnár • György Kriska

Atlas of Animal Anatomy and Histology

Springer

Péter Lőw
Department of Anatomy,
Cell and Developmental Biology
Institute of Biology
Eötvös Loránd University
Budapest
Hungary

Kinga Molnár
Department of Anatomy,
Cell and Developmental Biology
Institute of Biology
Eötvös Loránd University
Budapest
Hungary

György Kriska
Group for Methodology in Biology Teaching
Institute of Biology
Eötvös Loránd University
Budapest
Hungary

Danube Research Institute
Centre for Ecological Research
Hungarian Academy of Sciences
Budapest
Hungary

General professional reviewer

Zsolt Kovács
Department of Zoology
Faculty of Sciences and Technology
University of West Hungary
Szombathely
Hungary

Additional material to this book can be downloaded from http://extras.springer.com

ISBN 978-3-319-25170-7 ISBN 978-3-319-25172-1 (eBook)
DOI 10.1007/978-3-319-25172-1

Library of Congress Control Number: 2016931371

Springer Cham Heidelberg New York Dordrecht London

Printed on acid-free paper

Springer International Publishing AG Switzerland is part of Springer Science+Business Media (www.springer.com)

Preface

The purpose of this book is to provide an introduction to comparative anatomy and histology for biology undergraduates and for all those who are interested in the internal structure of animals. The information is presented in the form of colour photographs of step-by-step dissection stages integrated with histological sections of actual organs. A specialty of this atlas is that it contains only high-quality, accurate, and attractive photographs, not idealised line drawings. Dissection plays an important part in understanding the anatomy of an animal, and this book has been designed to make full use of the wealth of information made available through dissection. The accompanying text aims to outline the evolutionary and functional aspects of the anatomy revealed in the photographs. Our book encourages and facilitates active and self-directed learning by the students so that instructors can teach more effectively and efficiently. This manual emphasises dissection procedures that preserve as many structures as possible for later review of the entire specimens. Every effort has been made to give clear, lucid descriptions and instructions, and enough background material has been included to create interest in and understanding of the subject matter.

The animals dissected in this book have been chosen as representative examples of six invertebrate phyla and four classes of vertebrates. This book offers step-by-step illustrations and instructions for dissecting a roundworm, earthworm, snail, mussel, crayfish, cockroach, crucian, frog, chicken, and rat. The types included are commonly studied in undergraduate zoology courses. They can be used also as a guide to dissection of other animals in the same group. Dissections range from beginning to advanced and discuss the digestive, circulatory, respiratory, excretory, reproductive, and nervous systems. Skeletal material of vertebrate animals is also included to show the supporting framework of the body and its development during evolution.

Another valuable aspect of this atlas is that it features large-size, full-colour histological micrographs, with labels and legends that draw attention to details of microanatomy of the most important organs. The histological descriptions follow the anatomical pictures and explanation of an actual organ, and they are highlighted with a coloured background. In this way, students can correlate microscopic structures with the gross composition. Clear histological explanations give details of how tissues are structured and how they work. Students will learn to recognise different types of tissues easily. The detailed photographs enable the reader to gather microanatomical knowledge even in the lack of prepared light microscopic sections or microscopic facilities.

The digital annex of the book includes slide-shows and interactive tests that can be used to check the knowledge. A special item of the software is a stereoscopic (3D) application enabling to visualize three-dimensional (anaglyph) pictures on a monitor or by a projector. Anaglyph pictures should be viewed through red-cyan glasses. The slide-shows are also available on-line at http://bszm.elte.hu/anatomy/, optimized for mobile browsers.

Budapest, Hungary

Péter Lőw
Kinga Molnár
György Kriska

Points for Successful Dissection

It is always important to perform a dissection in an appropriate lab under the guidance of an experienced instructor. Do not do anything uncertainly; wait for specific instructions in the lab. Dissection is both a skill and an art. A good dissection requires time and patience. Always prepare for a dissection in advance, learn the structures you want to find, and work deliberately. Make small cuts and do not remove a piece of tissue unless you know what it is. Each dissection chapter in this book includes background information about the sample animal, availability and proper, species-specific anaesthesia of the animal.

The dissections should be performed in a wax-bottomed dish using small pins for attachment and display. Most of the structures described in this dissection guide will be best viewed with the aid of a stereomicroscope (dissecting microscope) or a hand lens. Dissecting tools will be used to open the body of the animal and unfold the structures. Learn the techniques of working with these instruments. The tools are very sharp, use them properly and be careful not to injure yourself. The development of good abilities at dissection desponds upon practice and, above all, patience. A comprehensive dissecting kit (Fig. 1) includes the following tools for almost all types of knacks:

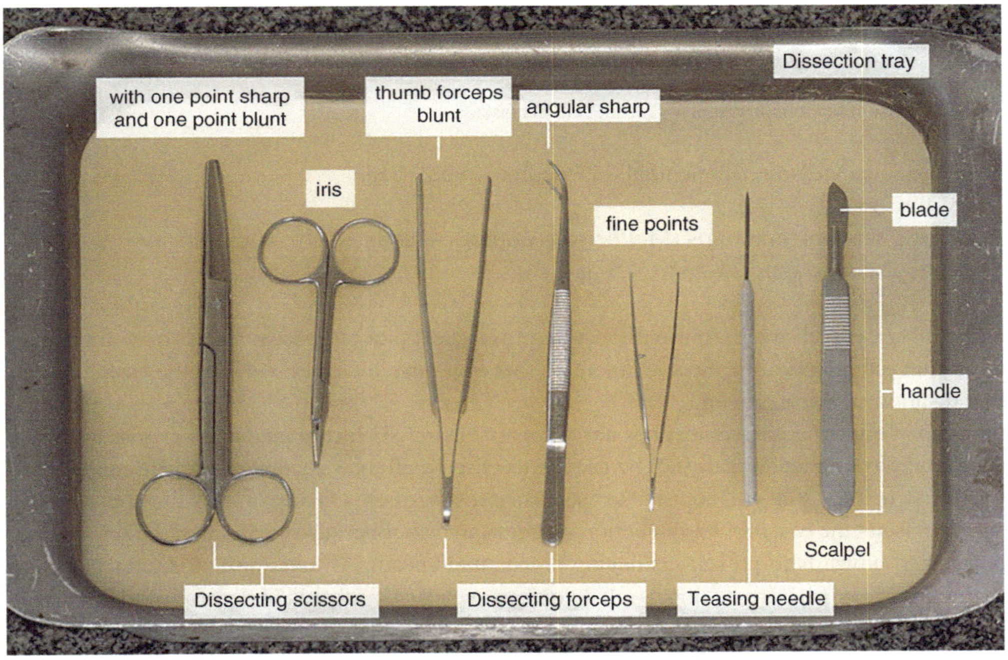

Fig. 1 A comprehensive dissecting kit arranged in a wax-bottomed dissecting dish

Dissecting scissors with one point sharp and one point blunt, 5.5 in.
Dissecting scissors, iris, 4.5 in.
Thumb forceps blunt, 5.5 in.
Forceps angular sharp, 5 in.

Forceps fine points, 4.5 in.
Teasing needle straight with metal chuck
Scalpel handle No. 4
Scalpel blade No. 22

Although these instruments serve the requirements of nearly all kinds of dissecting situations for special tasks and fine, elaborate work, we recommend some further tools (Fig. 2):

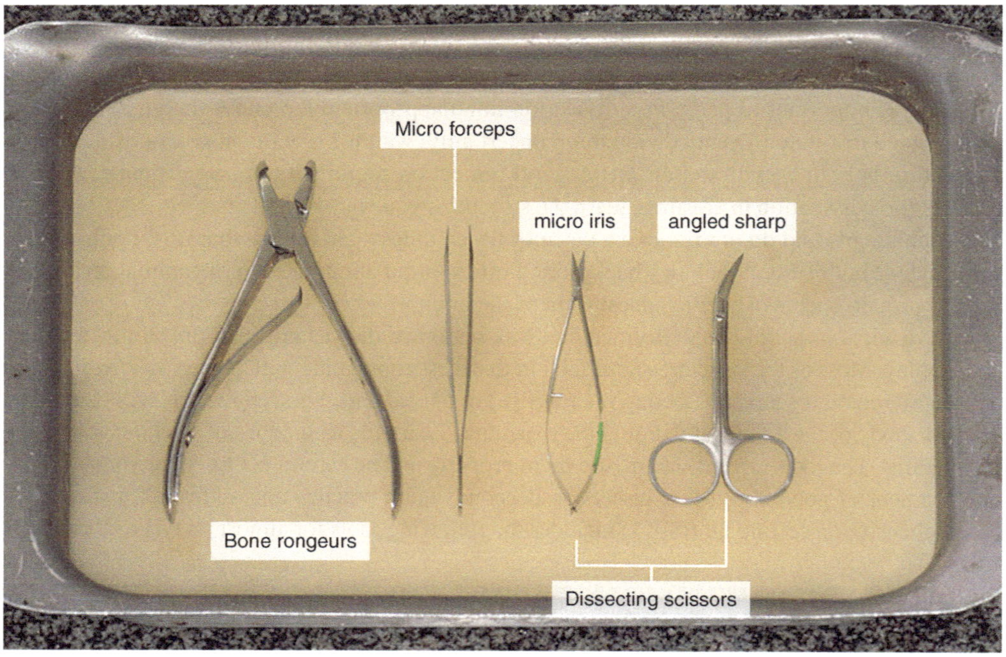

Fig. 2 Special dissection instruments for meticulous tasks

Bone rongeurs (Adson, Blumenthal, or Friedman type), 6 in.
Micro forceps
Dissecting scissors, micro iris (McPherson-Vannas), straight, sharp, 4.5 in.
Dissecting scissors, angled sharp, 4.5 in.

Finally, it is well worth to use dissecting pins (insect pins) to position parts as you proceed with your examination of the specimen, so that you have a clearer view of the structure and organisation of the organism.

Keep in mind that dissecting does not mean "to cut up"; in fact, it means "to expose to view". Careful dissecting techniques will be needed to observe all the structures and their connections to other structures. You will not need to use a scalpel very often. On the contrary to popular belief, a scalpel is not the best tool for dissection. Scissors are better because the point of the scissors can be pointed upwards to prevent damaging organs underneath. Always raise structures to be cut with your forceps before cutting, so that you can see exactly what is underneath and where the incision should be made. Never cut more than is absolutely necessary to expose a part. Sometimes the so-called blunt dissection is the most appropriate when you only tear connective tissue structures with forceps to reveal an underlying compact organ and do not cut anything.

When completed, clean up your dissection. Dispose of your materials according to the directions from your instructor. Pour your excess liquid into the sink and wrap the body parts in a paper towel before throwing them in the carcass container. Never dispose the body parts into ordinary communal waste. Immediately after use, rinse instruments under warm or cool running water to remove all blood, body fluids, and tissue. Dried soils may damage the instrument surface and make cleaning very difficult. Do not use hot water as this will coagulate proteinous substances. Clean up your work area and wash your hands before leaving the lab.

Histological Methods

Histological Sections

Histology is the study of the microscopic anatomy of cells and tissues of animals (or plants). During the routine procedure, the organs are fixed to prevent decay and embedded in paraffin (paraplast) to give support for cutting very thin (2–5 μm thick) sections. The sections are placed onto microscope slides and stained with histological stains, then covered with a coverslip and mounting medium for preservation. Histological slides are examined with light microscope.

Histological Stains

Histological stains are used to increase the typically minor differences in light refraction of biological samples. The procedure is based on the variances in binding of histological stains by tissue and cell components.

HE (haematoxylin – eosin) **stain:** It provides a general overview – haematoxylin stains the nucleic acids, and eosin stains the cytosol and the extracellular matrix.
Azan (azocarmine – aniline blue) **stain:** It provides a general overview – azocarmine stains the cell nucleus and the cytoplasm, aniline blue stains the connective tissue matrix and fibres and some mucous secret.
PAS (Periodic acid-Schiff reaction) **stain:** This reaction is used to detect structures containing a high proportion of carbohydrate macromolecules (glycoproteins, glycolipids, and polysaccharides). The reaction gives a purple-magenta colour typically in mucus gland cells, connective tissue, and basement membrane.

Semithin Sections

Plastic (epoxy resin) embedding is commonly used in the preparation of material for electron microscopy. Semithin sections (0.8–1 μm) are cut using glass knives. The sections are stained with toluidine blue and examined using a light microscope.

Important Technical Terms

Here we explain compass points of anatomy. Many of these are taken from Latin or Greek languages, and each has a very specific meaning. It is really important to understand the basic terms, which are used throughout the anatomical and histological descriptions.

Frontal plane: It is a vertical plane at right angle to median plane. If you draw a line from one ear to another from above the head and then divide the whole body along this line, the plane formed will be frontal plane. It is also known as coronal plane.
Median or mid-sagittal plane: This is the plane which divides the body into equal right and left halves.
Oblique plane: Any plane other than the above described planes will be oblique plane.
Sagittal plane: It is any plane parallel to the median plane. This plane divides the body into unequal right and left halves.
Transverse plane: It is the horizontal plane of the body. It is perpendicular to both frontal and median planes.

Directional terms describe the positions of structures relative to other structures or locations in the body:

Anterior: Towards the head end (e.g. the oesophagus is located anterior to the stomach)
Caudal: Away from the head, towards the tail end of the body
Cranial: Towards the head end of the body
Distal: Away from or farthest from the middle line of an organism or from the point of attachment (e.g. the hand is located at the distal end of the forearm)
Dorsal: Towards the back or upper part of the animal
Inferior: Lower
Lateral: Situated at the side away from the midline of the body (e.g. the little toe is located at the lateral side of the foot)
Longitudinal: Lengthwise; along the length of the body
Medial: Towards the midline of the body (e.g. the middle toe is located at the medial side of the foot)
Median: Along the middle of the long axis
Periferal: Referring to parts away from the centre
Posterior: Facing towards the tail end (e.g. the pelvic girdle is located on the posterior end of the backbone)
Proximal: Towards or nearest to the middle line of the organism or the point of origin of a part (e.g. the proximal end of the femur joins with the pelvic girdle)
Sagittal: along or parallel with the middle plane of the body
Superficial: On or near the surface
Superior: Upper
Transverse: Lying across or between or at right angles to the longitudinal axis
Ventral: Towards the abdominal surface

Acknowledgements

The authors wish to thank the many people who helped in the preparation of this book. In particular, thanks are due to Dr. György Csikós, Viktor Kis, Sarolta Pálfia, and Zsolt Pálfia of the Department of Anatomy, Cell and Developmental Biology, Eötvös University, Budapest, Hungary for their prepared specimens. We are indebted not only to them but also to Dr. Zsolt Kovács (Department of Zoology, Faculty of Sciences and Technology, University of West Hungary, Szombathely, Hungary) who revised the manuscript. Any inaccuracies remaining are, of course, our own responsibility. We are grateful for the excellent technical assistance to Eszter Papp. We thank Monika Truszka for the first-rate histological work from embedding to outstanding sections and brilliant staining. We thank András Barta for creating the on-line version of the slide-shows. We would also like to thank the cheerful and hardworking production team at Springer Verlag whose encouragement sustained us on several occasions.

P. Lőw
K. Molnár
G. Kriska

Contents

Part I

Invertebrates

Examination of a Hydra

1

The cnidarian body consists of a central blind sac, the *coelenteron* (gastrovascular cavity), enclosed by a body wall comprising two epithelia, the outer *epidermis* and the inner *gastrodermis* (Fig. 1.1). A gelatinous connective tissue layer, the *mesolamella* (mesogloea), lies between the two epithelia. The *mouth* opens at one end of the coelenteron and marks the oral end. The mouth is at the tip of a process, the *hypostome* that elevates it above the oral surface. The opposite pole is the aboral end forming the *pedal disc*. The imaginary line connecting the oral and aboral poles is the axis of symmetry around which the radial symmetry of the body is organised. The mouth is usually surrounded by a ring of hollow *tentacles*, which are well endowed with *cnidocyte* batteries (white spots in Fig. 1.1).

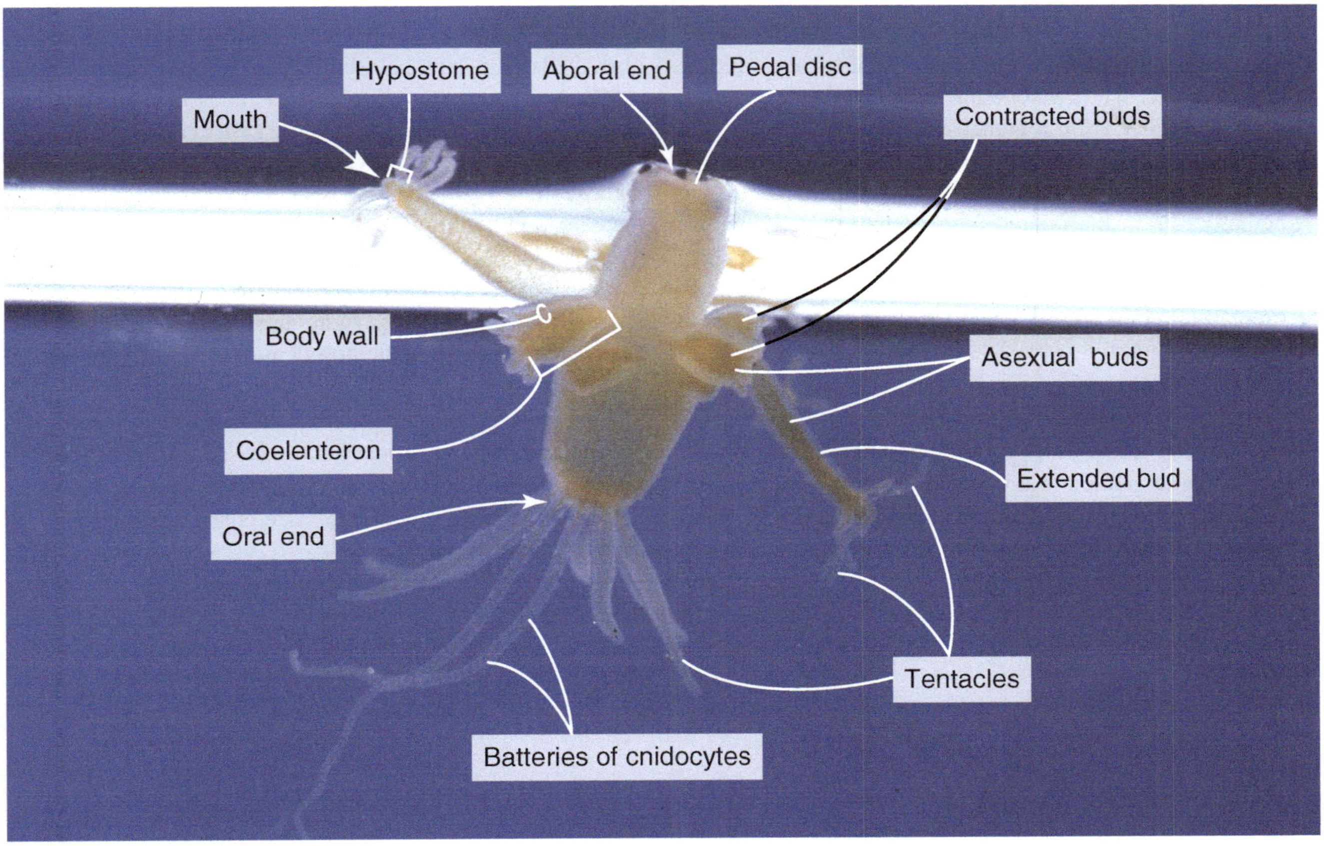

Fig. 1.1 A living hydra (*Hydra vulgaris*) attached to the water surface by its pedal disc. There are at least 11 buds on the parent animal, the size of which is 3 mm in this half way extended state

P. Lőw et al., *Atlas of Animal Anatomy and Histology*, DOI 10.1007/978-3-319-25172-1_1

Cnidarians are chiefly marine but the well-known *Hydra* is an exception. The *Hydra* is found in pools, quiet streams and spring ponds, usually on the underside of the leaves of aquatic vegetation. All cnidarians are carnivores feeding on live prey which they usually capture using tentacles armed with cnidocytes. Digestion occurs in the coelenteron which is typically equipped with ciliated canals for distribution of partly digested food. Cnidarians are ammonotelic, and diffusion across the body and tentacle surface eliminates the ammonia from the body. Gas exchange is across the general body surface. The nervous system is a plexus of basiepithelial neurons serving sensory and motor systems. Most cnidarians are gonochoric. Asexual reproduction is via asexual *buds* that form on the parent animal. They are small hydras that will separate from the parent and adopt an independent existence. *Hydra* does not form colonies. *Hydra vulgaris* is a freshwater species in which the medusoid generation is absent and the polyps are solitary. Polyps are small, about 1–5 mm in length, when contracted and up to 15 mm elongated. Slides can be used to supplement, or if necessary replace, the study of living specimens.

The body wall of hydra is composed of three layers: the outer epidermis, the inner gastrodermis and the middle mesolamella (ML) (Fig. 1.2, left). The two epithelial layers are formed by *epitheliomuscular* (myoepithelial) *cells* (EMCs) together with some additional cell types. All of them rest on a basement membrane attached to both sides of the *mesolamella* (mesogloea), which give a support for them. The visible thickness of epithelial layers in the section depends on the contraction state of the animal at the time of fixation. Epidermis synthesises a thin, protective cuticle, which detaches from epidermis during the histological procedure (Fig. 1.2, left). The gastrodermis surrounds a *gastrovascular cavity* (GVC). On a higher magnification, several cell types become identifiable in both layers (Fig. 1.2, right).

The *epidermis* contains dark particles: these are the nematocysts in the characteristic stinging cells of cnidarians, the *cnidoblasts* (CB, nematoblasts) and cnidocytes (nematocytes) (Fig. 1.2, right). The *cnidocyst* (nematocyst) is an explosive organelle, which, upon proper stimulation, inverts and ejects a slender, often barbed and toxic filament in the direction of prey or predator. Cnidoblasts originate from *interstitial cells* (ICs) which are in basal position and have large, round, euchromatic nuclei. They are mitotically active stem cells and give rise to cnidoblasts and neurons. *Neurons* have three main types: ganglion cells are connected into a network running on the basal membrane to form a diffuse nervous system. It contains neuroendocrine cells as well. Sensory cells reach the surface for taking up stimuli. Nerve cell types are not distinguishable in our section. Epitheliomuscular cells have an independent self-renewal population. Their basal portions contain myofibrillary bundles running parallel with longitudinal axis of the body. Their central portions contain the nucleus.

The *gastrodermis* is made of epitheliomuscular and gland cells. The latter have two types. *Mucous gland cells* (MGCs) contain empty-looking vacuoles in their apical domain and secret mucous into the gastrovascular cavity. *Enzymatic gland cells* (EGCs) synthesise enzymes and their secretory granules become visible if stained. The flagellated epitheliomuscular cells have digestive (nutritive) function (DC): they phagocytose food particles partly digested extracellularly and finish their digestion intracellularly (Fig. 1.2, right, arrowhead). The undigested remnants are exocytosed into the gastrovascular cavity.

Lots of empty spaces with various sizes can be seen in the epidermal and gastrodermal layers. These are intracellular vacuoles of epitheliomuscular cells, which serve as buffer spaces for the enormous size changes during contraction and elongation cycle.

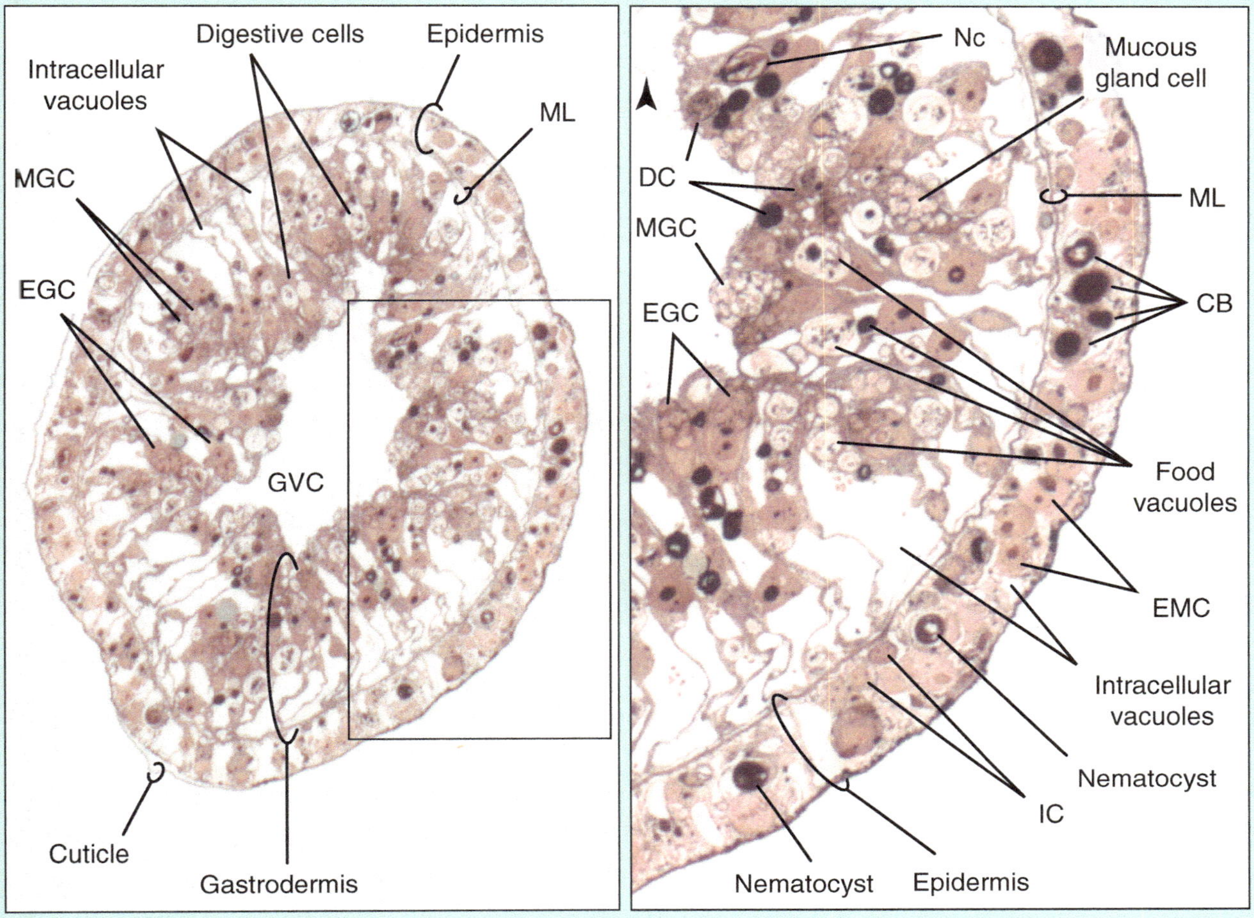

Fig. 1.2 Whole-body cross section of a hydra (*Pelmatohydra oligactis*, silver staining). The *right panel* reveals, at a higher magnification, the cell types of the body wall from the area enclosed in the *left panel*. *Arrowhead* flaggella of epitheliomuscular cells, *CB* cnidoblasts, *DC* digestive (nutritive) cells, *EGC* enzymatic gland cells, *EMC* epitheliomuscular cells, *GVC* gastrovascular cavity, *IC* interstitial cells, *MGC* mucous gland cells, *ML* mesolamella, *Nc* nematocyst in the gastrodermis, that was extruded from the epidermis during the prey catch, but did not open, so the filament is well visible inside (Section is made by Sarolta Pálfia; courtesy of Zsolt Pálfia)

Examination of a Planarian

2

The simplest animals that are bilaterally symmetrical and triploblastic (having three germ layers) are the flatworms (Platyhelminthes). Flatworms have no body cavity (acoelomate) and lack an anus. One of their groups is the freshwater triclads (*Tricladida*), or planarians. They are large free-living flatworms which are commonly found on the underside of stones or submerged leaves or sticks in freshwater springs, ponds, and streams. Planarians are mobile and use cilia on their ventral surface to glide over surfaces (Fig. 2.1).

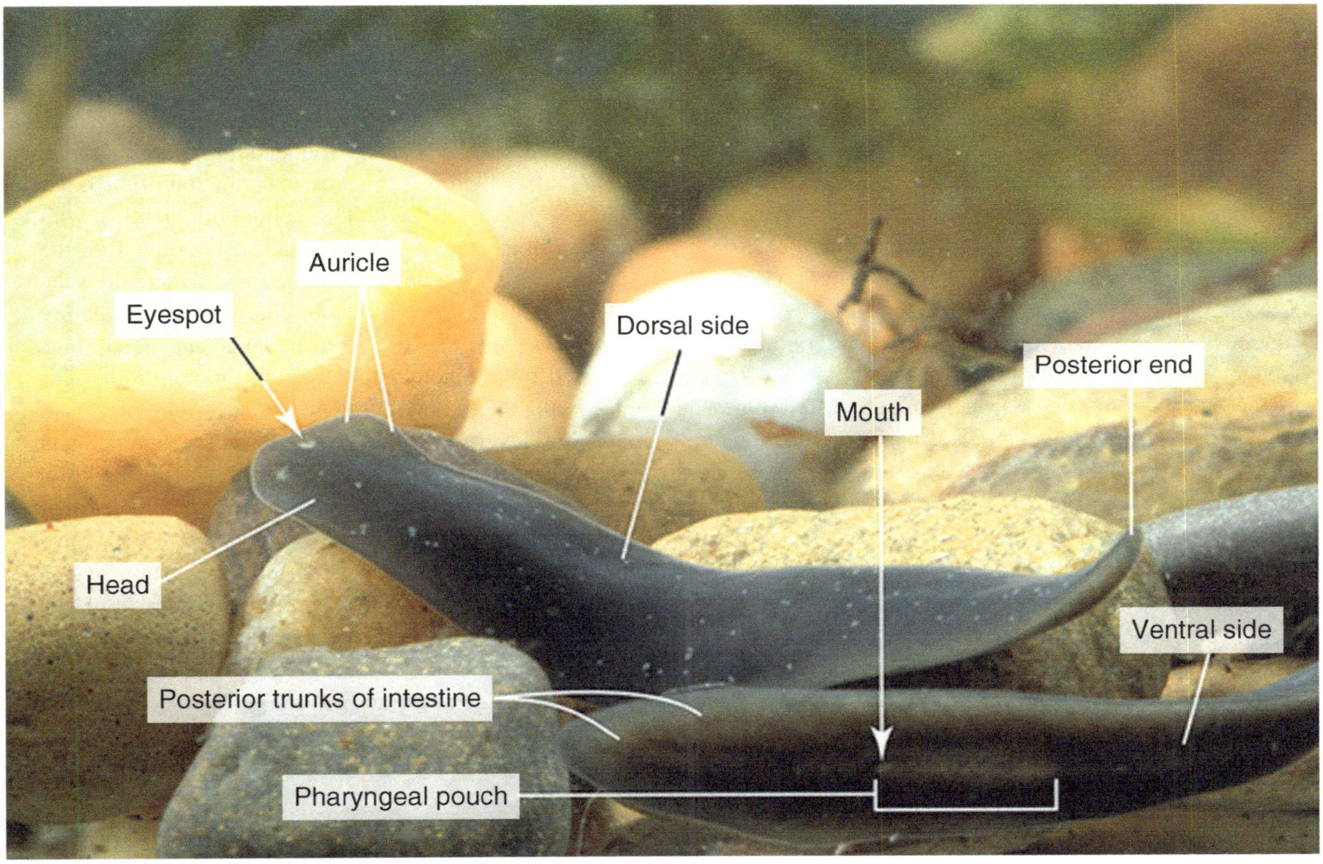

Fig. 2.1 Two living black planarians (*Dugesia lugubris*)

P. Lőw et al., *Atlas of Animal Anatomy and Histology*, DOI 10.1007/978-3-319-25172-1_2

Planarians can have different pigmentation such as light brown, dark brown, black or white. The characteristic planarian triangular *head* has two *auricles* and two light-sensing *eyespots*. Planarians are predators and scavengers and eat live or dead animals using their muscular retractable *pharynx* which can extend out of the *mouth* opening on the ventral side up to half of their body length. Planarians have very simple organ systems: The digestive system consists of a mouth, a pharynx and a three-branched *intestine* which makes planarians referred to as triclads. The digestion occurs in the intestine after the food has been sucked through the pharynx. The mouth is the only opening in the gut, so undigested food must also exit the body through the mouth. This highly branching gut system is called *gastrovascular system* as it unites the functions of the digestive and circulatory systems. Planarians do not have a skeletal, circulatory or respiratory system. Oxygen and carbon dioxide are transported into and out of individual cells by simple diffusion. The nervous system is made of a small brain beneath the eyes (the *cerebral ganglia*) which is connected to two long parallel *ventral nerve cords* running along the body to the tail. The two cords are connected by transversal nerves. The *auricles* contain chemoreceptors that are used to find food. The eyespots are connected to the cerebral ganglia and are used to detect and avoid sunlight (negative phototaxis) but do not detect images.

The *body wall* of a planarian is formed by epidermis and three muscular layers. Epidermis on the dorsal and ventral sides shows some differences. Most frequent cell type is ciliated in the ventral epidermis (VE), whereas the dorsal epidermis (DE) seems to be non-ciliated (Fig. 2.2, upper left, lower left, lower right).

The *epidermis* (E) on the surface is abundant in endo- and subepithelial (parenchymal) unicellular gland cells. *Endoepithelial glands* containing mucous granules (MG) seem to be swollen, and their vacuoles become empty during the microtechnical procedure, so these cells can be easily identified (Fig. 2.2, lower left). They secrete viscous mucus to create a thick coating on the surface. *Parenchymal gland cells* (PG) have a long neck region passing through epidermis to reach the surface. They produce *rhabdites* (RB), which are secretory granules with rod or spherical shape (Fig. 2.2, lower left and right). They bud from a Golgi-derived vacuole. Many types of rhabdites have been documented, but their functions are not yet clarified: they may serve as protective and repellent substances or as territorial markers. They are made of proteinaceous material featured by acidophil staining. There is a characteristic gland strip on the lateral "margin" of the animal called *marginal* (*adhesive*) *glands* (AG) (Fig. 2.2, lower right). Here groups of subepithelial (parenchymal) glands secret adhesive and releasing material onto the surface to adhere and release from a substrate several times within a second.

Musculature is composed of three layers. Outer layer is formed by circular muscle fibres (CML), inner layer contains longitudinal muscle fibres (LML) and there is an intermediate layer of radial (diagonal) muscle fibres (RM) between them. Several dorsoventral muscle bundles (DVM) can be seen between the dorsal and ventral side – they maintain the flattened shape of the animal. Body cavity is occupied by parenchymal tissue (P) embedding mid-gut branches and nervous and genital system. *Pharynx* in resting state is founded in the *pharyngeal pouch* (PP), which is formed by invagination of the outer surface – so it is lined with thin epithelium identical with the epidermis (Fig. 2.2, upper left). It is ciliated on the pharyngeal surface, but non-ciliated and flattened on the surface of the pharyngeal pouch. Pharyngeal musculature is well developed and ordered in outer and inner rings separated by parenchyma. Both rings contain longitudinal, circular and radial muscle layers. *Mid-gut* gives three main and several smaller branches in the parenchyma (Fig. 2.2, upper left, asterisks). Its wall is composed of a tall epithelial layer with gland cells (GC) secreting enzymes and digestive (nutritive muscular) cells (DC) for phagocytosing partially digested food. Digestion begins extracellularly and it is completed intracellularly. Indigestive remnants are exocytosed into mid-gut lumen. Section profiles of the *ventral nerve cord* (VNC) appear in the ventral side of the animal as lighter tissue islands in parenchyma (Fig. 2.2, upper left).

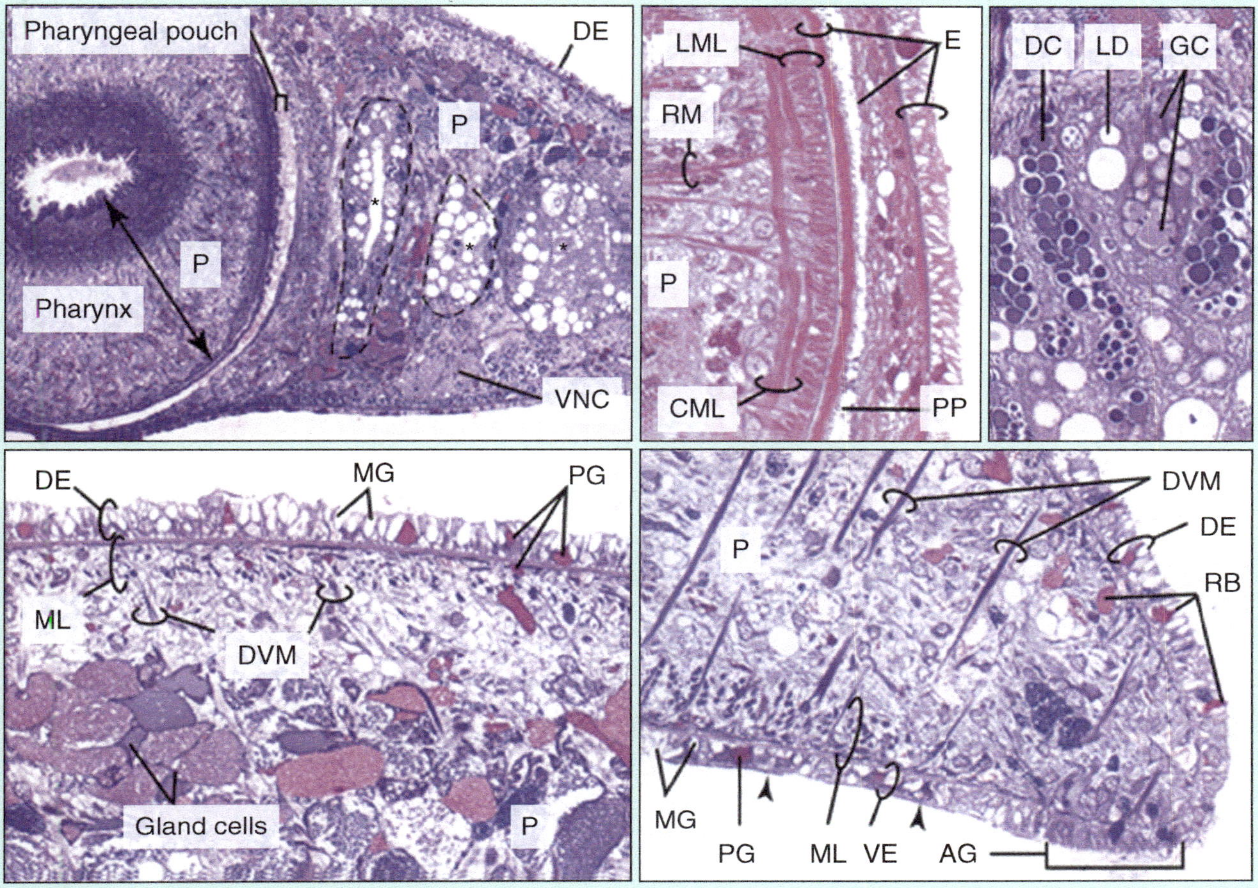

Fig. 2.2 Whole-body histological cross section of a planarian (semi-thin section, cresyl violet staining). *Asterisks* mid-gut branches, *arrowheads* cilia, *AG* adhesive (marginal) glands, *CML* circular muscle layer, *DC* digestive (nutritive muscular) cell, *DE* dorsal epidermis, *DVM* dorsoventral muscles, *E* epidermis, *GC* gland cell, *LD* lipid droplet, *LML* longitudinal muscle layer, *MG* mucous granules, *ML* muscle layers, *P* parenchyma, *PG* protrusions of subepithelial gland cells, *PP* pharyngeal pouch, *RB* rhabdites, *RM* radial bundles of muscles in the pharynx, *VE* ventral epidermis, *VNC* ventral nerve cord

The *eye of planarians* is a cup-shaped organ immersed in the parenchyma (Fig. 2.3). The cup is made by pigment cells forming an epithelial layer. The *pigment cell cup* (PC) has an opening which is oriented laterally. Light may enter the cup only through this hole because pigment cells absorb the light coming from any other directions. Orientation of photoreceptive projections is the opposite of arrival of the light – this eye is an inverse type. On its morphology this eye is suitable for sensing the direction and intensity of light for the purpose of choosing the shady places (planarians show negative phototaxis). Nerve projections of sensory cells enter the *cerebral ganglion* (CG).

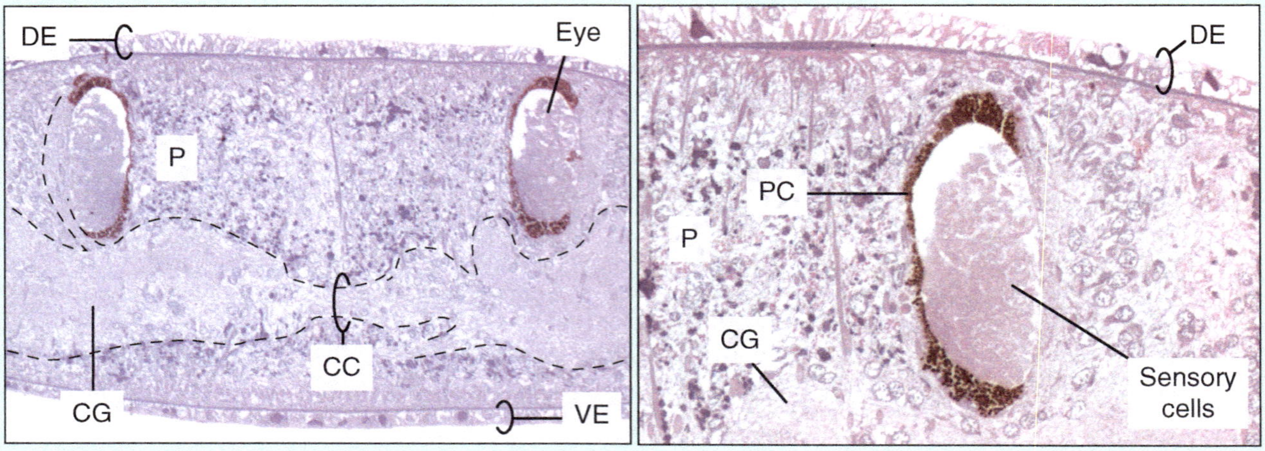

Fig. 2.3 Histological section of the planarian eye (semi-thin section, HE staining). *CC* cerebral commissure, *CG* cerebral ganglion, *DE* dorsal epidermis, *P* parenchyma, *PC* pigment cell cup, *VE* ventral epidermis

3 Dissection of a Roundworm (*Ascaris suum*)

- *Availability:* Specimens preserved in alcohol are available at biological supply companies. Cross section slides are also offered commercially. Ascaris eggs are extremely resistant to chemical treatment. Although it is unlikely, some eggs may survive immersion in preservatives for short periods. To avoid ascariasis (a disease caused by the parasitic roundworm; see life cycle at the end of the chapter), you should keep your hands away from your mouth and nose while performing this dissection and wash your hands afterwards. Put on a laboratory coat and make sure you handle all specimens with rubber gloves.

The roundworms (nematodes) are an extensive group with worldwide distribution. They inhabit terrestrial, marine and freshwater environments and are found in almost all moist habitats. The taxon includes numerous plant and animal parasites, many of which are of medical or agricultural importance, but the majority are free living (non-parasitic). Most roundworms are long, slender and almost featureless externally, tapered at both ends, and round in cross section. *Caenorhabditis elegans* is the most extensively studied roundworm. It is a free-living nematode, 1 mm in length and transparent; it can be cultured in a laboratory. It is an organism where it is possible to identify every cell as it develops and to trace its lineage. The genome of *C. elegans* was the first invertebrate genome to be sequenced. Genes controlling programmed cell death were also discovered in *C. elegans*. For laboratory studies of roundworm anatomy, however, *Ascaris suum*, the pork roundworm (Fig. 3.1), is convenient because of its large size (lengths up to 40 cm) and availability.

P. Lőw et al., *Atlas of Animal Anatomy and Histology*, DOI 10.1007/978-3-319-25172-1_3

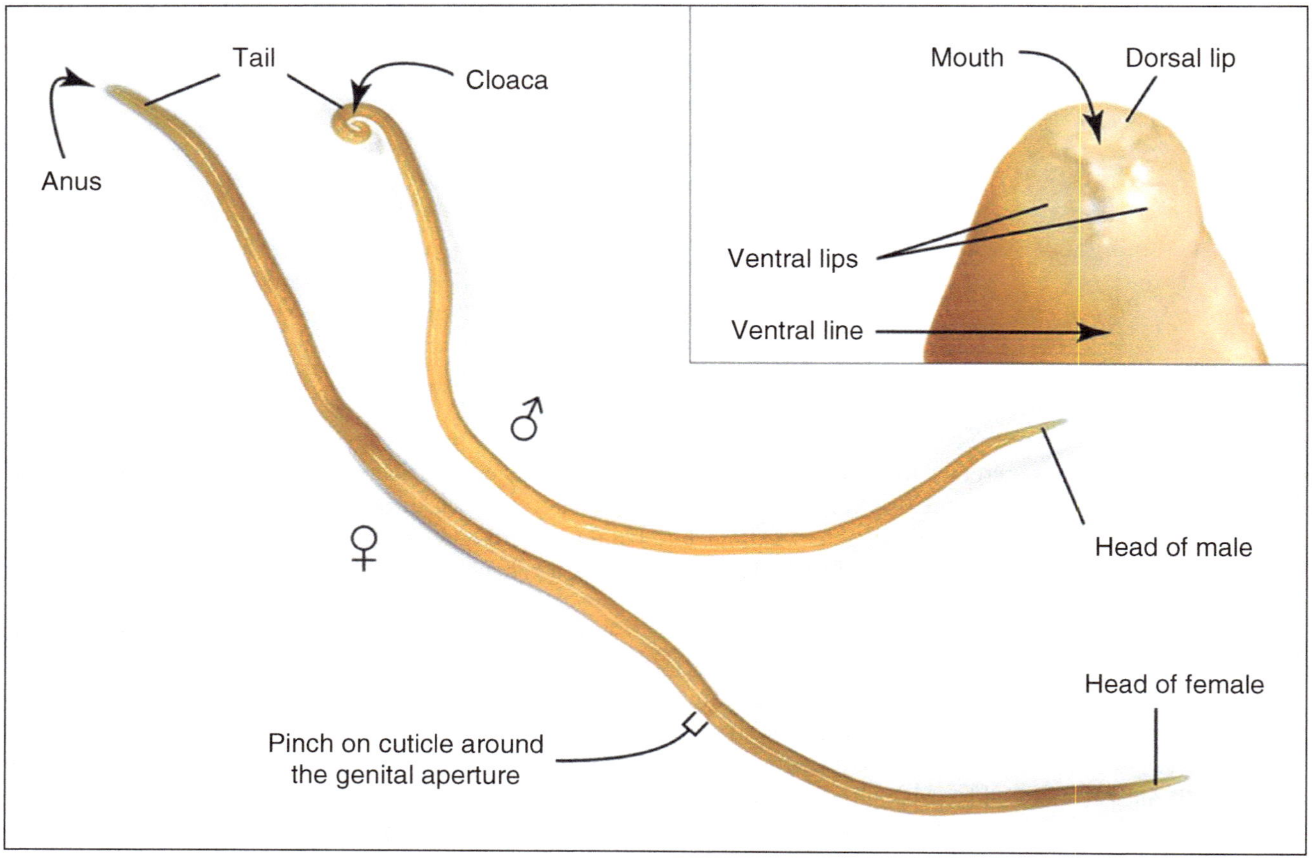

Fig. 3.1 External views of a male and a female roundworm (*Ascaris suum*). *Inset*: the head region enlarged

The body wall of preserved worms is reasonably tough, but the internal organs are extremely brittle and must be handled very carefully. The dissection should be performed in a large wax-bottomed dish using small insect pins to hold the body wall. The dissection is best conducted with a dissecting microscope. Place an adult *Ascaris* in the dissecting pan. Examine the external appearance carefully, using a hand lens to study the lips, genital aperture and anus (Fig. 3.1). In both sexes, the mouth is terminal at the anterior end, but the posterior end has no terminal opening. Viewed head-on with the help of a hand lens, the mouth can be seen to be surrounded by three small lips (Fig. 3.1, inset). One of the lips is dorso-median in position, whereas the other two are ventrolateral. The subterminal anus of both sexes is located slightly anterior to the posterior tip of the worm (Fig. 3.1). It is a transverse ventral slit and is the best landmark for recognising the ventral surface.

Look at the surface of the worm with the dissecting microscope or a hand lens and note that it is firm and resists deformation. It is covered with a thick proteinaceous cuticle which plays an important role in containing the high hydrostatic pressure of the body fluid. Look for the characteristic ornamentation of the cuticle, which in this species consists of fine circumferential ridges (Fig. 3.2). These ridges do not refer to inner structures; the animal is unsegmented.

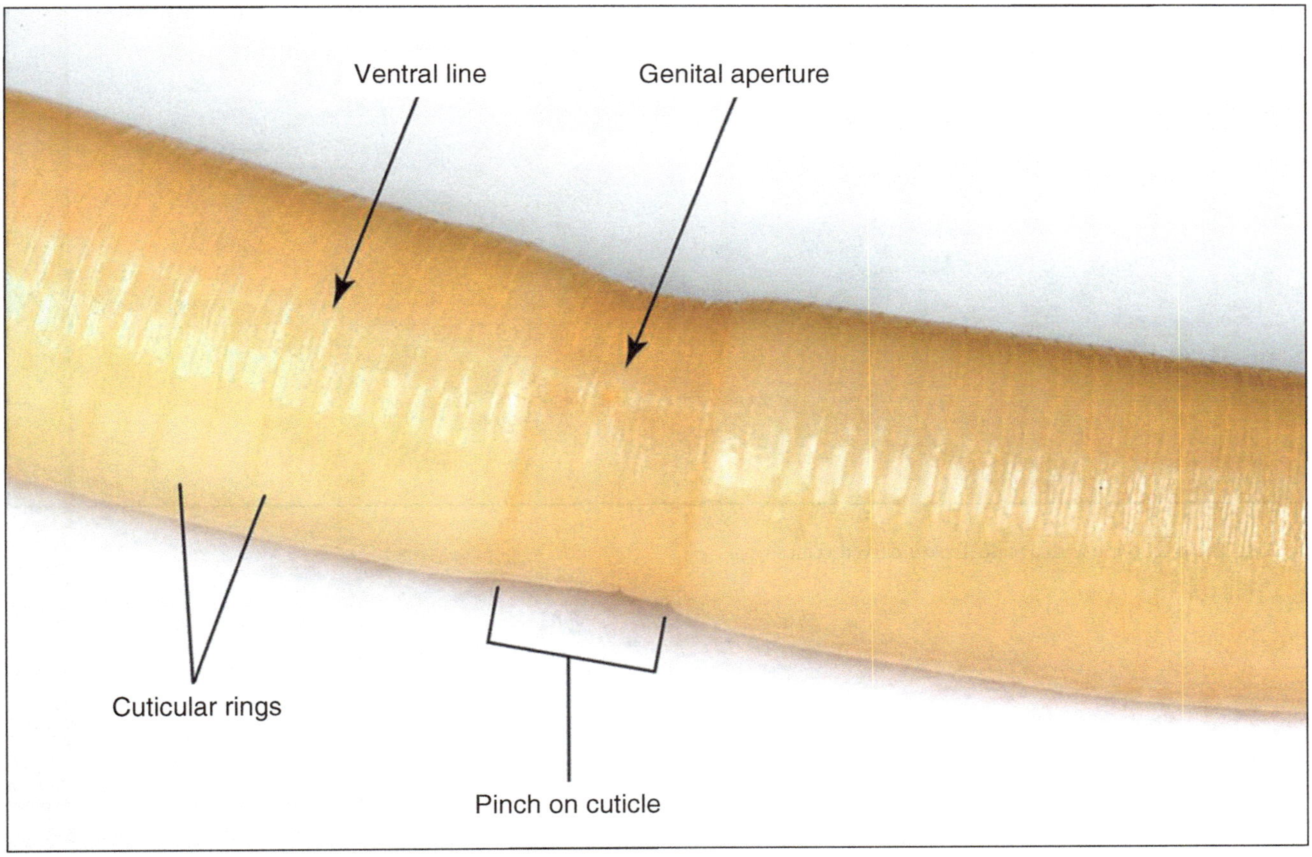

Fig. 3.2 The female genital pore is in the middle of a pinch on the cuticle on the ventral side of the body

Determine the sex of your specimen. *Females*, which run 20–40 cm in length, are more numerous and are larger than males, which average 15–30 cm in length (Fig. 3.1). The female genital aperture, known as the *vulva*, is located on the midventral line about 1/3 of the animal's length from the mouth in the middle of a pinch on the cuticle (Fig. 3.2). The female reproductive system opens to the exterior independently of the gut.

The posterior end of *males* is curved ventrally and looks like a shepherd's crook (Fig. 3.3). The posterior end of females is not noticeably curved.

Fig. 3.3 The posterior end of a male *Ascaris* curved ventrally

The four longitudinal *hypodermal cords* in the body wall are visible from the exterior as thin, pale stripes (Fig. 3.2). These are the dorsal, ventral and two lateral cords. They are faint, but discernible with good light. The two lateral cords are easier to see. Identifying these structures, you can position the worm in the dissecting pan with its ventral side down. Males must be rotated a little to accommodate the curl of the tail.

Fix the worm with two pins on each end. Be careful piercing the body wall as the high-pressure body fluid might gush out. Using a small pair of scissors, cut up the dorsal midline. Do your best to keep the incision on this line. Extend the cut forwards to the lips and backwards to the level of anus. Pin the cut edges of the body wall to the wax using insect pins slanting the pins outwards to allow room for dissection. Handle the internal organs, especially the gut, carefully because they are very delicate and break easily. Opening the middle region of the worm is a bit more difficult because it is packed with the reproductive system (Fig. 3.4). Finally, cover the specimen completely with water.

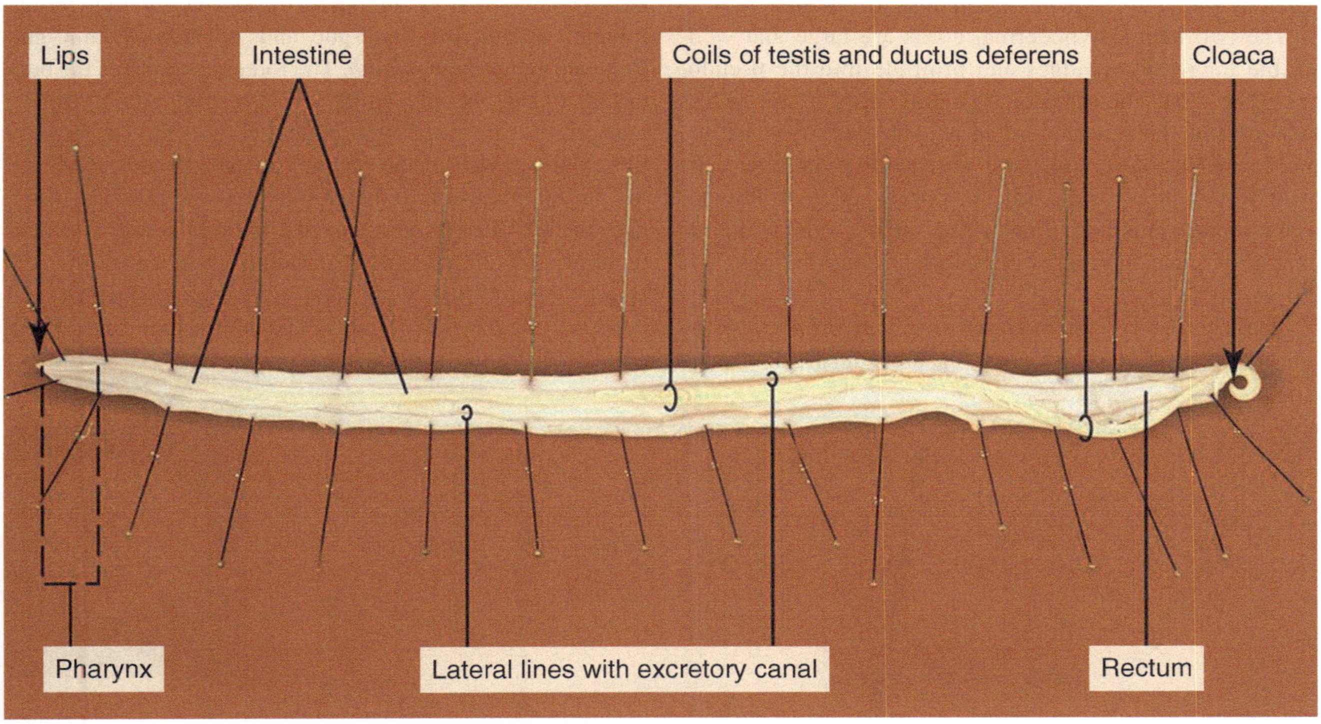

Fig. 3.4 Internal structure of male *Ascaris*

The heavy, transparent *cuticle* is the outermost layer. Immediately under the cuticle is the inconspicuous, thin epidermis. This is called *hypodermis* in *Ascaris* as it is under a very thick cuticle (Fig. 3.6). Inside the hypodermis is a thick, white sheath of *longitudinal muscles* composed of a single layer of cells which protrude into the *pseudocoel* (Figs. 3.5 and 3.6 upper left, bottom right). The pseudocoel, or primary body cavity, is filled with fluid under a high pressure. Virtually all other organs are affected by this pressure and must be able to function under its influence. The pressure maintains the body shape and acts as a *hydrostatic skeleton* against which the body wall muscles act to accomplish locomotion.

The two lateral hypodermal cords are large and well-visible longitudinal ridges and protrude into the pseudocoel (Fig. 3.5). The dorsal and ventral cords are much less evident and the dorsal cord is usually destroyed by the middorsal incision. Push the surrounding muscle cells aside to see the ventral cord. The dorsal and ventral hypodermal cords include longitudinal nerve cords and an excretory canal is present in each lateral cord (Figs. 3.5, 3.6, and 3.8).

The locomotory system comprises the hydrostatic skeleton (the pressurised pseudocoel), the antagonistic dorsal and ventral longitudinal muscle fields of the body wall and the elastic cuticle, which contains the hydrostatic pressure and opposes the longitudinal muscles. When one muscle field contracts, the opposite side of the body elongates to relieve the hydrostatic pressure. Alternate contractions of dorsal and ventral muscle fields result in sinusoidal waves in the dorsoventral plane passing along the length of the body. If living nematodes, like *Caenorhabditis elegans*, are available in the laboratory, place a culture in a Petri dish in an inverted microscope and observe their motion.

The gut is a long, straight tube running from the mouth to anus (Fig. 3.4). It is composed of an anterior, ectodermal foregut, endodermal mid-gut and ectodermal hind-gut. Relocate the terminal *mouth*. The foregut comprises the buccal cavity and *pharynx*, which, consistent with their ectodermal origins, are lined with cuticle (Fig. 3.5). The heavily muscularised wall of the pharynx is used to suck food into the gut in opposition to the high hydrostatic pressure of the pseudocoel. The pharynx is round in cross section. At rest, its lumen is collapsed and is triradiate (Y-shaped) (Fig. 3.5, inset). When filled with food, the lumen expands and becomes circular. The lumen is dilated by contraction of the radial muscles in the pharyngeal wall.

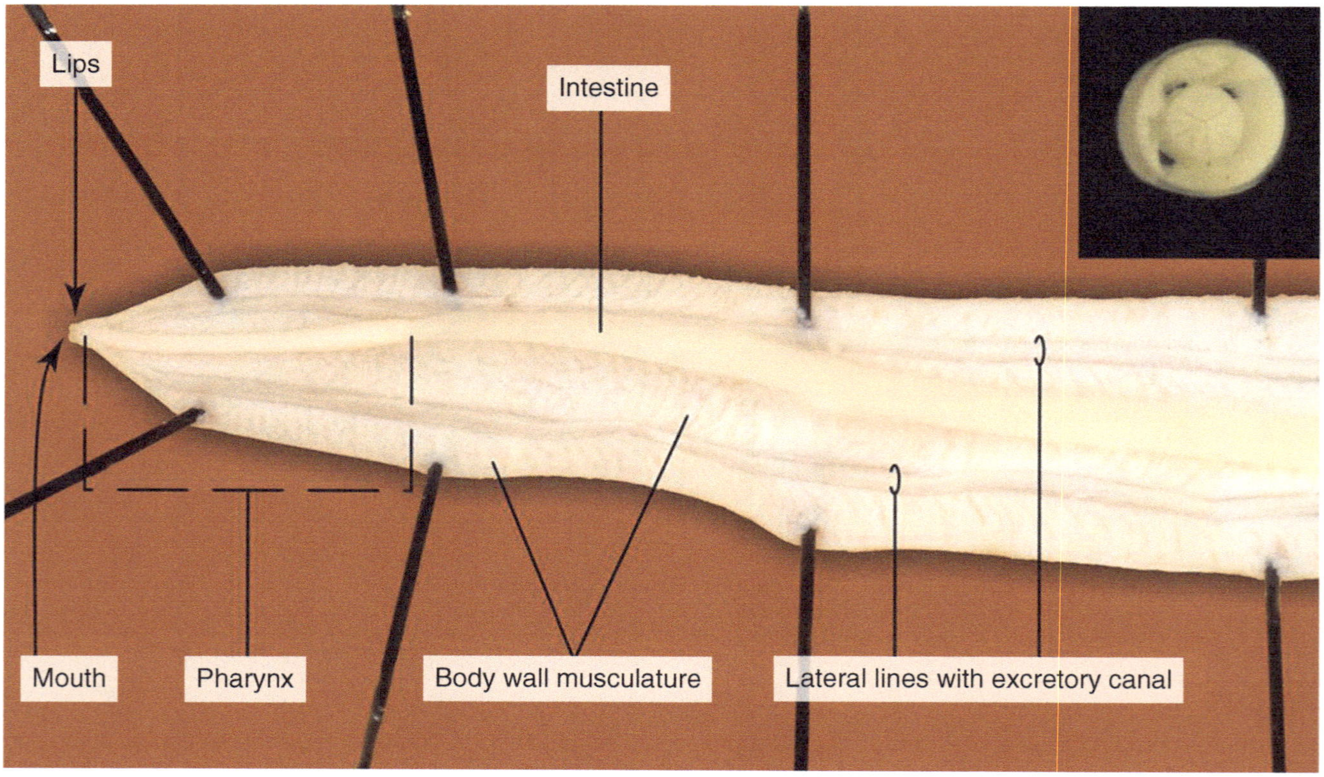

Fig. 3.5 Details of the anterior internal structures of male *Ascaris*. *Inset*: Transverse section of *Ascaris* at the level of the pharynx

The mid-gut, or *intestine*, begins immediately posterior to the pharynx (Figs. 3.4 and 3.5). It is a dorsoventrally flattened, ribbon-like tube. The intestine is the region of hydrolysis and absorption. *Ascaris* lives mainly on monomers (simple sugars and amino acids) from the intestinal contents of its host. These are absorbed by the microvilli of mid-gut epithelium. The intestine extends posterior to join the short ectodermal hindgut, or *rectum* (Fig. 3.4). In females, the rectum is difficult to differentiate from the intestine, but in males the rectum is a *cloaca* which receives the male gonoduct and the intestine before opening to the exterior (Fig. 3.4). Being ectodermal, the rectum is lined with a cuticle.

Study prepared stained slides of the posterior half of the worms containing the female or male reproductive systems. Orienting these sections is sometimes difficult. The lateral hypodermal cords are much larger than the dorsal and ventral cords and can be used to distinguish lateral from dorsoventral. Distinguishing dorsal from ventral is difficult, but the best landmark is the gut, which is usually (but not always) in the dorsal half of the pseudocoel.

The outermost layer of the body wall is the thick *cuticle*. It is a nonliving extracellular secretion. Below the cuticle is the thinner syncytial *hypodermis* (epidermis) which secretes the cuticle (Fig. 3.6, top and left bottom). (Syncytium is a multinucleated cytoplasmic mass.)

Together the hypodermis and cuticle make up the integument. The hypodermis has four groove-like evaginations called as *hypodermal cords*. The dorsal and ventral cords are small but can be found by careful inspection (Fig. 3.6, top left). The dorsal and ventral longitudinal nerve cords are usually visible in the dorsal and ventral hypodermal cords, respectively. The lateral hypodermal cords are large and easily located (Fig. 3.6, top right). The inconspicuous excretory ducts are in these cords. Note that the body wall at lateral lines is very thin because of missing muscle cell arms in the vicinity of these cords (Fig. 3.6, top left). Hypodermal cords divide the wall musculature into four stripes.

The thickest part of the body wall is the *longitudinal muscle layer* (Fig. 3.6, top left and right, bottom left). This is a single layer of large cells which bulge far into the body cavity and occupy much of it. Each muscle cell comprises an obvious peripheral and a less evident central portion. The *contractile portion* of the muscle cell sits on the inside of the basal membrane of the hypodermis and contains the contractile myofibres of the cell. It is easily recognised because the fibres stain dark pink and form a thick outline around this portion of the cell. The nucleus and most of the cytoplasm (or sarcoplasm), however, are in a large, but less conspicuous, bulging, the *cell body or sarcoplasmic region* (SR) that extends deep into the pseudocoel (Fig. 3.6, top left and right, bottom left). This region contains glycogen granules as energy store (cell body region appears almost empty in sections, because of not adequate preservation and dissolving of glycogens by the routine histological procedure). In nematodes, the axons of motor neurons do not exit the central nervous system (nerve cords) and do not approach the muscles. Narrow sarcoplasmic projections, or *arms*, arise from the apical ends of the muscle cells and run to a dorsal or ventral nerve cord to synapse with neurons confined to the cord (Fig. 3.6, top left and right, bottom left).

The dorsoventrally flattened *gut* is observable in the middle region. At high power, you can see that the intestinal walls are composed of a simple columnar epithelium of very tall cells. Unlike the ectodermal foregut, the mid-gut wall consists solely of a simple columnar epithelium and its basal membrane. There is no associated muscle, connective tissue or mesothelium. The basal ends of the cells rest on a basal membrane. The basal membrane separates the epithelium from the pseudocoel. The apical ends of the epithelial cells are microvilliated and form an absorptive brush border which is visible as a dark line around the mid-gut lumen (Fig. 3.6, bottom right). The pseudocoel is bounded on the outside by somatic musculature, which is mesodermal, and on the inside by the mid-gut epithelium, which is endodermal.

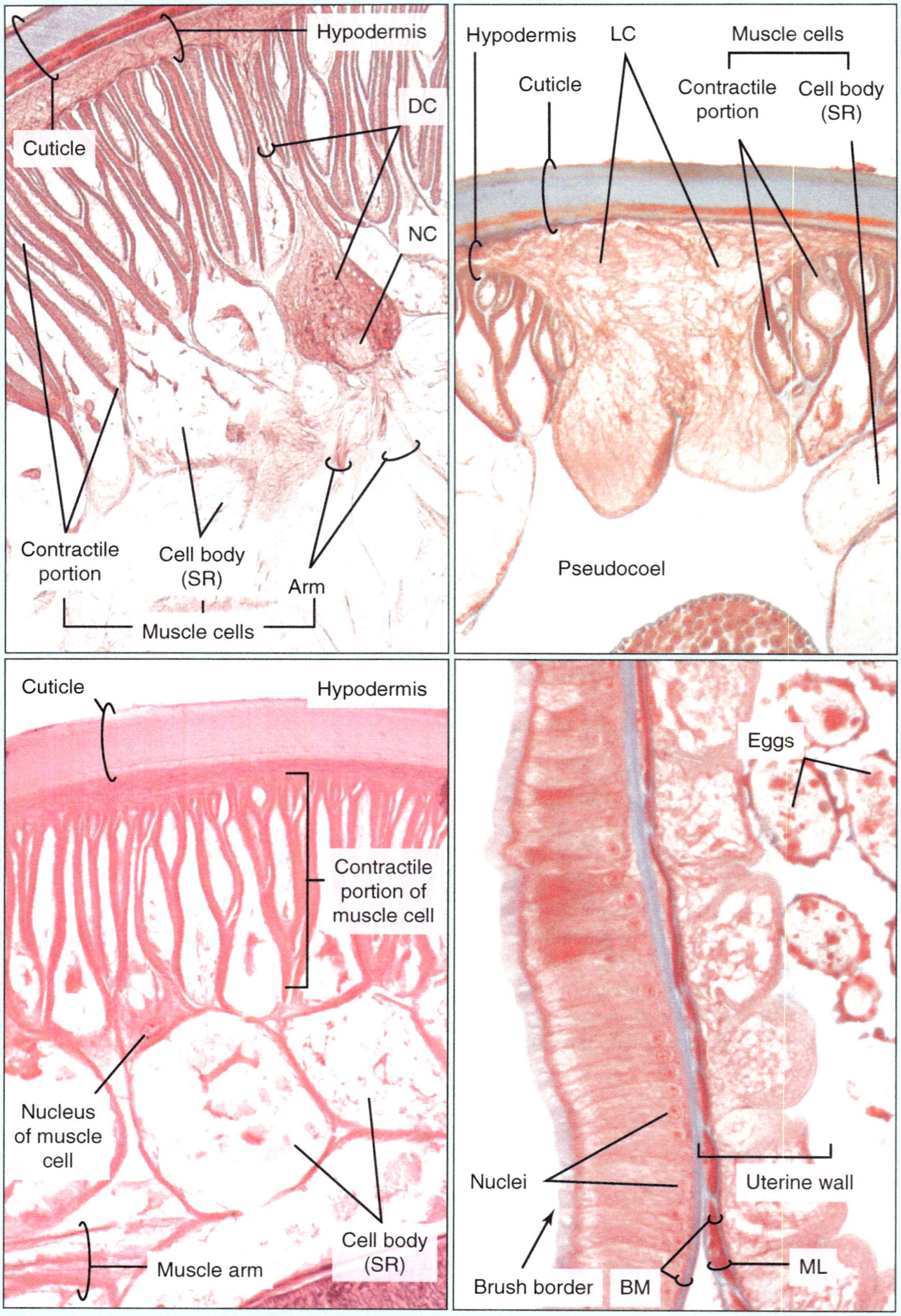

Fig. 3.6 Histological cross sections of *Ascaris* body wall (*top* Azan, *bottom left* HE) and gut (*bottom right* Azan). *BM* basal membrane, *DC* dorsal hypodermal cord, *LC* lateral hypodermal cord, *ML* muscle layer, *NC* nerve cord, *SR* sarcoplasmic region

Ascaris has *no circulatory system* as the unpartitioned pseudocoel makes a haemal system unnecessary. Transport is accomplished by movement of the pseudocoelic body fluid.

The *excretory system* consists of an enormous H-shaped canal system contained within a single cell. The longitudinal canals are located in the lateral hypodermal cords and extend over the entire length of the worm (Figs. 3.4 and 3.5). The two longitudinal canals connect with each other via a transverse canal near the anterior end of the worm. A short excretory duct leads from the transverse canal to the excretory pore on the anterior ventral midline. The system is thought to be chiefly osmoregulatory. The excretory canal system is difficult to observe in gross dissection of preserved whole specimens. The excretory pore is located immediately posterior to the mouth on the ventral midline, but it is difficult to find.

Study of the nervous system of *Ascaris* requires specially prepared material and will not be attempted.

The *reproductive systems* in both male and female are long tapered tubes lying coiled in the pseudocoel (Figs. 3.7 and 3.9). The solid upper ends are the gonads, ovaries or testes. The hollow, larger regions are specialised for transport and storage of gametes. You should be familiar with both sexes. Study the reproductive system of your specimen and then look at a dissection of the opposite sex as well. Using forceps and a teasing needle gently disentangles the intestine and oviduct/ductus deferens from the much-coiled ovaries/testis. Float rather than pull the strands apart. It is possible to disentangle the ovaries/testis completely so that the extreme length and graded thickness can be seen, but this is very difficult and time consuming and is not really worth the effort.

Female reproductive system is a Y-shaped organ consisting of two tubes, each with an *ovary*, *oviduct* and *uterus* forming an arm of the Y. The two arms join to form a common (unpaired) *vagina* which is the stem of the Y. The vagina opens to the outside at the *vulva*. It is convenient to trace the system backwards beginning at the gonopore, but keep in mind that female gametes travel in the opposite direction (Figs. 3.7 and 3.8).

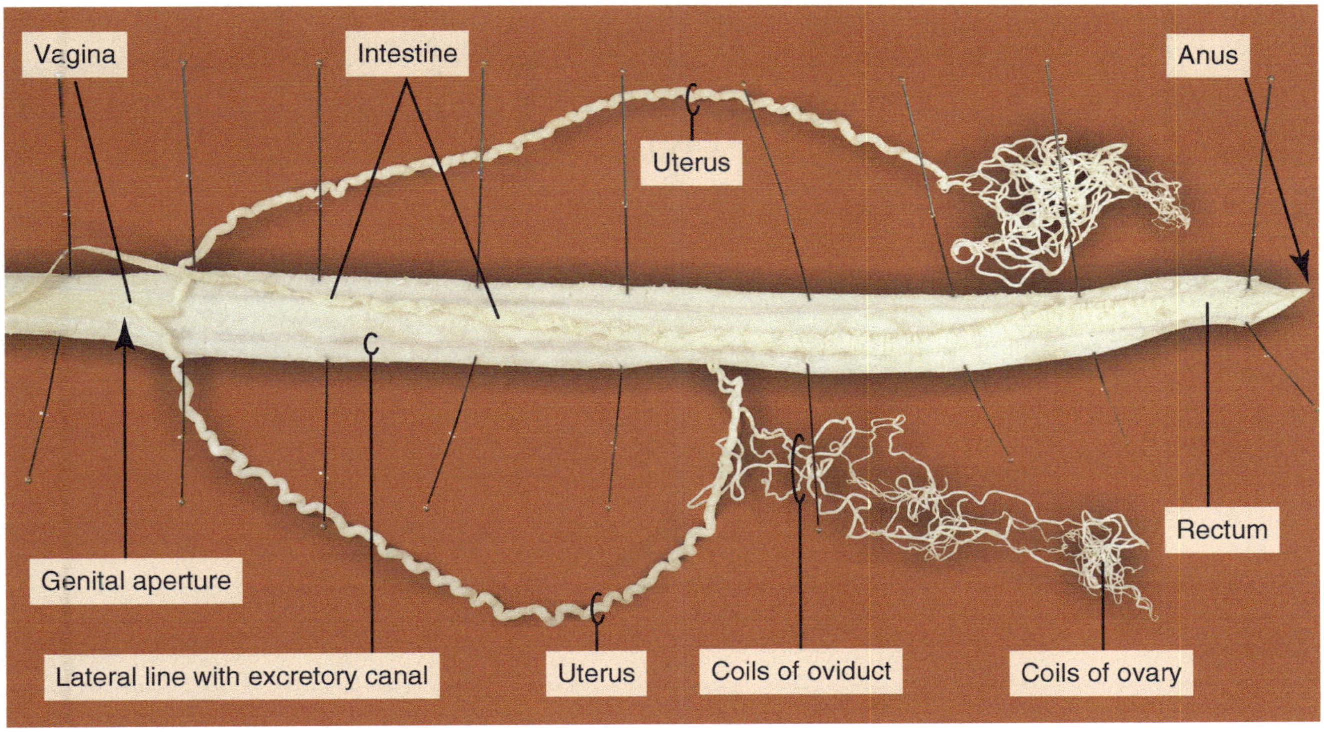

Fig. 3.7 Dorsal dissection of a female *Ascaris*. The reproductive system has been moved to the side and untangled for clarity

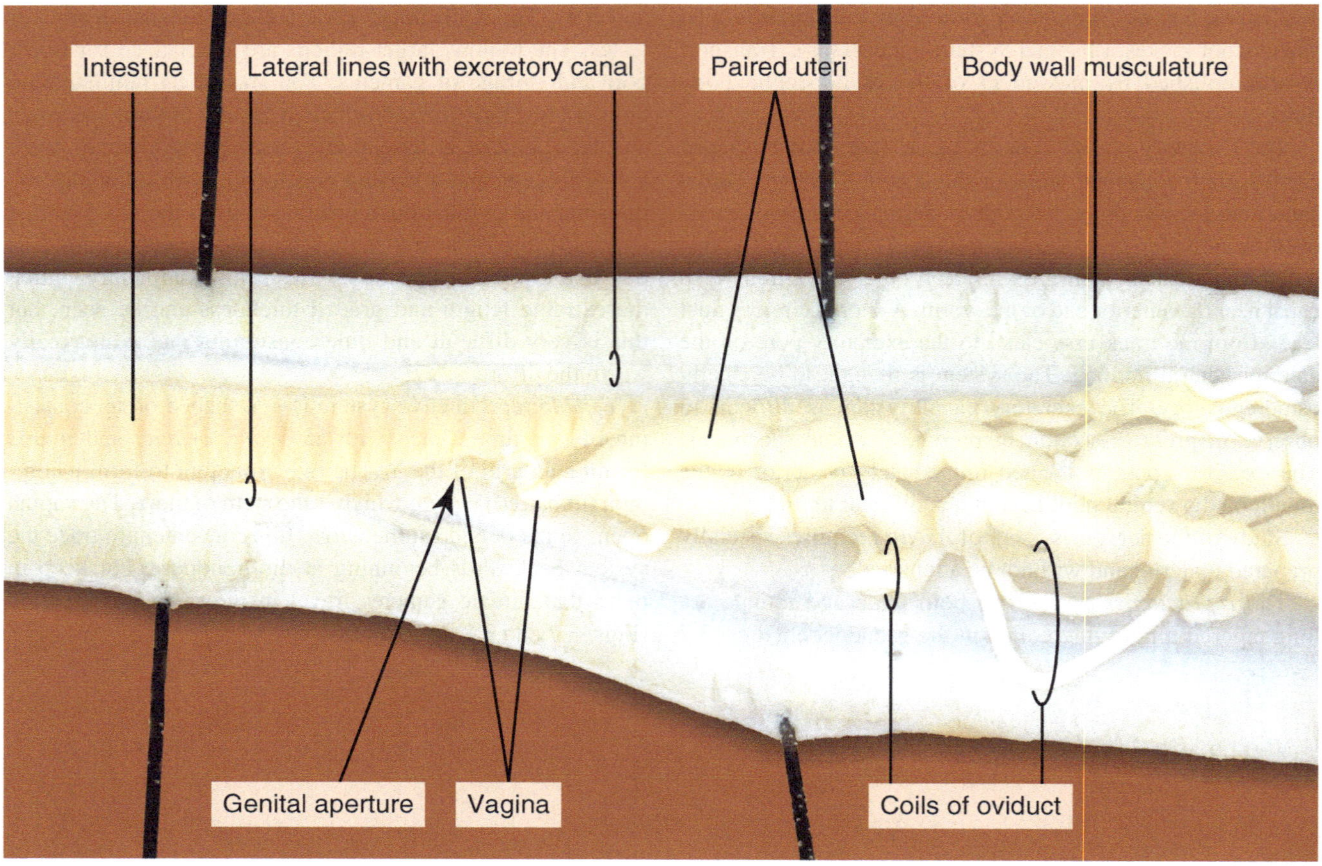

Fig. 3.8 Details of the reproductive system of female *Ascaris*

The ovaries are solid, not tubular, and form a mass of small-diameter threads in the middle of the worm. Oviducts have a visibly wider diameter and fertilisation of mature eggs occurs here. The widest parts are the uteri where fertilised eggs enter. Here, they receive a chitinous eggshell and undergo embryonic development.

The *male* reproductive system is essentially a single, long tube (Fig. 3.9).

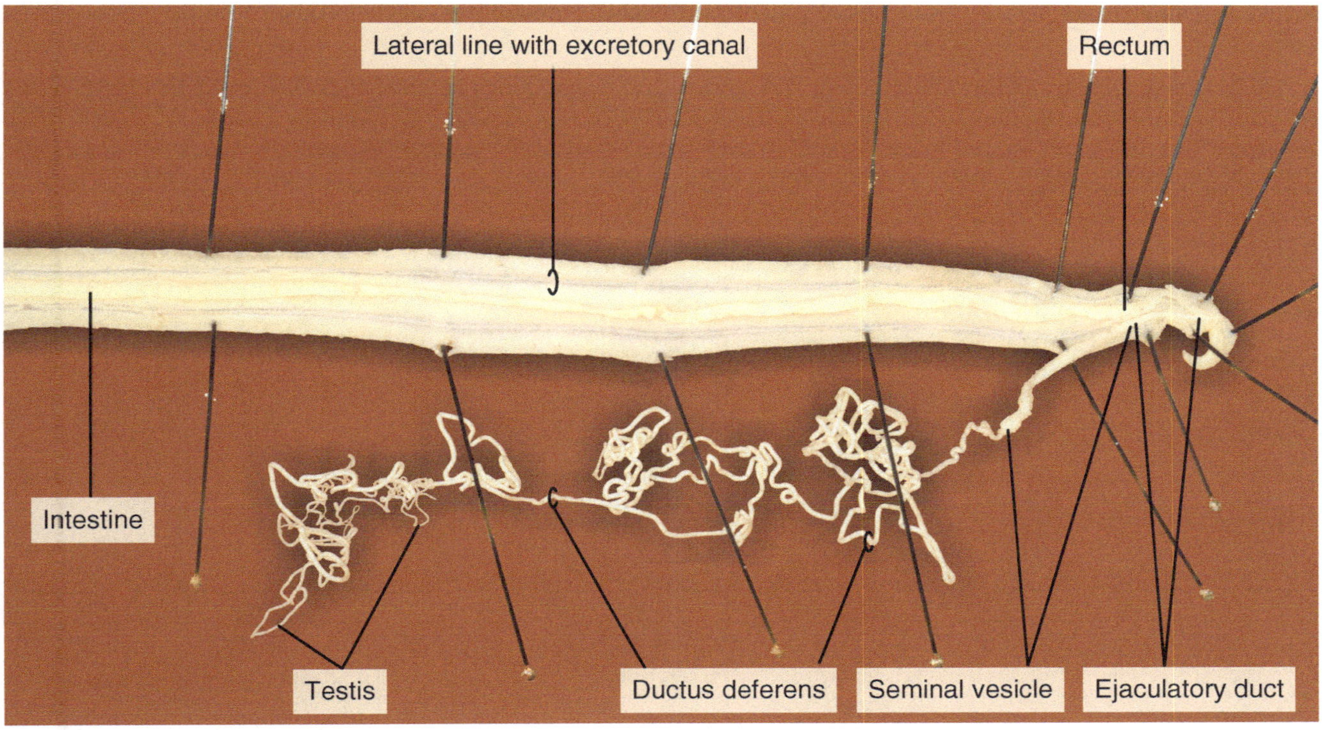

Fig. 3.9 Dorsal dissection of a male *Ascaris*. The reproductive system has been moved to the side and untangled for clarity

The solid free upstream end of the tube is the threadlike *testis*, which continues as a thicker *ductus deferens*, or sperm duct. Both are much coiled. The ductus deferens connects with the wider *seminal vesicle*, which empties by a short, muscular *ejaculatory duct* into the cloaca. The male reproductive system does not have its own external gonopore; it has a common exit with the gut called *cloaca*. Two protrusible chitinous *penial seta* or copulatory spicules are located beside the cloaca (Fig. 3.10). Spicules secreted by and contained in spicule pouches may be extended through the cloaca. The spicules are used to hold the vulva of the female open during copulation and fix the mating partners.

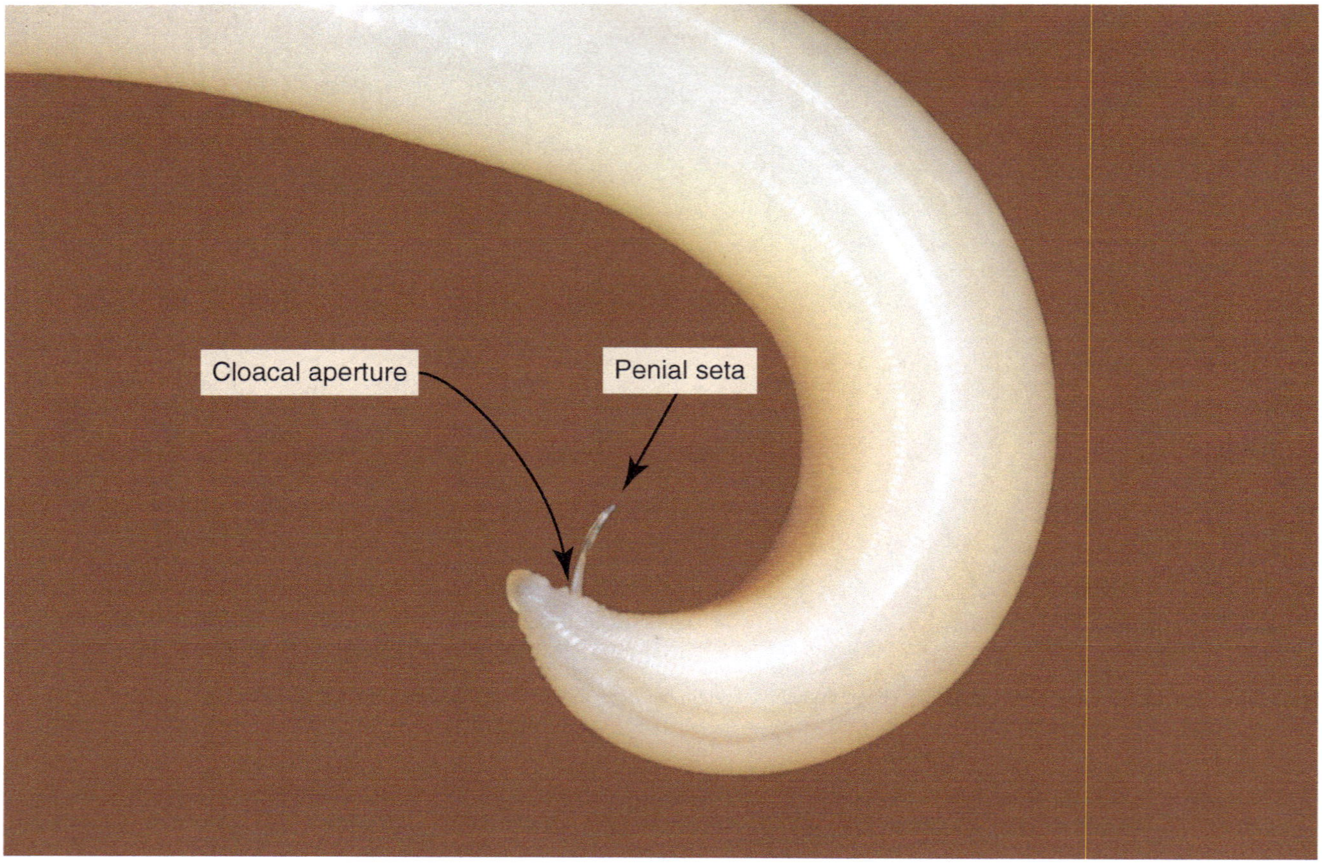

Fig. 3.10 One of the two penial setae (copulatory spicules) of a male *Ascaris*

Study the cross section of the reproductive system of a female *Ascaris* (Fig. 3.11). It should be recognisable by its two large egg-filled uteri. The twisting coiled nature of the reproductive system results in multiple sections through the ovaries and oviducts, but not the uteri. The smallest diameter sections are of the *ovary*. These are easily recognised because they are wheel shaped and solid, whereas the oviducts and uteri are hollow. The ovary has two typical compartments (Fig. 3.11, left). The proximal part is a *germinative zone*: it contains proliferating oocytes and their daughter cells in irregular pattern. Differentiating cells are pushed into a *growth zone*, where they form a connection to a common cytoplasmic cord named *rachis* in central position. Each cell sits on the thin ovarian wall and reaches the central rachis, which is essential for their development (perhaps for uptake yolk and growth factors). Matured oocytes get into the oviduct separately. The ovary and oviduct are surrounded by thin epithelia which are often pulled away from the germinal cells leaving a white space between them. This space is an artefact (Fig. 3.11, right). The *oviduct* is a little larger in diameter and is hollow (it has no rachis). The *uteri* are much larger than either oviducts or ovaries. Its wall is lined with squamous, thick epithelium, which forms several folds in upper part of the organ. This partitioned cavity is ideal for the storage of the mating partner's sperm (Fig. 3.11, right). The fertilisation takes place here. The eggs get into the lower part of the uterus, where embryogenesis begins and a spiny capsule is formed around them. The uterus contains shelled "eggs" in all stages of embryonic development. The uterine wall is well muscularised (Figs. 3.6, bottom right and 3.11, right).

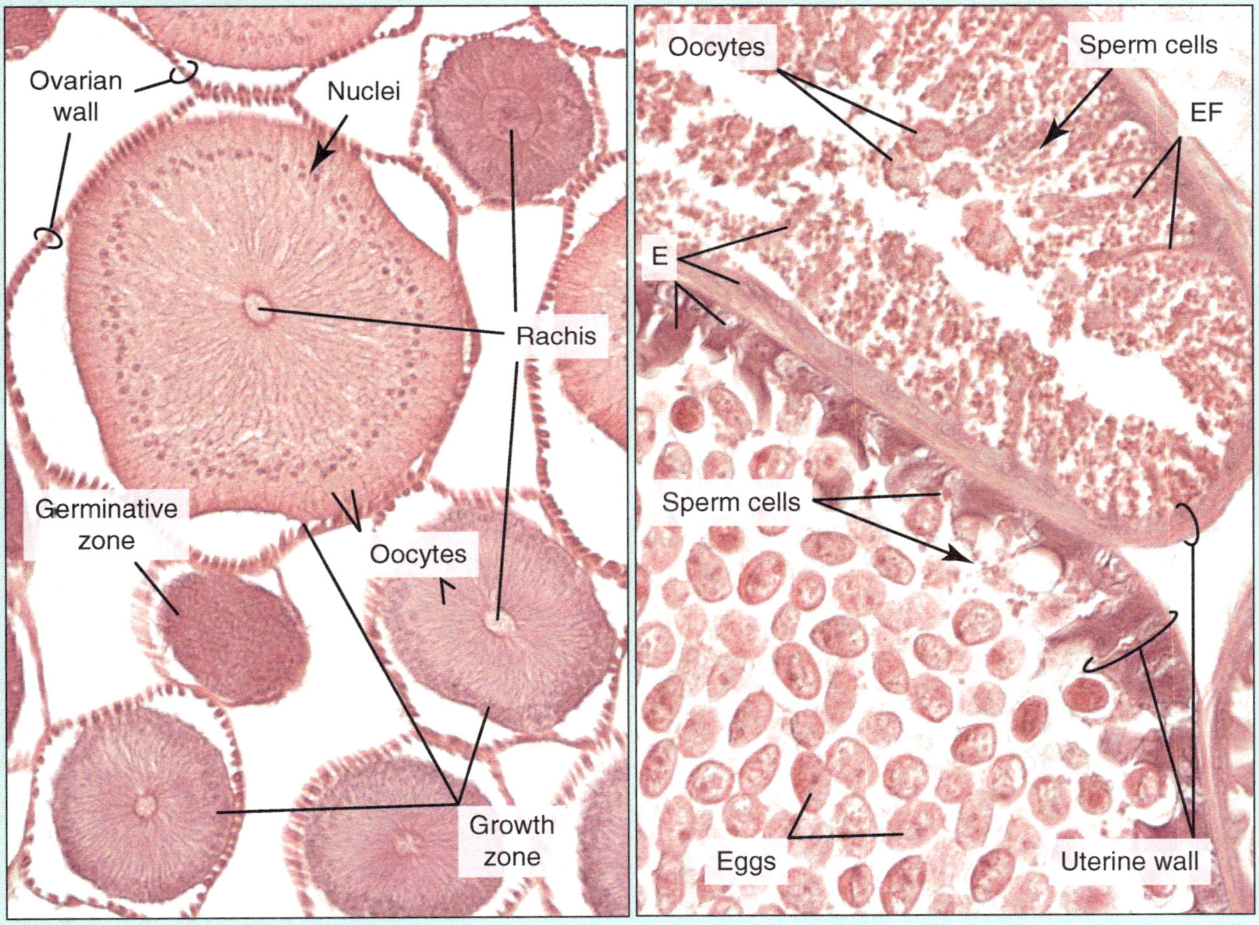

Fig. 3.11 Histological cross sections of the gonads of a female *Ascaris* (HE). *E* epithelium, *EF* epithelial folds

Study the cross section through a male *Ascaris* (Fig. 3.12). Male cross sections are usually smaller in diameter than female. The proximal, upstream end of the tube is the *testis* which is small in diameter and enclosed by an epithelium. Testis has two compartments: proximal one is the germinative zone and distal one is the growth zone. Unlike the ovary, testis does not have rachis. The germinative zone is filled with small, spherical primordial germ cells and has no lumen. Mitotic divisions of the germ cells produce spermatogonia which move downstream to undergo spermatogenesis in the growth zone. Several sections through the testis may be present. The next region of the male tube is the *ductus deferens*. It is slightly larger in diameter than the testis and is also enclosed by a thin epithelium. Its interior is filled with round, atypical spermatozoa (they have pseudopodia for crawling). There should be several sections through the ductus deferens in the pseudocoel of your specimen (Fig. 3.12, right).

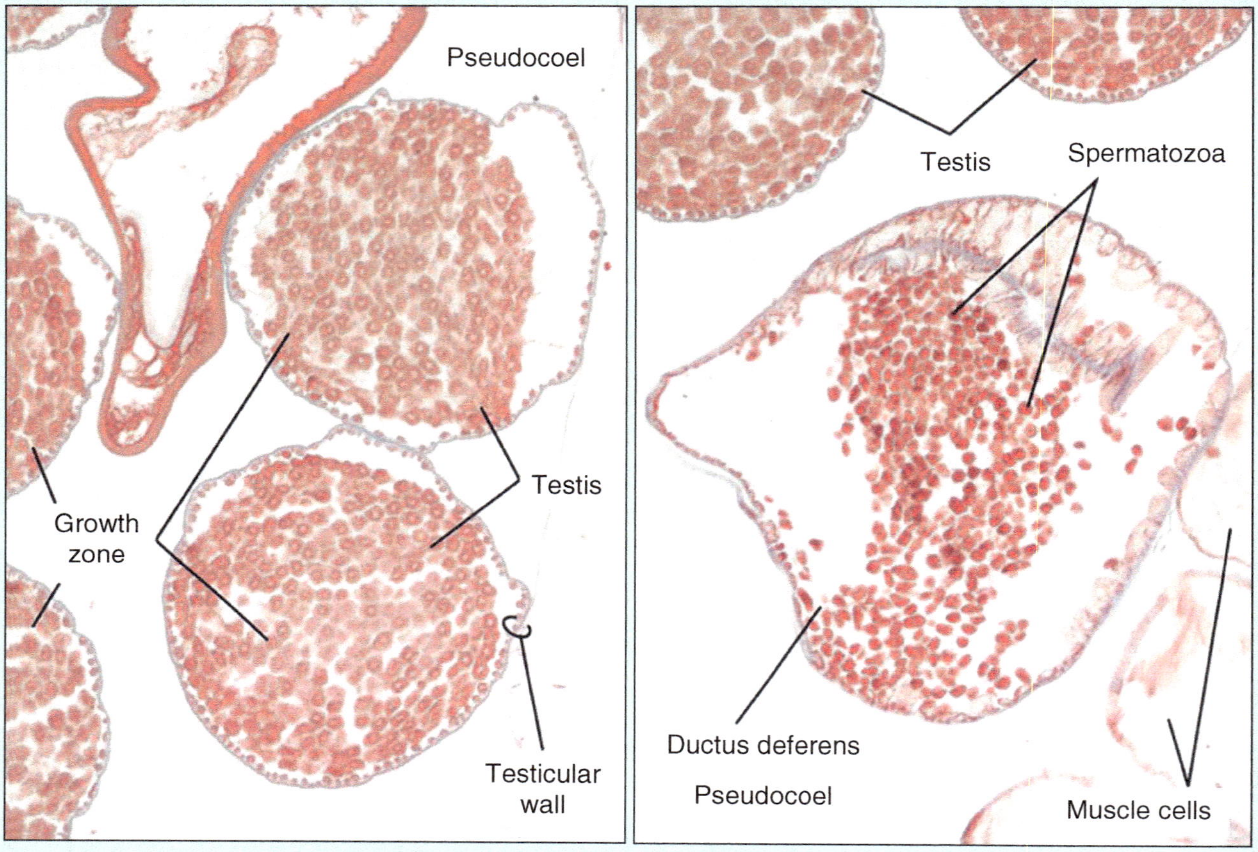

Fig. 3.12 Histological cross sections of the gonads of a male *Ascaris* (Azan)

The nematode life cycle involves only one host which becomes infected when it ingests *Ascaris* eggs in its food or water. These hatch in the intestine and larval worms migrate to the liver where they enter the host's haemal system. They are carried in the blood to the lungs where they enter the lumen of the alveoli. From here, they crawl to the pharynx and then follow the gut lumen to return to the small intestine where they mature into adult roundworms and feed on chyme. The larvae undergo four moults to become adults. After maturation, copulation occurs and females produce and release shelled eggs which leave the host in the faeces.

Safety Unpin the dissected specimen, and using forceps, place it into the disposal container. Wash all dissecting equipments with soap and water. Wipe down lab surface. Finally, wash your hands with hand sanitizer!

Dissection of the Earthworm (*Lumbricus terrestris*)

4

- *Availability*: Earthworms are found all over the earth. They prefer moist rich soil that is not too dry and sandy. Among earthworms there are many species which grow large enough to be used for dissection. There are a number of differences in the number and position of the internal organs which may affect the dissection. It is therefore important to identify the specimen to be used and not to assume that it is a *Lumbricus terrestris* as described here. Earthworms are chiefly nocturnal and come out of their burrows at night. They can be easily found during warm, moist nights of spring and early summer by searching with a flashlight around a rich soil. They can be collected by day instead when a bucketful of water is poured out at the same rich soil. They come out of their burrows because of the flood.

  Place a live earthworm on a sheet of paper and observe the mechanics of crawling. Its body wall contains well-developed layers of circular and longitudinal muscles. Notice the processing peristaltic waves of their alternate contraction as the animal crawls. The setae help in providing holding power, when the worm is burrowing. You can hear their scratching noise on the paper.
- *Anaesthesia:* Earthworms can be anaesthetised submerged in 10 % ethanol for 15 min. Alternatively the worms may be killed with chloroform provided that the liquid is placed on a cotton wool and not allowed to come into contact with the worms. Wash the specimen thoroughly with water.

The earthworm belongs to a group of animals called annelids (segmented worms). The body of an annelid is usually divided internally and externally into well-defined *segments* which are separated from each other by septa, or dividing walls. Except for the tail and head regions, all segments are essentially alike. Other members of this group include the clamworms and tubeworms, which live in the ocean, and the leeches.

Examine the earthworm externally using a dissecting microscope or a magnifying glass as necessary. Identify the dorsal side, which is the worm's rounded top, and the ventral side, which is its flattened bottom. The anterior end of the animal is more cylindrical and usually more pointed than the flattened posterior end (Fig. 4.1).

P. Lőw et al., *Atlas of Animal Anatomy and Histology*, DOI 10.1007/978-3-319-25172-1_4

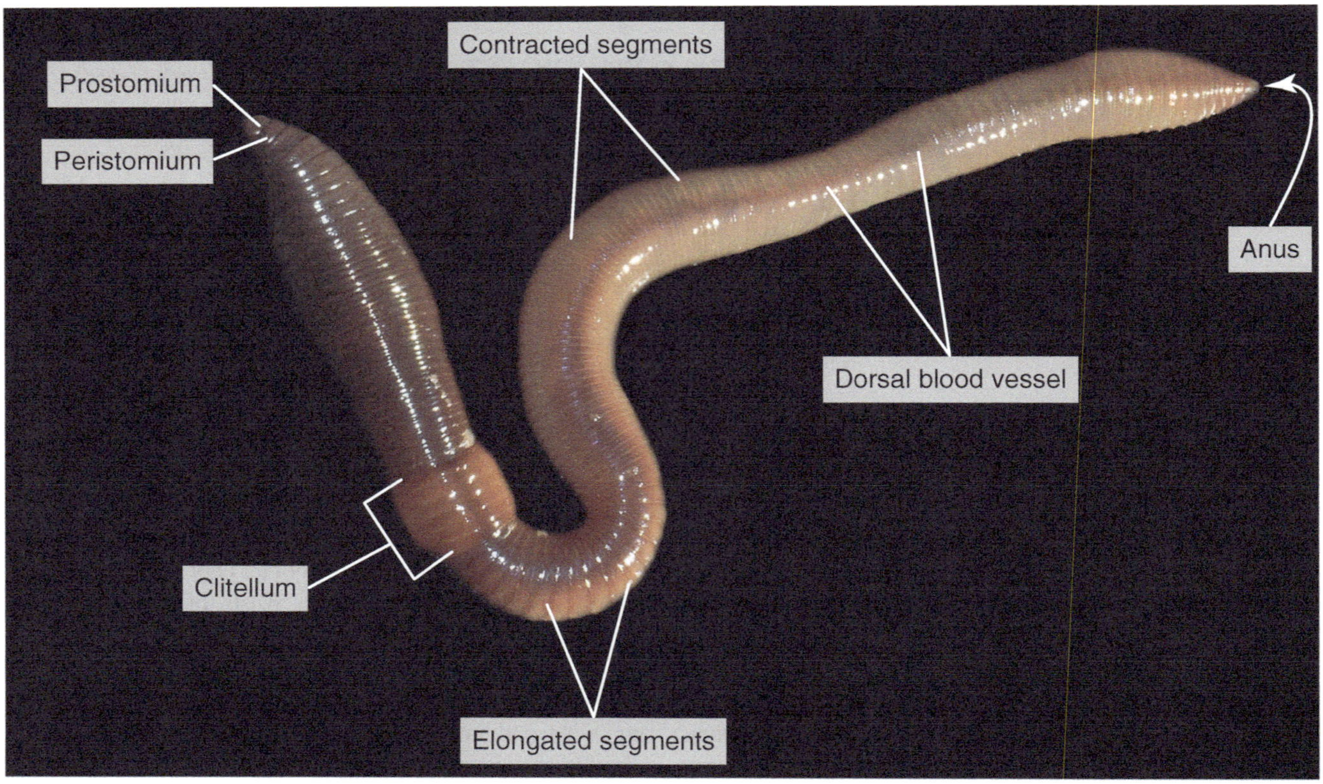

Fig. 4.1 Dorsal view of a live earthworm

The first segment is the *peristomium*. It bears the *mouth*, which is overhung by a fleshy lobe, the *prostomium* (Figs. 4.1 and 4.2). The head of the earthworm, lacking in specialised sense organs, is considered degenerate and is not a truly typical annelid head. Find the *anus* in the last segment. Adult (sexually mature) earthworms have a distinct, conspicuous, saddle-like swelling called a *clitellum* on the dorsal surface (Fig. 4.1). Young or juvenile worms do not have one. It is often white or orange in colour. It is located about one-third of the way down the earthworm and generally extends from segment 33 to 37. The clitellum produces a mucus sheath used to surround the worms during mating and is responsible for making the *cocoon* within which fertilised eggs are deposited.

Turn the worm ventral side up, as shown in Figs. 4.2 and 4.3. The ventral surface of the earthworm is usually a lighter colour than the dorsal surface.

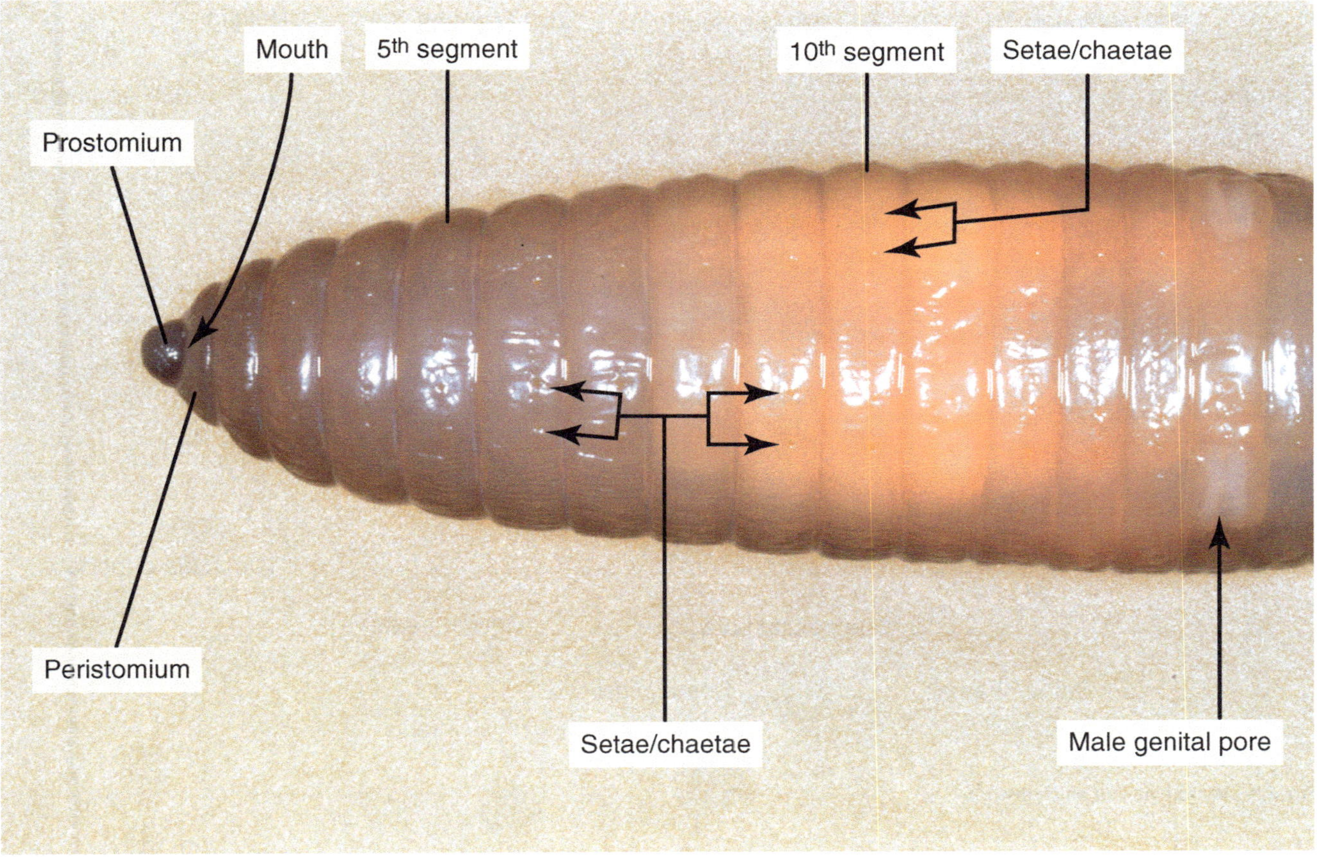

Fig. 4.2 Ventral view of the anterior end of an earthworm

Run your fingers over the ventral surface of the earthworm's body; you should be able to feel bristlelike projections used by the worm to prevent slipping. Observe the worm's *setae* (chaetae) with a dissecting microscope or a magnifying glass (Fig. 4.2). They are minute spines, of which four pairs are located on every segment except the first and last one. Locate and identify the external structures of the reproductive system on the ventral side of the worm (Fig. 4.3).

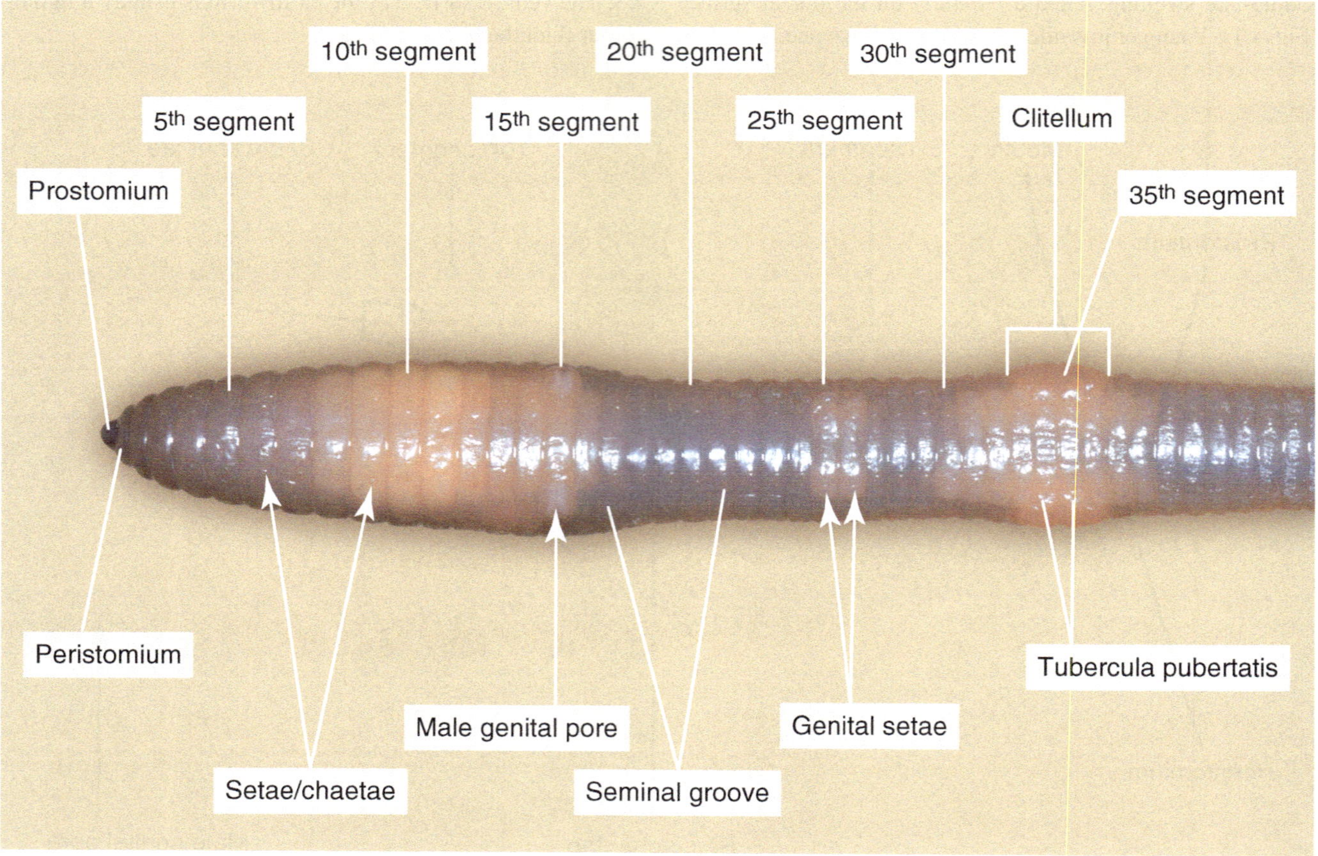

Fig. 4.3 Ventral view of an earthworm

These monoecious organisms have *male genital pores* on the ventral surface of segment 15 (Figs. 4.2 and 4.3). These are conspicuous openings of the sperm ducts from which spermatozoa are discharged. They have a pair of long *seminal grooves* extending between the male genital pores and the clitellum. These guide the flow of spermatozoa during copulation. The small female genital pores are inconspicuous on ventral side of segment 14. Here the oviducts discharge eggs. Also hard to observe are the openings of two pairs of the spermathecae in grooves between segments 9 and 10, and 10 and 11. Another key structure found on the clitellum is the *tubercula pubertatis*, an additional feature used to identify mature earthworms (Fig. 4.3). The tubercula pubertatis are glandular swellings located on both sides of the clitellum. The genital tumescences are areas of modified epidermis that do not have distinct boundaries. Here follicles of *genital setae* open. The pattern and location of the genital setae are also important clues to identifying different species of earthworms.

The *body wall* of the earthworm has three main layers: epidermis and two muscle laminae. The *epidermis* (E) is made of tall, columnar epithelial cells (Fig. 4.4). They secrete a thin cuticular layer onto the surface. There are different *gland cells* (GC) among columnar cells, which produce mucous layer protecting against desiccation (Fig. 4.4, top left). Musculature is divided into outer circular (CM) and inner longitudinal layers (LM), separated by a connective tissue septum. Feather-like pattern of longitudinal muscle layer showed in cross section is very characteristic (Fig. 4.4, top left and bottom left). Routine histological protocol causes dehydration and shrinking of muscle cells, so they are separated in a feather-like pattern. Musculature is partitioned into stripes by repetitive septa. Closing layer of body wall is the parietal peritoneum. *Clitellum* is a saddle-like swelling on the body. Huge amounts of gland cells develop in its epithelial layer, so these glands become subepithelial. They show regional distribution and different staining (Figs. 4.4, top right and 4.13). Clitellar glands produce a mucus sheath that surrounds the worms during mating and are responsible for secreting the cocoon, within which fertilised eggs are deposited. *Setae* are bristles formed by setal sacs, which are invaginations of the epidermal layer. They are manipulated by small muscles at their bases. Retractor and protractor muscles are attached to them for directing their movement (Fig. 4.4, bottom left and right).

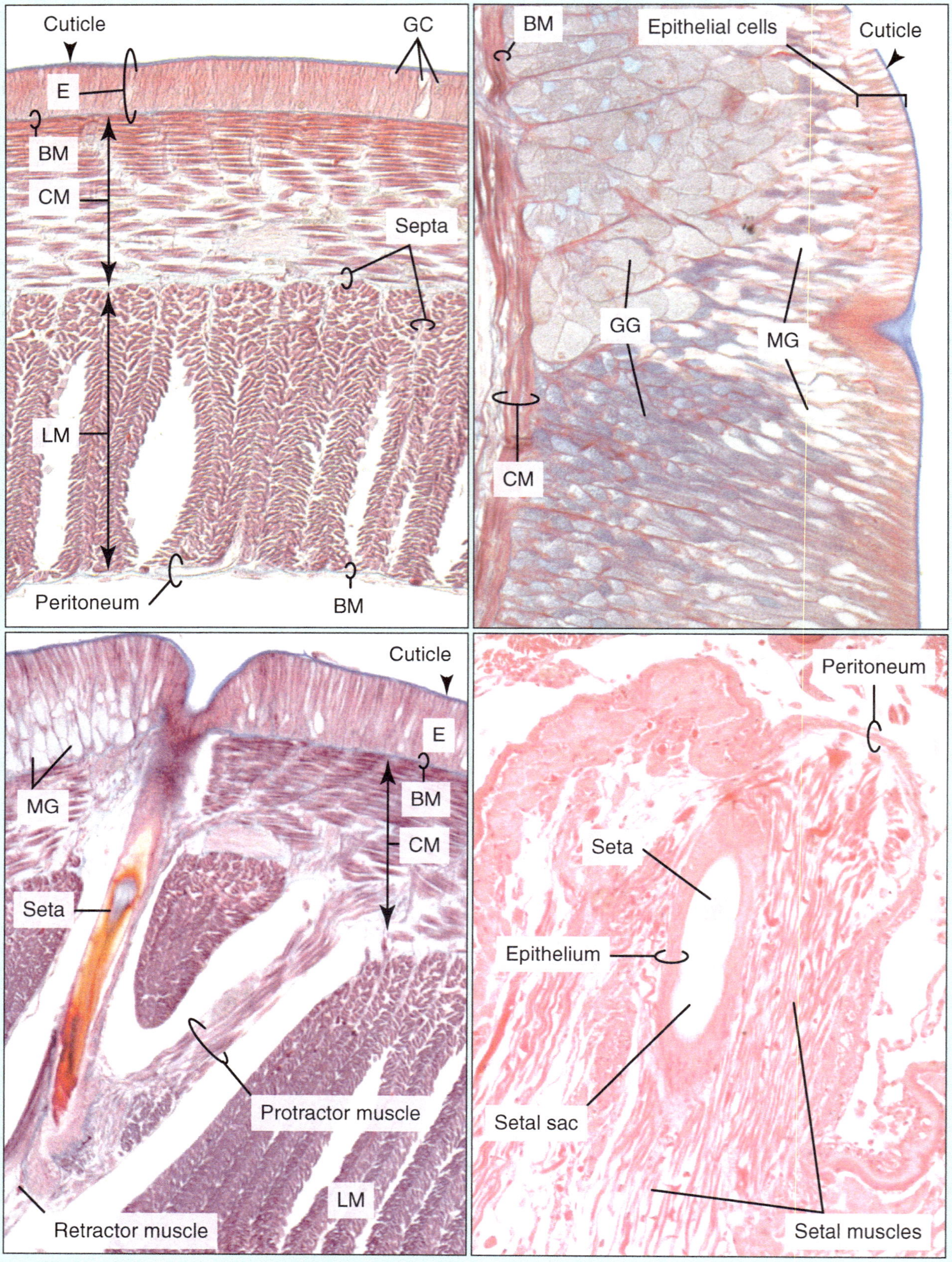

Fig. 4.4 Cross sections of the body wall of the earthworm (*top* and *bottom left*: Azan, bottom right: HE). *BM* basal membrane, *CM* circular muscle layer, *E* epidermis, *GC* gland cells, *GG* granular glands, *LM* longitudinal muscle layer, *MG* mucous glands

After observing the external features, start the dissection of the earthworm. The dissection should be performed in a wax-bottomed dish using small pins for attachment and display. The dissection of the earthworm is much improved when it follows the dissection of the roundworm *Ascaris*. Compare the structures of *Ascaris* and the earthworm to see how the additional features of the earthworm make it more complex. The major points are: *Annelids* are segmented and have a true coelom, while roundworms are not segmented and have a pseudocoelom. *Annelids* have a circulatory system, while roundworms have not. *Annelids* have a more complex digestive, excretory and nervous system than roundworms.

Reanaesthetise the earthworm in 10 % ethanol if necessary. Turn the worm dorsal side up. Holding it in one hand, make a small slit in the body wall in the middorsal line in the region of the clitellum. Be very careful not to cut deeply (Fig. 4.5).

Fig. 4.5 Holding the worm in one hand, make a small slit in the body wall in the middorsal line in the region of the clitellum

Keeping the points of the scissors well up, cut forwards as far as the prostomium (Fig. 4.6).

Place earthworm in the dissecting tray ventral side down and start to pin it open (Fig. 4.7).

Fig. 4.6 Cut forwards as far as the prostomium, keeping the points of the scissors well up

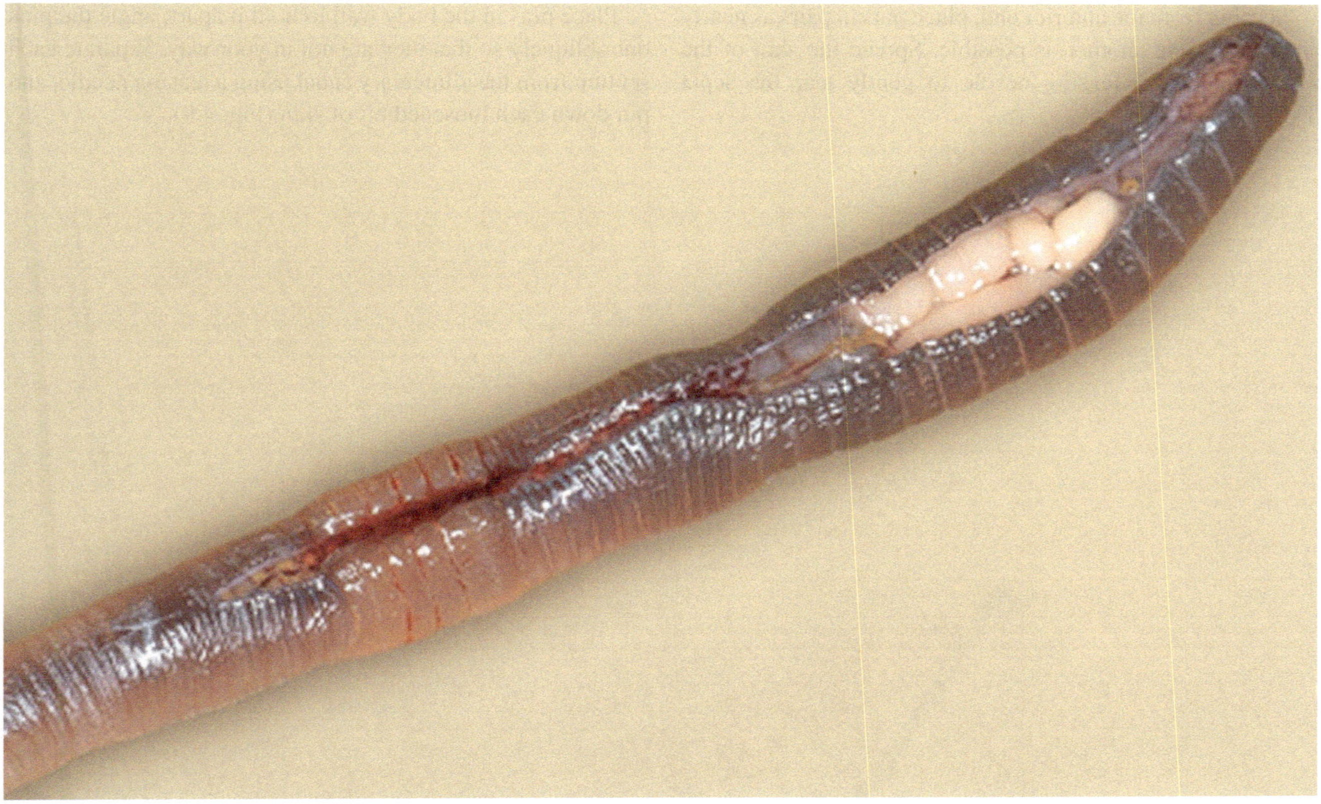

Fig. 4.7 The anterior end with the finished dorsal cut before pinning out

Starting from the anterior end, place pins in pairs as nearly opposite to one another as possible. Spread the skin of the worm out; use a teasing needle to gently tear the septa (Fig. 4.8).

Place pins in the body wall to hold it apart; angle the pins out obliquely so that they are not in your way. Separate each septum from the alimentary canal using a teasing needle, and pin down each loosened bit of skin (Fig. 4.9).

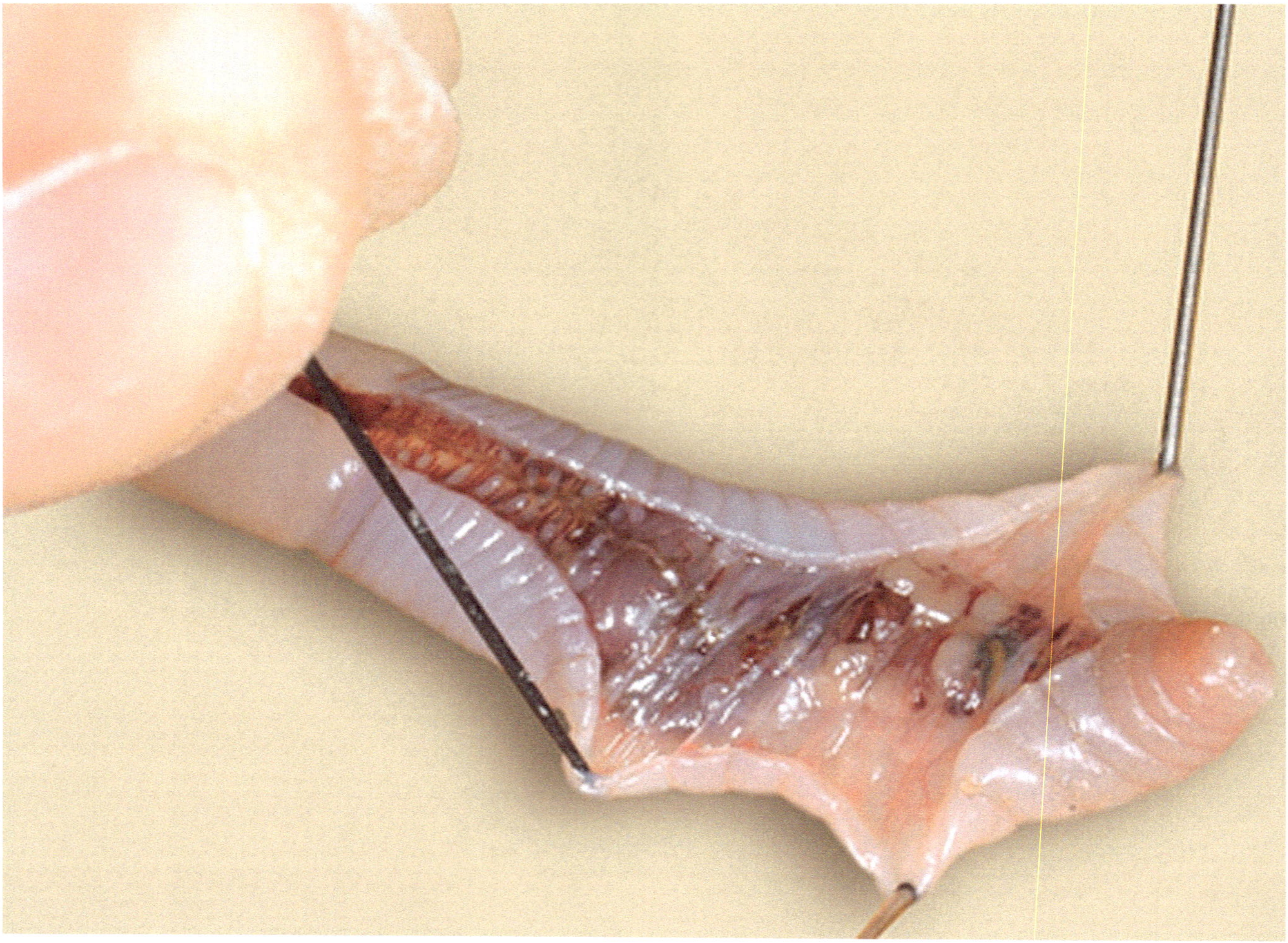

Fig. 4.8 Place pins in pairs as nearly opposite to one another as possible starting from the anterior end

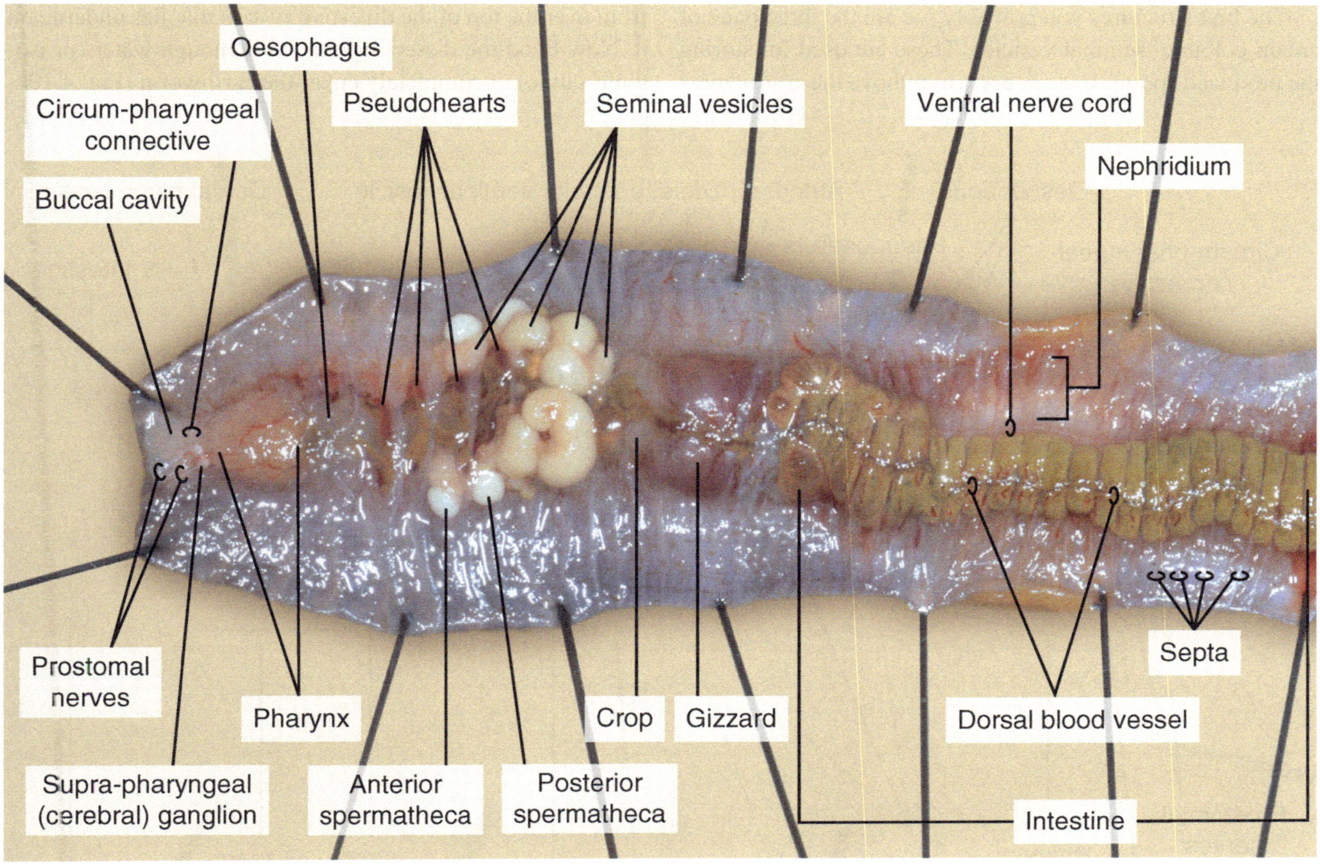

Fig. 4.9 Opened and pinned out earthworm (before the water cover)

The first structures you probably see are the three pairs of cream-coloured seminal vesicles. These are used for storing the produced sperm. Use tweezers to remove these structures from over the top of the digestive system that lies underneath it. Now flood the dissecting tray with enough water (or isotonic saline) to completely cover the earthworm (Fig. 4.10).

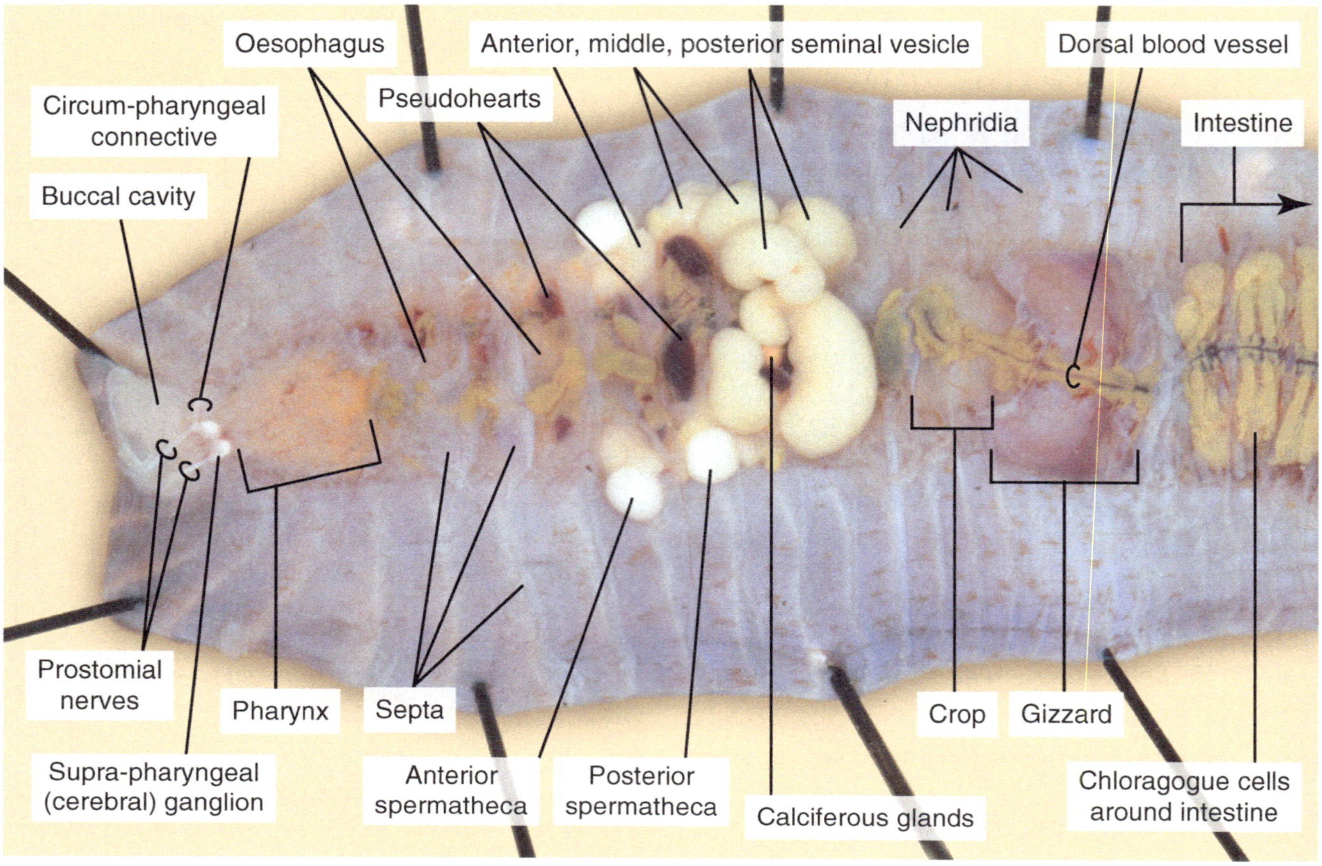

Fig. 4.10 The internal structure of an earthworm (the specimen is completely covered with water)

The earthworm is an example of a foraging herbivorous annelid, obtaining food by eating its way through the soil and extracting nutrients from the soil as it passes through the digestive tract. The earthworm takes in a mixture of soil and organic matter through its mouth, which is the beginning of the digestive tract. Identify the *mouth*; the first part after the mouth is the muscular *pharynx*, attached to the body wall by dilatator muscles for sucking action (Fig. 4.10). The muscles are torn by the dissection and give the pharynx a hairy appearance. The slender *oesophagus* leads from the pharynx to the large thin-walled *crop*, which serves for temporary food storage. The oesophagus is hidden by the pseudohearts and seminal vesicles. Three pairs of yellowish, *calciferous glands* (Morren's glands) lie on either side of the oesophagus, usually partly concealed by the seminal vesicles (Figs. 4.10, 4.11 and 4.16). They are believed to remove excess calcium and carbonates from the blood taken in with the soil. These ions are accumulated as calcite crystals. The crop is followed by the muscular *gizzard* (Figs. 4.9 and 4.10). Gently press on the crop and gizzard to test their firmness. While the crop is soft and thin, the gizzard is muscular (soil is ground up and churned within the gizzard). The gizzard leads to the *intestine* (mid-gut) which is straight and in which both digestion and absorption occur (Fig. 4.10). Yellowish-brown *chloragogue cells* cover the intestine and the dorsal blood vessel (Figs. 4.10 and 4.12). They store glycogen and lipids and have other functions as well, similar to those of vertebrate liver. Undigested material is voided through the *anus* (Fig. 4.1).

The earthworm has a closed circulatory system. Locate and identify the five pairs of *pseudohearts* (or "aortic arches") over the oesophagus (Fig. 4.11).

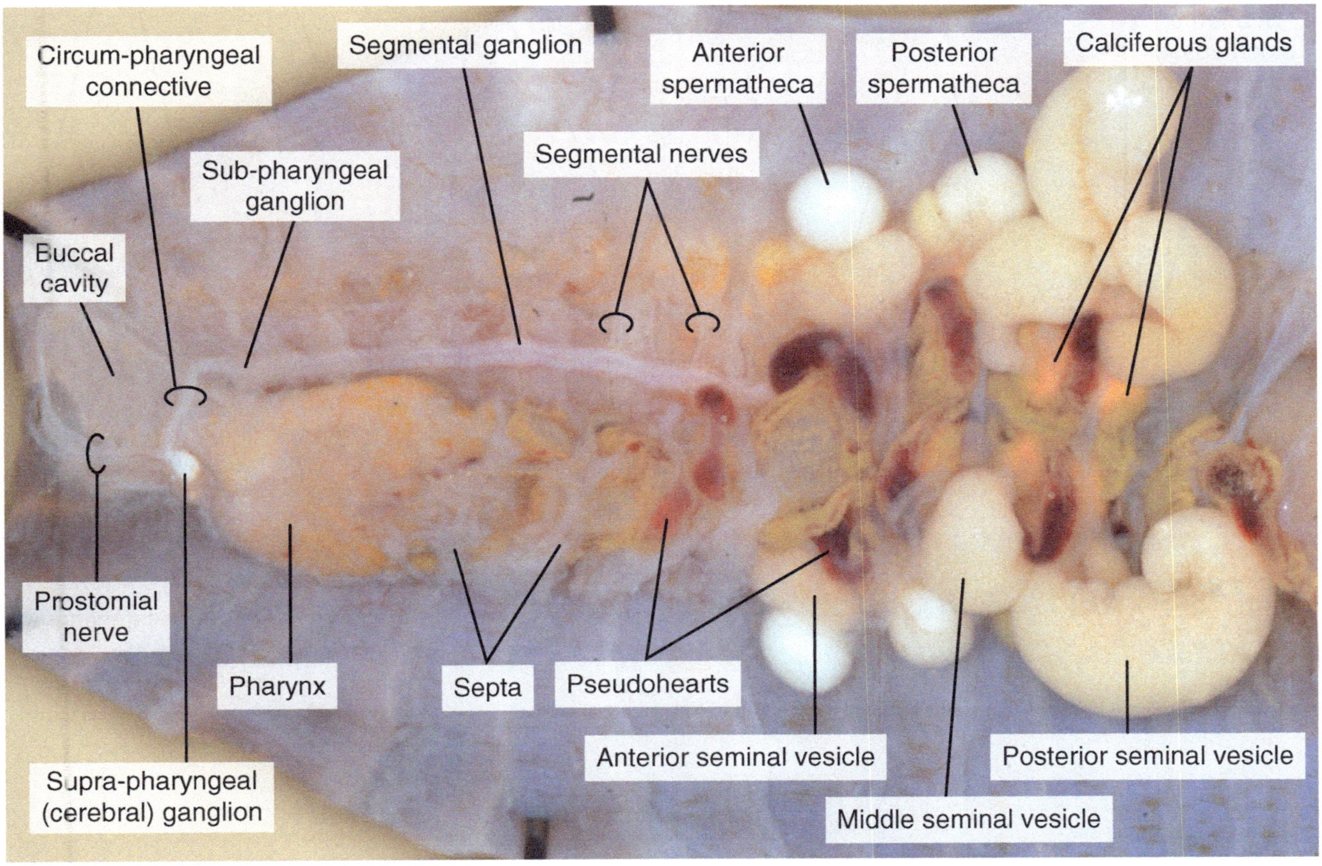

Fig. 4.11 Organs in the anterior end of an earthworm

They are the pumping organs of the circulatory system. Carefully tease away the tissues to expose the arches of the pseudohearts. Then find the *dorsal blood vessel*. Look for smaller blood vessels that branch from the dorsal blood vessel. The dorsal blood vessel appears as a dark brownish-red vessel running along the intestine. The *ventral blood vessel* is opposite the dorsal blood vessel and cannot be seen at this time because the digestive system covers it (Fig. 4.12). Retract the digestive tract. Lift up the ventral nerve cord from the ventral wall. Note the *subneural blood vessel* clinging to its lower surface and a pair of *lateroneural blood vessels*, with one located on each side of the nerve cord (Fig. 4.12). Circulatory fluids travel from the pseudohearts through the ventral blood vessel to capillary beds in the body. The blood then collects in the dorsal blood vessel and re-enter the pseudohearts.

The earthworm has no respiratory organ (gills or lungs). Gases are exchanged between the circulatory system and the environment through the moist skin.

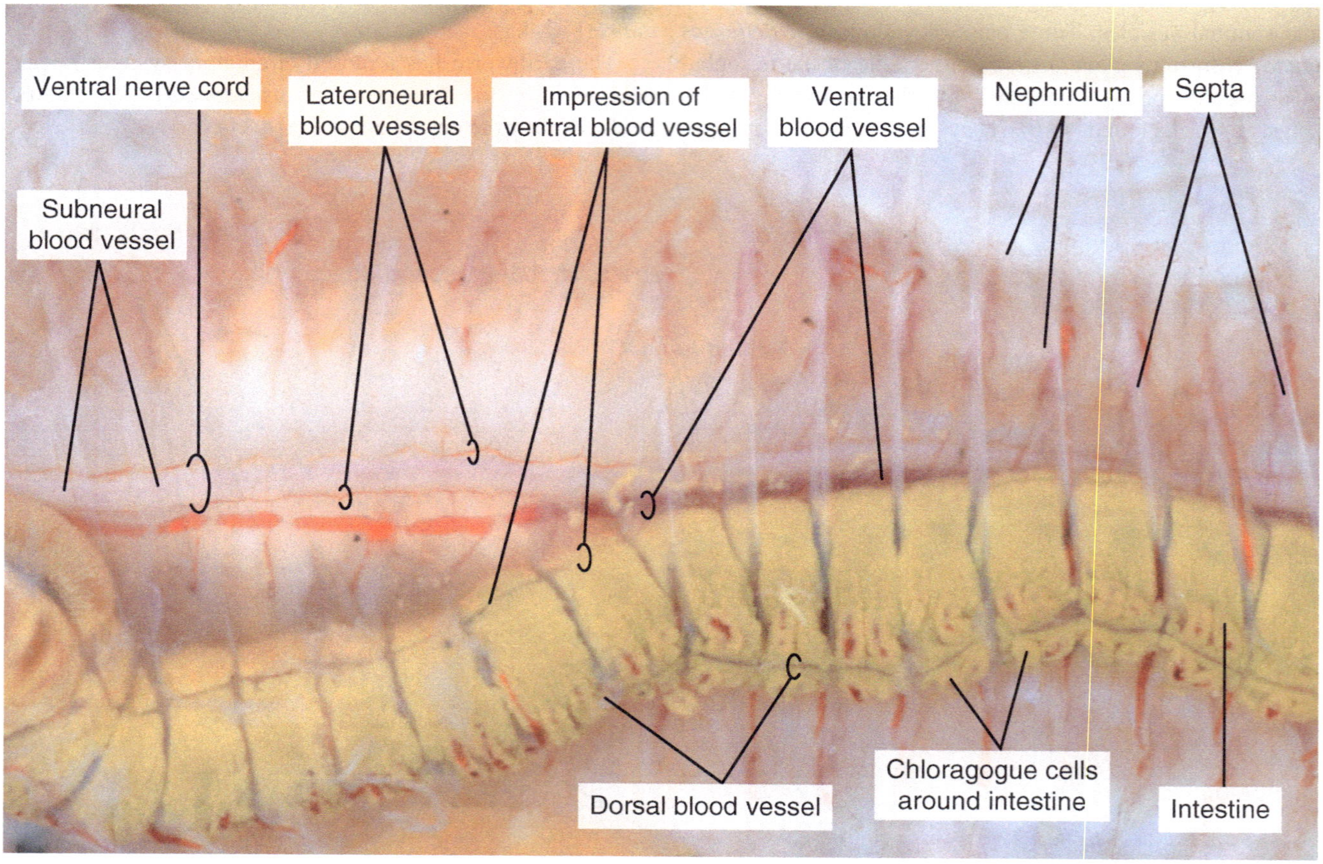

Fig. 4.12 The mid-gut (pulled partly to the left) and the underlying ventral nerve cord

The earthworm's excretory organs are tiny, tubular *nephridia* (metanephridia). They appear as tiny white fibres on the dorsal body wall. They are found in pairs in each body segment except the first three and the last one. Locate some nephridia (Figs. 4.10 and 4.12). Each nephridium begins with a ciliated, funnel-shaped *nephrostome*, which projects through the posterior septum of the segment and opens into the next segment. Coelomic fluid is drawn by ciliary activity into the nephrostome and then flows through the narrow convoluted tubule where ions are reabsorbed. The urine, containing wastes, collects in the bladder, which empties to the outside through a *nephridiopore*.

Mid-gut of the earthworm provides a large surface for enzyme production and nutrient absorption by forming a deep, folded invagination called *typhlosole* (Fig. 4.13, top left). Whole inner surface of the mid-gut is lined by tall, columnar epithelium bearing a brush border. Intestine wall is supported and moved by thin circular muscle layer (CM) and longitudinal muscles (LM) – latter rather form a network than a continuous layer (Fig. 4.13, top left). The outer surface of mid-gut is covered by modified visceral peritoneal cells, named as *chloragogue cells* (ChC) (Fig. 4.13, top left). They have essential role in keeping the homeostatic equilibrium of blood. They can be recognised by their foamy cytoplasm and strange, yellowish colour. The earthworm has a closed *circulatory system*, which consists of a continuous network of endothelial-lined blood vessels. Main vessels have their own musculature by which blood is pumped. Pseudohearts ("hearts") are repetitive vessels connecting dorsal and ventral (subintestinal) main vessels. They have well-developed circular (CM) and longitudinal muscle layers (LM) (Fig. 4.13, top right). Annelids are the simplest organism to have a true *coelom* (Fig. 4.13, top left and right, and bottom left). A coelom (see-lum) is a fluid-filled cavity containing *coelomocytes* (CC), lined with mesodermal tissue. It helps to protect organs, aids in digestion and movement and provides space for the circulatory "pumping". A separate coelomic sac is found in every segment of the earthworm. *Nephridia* are segmentally repeated, paired excretory organs of the earthworm. It begins as a ciliated funnel which propels the coelomic fluid into a folded nephridial tubule (Fig. 4.13, bottom left and right). It has morphologically different divisions on the basis of the epithelial lining. The tubule is accompanied by a capillary network for reabsorption necessary materials from urine. Ciliated distal part of the tubule perforates the body wall to open onto the surface (Fig. 4.13, bottom right).

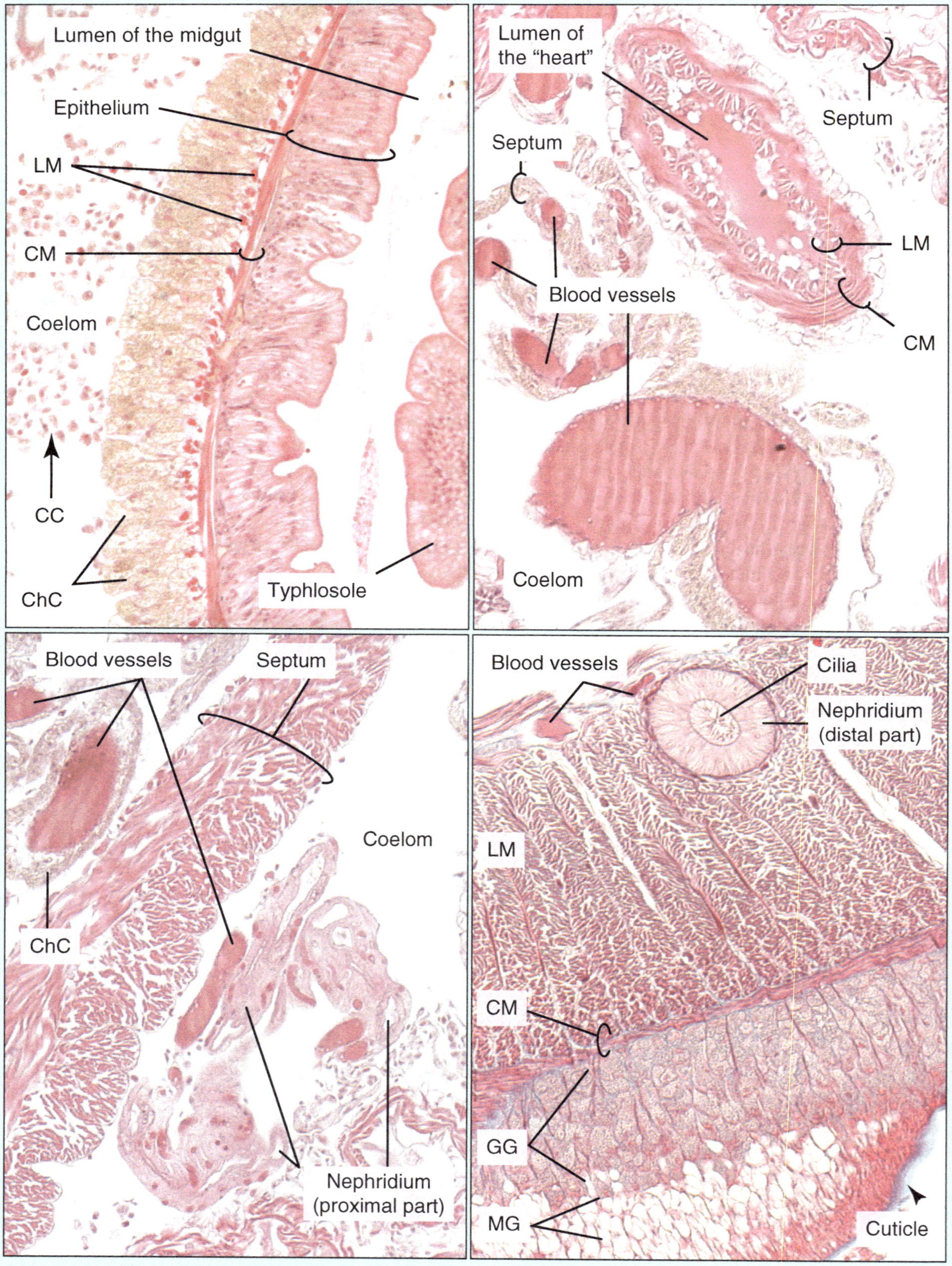

Fig. 4.13 Histological sections of the digestive, circulatory and excretory systems (*top* and *bottom left*, HE; *bottom right*, Azan). *CC* coelomocytes, *ChC* chloragogue cells, *CM* circular muscle layer, *GG* granular glands of the clitellum, *LM* longitudinal muscle layers, *MG* mucous glands of the clitellum

Find the small pair of white *supra-pharyngeal* or cerebral *ganglion* (the brain) lying on the anterior dorsal end of the pharynx and partially hidden by dilatator muscles (Figs. 4.11 and 4.14). (If you can't find it, it is probably because it was destroyed when you cut the worm.)

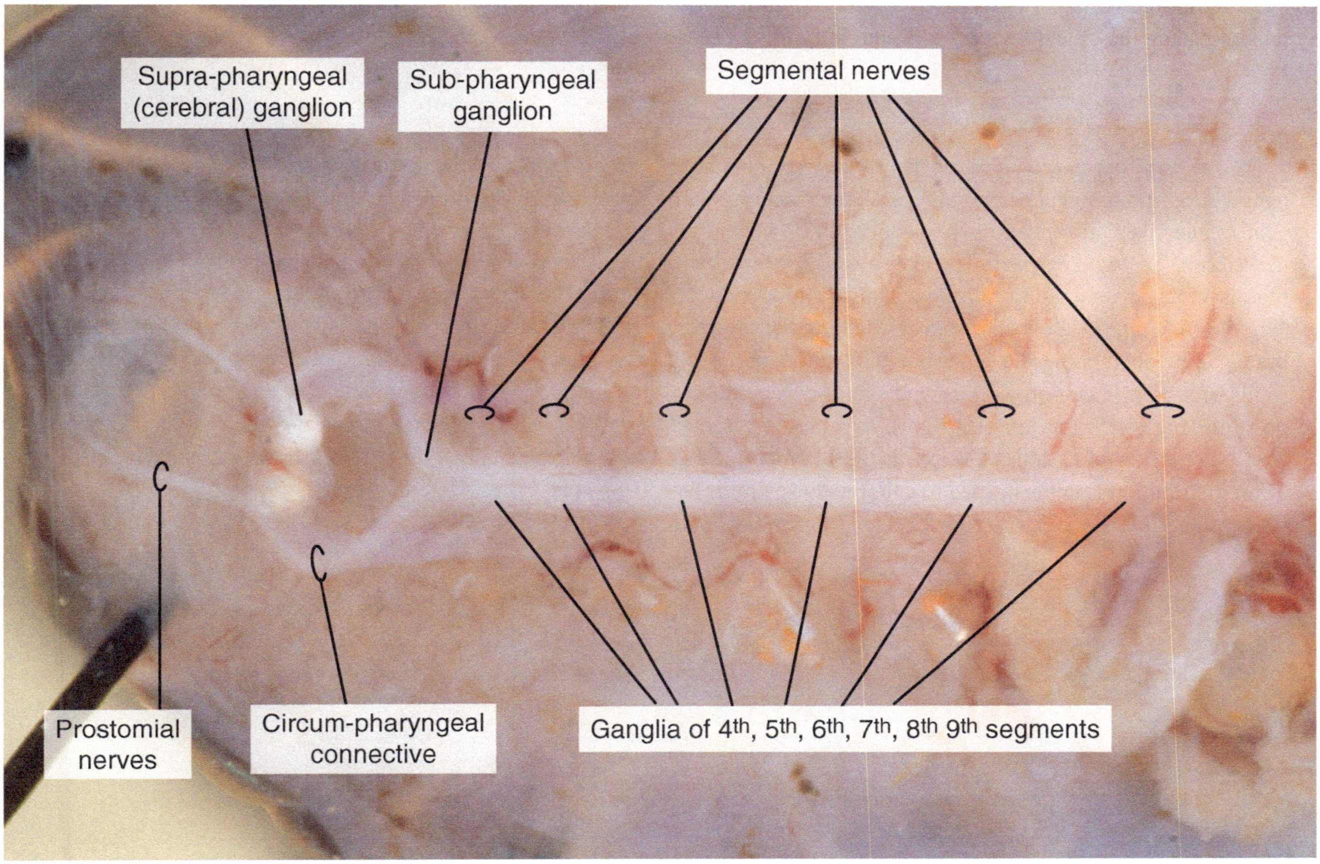

Fig. 4.14 The dissected anterior end of the ventral nerve cord

Observe the delicate white *prostomial nerves* from the cerebral ganglia to the prostomium and a pair of *circumpharyngeal connectives*, extending from the ganglia and encircling the pharynx to reach the *sub-pharyngeal ganglion* under the pharynx (Fig. 4.14). Locate the *ventral nerve cord* by pushing aside the digestive tract and searching for a white string-like structure that runs the length of the worm and attaches to the sub-pharyngeal ganglion. Each segment contains a slight enlargement, or *ganglion*, which is a mass of tissue containing many nerve cells. Lateral *segmental nerves* branch from each ganglion to innervate the segments (Fig. 4.14).

The *central nervous system* of the earthworm contains paired supra-pharyngeal (cerebral) ganglia, circumpharyngeal connectives, sub-pharyngeal ganglia and the ventral nerve cord. The whole system is encapsulated by a connective tissue layer containing muscle cells and blood vessels (Fig. 4.15). *Ganglia* show characteristic features of invertebrate ganglia: cell bodies have peripheral position, and neuropil composed of cell projections and synapses has central position (Fig. 4.15, top left and right). In addition to neurons, supporting glial cells develop in the nervous tissue. Their distribution is not restricted to ganglia; these tiny cells can be found in interconnecting elements as well. Ventral nerve cord has paired segmentally repetitive ganglia interconnected by transverse *commissures* (these are not visible on the sections) and longitudinal *connectives* (Fig. 4.15, top right). The ventral nerve cord is accompanied by subneural (SV) and lateroneural blood vessels (LV), one medial and two lateral giant fibres. *Giant fibres* are nerves with the fastest conductivity among invertebrates: median one transmits impulses to the posterior direction, whereas lateral ones to the anterior direction. They have separate capsules (Fig. 4.15, bottom left and right).

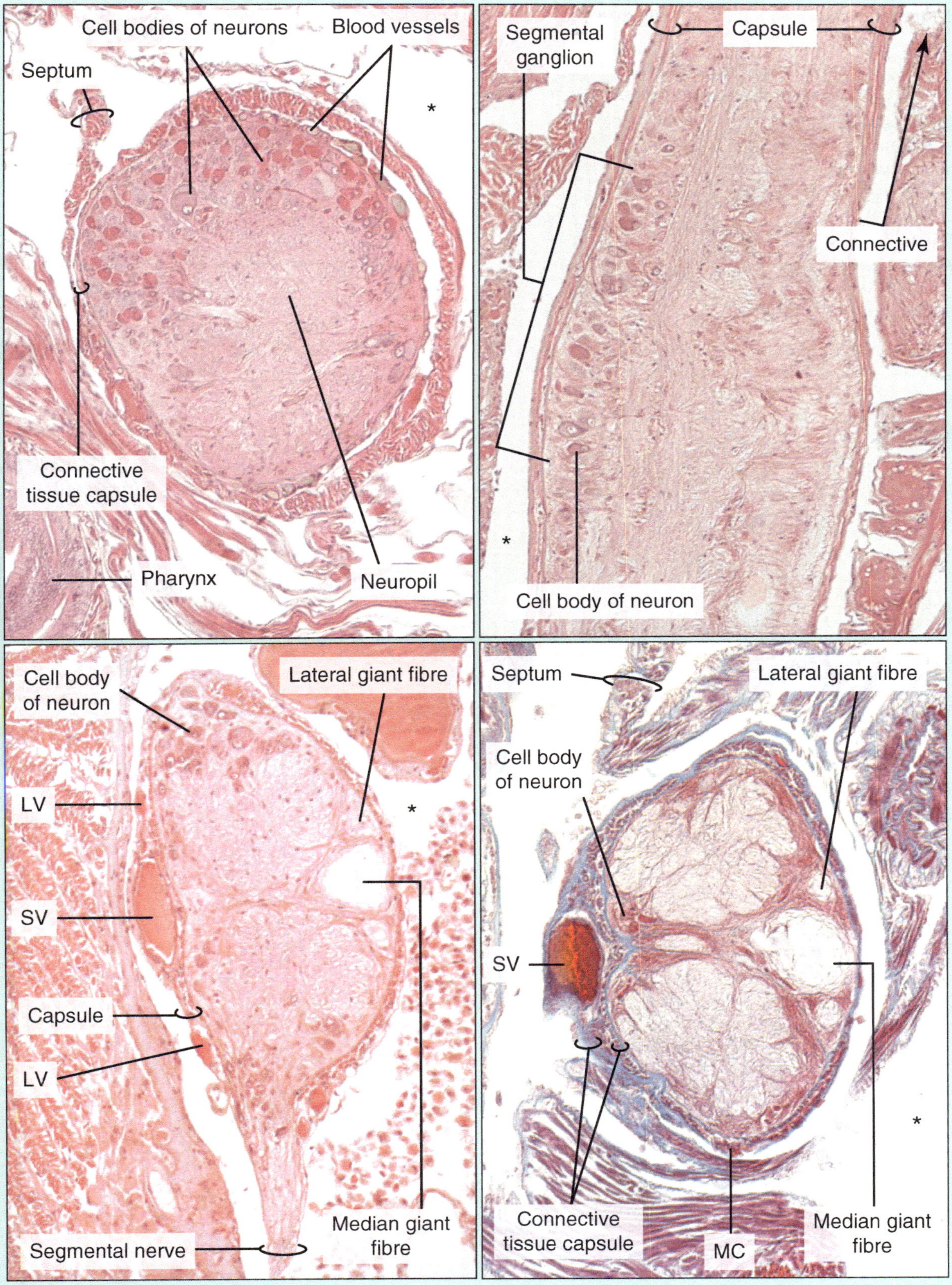

Fig. 4.15 Histological cross sections and a longitudinal section (*top right*) of the ventral nerve cord (*top ones* and *bottom left*, HE; *bottom right*, Azan) *asterisks* coelom, *LV* lateroneural blood vessel, *MC* muscle cells, *SV* subneural blood vessel

The earthworm is monoecious: it has both male and female organs in the same individual, but cross fertilisation occurs during copulation. First, consider the male organs (Figs. 4.11 and 4.16).

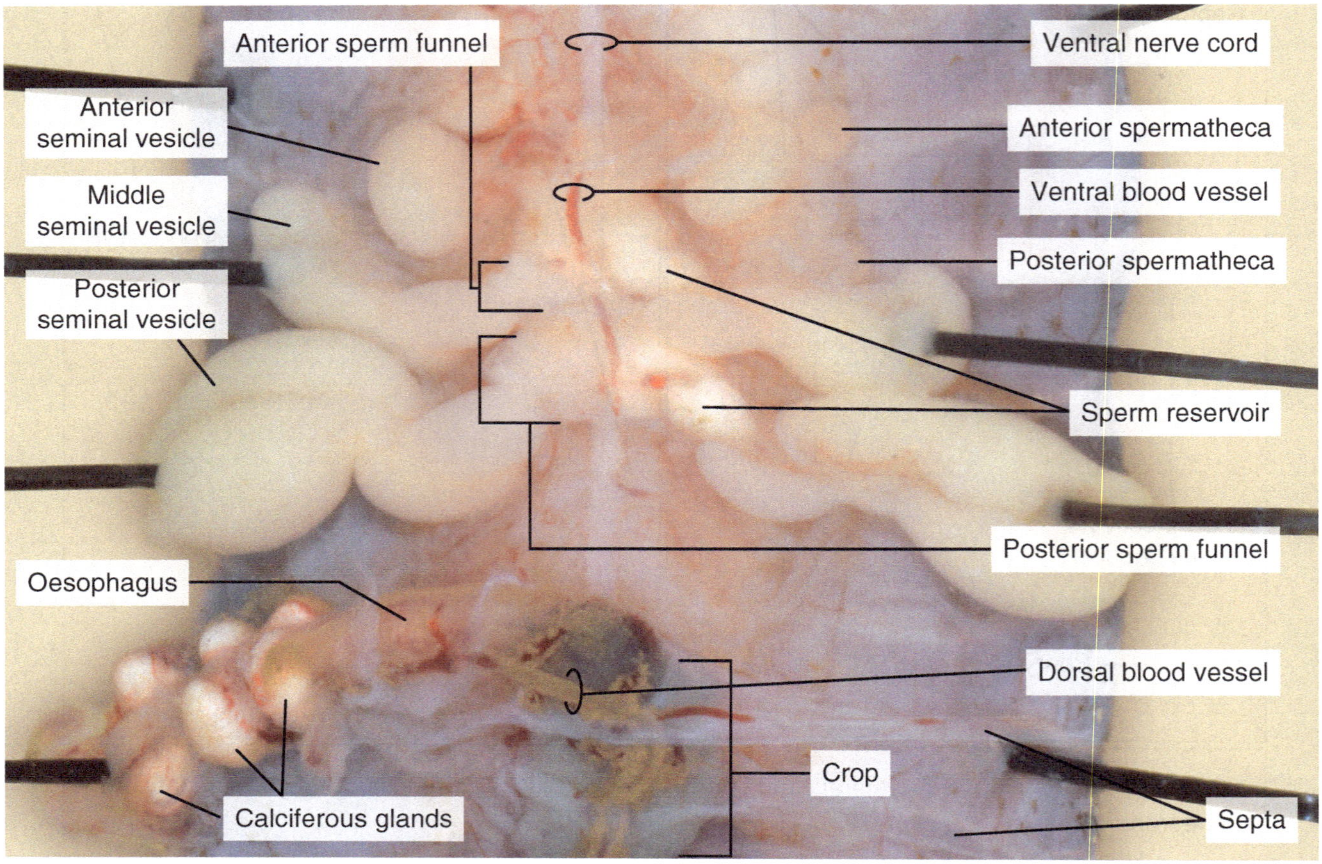

Fig. 4.16 Structures of the male part of the reproductive system. Note the oesophagus is cut and folded out to the left; the oesophageal calciferous glands are well visible on the two sides of it

The three pairs of *seminal vesicles* (sperm sacs in which spermatozoa mature and are stored before copulation) are attached to segments 9, 11 and 12. They lie close to the oesophagus. To find these organs, you will again have to push aside some parts already dissected. The two pairs of *testes* are housed in special sperm reservoirs in the ventral part of the seminal vesicles. *Sperm funnels* collect mature spermatozoa, and then through the *efferent ducts* they are transferred into the two small vas deferens or *sperm ducts* which connect with the male genital pores in segment 15. These structures are too small to be found easily. Sperm funnels are likely to be found if the dorsal wall of the seminal vesicles is removed (Fig. 4.16).

The female organs are also very small. The two pairs of round, glistening white seminal receptacles or *spermatheca*, easily seen in segments 9 and 10, store spermatozoa after copulation (Fig. 4.11). It is much harder to find and identify the pair of *ovaries* in segment 13. The paired *oviducts* with ciliated funnels that carry eggs to the female genital pores in the next segment will probably not be seen.

During mating, sperm from one worm travel along the seminal grooves to the spermatheca of another worm. Fertilisation of the eggs takes place outside the body in the cocoon, as it moves forward over the body, picking up the eggs of one worm and its mating partner's sperm from the spermathecae.

Earthworms are hermaphrodite animals. Their testes are found in a separated cavity of the seminal vesicles (Fig. 4.17). The *testis* contains mitotically active spermatogonia and spermatocyte groups, named morulae upon their shape. A *morula* is formed by interconnected sister cells. These cell groups break off the testis and fall into the cavity of the *seminal vesicle*; their development continues here. A morula has a central, common cytoplasmic mass, called cytophore, which is encircled by developing spermatocytes and spermatids (Fig. 4.17, bottom left insert). Every cell is in the same developmental stage in a morula. Differentiated, flagellated spermatozoa detach the cytophore and get into the highly folded, ventral part of the seminal vesicle, the so-called *sperm reservoir* (Fig. 4.17, bottom left). The coiled and ciliated *efferent duct* collects them during mating and transfers them towards the male genital pore. From here through the seminal groove, they get into the mating partner's seminal receptacle or spermatheca. The *spermatheca* has a round profile with a thick wall (Fig. 4.17, bottom right). It contains only morphologically fully developed, flagellated spermatozoa forming a whirl-like pattern.

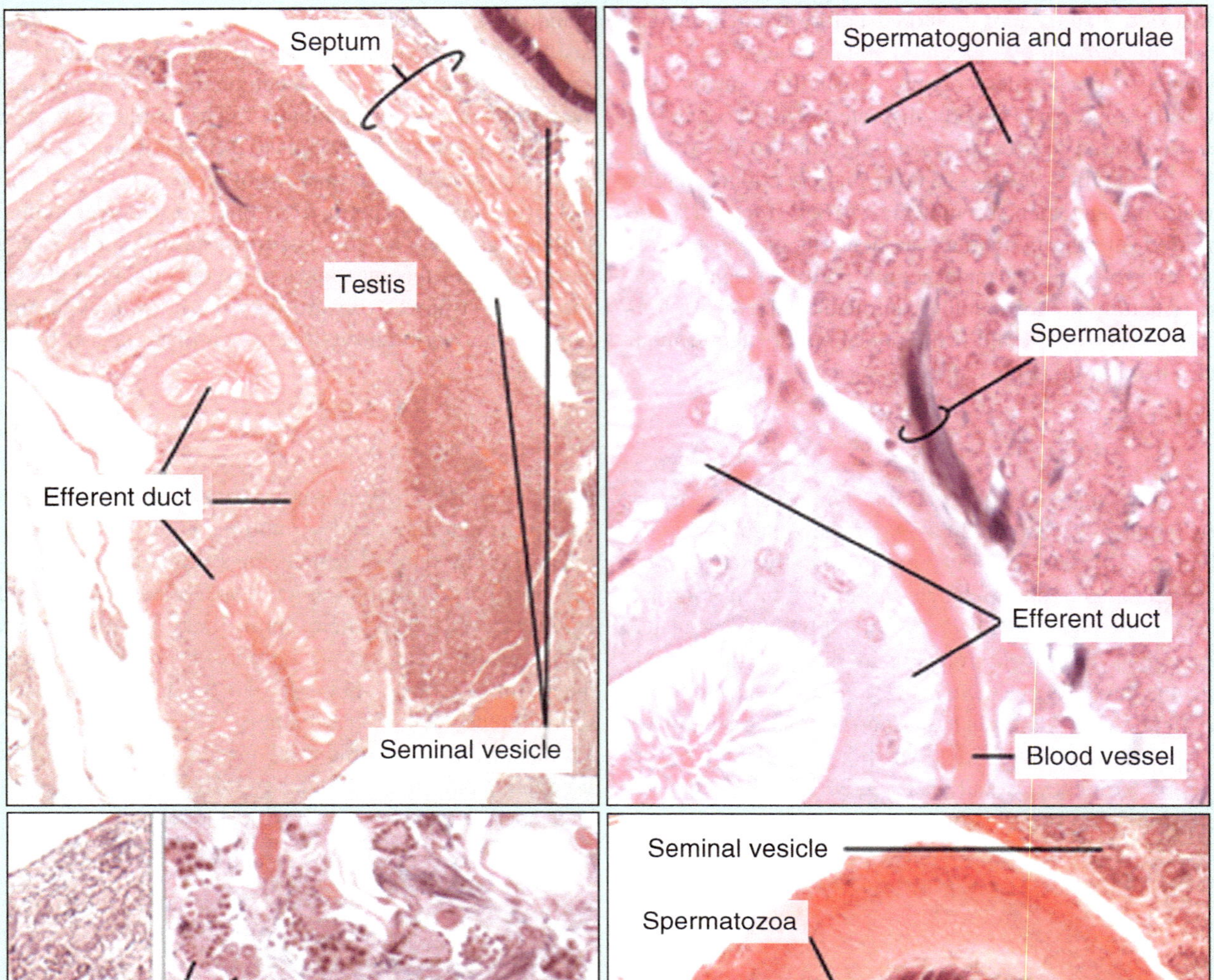

Fig. 4.17 Histological sections of the reproductive system (HE); *top*, testis and efferent duct in two magnifications; *bottom left*, seminal vesicle; *bottom right*, spermatheca

5 Dissection of a Snail (*Helix pomatia*)

- *Availability*: Roman snails (*Helix pomatia*) like places which are dark and damp. In spring and autumn, they are most active and easy to collect. When it's dry or cold, they seal themselves up; they hibernate in winter and aestivate in summer. During these periods giant Ghana snail (*Achatina* sp.) can be purchased from zoos instead. Different species of snails differ slightly internally. The instructions given here should enable the student to dissect any type of pulmonate (air breathing) snail. Examine the snail alive and watch it crawl. Place it on a sheet of glass, watch it from the other side and observe foot muscle contraction waves. Study the external appearance, noticing especially the head, tentacles, collar and opening of mantle (Figs. 5.1 and 5.2).

Fig. 5.1 Roman snail (*Helix pomatia*)

P. Lőw et al., *Atlas of Animal Anatomy and Histology*, DOI 10.1007/978-3-319-25172-1_5

Fig. 5.2 Head of a living Roman snail

- *Anaesthesia*: The complete procedure takes about 24 h, so it should be started the day before the dissection. Put the snail in a screw-cap jar and pour over up to the rim boiled and re-cooled water. It is important not to leave any bubbles above water. Boiling is necessary in order to remove oxygen from water. Check if the snail is completely dead before dissection by pricking it with a teasing needle. The slightest reaction shows that it's still alive. In this case inject some 4–6 % (w/v) $MgCl_2$ solution into the body cavity. Once the edge of the mantle is everted and detached from the shell so that the outer surface of the mantle is visible (Fig. 5.4, right) the snail is dead. The process of anaesthesia can be speeded up by dissolving 4 % (w/v) chloral hydrate or $MgCl_2$ in the boiled water. By this method snail takes up water so the head and foot remain out of the shell. It secretes a lot of mucous, which should be washed away.

Examine thoroughly the snail before starting the dissection (Fig. 5.3). At the anterior end, find the *mouth* with the four *lips* around and below that the opening of the duct of mucous *pedal gland*. Identify two pairs of *tentacles* on the dorsal side: a pair of anterior, shorter tentacles and a pair of posterior, longer tentacles. On the latter *eyes* appear as little black dots.

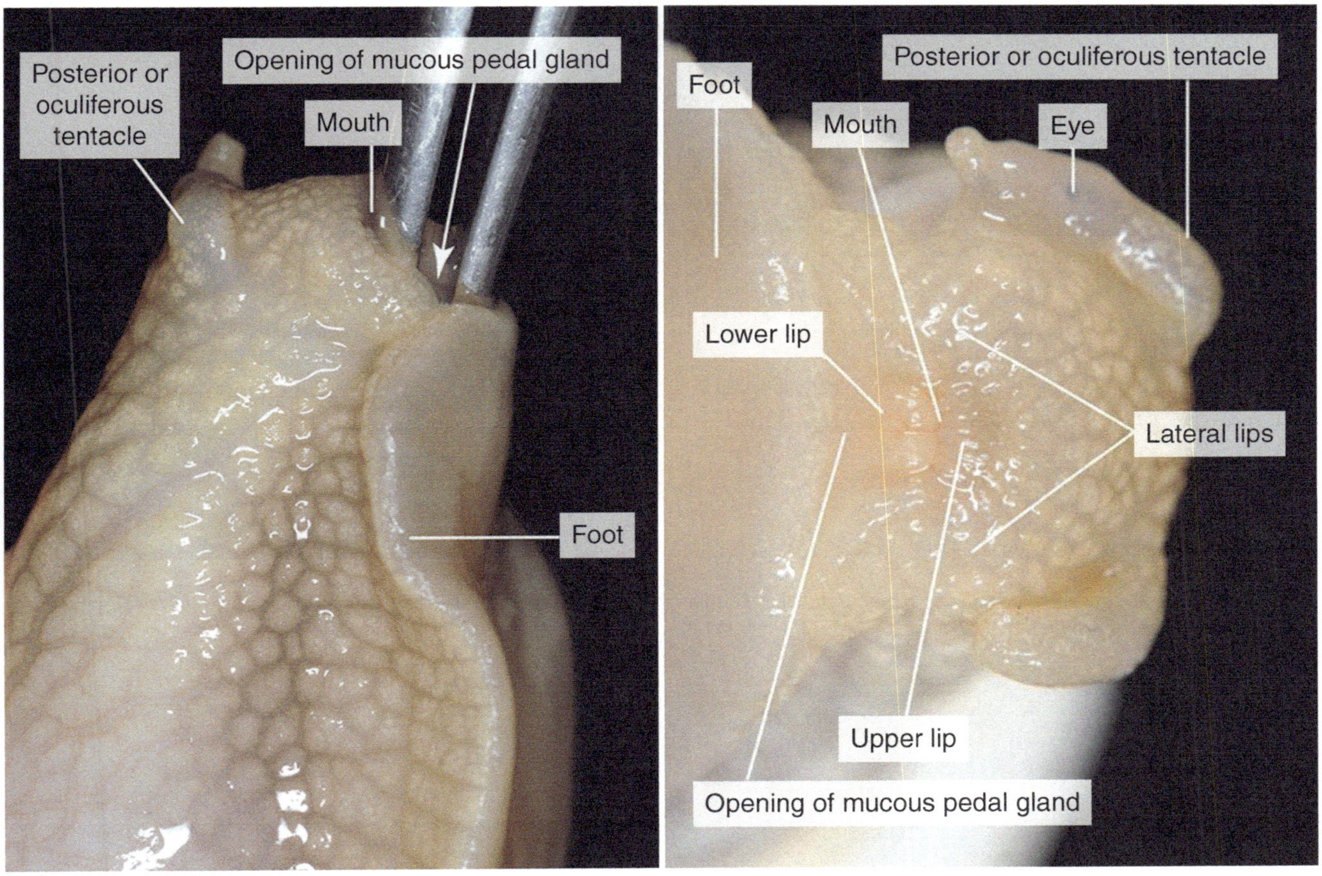

Fig. 5.3 Head region of snail. *Left panel:* right side view of the head. The opening of the duct of mucous pedal gland is between the head and the edge of foot; the tweezers are in it. *Right panel*: front view of the head

The *genital aperture* is situated halfway between the right posterior tentacle and the edge of the foot. The opening of the pulmonary chamber or mantle cavity is positioned on the *edge of mantle* or collar. This serves as an airway, but the secondary urinary duct and the rectum opens here as well (Fig. 5.4).

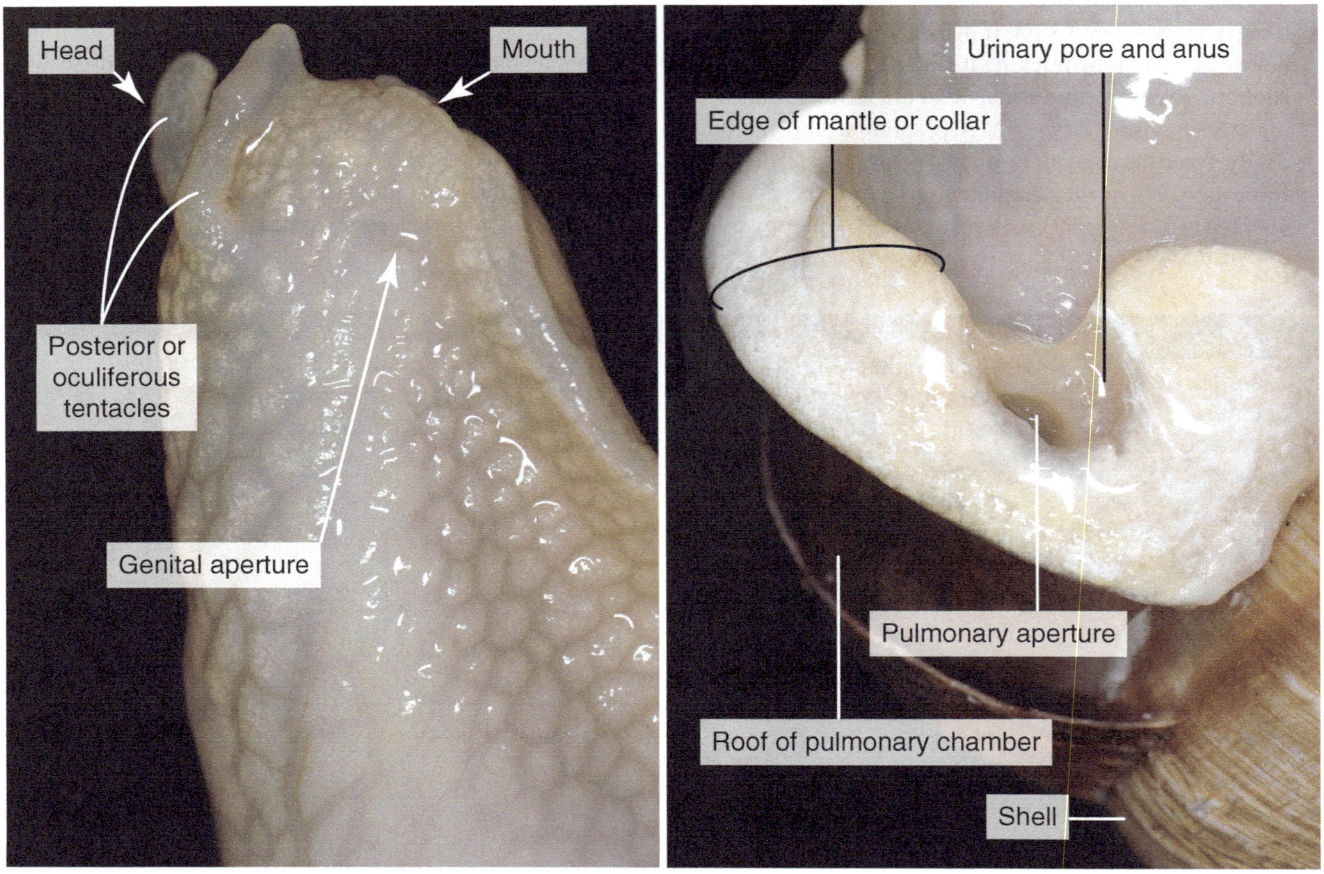

Fig. 5.4 Openings of the snail are situated on the right side. *Left panel*: genital aperture on the right side of the head. *Right panel*: pulmonary aperture on the edge of mantle

The light microscopic sections made from the *edge of mantle* show a typical structure and pattern (Fig. 5.5). The surface is covered by the *epidermis* which is composed of a simple columnar epithelium. The underlying *connective tissue* contains several *unicellular glands*, which produce a calcareous mucous layer onto the surface. Just beneath the surface, the *mucous glands* (UMG) have scummy cytoplasm because the secretion is lost during the histological preparation. The so-called *protein glands* (UPG) contain large amount of protein which show homogenous appearance. The *calciferous glands* (UCG) have a granular content. This pattern is visible on all sections stained either with haematoxylin-eosin or Azan or PAS-alcian blue. In the latter case, the blue or purple colour of the glands and the colour of the mucous layer depend on the pH of the secreted material (Fig. 5.5, right).

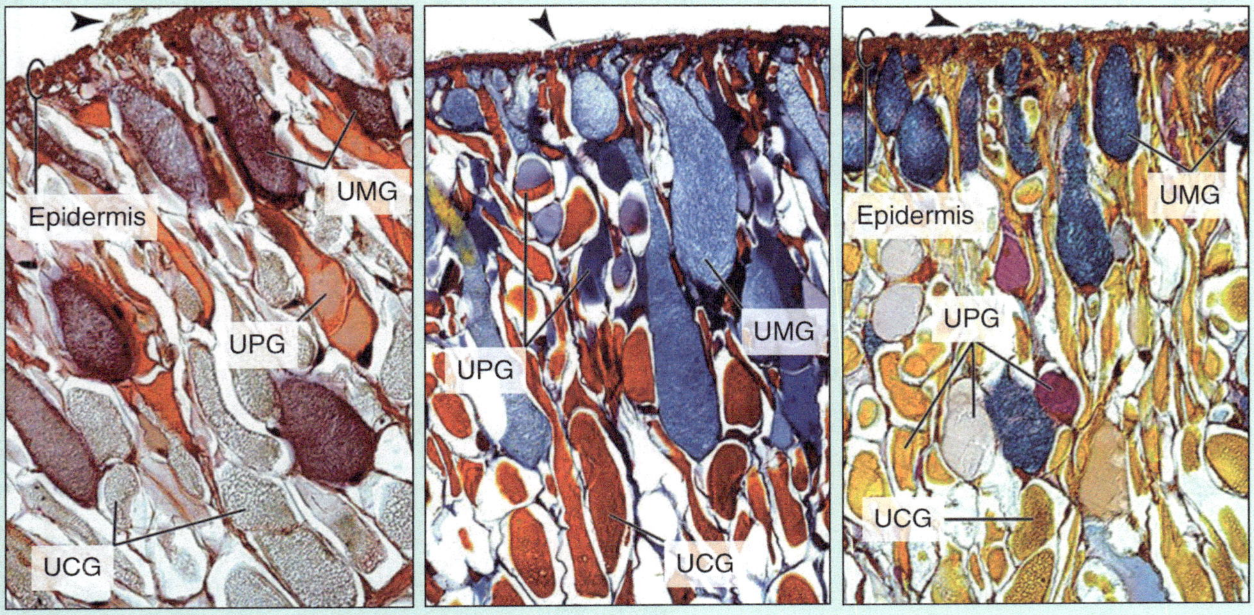

Fig. 5.5 Histological sections of the edge of the mantle. (*Left*, HE; *middle*, Azan; *right*, alcian blue-PAS staining) *arrowheads* secreted mucous, *UCG* unicellular calciferous glands, *UMG* unicellular mucous glands, *UPG* unicellular protein glands

The dissection of the snail is simple in principle as it consists only of removing the shell, then opening the mantle cavity, cutting the dorsal body wall and disentangling the various internal organs. It is, however, difficult to describe because the varying amount of contraction of the individual specimen alters the placing of the parts. As the first step of dissection, remove the shell with a strong pair of tweezers. Go around the whorls, and proceed gradually (Fig. 5.6).

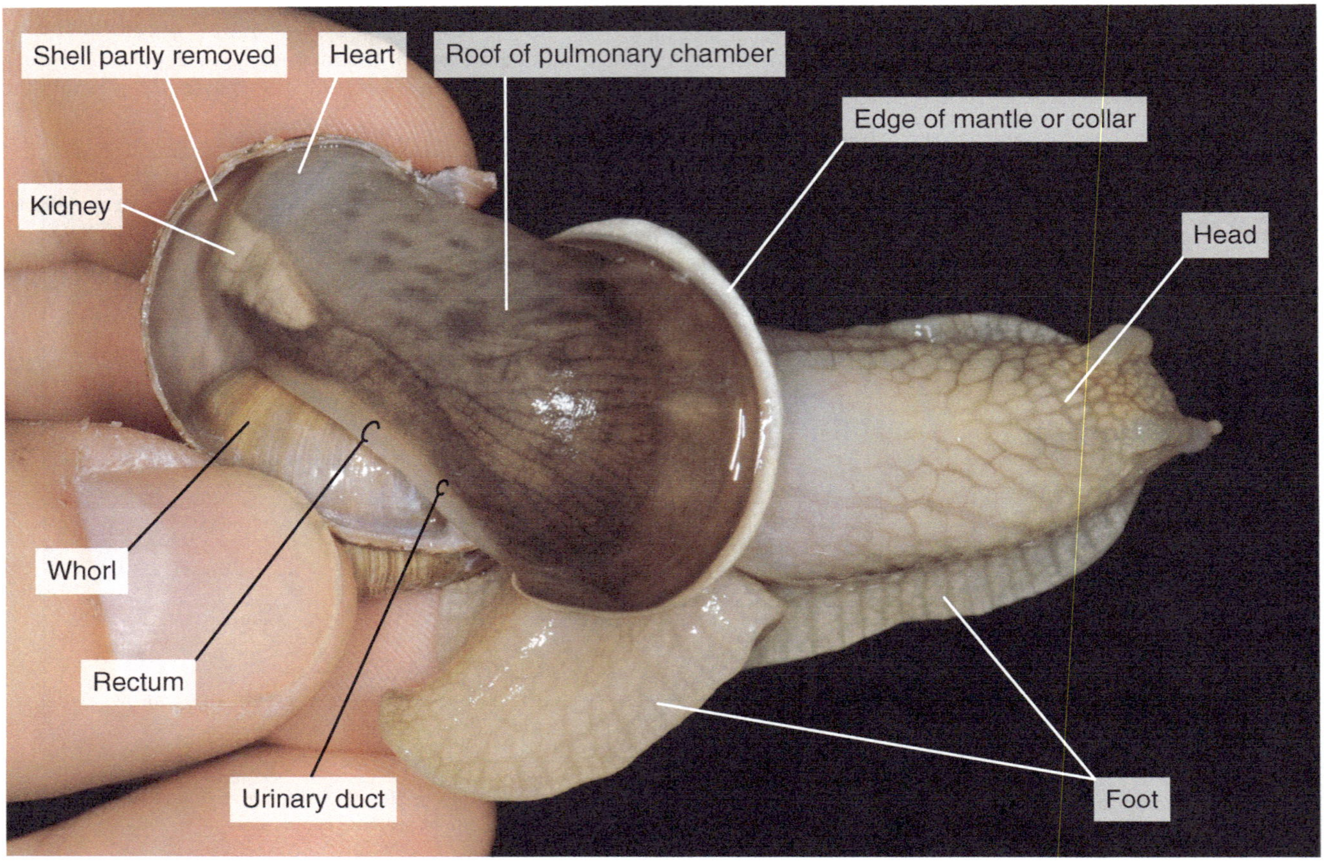

Fig. 5.6 Snail with shell partly removed. The organs are visible through the transparent epidermis

Prise away the pieces as you break them off, leaving only the central parts round the columella. Be very careful not to damage the coiled part of the visceral mass. Look for the *columellar muscle* (Fig. 5.18) and push it off from the shell with the tip of your tweezers. Remove the remainder of the shell. The visceral mass slides out from the last whorls. Finally wash away the mucous and splinters of the shell (Fig. 5.7).

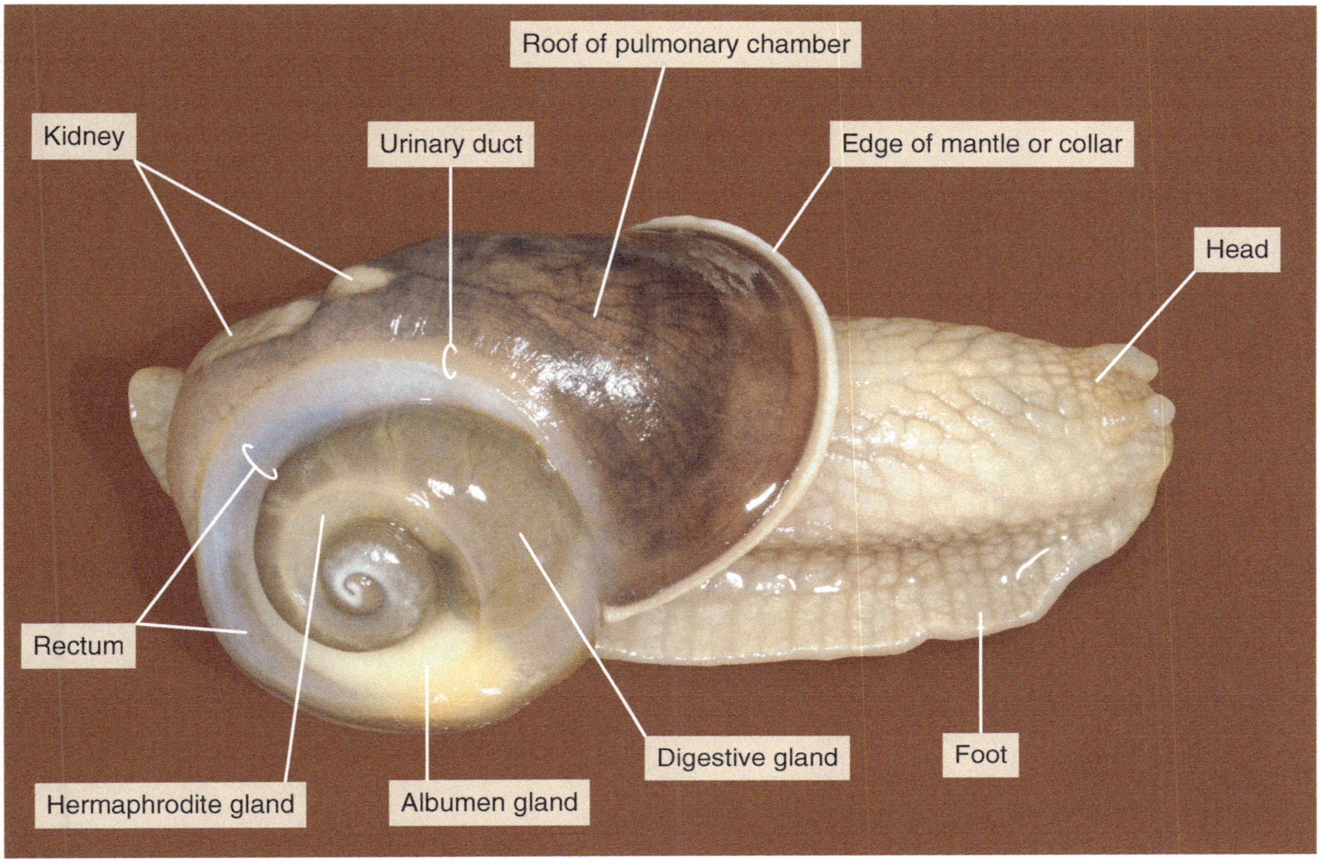

Fig. 5.7 Snail after removal of the shell. Note some organs are visible through the transparent epidermis of the visceral mass

Three cuts are needed to reveal the internal organs: for the *first*, hold the snail in one hand, insert one blade of a pair of scissors through the *pulmonary aperture* (pneumostome) and cut the mantle away from the body wall. The line of attachment of the mantle to the body wall is under the collar, not along its edge (Fig. 5.8).

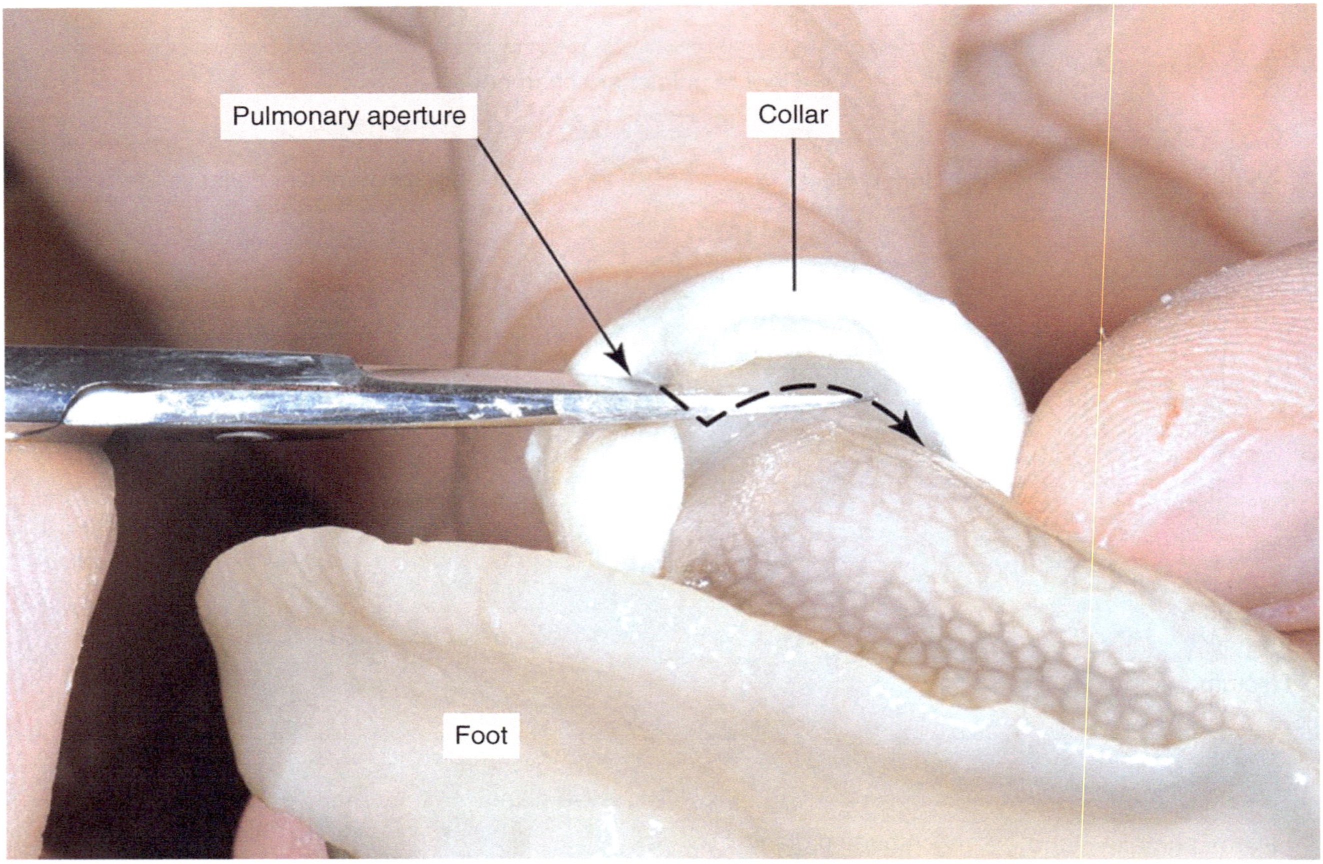

Fig. 5.8 The first cut of the dissection, as indicated by the *dashed line* under the edge of mantle

For the *second* cut, alter the position of grip, start again from the pulmonary aperture and cut in the opposite direction in order to free the rest of the edge of the mantle (Fig. 5.9).

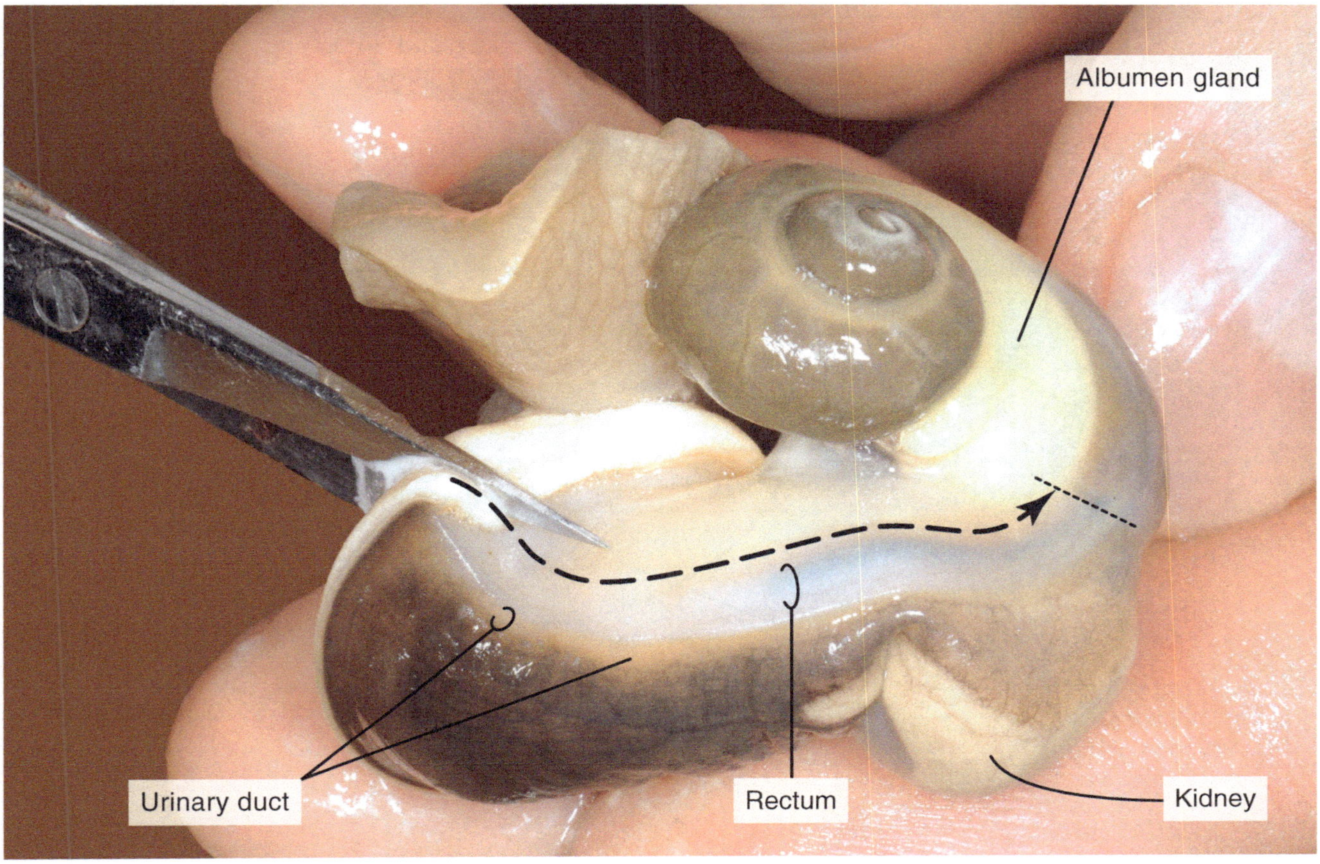

Fig. 5.9 The second cut follows the rectum to the highest level of the pulmonary chamber (indicated with *dashed line*)

Continue the cut along the wall of the visceral mass up to the albumen gland and beyond to the hermaphrodite gland (ovotestis) in such a way that the rectum remains attached to the loosened part (Fig. 5.10).

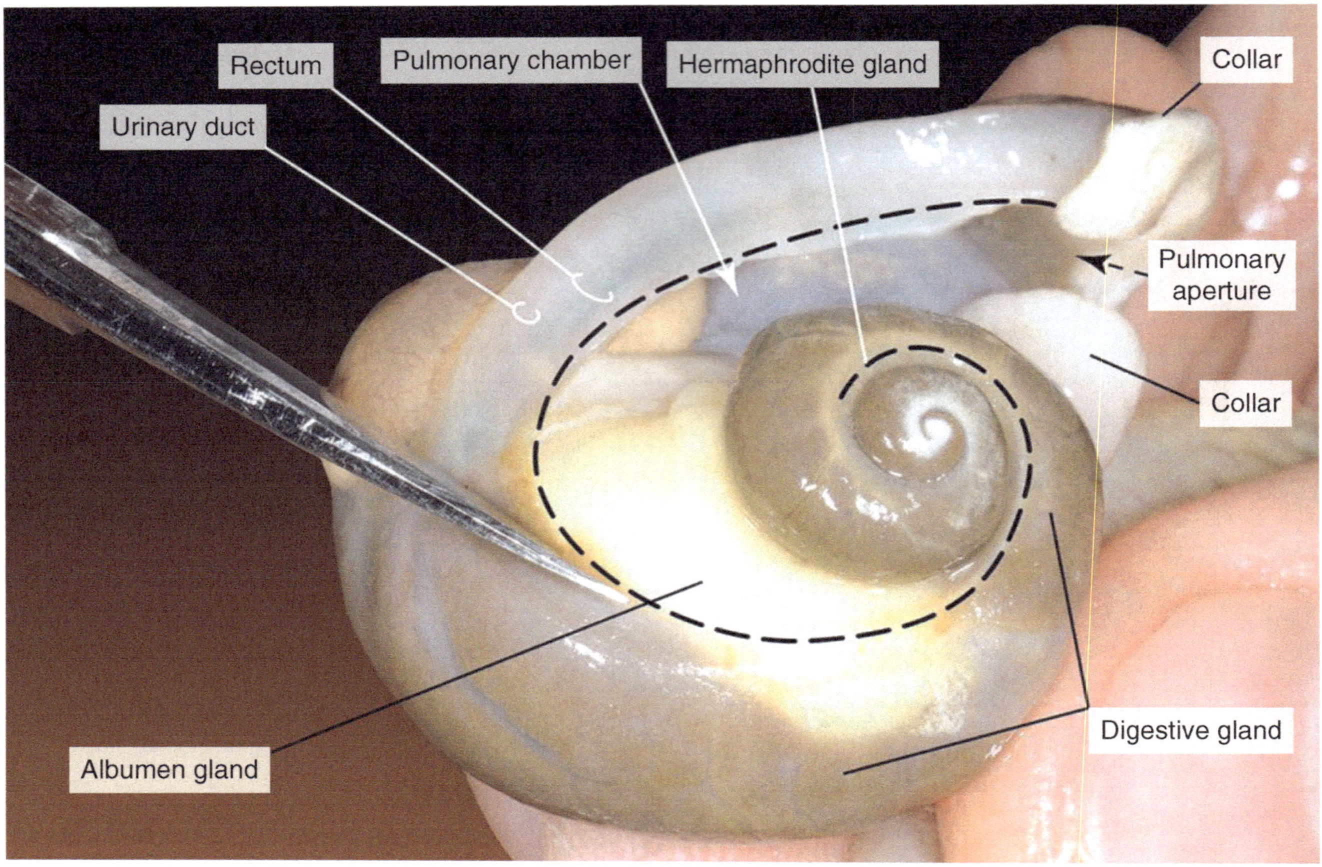

Fig. 5.10 The second cut continues on the wall of the visceral mass (indicated with *dashed line*)

Then using a pin on either side of the head and another through the posterior part of the foot, attach the snail to the wax of a dissecting dish leaving space on both sides. Turn back the mantle on the left hand side, and pin it as shown on Fig. 5.11.

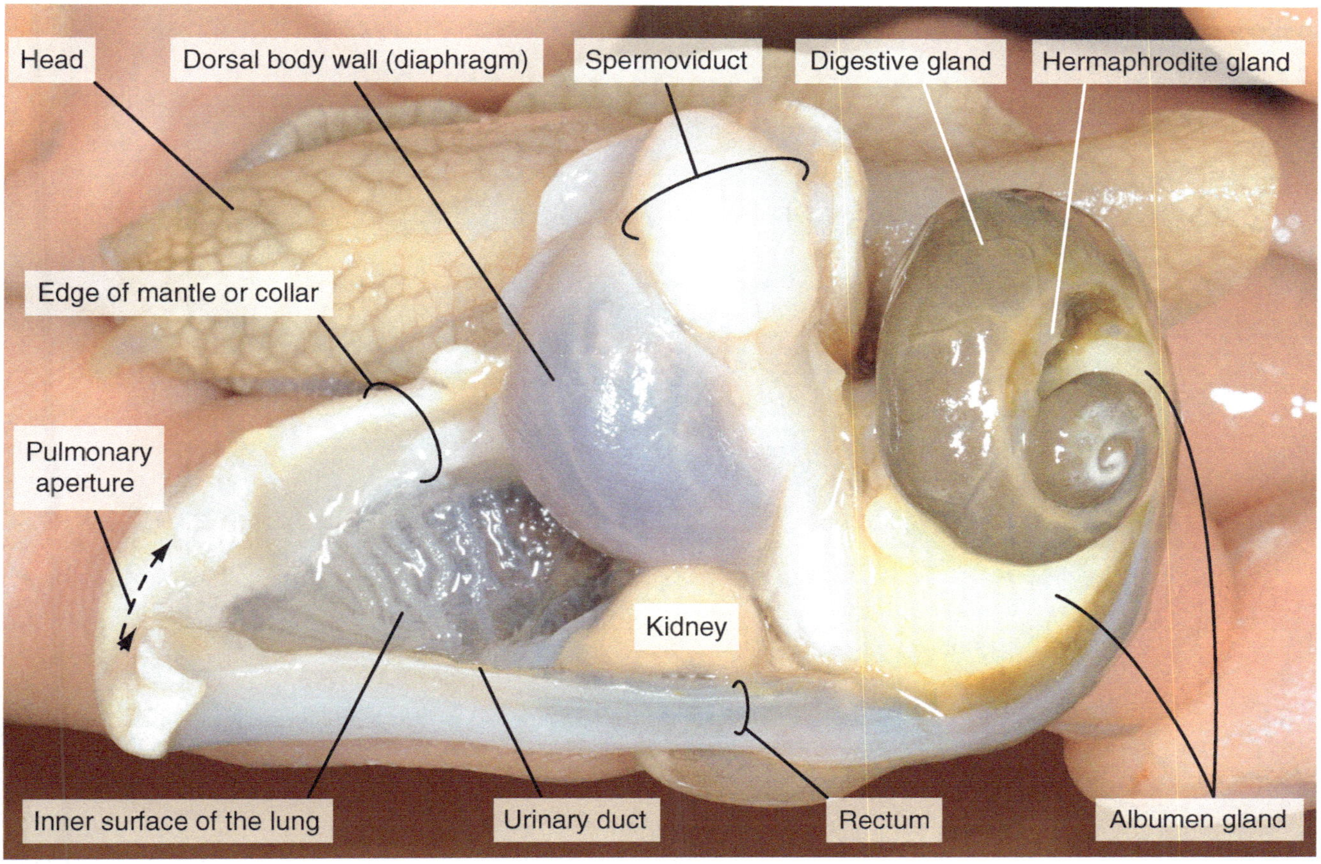

Fig. 5.11 Organs of the snail after the first two cuts

For the *third* cut, cut from the visceral mass forwards through the body wall along the dorsal midline to the dorsal lip of the mouth as indicated by the dashed line on Figs. 5.12 and 5.13.

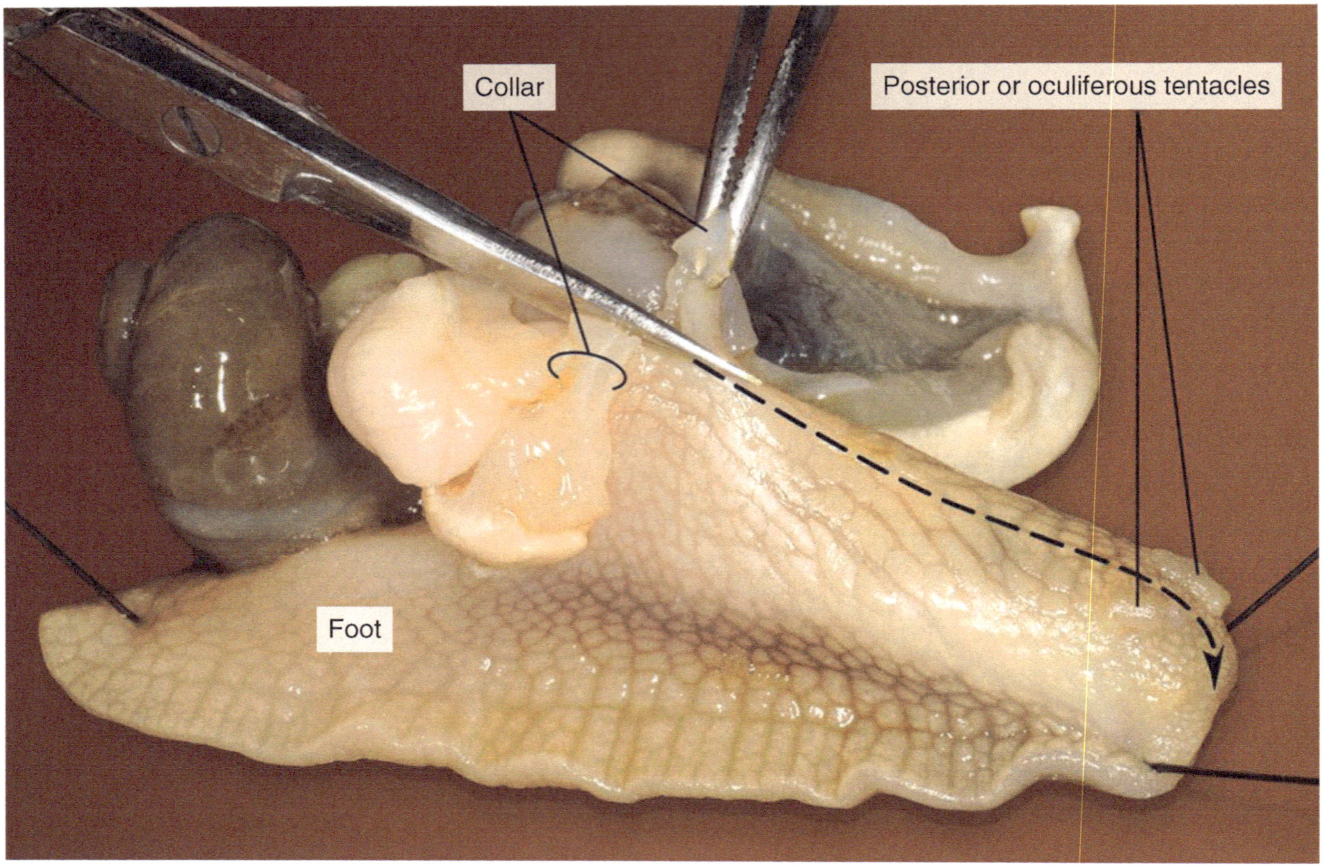

Fig. 5.12 The direction of the third cut (indicated with *dashed line*)

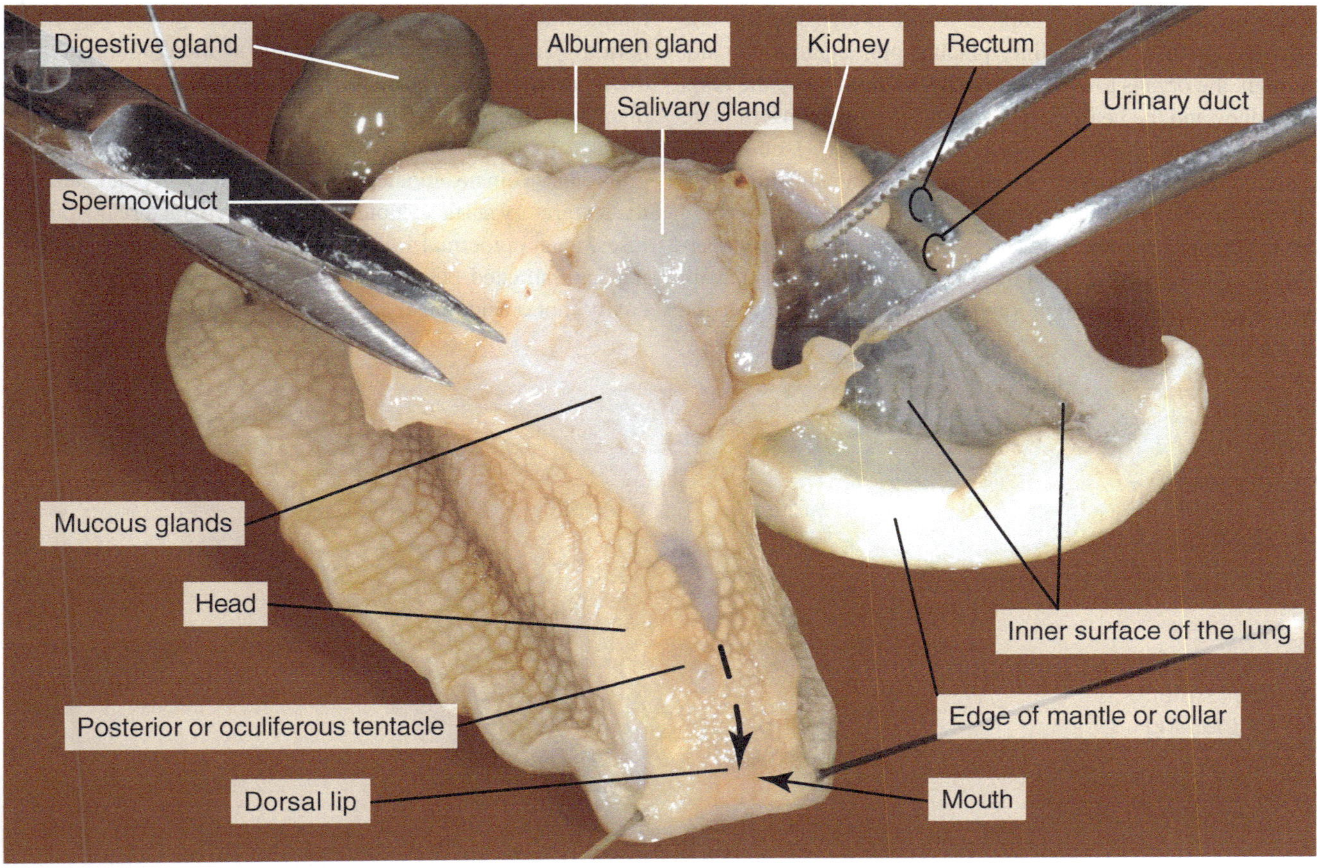

Fig. 5.13 The line of the third cut runs to the mouth

Cover the specimen completely with water. Display the structure as fully as possible, placing pins against, but not through, the organs to hold them in position. Identify the lung, the kidney and the heart. Snails are one of the few invertebrate groups to successfully invade land. In dry-land snails (*Pulmonates*), the mantle cavity has become an air-filled *lung*. The wall of the mantle cavity is heavily vascularised. The snail can exchange air in the lung across the small pulmonary aperture by contraction and relaxation of the dorsal body wall muscles, called diaphragm (named upon analogy with the mammalian diaphragm) (Fig. 5.11). During inspiration the bottom of the mantle cavity descends as the diaphragm contracts and the lung expands. In the course of expiration, the diaphragm relaxes, the dorsal body wall ascends and the lung shrinks. Examine the surface of the lung; the plexus of the pulmonary vein invaginates into the mantle cavity and enlarges the respiratory surface. Note that mantle, characteristic to molluscs, is an epidermal fold and mantle organs (heart and kidney) are situated between the two epithelia inside the body cavity (not outside the mantle wall) (Fig. 5.14).

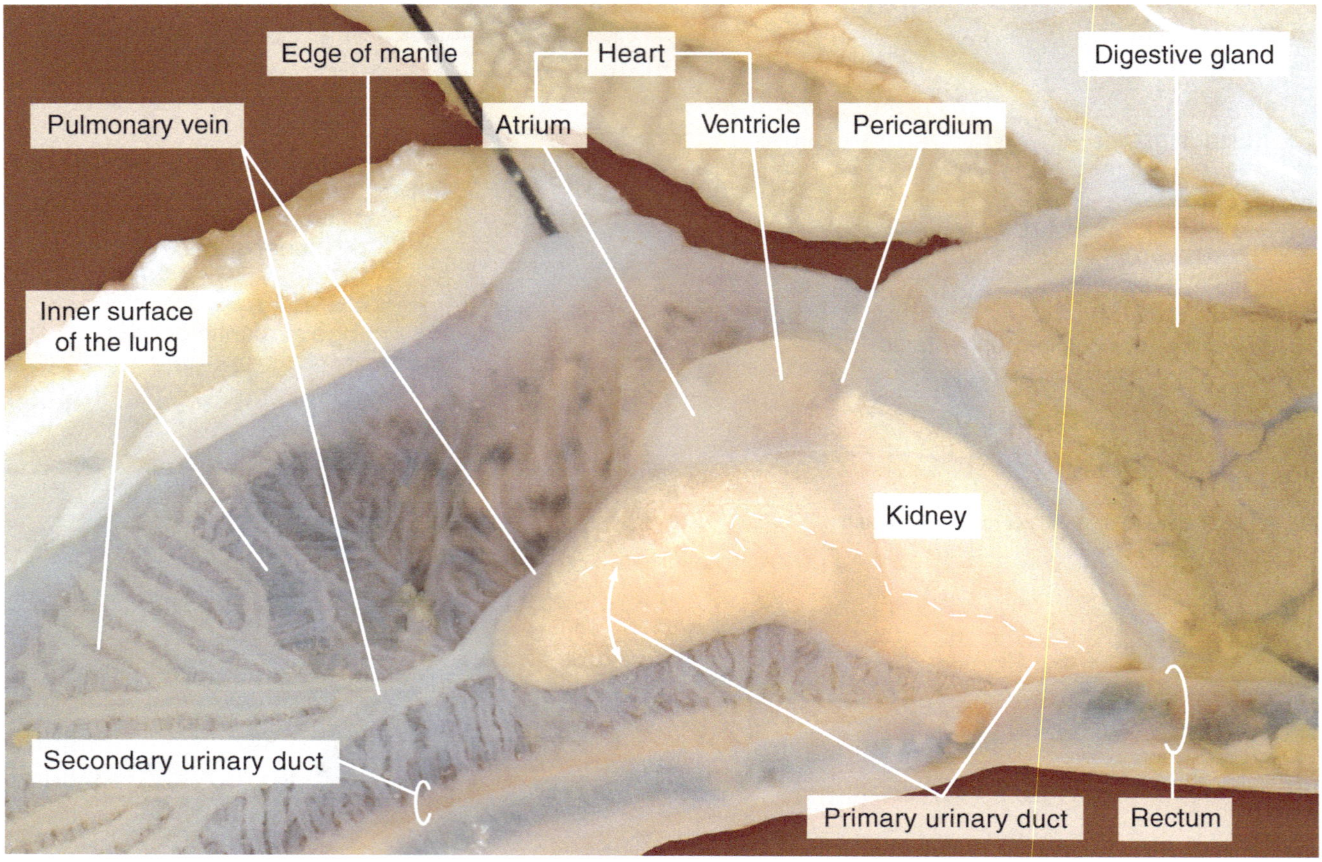

Fig. 5.14 Snail's organs in the mantle wall. The *dashed line* indicates the boundary between the kidney and the primary urinary duct

Snails have an open circulatory system with a simple two-chambered *heart* covered with a *pericardium*. The heart pumps the *haemolymph* (blood in an open circulatory system) into the *aorta* and then into smaller arteries. From these it empties into the body cavity which is called *haemocoel*. The haemolymph is bluish due the haemocyanin, a copper-containing oxygen-carrier protein. The haemolymph passes from the body cavity into the vessels surrounding the lung (venous circle) then into the *pulmonary plexus*, where O_2 is absorbed and CO_2 is released. The haemolymph returns the heart via the *pulmonary vein* (Figs. 5.14 and 5.15).

The *kidney* of the snail lies as a yellowish, triangular organ in the rear segment of the mantle close to the pericardium (Fig. 5.14). The kidney consists of two parts: the *kidney cavity* where the excrements are secreted in the form of yellow granules and the *primary ureter* which is coiled up forming a compact organ. This joins then to a second part, the *secondary ureter*, which runs parallel with the rectum (Fig. 5.14). Since water conservation on dry land is principal, excretory system of snails has become adapted to reduce water loss. Nitrogenous wastes are converted into uric acid which is then excreted as a crystalline solid waste, like that of birds and reptiles. Uric acid is insoluble and so does not use water when it is excreted; this is important to help conserve water.

Inject some toluidine blue solution or any other dye (ink, carmine suspension) into the heart or the pulmonary vein. Press the dye slowly and carefully out of the syringe. The dye infiltrates the network of vessels and makes the system visible (Fig. 5.15).

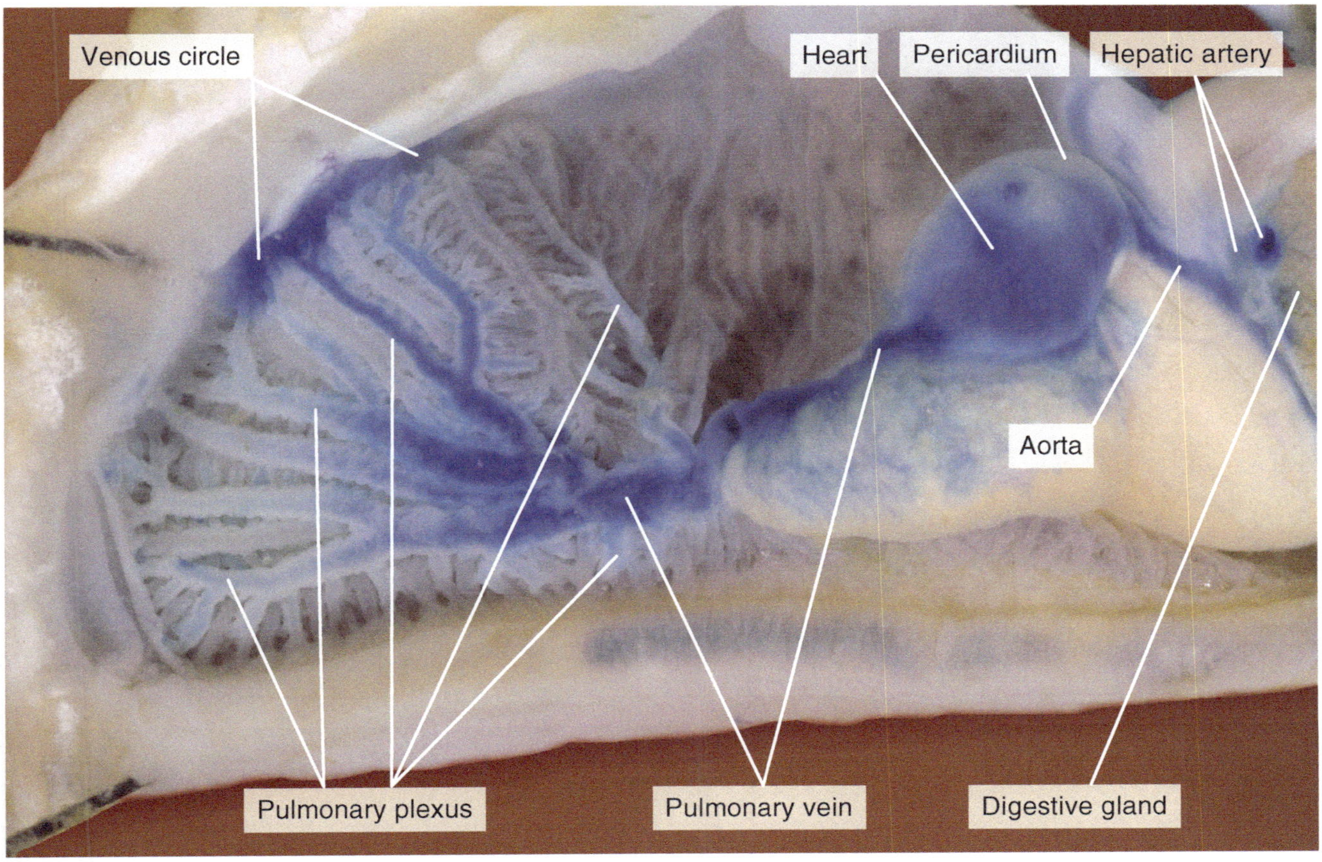

Fig. 5.15 Organs in the mantle wall of the snail (vessels and heart filled with toluidine blue stain)

The *lung* is formed by the inner surface of the mantle, where the *trabeculae* of the pulmonary plexus (TPP) are clearly visible (Fig. 5.16, top left). This inner, wavy surface is covered by a *respiratory epithelium* (RE), while the outer surface facing the shell is covered by the epidermis. The scaffold of the trabeculae is composed of connective tissue in which there are *lacunae*. These latter are prominent just beneath the respiratory epithelium. Examining the trabeculae of the pulmonary plexus, we can identify the branches of the *pulmonary vein* (PV, Fig. 5.16, top right). These vessels have a muscular wall comprising a circular (CML) and a longitudinal muscle layer (LML) and they are filled with haemolymph. Following the circulatory system which is indicated by the blood cells, we can reach the respiratory epithelium, a thin layer formed by flattened cells (Fig. 5.16, bottom). The inner surface of the lung is covered by a *mucous layer* (ML) which has an important role in the gaseous exchange.

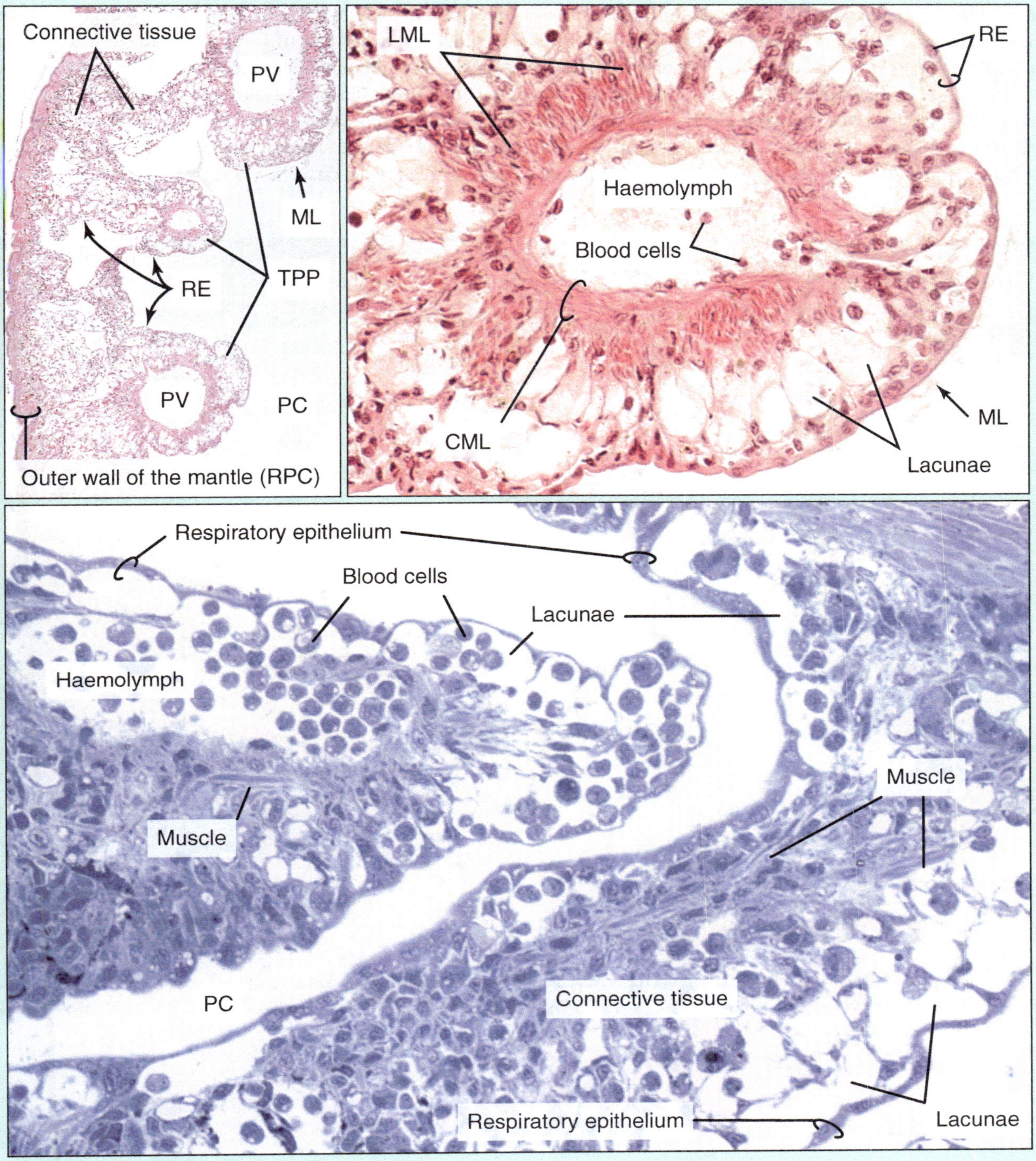

Fig. 5.16 Histological sections of the lung (*top micrographs*, HE; *bottom*, semi-thin section, toluidine blue staining). *CML* circular muscle layer, *LML* longitudinal muscle layer, *ML* mucous layer, *PC* pulmonary chamber, *PV* pulmonary vein, *RE* respiratory epithelium, *RPC* roof of the pulmonary chamber, *TPP* trabeculae of pulmonary plexus

Dry-land snails (*Pulmonates*) are hermaphrodites. The animals have only one gonad, the *hermaphrodite gland* (ovotestis) which produces both sperm and egg cells. It opens into the *hermaphrodite duct* connecting it with the *fertilisation pouch* (not visible from the outside). Cut out the hermaphrodite gland from the digestive gland's substance with a pair of small scissors carefully; do not tear it off from its coiled duct (Fig. 5.17, left). Find the *albumen gland* and the beginning *spermoviduct* on the other end of the hermaphrodite duct. Displaying the gonads, it is useful to grab the spermoviduct with our pair of forceps and move it to the right together with the other attached glands (Fig. 5.18). The best way to perform this part of the dissection is to use two pairs of tweezers and separate the organs without any cuts, just tearing the connective tissue pieces connecting them. This is the so-called *blunt dissection* (Fig. 5.17, right).

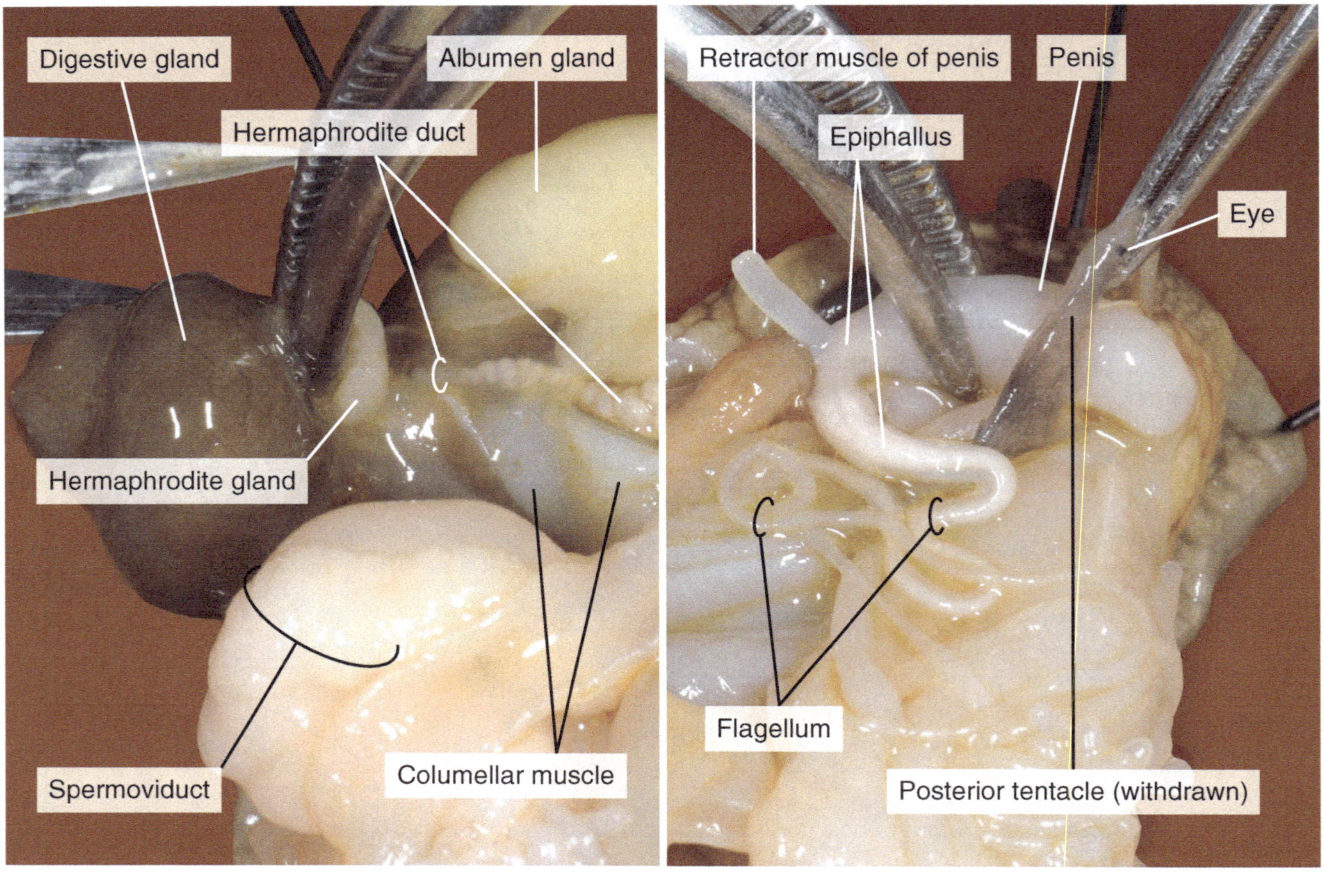

Fig. 5.17 Dissection of the gonads. *Left panel*: cut out the hermaphrodite gland from the digestive gland's substance with a pair of small scissors carefully; do not tear it off from its coiled duct. *Right panel*: locate the retractor muscle of the penis and cut it, then pull both organs under the right posterior tentacle

The spermoviduct is a thick, white canal. It is divided functionally, but not morphologically, into male and female channels. An open groove connects the *sperm duct* (vas deferens) and *oviduct*. Further on, the sperm duct separates and goes towards the *penis* and the oviduct towards the *vagina*. Notice the *retractor muscle of the penis* and cut it, then pull both organs under the right posterior tentacle (Fig. 5.17, right). Sperm cells are transferred in a *spermatophore*, which is produced in the *epiphallus*, while the spermatophore's tail is formed by the *flagellum* (Fig. 5.17, right). The penis and vagina are both connected to the common *genital atrium* that opens through a common *gonopore*. Duct of *spermatheca* opens from the oviduct (Fig. 5.18).

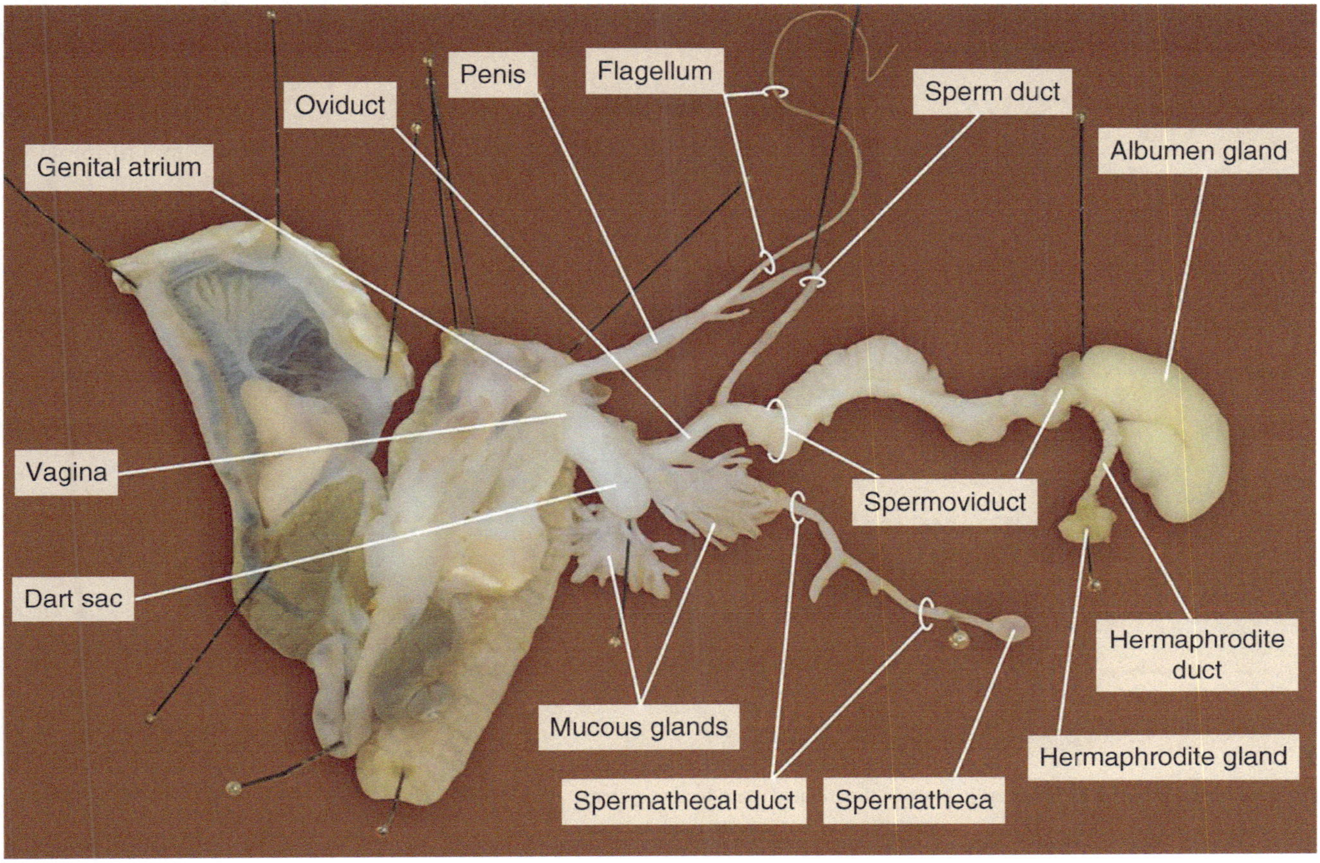

Fig. 5.18 The reproductive organs of the Roman snail

Grip the spermatheca carefully and lift away from the digestive gland, tearing the connective tissue around it. It needs more attention as it is attached firmer, so it might be lost.

The *love dart* (spiculum amoris) is produced and stored in the *dart sac* (stylophore) (Fig. 5.19). The *mucous glands* (finger-shaped glands) lay behind the dart sac (Figs. 5.18 and 5.19).

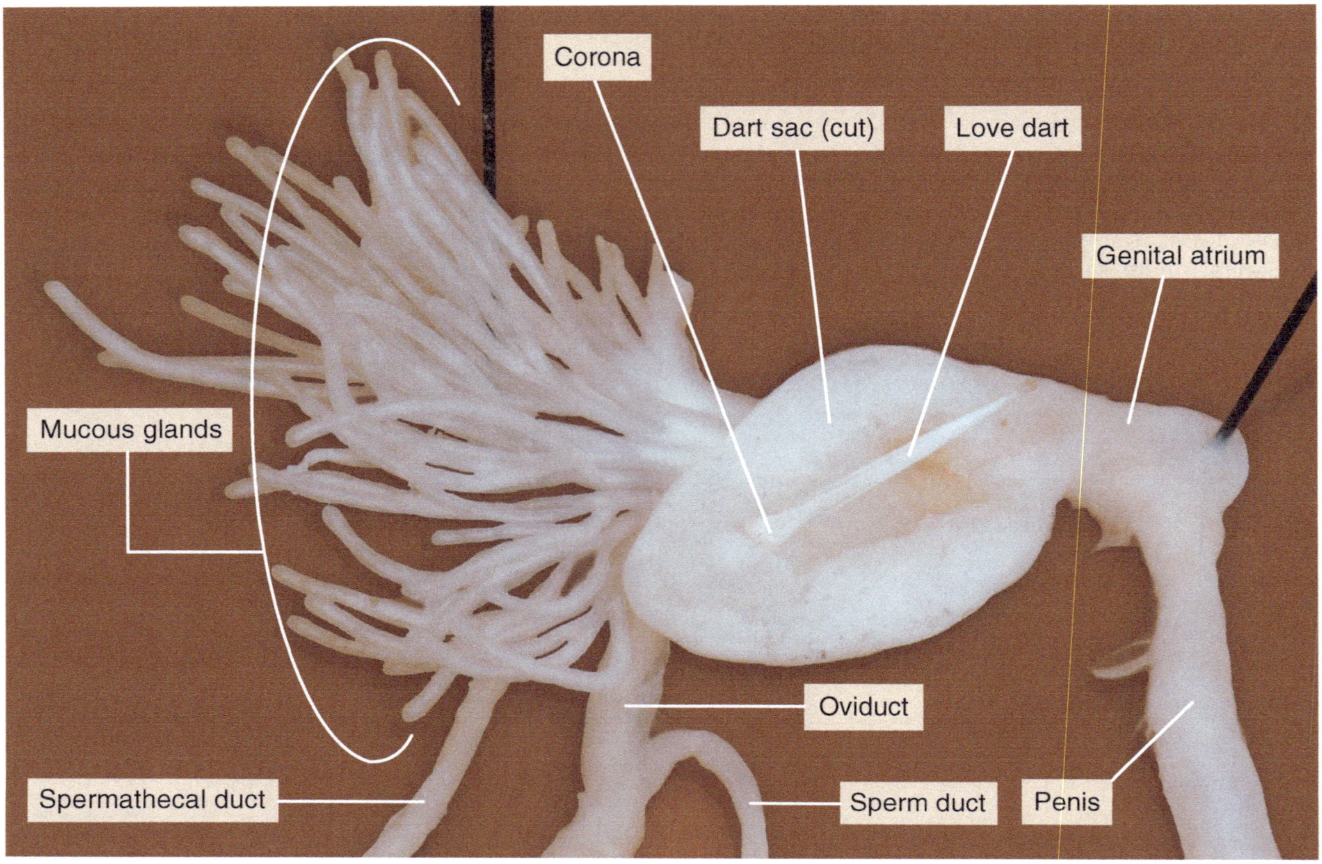

Fig. 5.19 Cut dart sac with love dart in place in the isolated reproductive system

Cut the dart sac and find the extremely fragile, calcareous love dart. Note the corona at its posterior end and the four blade-like vanes along the shaft (Fig. 5.20). The function of the love dart and the mating habits of snails are described later.

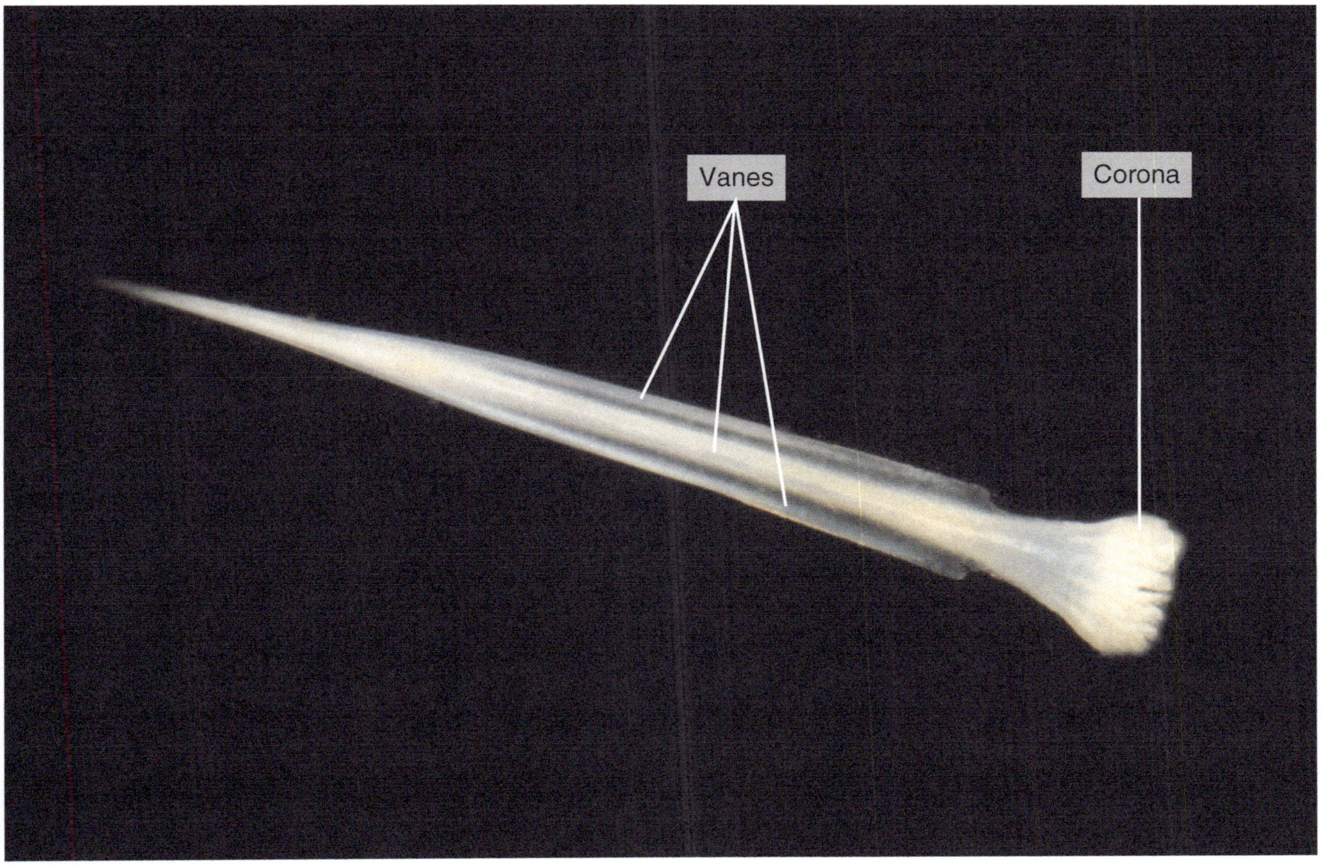

Fig. 5.20 The isolated love dart is only about half a centimetre long

The *hermaphrodite gland* produces both egg cells (oocytes) and sperm cells (spermatocytes). The organ is formed by several finger-shaped follicles, which are lined by *follicular epithelium* (Fig. 5.21).

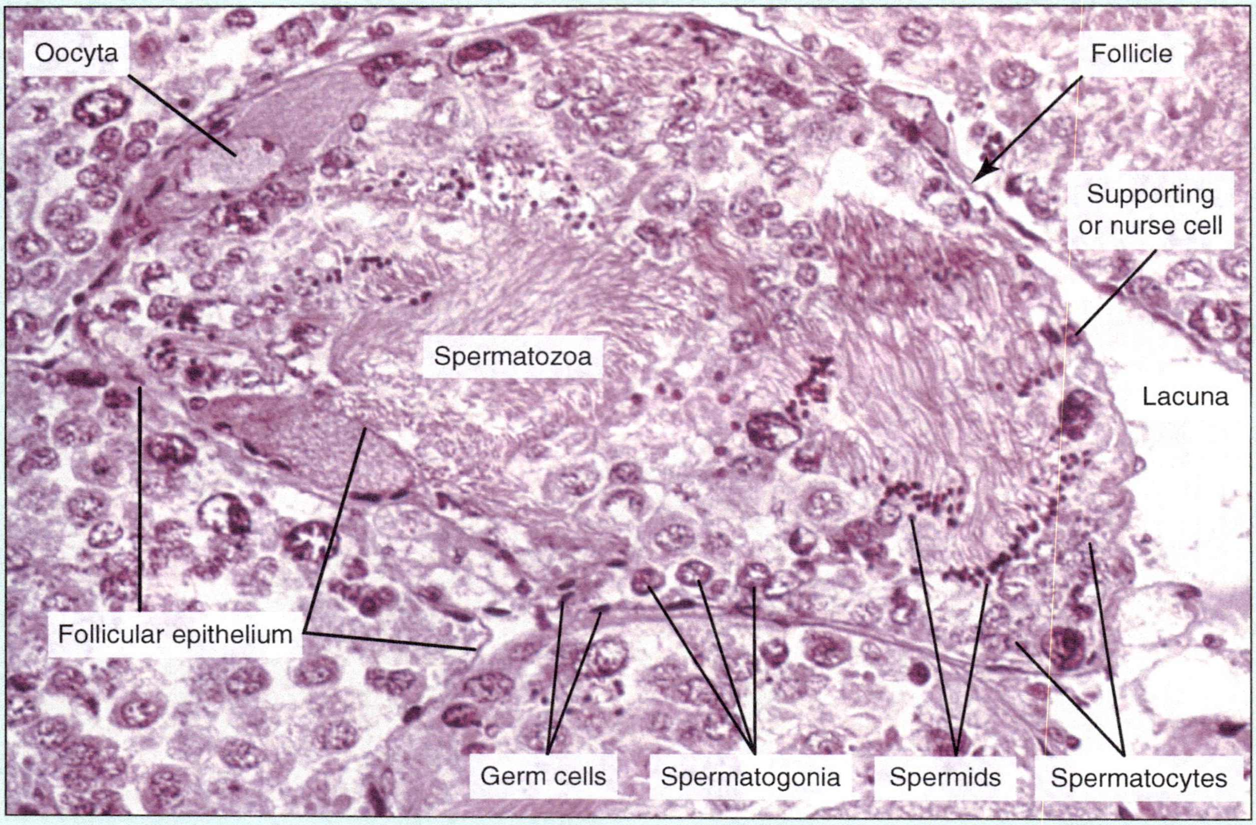

Fig. 5.21 Histological section of the hermaphrodite gland (HE)

This epithelial layer segregates the *developing oocytes* and *spermatocytes*. The oocytes have huge nuclei with several nucleoli and their granular cytoplasm contains large amount of yolk. The stages of the spermatogenesis are easily distinguishable on the basis of the cell morphology and position. The *germ cells* are small cells with dark nuclei in the neighbourhood of the follicular epithelium. Their descendants, *spermatogonia* and *spermatocytes*, are larger, round cells. The smaller *spermids* enter the morphogenetic-stage, loose cytoplasm and attach to the supporting cells. The supporting cells feed the developing sperms and help their maturation. The mature spermids (*spermatozoa*) accumulate in the central part of the follicles (Fig. 5.21).

The Roman snail performs a curious precopulatory ceremony. The two animals approach each other with their genital atria everted. Each fires the dart by a forceful eversion of the dart sac which penetrates deeply into the internal organs. Mucous glands produce a sort of lubricant that is deposited on the dart before shooting, making it easier to push out the love dart and receive the penis. The function of the love dart is to transfer hormones which stimulate the reception of sperm cells. During mating the penis extends by haemolymph pressure; it withdraws with a penis retractor muscle. The penis is used to transfer the spermatophore; the mating partners mutually exchange sperm. When a spermatheca tract diverticulum is present, the spermatophore is transferred into that of the partner, after which the snails separate. The spermatophore is transported through the spermatheca tract into the spermatheca, where it is digested. The freed sperm cells swim back to enter the female tract and find their way through the vagina and oviduct up to the fertilisation pouch. There they fertilise the eggs. In hermaphroditic animals, like the Roman snail, eggs and sperms mature in different time because of hormonal control and this rules out self-fertilisation. In the ovotestis of the Roman snail, sperms mature first then, after mating and the digestion of the spermatophore, the eggs. The albumen gland and the wall of spermoviduct coat the fertilised eggs with several layers as they descend, finally with a calcareous cover. Eventually the snail lays the eggs.

The Roman snail is an herbivore; it takes in larger pieces of plant material or uses the *radula*, a ribbon-like rasping tongue, to fragment pieces of food (Fig. 5.23). The radula is covered with fine horny curved teeth (Fig. 5.25). The radula is rubbed back and forth against the horny jaws located on the dorsal wall of buccal cavity and tears off small pieces of vegetation in a similar manner to the action of a rasp. Start the dissection of the digestive tract at the *mouth* (Figs. 5.22 and 5.23). Cut the dorsal side of the *buccal cavity* beginning from the mouth and examine the teeth of the radula with a magnifying glass or stereomicroscope (Figs. 5.24 and 5.25). After that there is no need to any further dissection; other parts of the alimentary canal are free to study (Fig. 5.22).

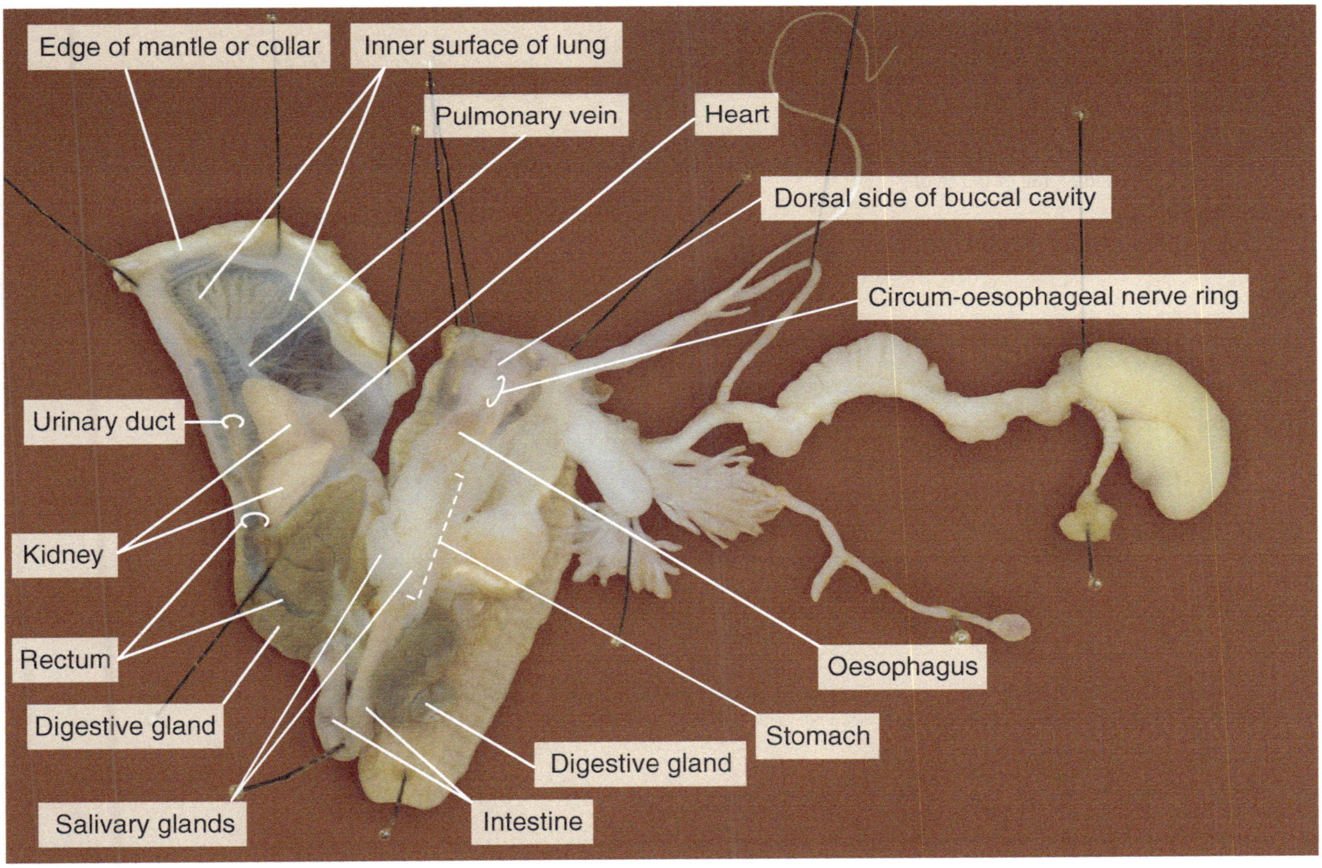

Fig. 5.22 Completely dissected snail in water cover

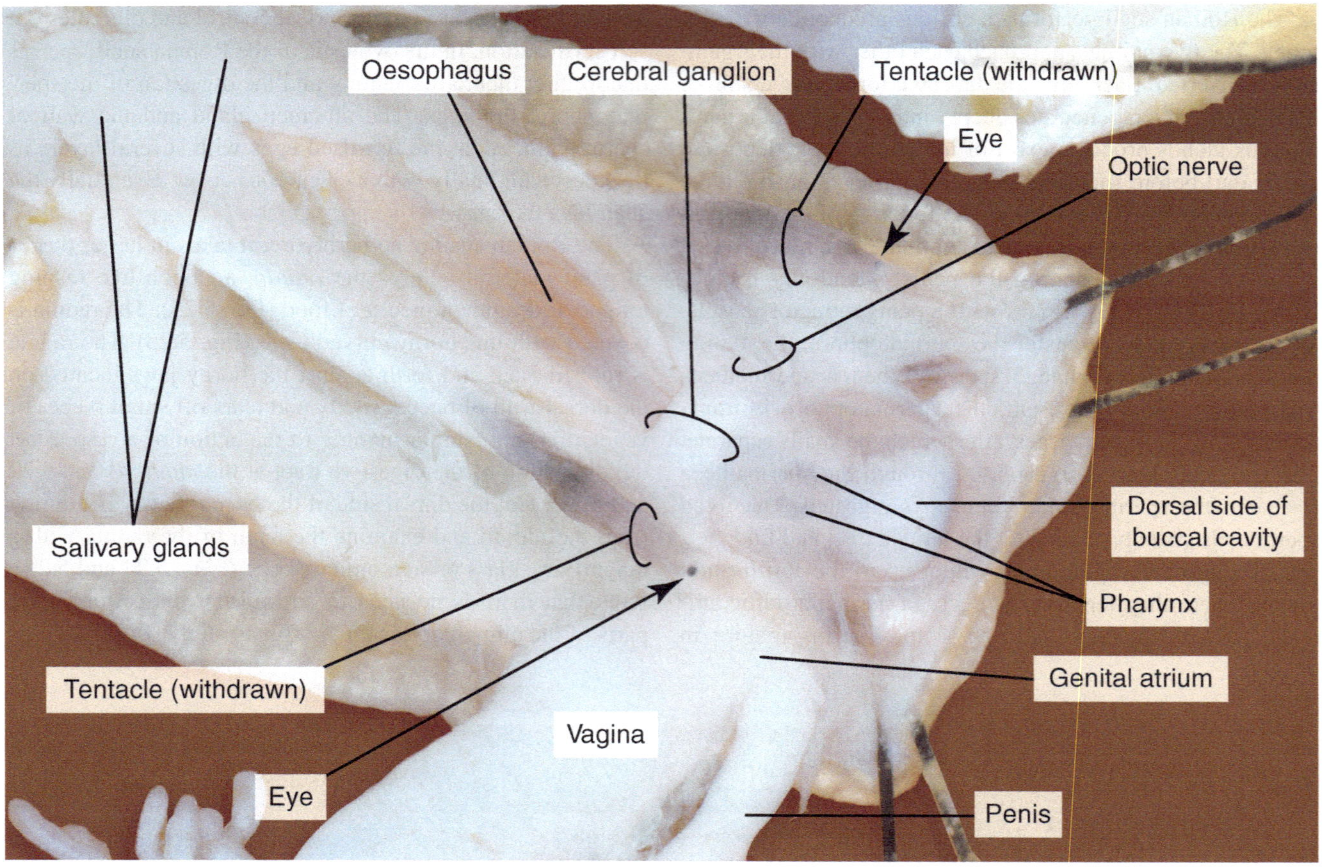

Fig. 5.23 Dissected head region of the snail

Near the opening of the mouth, *salivary glands* release digestive enzymes. The salivary glands' secretion moistens the food, thereby making easier it for the food to go into the oesophagus. The *oesophagus* ends in the stomach (Figs. 5.22 and 5.23). The *stomach* contains symbiotic bacteria which digest the cellulose in the plant matter. The *intestine* transports back large quantities of a brown digestive juice into the stomach. The *digestive gland* (hepatopancreas) fills up most of the space in the visceral sac (Fig. 5.22). The digestive gland consists of smaller and bigger follicles. A steady back and forth movement of the digestive juices between the stomach and intestine enhances the process of absorption of the food. The movement of the digestive juices is caused by the muscles of the digestive gland and cilia. The digested food flows over the digestive gland cells which absorb the food. The *rectum* starts at the visceral sac; it follows the edge of the kidney and runs parallel with the secondary ureter at the edge of the mantle (Figs. 5.14 and 5.22). It ends in the *anus* near the pulmonary aperture. In the rectum the solids are compressed and enveloped with a layer of slime and they leave the body.

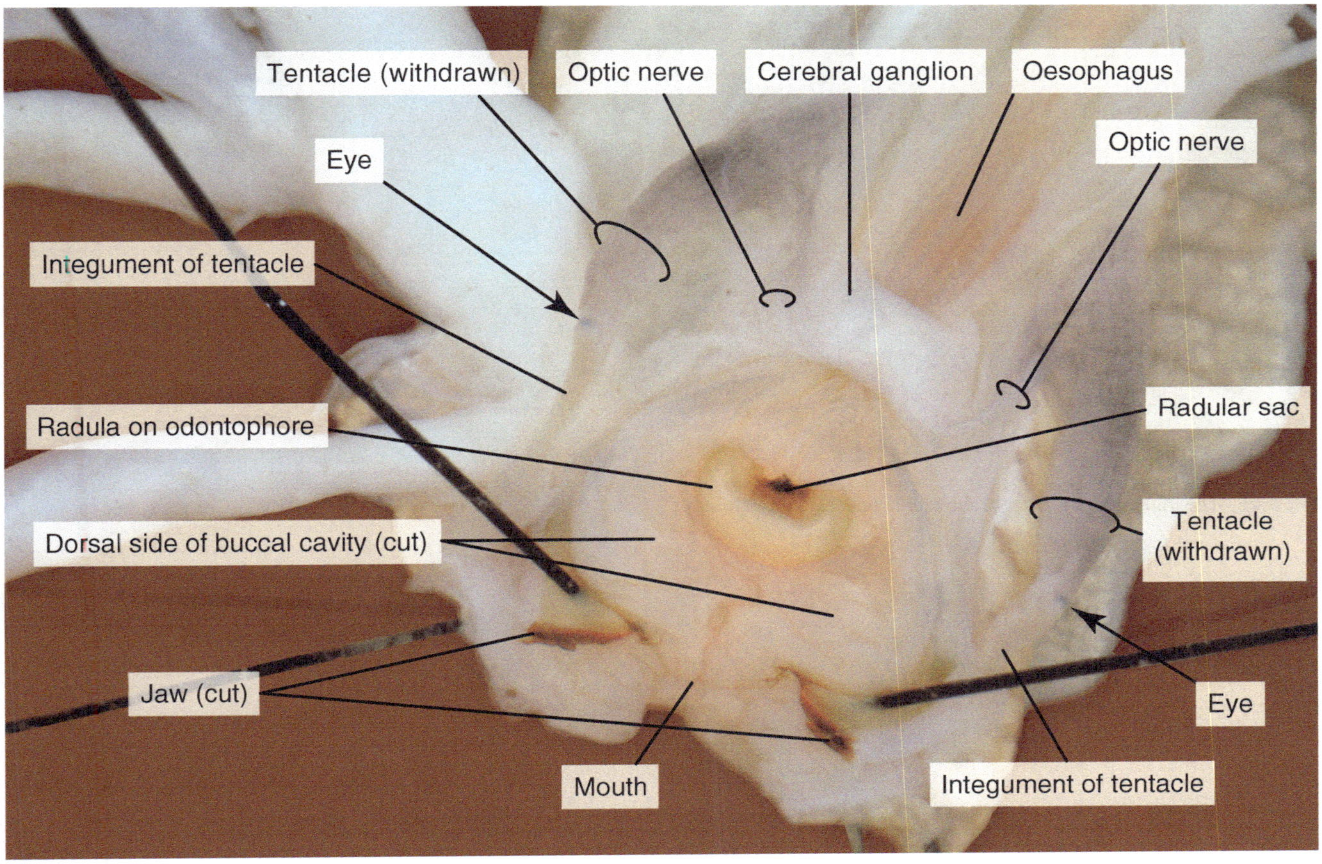

Fig. 5.24 Radula of the Roman snail displayed after opening the dorsal side of the buccal cavity

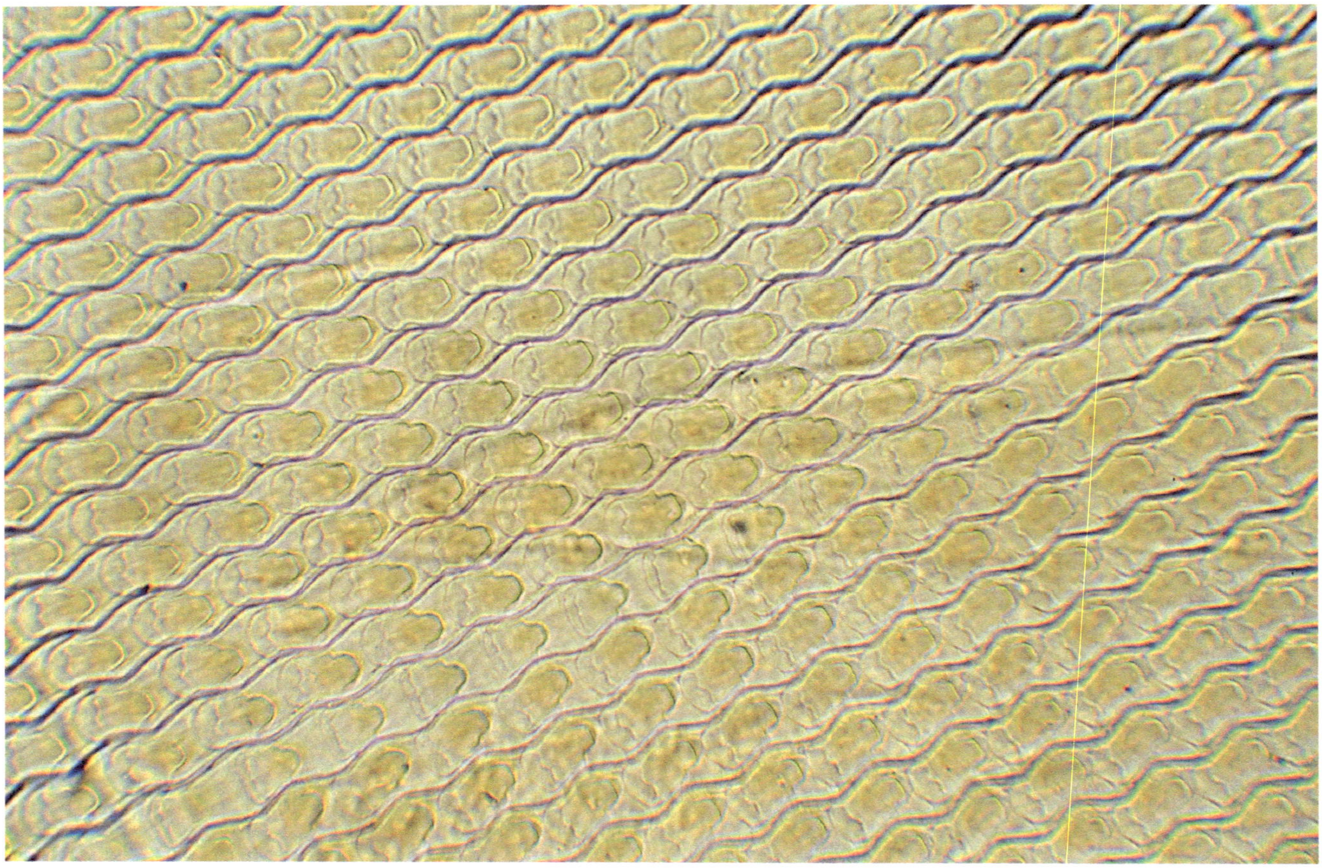

Fig. 5.25 The thin chitinous layer of the radula under light microscope

For the examination of the *central nervous system*, cut the pharynx and pull it out from the circum-oesophageal nerve ring. The central nervous system of snails consists of paired groups of nerve cells or *ganglia* (singular ganglion). The five pairs of ganglia lie near the oesophagus. The cerebral and buccal ganglia lie above the oesophagus, the others below it forming a more or less compact organ, the sub-oesophageal ganglion (Fig. 5.26). Numerous thick nerves emerge from the latter. The ganglia are connected lengthwise, and there are also connections across the body. They have all different functions.

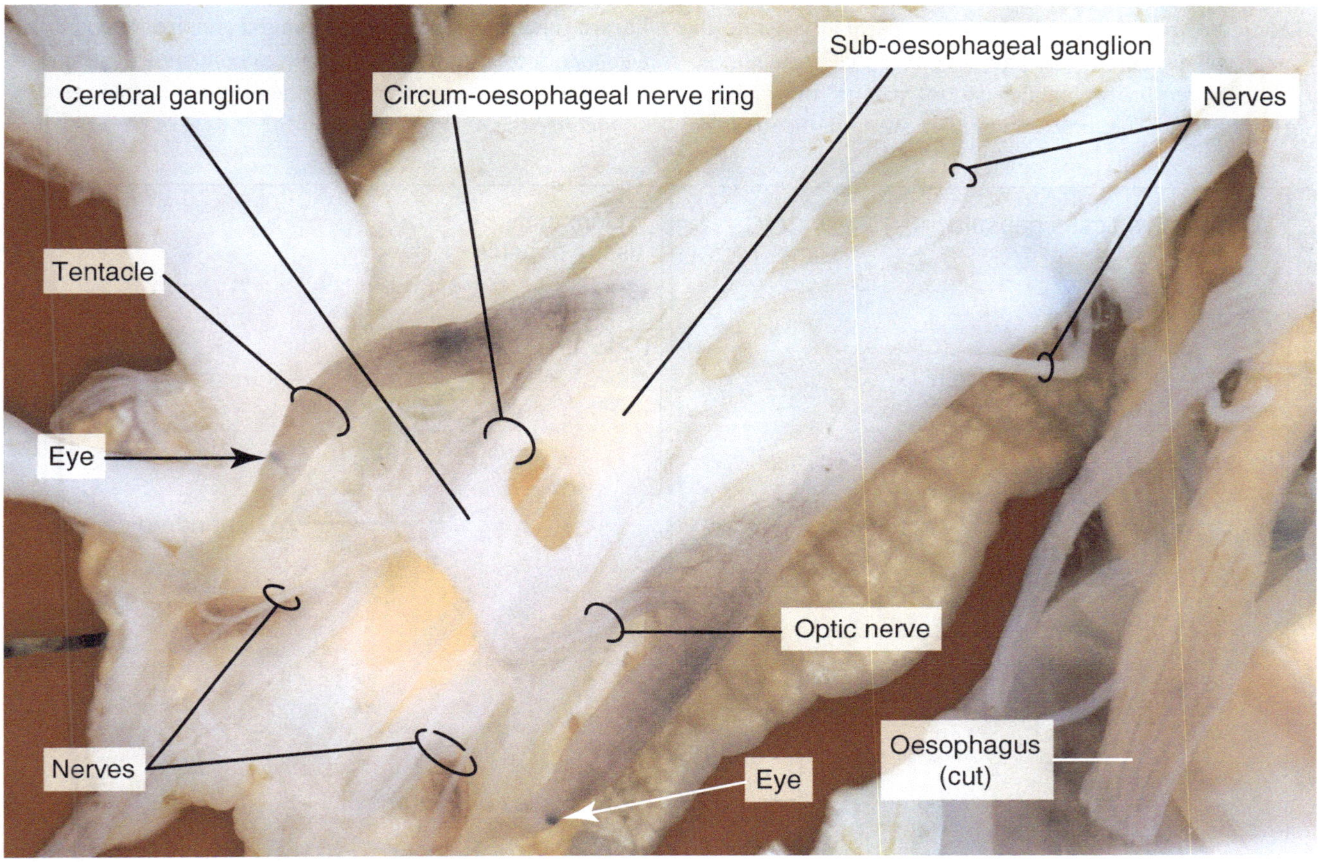

Fig. 5.26 The circum-oesophageal nerve ring with emerging nerves after removal of anterior part of the alimentary canal

The *cerebral ganglia* act as a "brain" for the sensory organs on the head (tentacles and eyes). Leaving the eyes the optic nerves run within the tentacles and then enter the cerebral ganglia (Fig. 5.26).

The histological section of the *cerebral ganglia* shows typical morphology of the invertebrate ganglion: the cell bodies of neurons are seen on the periphery, while the neuronal processes and the synapses that form the so-called *neuropil* are in the central part of the ganglion (Fig. 5.27). The tiny cell nuclei seen in the neuropil belong to glial cells. The snail CNS contains several *giant nerve cells* featured with unvarying positions and well-known functions. The cerebral ganglia are enveloped into a *connective tissue capsule* and have connections with the *sub-oesophageal ganglion* by the *circum-oesophageal connective* (COC, Figs. 5.26 and 5.27).

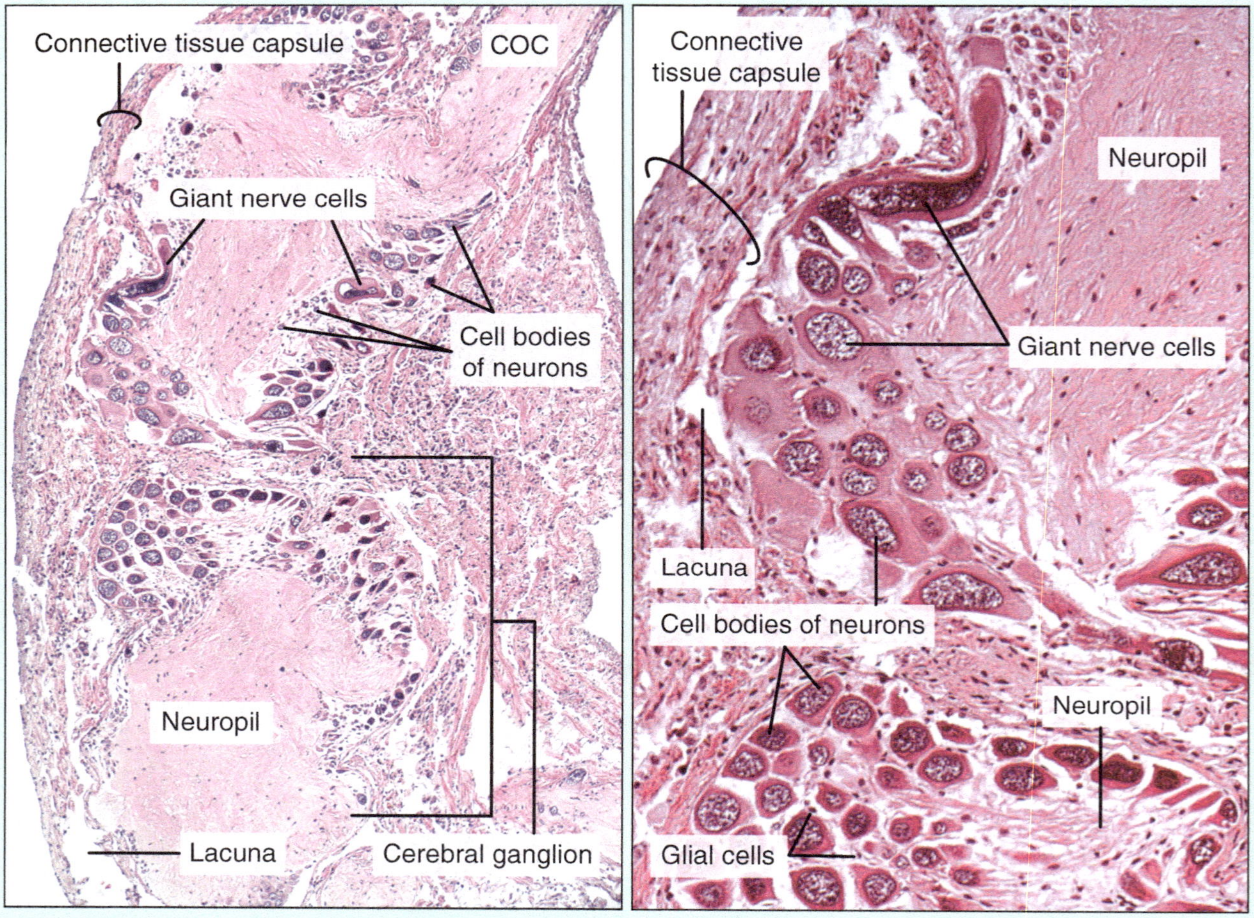

Fig. 5.27 Histological sections of both cerebral ganglia of the Roman snail (HE). *Right panel* shows detail of the *left*. *COC* circum-oesophageal connective

The *buccal ganglia* lie in front of the cerebral ganglia and control the mouth. The sub-oesophageal ganglion is covered with thick connective capsule, so its parts are hardly distinguishable without removing the connective sheet. It is composed of three paired and one unpaired ganglia. The *pedal ganglia* coordinate movement of the muscles in the foot; the *pleural ganglia* coordinate the respiratory movements of the mantle wall. The *parietal* (lateral) *ganglia* control the food uptake, respiratory movements and mating. A separate, not paired, ganglion is the *visceral ganglion* and it innervates the digestive system and the heart.

Eye of the Roman snail is a simple epithelial sphere under the surface of the body epidermis (Fig. 5.28). Parts of the snail's eye are named upon analogy with the vertebrate eye which has an entirely different development. The transparent *external cornea* (EC) is derived from the superficial epidermis. *Internal cornea* (IC) makes up the anterior part of the epithelial sphere. The rear part of the ball is called the *retina*; this contains the *photoreceptor cells* separated with *pigment cells*. That is why the snail's eye appears as a black dot at the tip of the tentacle. Inside of the sphere is filled with a high protein content fluid. It is thinner in the outer part, called *vitreous body*, and thicker in the middle, named *lens*. Unlike vertebrate eye, the retinal photoreceptor cells face the interior of the sphere, towards the incoming light. Such eye structure is called an everse-type eye (Fig. 5.28).

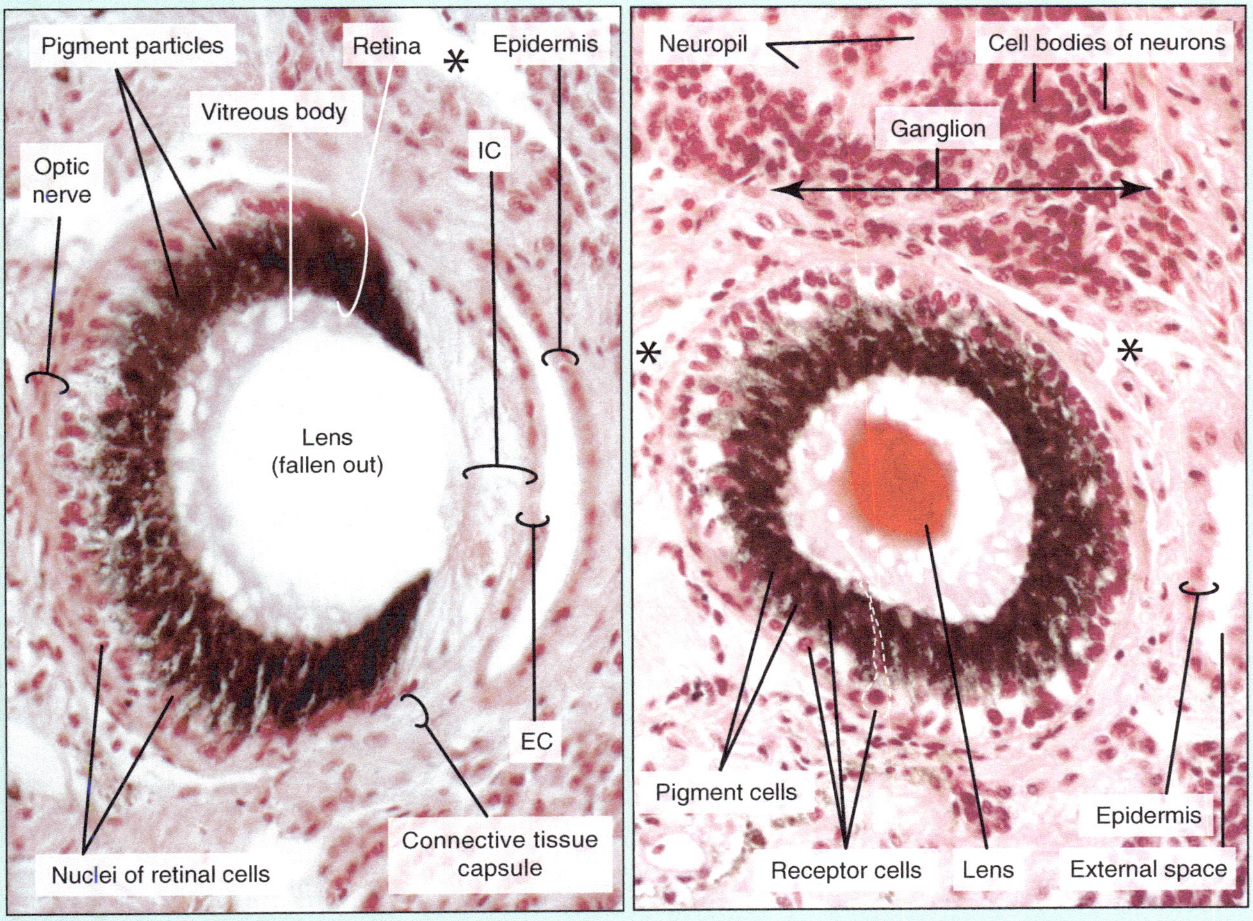

Fig. 5.28 Histological sections of the eye of the Roman snail in a withdrawn tentacle in two different planes (HE). *Asterisks* haemolymph spaces, *EC* external cornea, *IC* internal cornea

Dissection of a Freshwater Mussel (*Anodonta anatina*)

6

- *Availability*: Mussels live in a variety of freshwater habitats but are most prevalent in lakes, ponds, rivers, streams and canals. They are common in areas with muddy, silty or sandy bottoms and slowly flowing permanent water where they can be collected. Alternatively we can purchase mussels from companies supplying restaurants.
- *Anaesthesia*: Place the mussel in 4–6 % (w/v) $MgCl_2$ solution; it can be dissected within 15 min. Use 4 % (w/v) chloral hydrate solution instead to anaesthetize the mussel in 1 h time.

Place the mussel in a dissecting tray and determine the correct orientation: identify the anterior and posterior ends of it as well as the dorsal, ventral and lateral surfaces (Fig. 6.1).

Fig. 6.1 Lateral view of a freshwater mussel (*Anodonta anatina*)

P. Lőw et al., *Atlas of Animal Anatomy and Histology*, DOI 10.1007/978-3-319-25172-1_6

The mussel's external shell is composed of two hinged halves or *valves* (Fig. 6.1). The two valves of the mussel are attached by a *hinge ligament* on the dorsal side and are closed when necessary by the strong *adductor muscles*. The ventral side is free for the protrusion of the foot. Locate the *umbo*, a swollen hump at the anterior end of the valve. This is the oldest part of the mussel shell usually with periostracum rubbed off so that prismatic layer is exposed. Observe the shape of the shell and the concentric *lines of growth* (Fig. 6.1). The lines represent alternating periods of slow and rapid growth. The youngest part of the shell is the edge. Mussel shells carry out a variety of functions, including support for soft tissues, defence against predators and protection from desiccation.

The shell has three layers. In the pearly mussels there is an inner iridescent *nacreous layer* (mother of pearl, *hypostracum*) composed of calcium carbonate, which is continuously secreted by the mantle surface. This surface is not ciliated, so the animal is unable to rid itself of foreign objects that might get between the shell and the mantle. Instead the mantle secretes nacre around the foreign objects, forming a pearl. The middle layer, the *prismatic layer* (*ostracum*), consists of aragonite or calcite, chalky white crystals of calcium carbonate in a protein matrix. It is secreted by glands in the inner side of the outer fold at the edge of the mantle. The outer *pigmented layer* (*periostracum*) is composed of a protein called conchiolin, and its function is to protect the prismatic layer from abrasion and dissolution by acids (especially important in freshwater species where the decay of plant materials produces acids) (Fig. 6.1). It is secreted by the inner side of the outer fold at the edge of the mantle.

The dissection can be started if the adductor muscles do not contract on the straining of the valves; the slightest reaction shows that the mussel is still alive. Hold the specimen in the left hand with its dorsal side down and anterior end towards your right and gently prise the valves of the shell apart. Grip it so that your thumb spreads the shell slightly open. Do not attempt to force the valves too far apart as this will tear the muscles and possibly damage other parts at the same time. Locate the adductor muscles. Slide the scalpel on the layer of nacre and cut the muscles away from the right valve of the shell. Cut through the anterior adductor muscle, cutting as close to the shell as possible. Repeat this in cutting the posterior adductor muscle. Cut each muscle in turn, and when the valve is free, force it back and break the hinge. This leaves soft parts attached to the left valve.

Examine the inner surface of the right valve (Fig. 6.2).

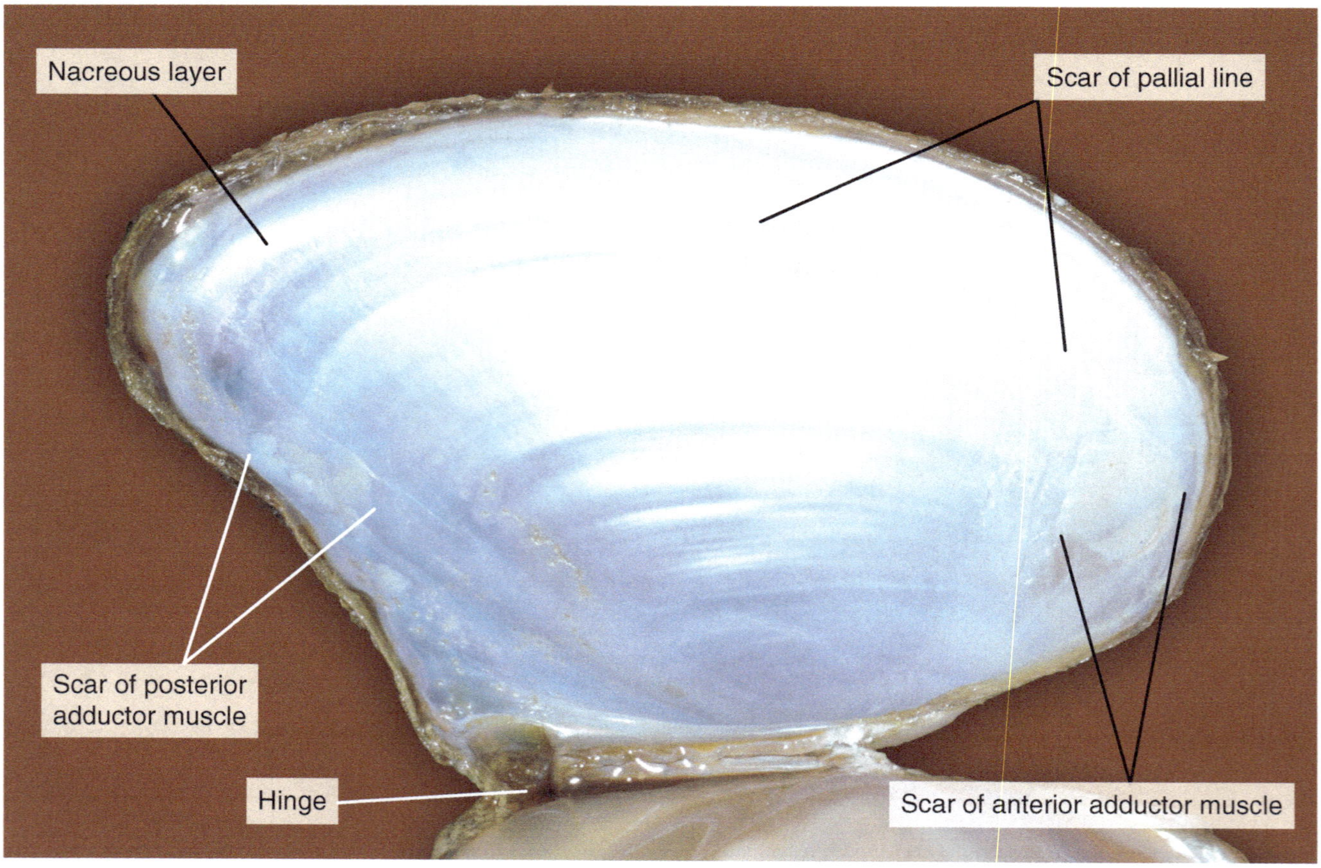

Fig. 6.2 Inner surface of the right valve of a mussel's shell with the iridescent hypostracum layer, showing muscle scars

Locate on the valve the scars of the *anterior* and *posterior adductor muscles*, which close and hold the valve together. Note that there are no muscles to open the shell. This is done by the elastic *hinge ligament*, which acts like a spring and forces the shells apart when adductors relax. Find the *pallial line*, where the pallial muscle of the mantle was attached to the valve (Fig. 6.2).

Examine the thin, semitransparent *mantle*, which lines both valves and covers all the soft tissues of the mussel (Fig. 6.3).

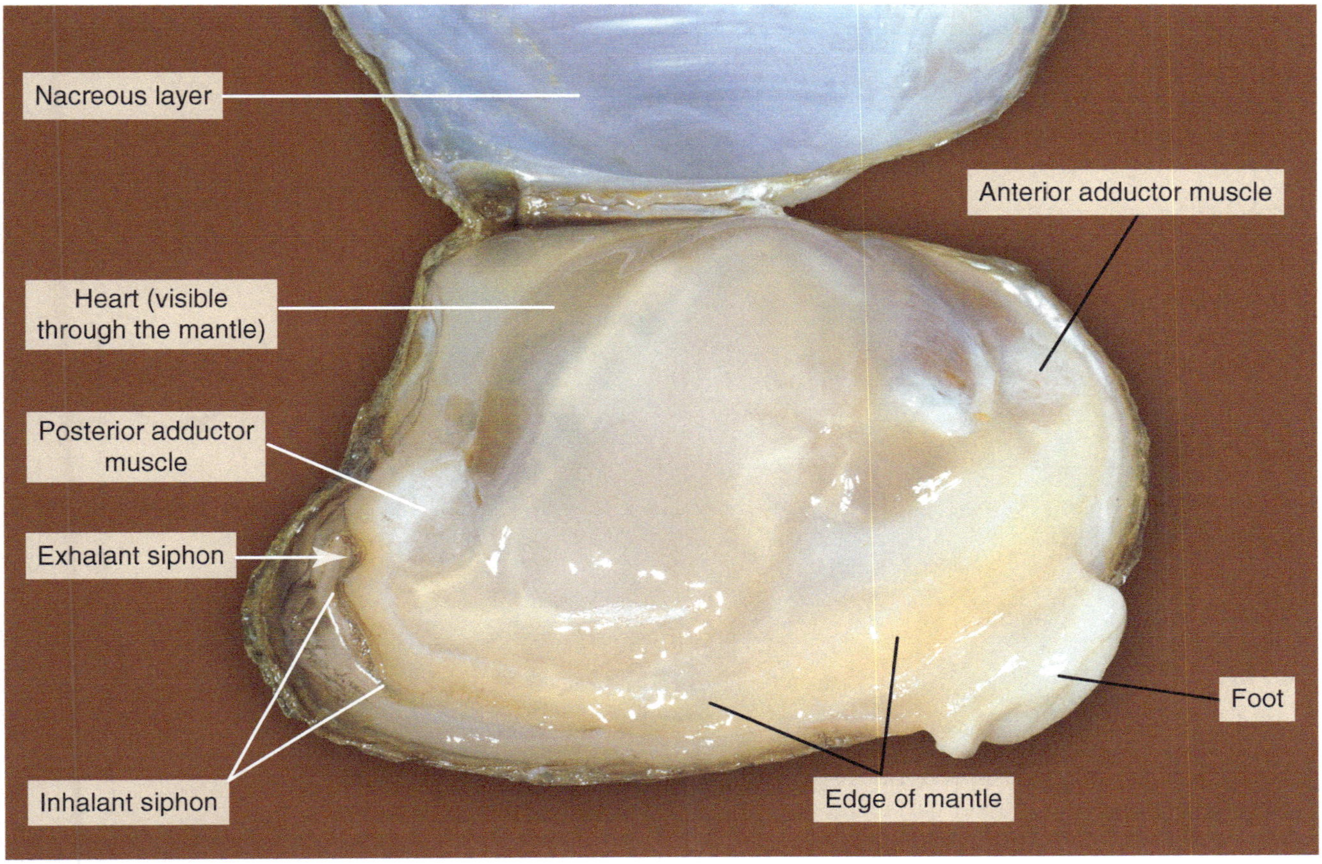

Fig. 6.3 Semitransparent mantle of the mussel and organs which show through it after opening the shell

Locate two openings on the posterior end of the mussel. Posteriorly the edges of the two mantles are thickened, darkly pigmented and fused together dorsally to form the ventral *inhalant siphon* that carries water into the mussel and dorsal *exhalant siphon* where wastes and water leave (Fig. 6.3). The apertures permit a continuous flow of water through the mantle cavity. Look at the free edge of the mantle and note that it has three lobes. The outer lobe secretes the prismatic and pigmented layers of the shell.

Lift up the mantle to expose the outer pair of gills and body mass beneath (Fig. 6.4).

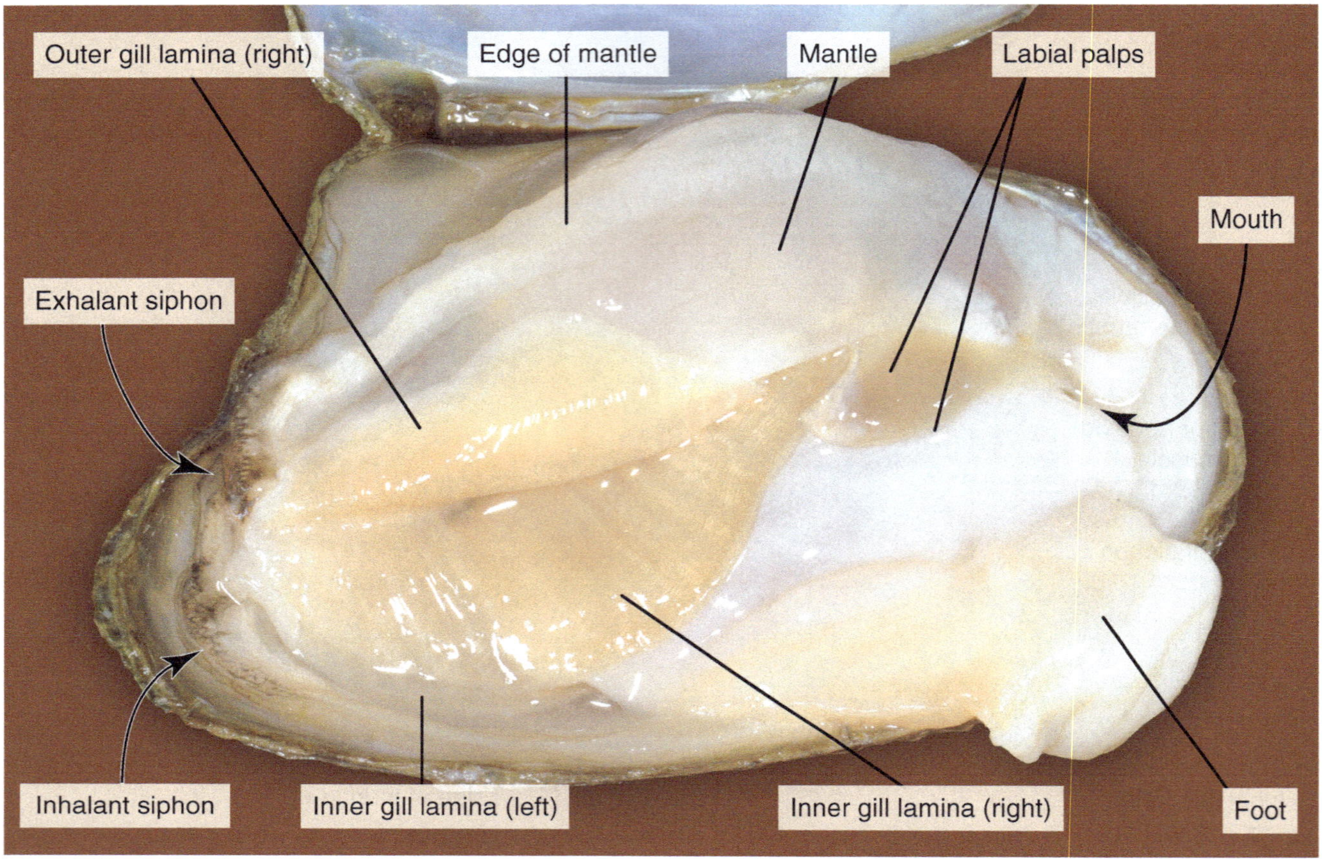

Fig. 6.4 The mantle folded up to expose the right pair of gills and body mass beneath

The entire space between the right and left lobes of the mantle is the *mantle cavity*. Cilia on the inner side of the mantle keep water flowing through the mantle cavity. The cilia beat constantly, pulling water in the inhalant siphon, through the ostia (see later) and into the water tubes and then the epibranchial canal. From here it is carried out through the exhalant siphon.

Like most bivalves, mussels have a large organ called *foot* (Figs. 6.3 and 6.4). In freshwater mussels, the foot is large, muscular and generally hatchet shaped. It is used to pull the animal through the substrate (typically sand, gravel or silt) in which it lies partially buried. Observe the muscular foot of the mussel, which lies ventral to the gills and the visceral mass (Fig. 6.4). The foot operates by a combination of muscular movement and hydraulic mechanisms. The mussel can extend or enlarge the foot hydraulically by engorgement with haemolymph and uses the extended foot for anchorage and then pulls the rest of the animal with its shell forwards. Furrows can be seen along banks and sandy/muddy patches of stream bed, where mussels have moved themselves along the bottom. Unlike their marine and estuarine cousins, they do not attach to structures. This allows them to move with retreating water levels and position themselves to the best feeding spots. It also serves as a fleshy anchor when the animal is stationary.

Lift the free part of the mantle that lined the right valve and carefully cut along the line indicated with scissors as close to the muscles and gill as possible (Fig. 6.5).

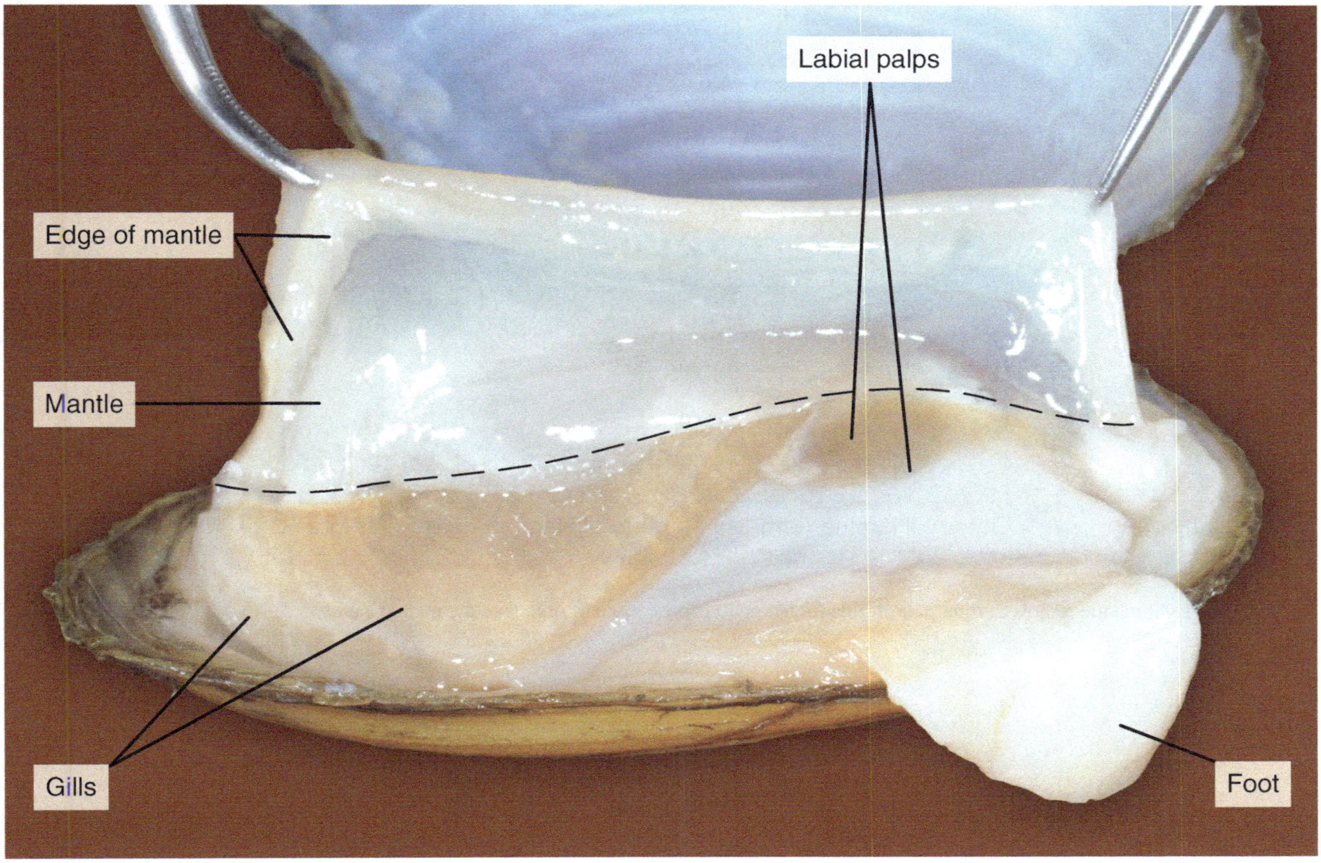

Fig. 6.5 The free part of the right mantle (lifted) and organs of the mantle cavity. *Dashed line* indicates the line of cut to remove the mantle

After removing this part of the mantle, you can see the *gills*, respiratory structures. Note that there are two gills of each side of the visceral mass, an outer and a somewhat larger inner one. If the outer gill is much thicker than the inner gill, the animal is probably a female in which the gill is serving as a brood chamber for eggs, embryos and larvae during the breeding season. Observe that only the dorsal margins of the gills are attached; the ventral edge hangs freely in the mantle cavity (Fig. 6.6).

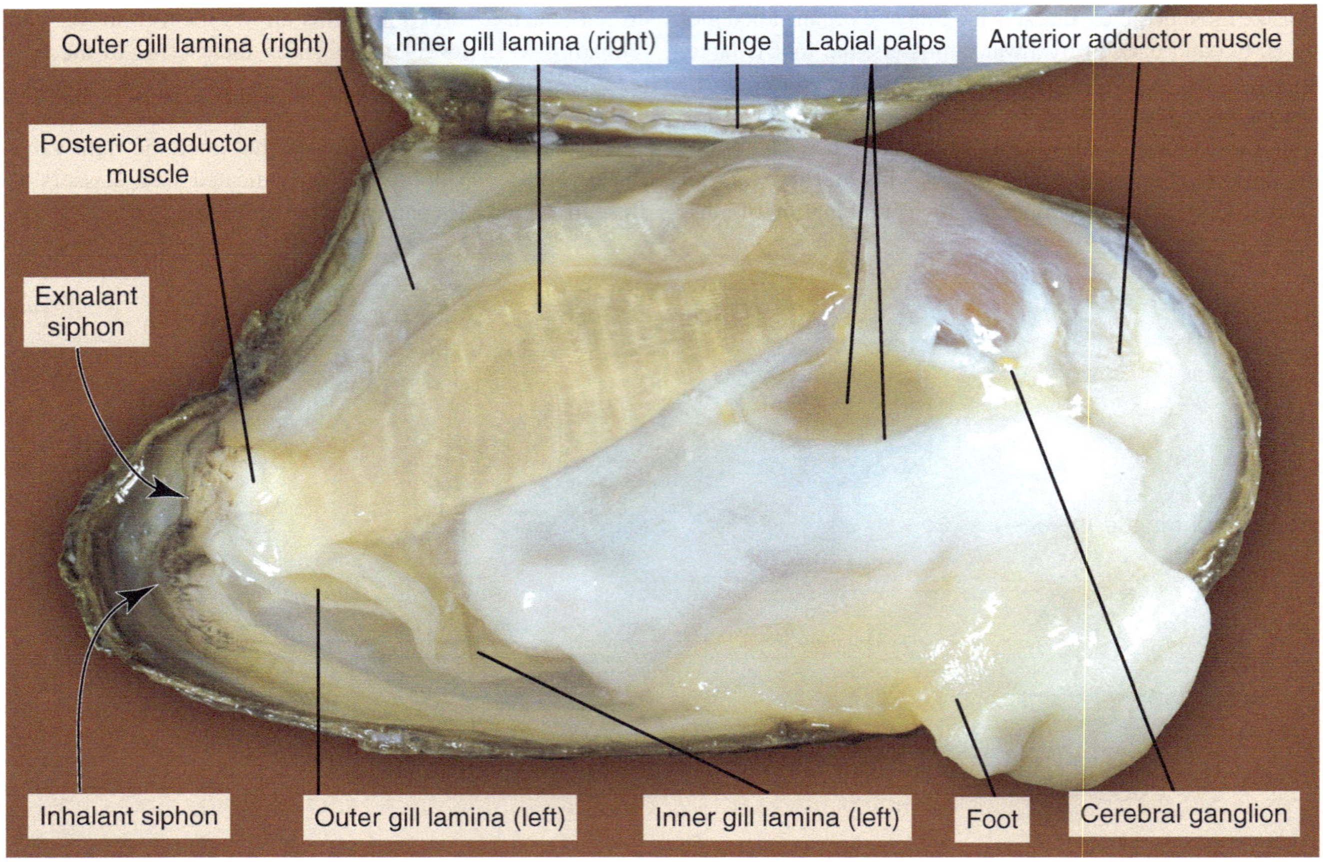

Fig. 6.6 The right mantle removed; right pair of gills folded up to expose more of organs of the mantle cavity

Sprinkle a few grains of powdered carmine onto the gill and see the current produced by the cilia. Mount a small portion of a gill and examine it under a light microscope. Notice that the transverse lines which are visible to the naked eye are not the gill filaments which are actually much finer and perpendicular to these lines (Fig. 6.7).

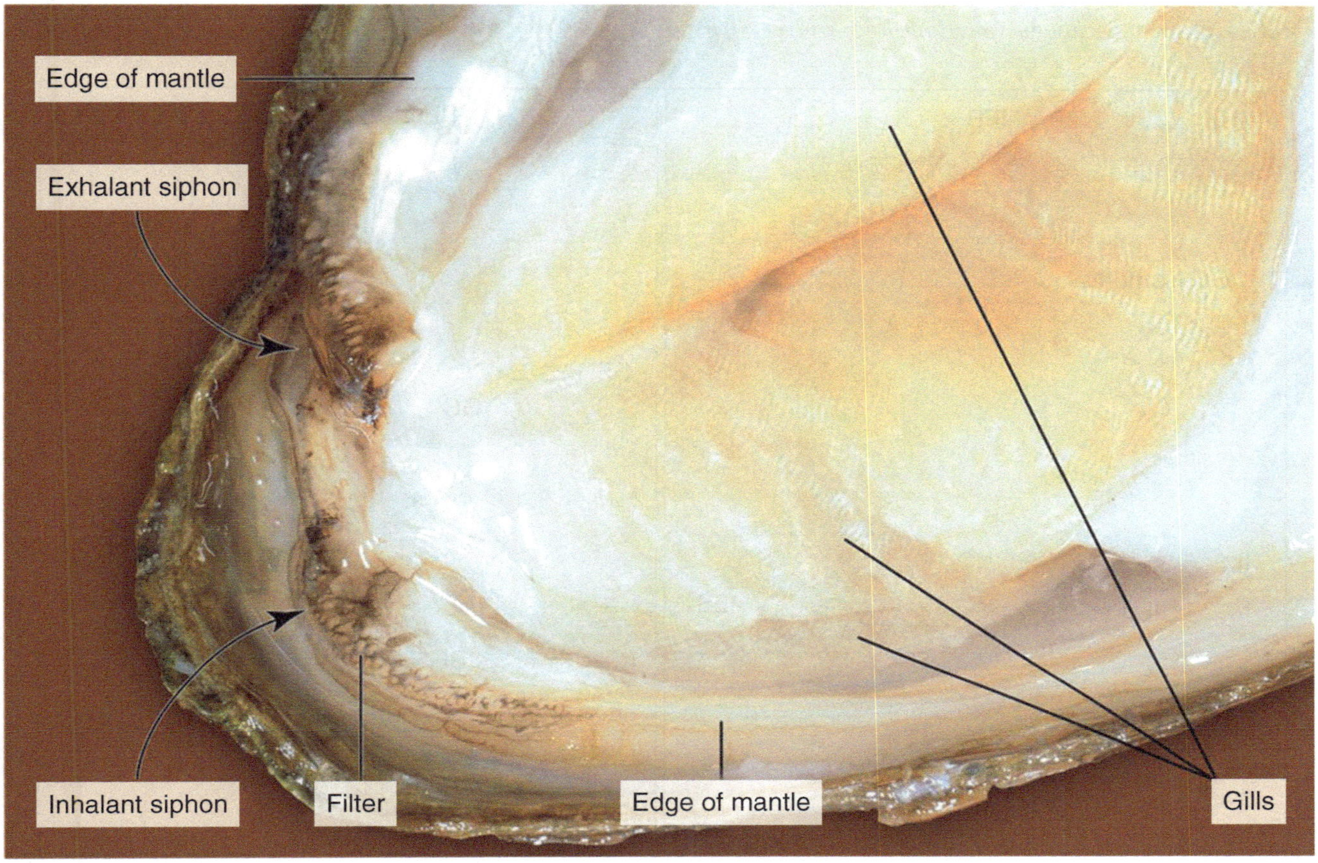

Fig. 6.7 Closer view of the right pair of gills. Notice the transverse lines; the gill filaments are actually perpendicular to these

Each gill is a double fold, with W profile in cross section of the whole animal. The "W" is composed of two V-like unit forming descending and ascending gill lamellae. The two lamellae of each gill are connected to each other by a series of thin partitions (*interlamellar bridges*, ILB), which divide the gill into many vertical spaces, the *water tubes* (interlamellar spaces, ILS). The *lamellae* are ridged in appearance, consisting of numerous parallel *gill filaments* (GFs) which are actually plates in three dimensions (Fig. 6.8).

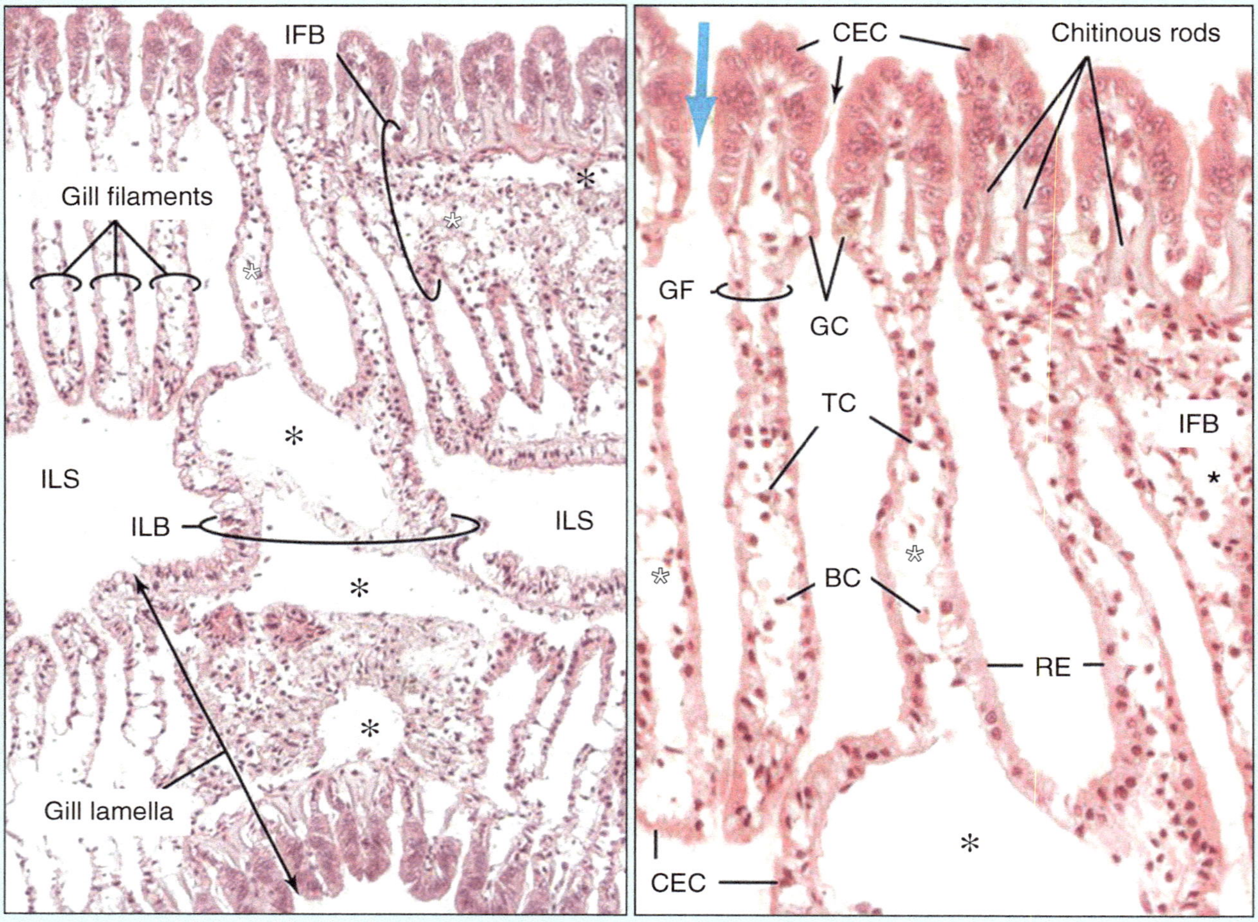

Fig. 6.8 Transverse cross section through a portion of the gill of a mussel (HE). *Black asterisks* large vessels, *white asterisks* sinusoids, *blue arrow* water current, *small black arrow* ostium, *BC* blood cells, *CEC* ciliated columnar epithelial cells, *GC* mucous gland cells, *GF* gill filament, *IFB* interfilamental bridge, *ILS* interlamellar space, *ILB* tube/interlamellar bridge, *RE* cuboidal respiratory epithelium, *TC* trabecular cells

Their cross section shows thin, long profiles with two ends. One end faces to the mantle cavity, whereas the other end faces to the interlamellar space. The surface located towards the mantle cavity is covered by *ciliated columnar epithelial cells* (CECs). There are groups of *mucous gland cells* (GCs) among them. This end of the filament (edge of the plate) is supported by *chitinous rods* (plates) lying just beneath the epithelium (Fig. 6.8). This reinforcement is essential: it works as a cutwater in the water stream. Lateral side of filaments covered by *respiratory cuboidal epithelium* (RE). Circulatory system gives vessels and smaller, irregular sinusoids into the gills. Vessels can be seen in interlamellar and *interfilamental bridges* (IFBs); sinusoids are seen in filaments. The latter are supported by *trabecular cells* (TCs) forming pillars. Water enters the tubes (interlamellar spaces) through innumerable small holes (*ostia*) between the filaments (Fig. 6.8). The water tubes connect dorsally with the *epibranchial canal*, which in turn empties to the outside through the exhalant siphon (Fig. 6.9).

Freshwater mussels are filter feeders; they feed on plankton and other microscopic creatures, sediment and organic debris which are free-floating in water. A mussel draws water in through its inhalant siphon into its mantle cavity by the actions of the cilia located on the inner mantle wall. Then the water is filtered through the small ostia on the gills. The water gets into the water tubes which connects to the epibranchial canal, then exits through the exhalant siphon (Fig. 6.9).

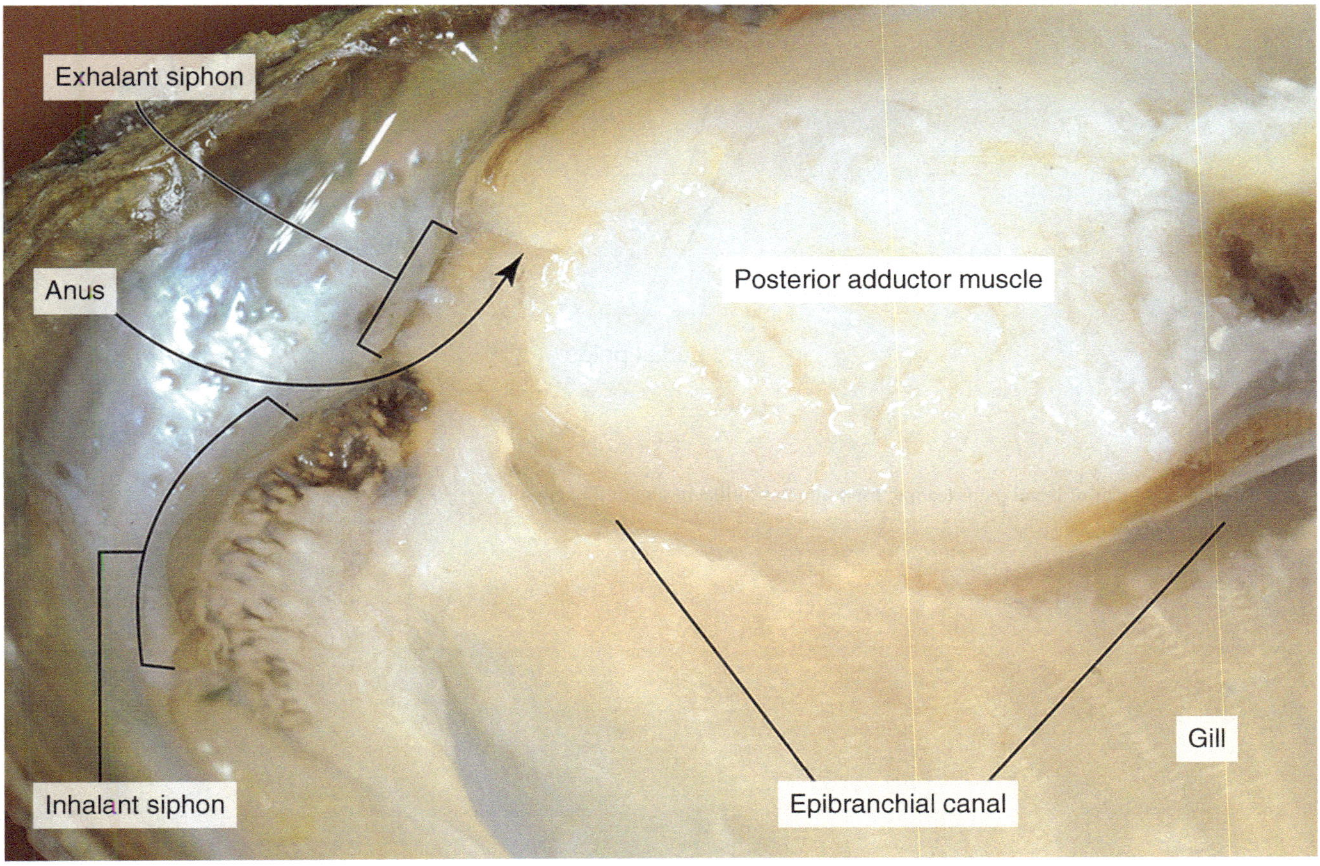

Fig. 6.9 The epibranchial canal is found dorsally to the gills leading to the exhalant siphon

Particles too large for the ostia are trapped in the mucus secreted on the surface of the gill lamellae. Cilia of the gill epithelial cells forward the mucus-trapped food particles to the ventral edge of the gills where they get into a groove with the ciliary tract. The food is then conveyed to the labial palps.

Locate the two pairs of *labial palps* (a pair on each side of the body); they are between the gills and the anterior adductor muscle (Figs. 6.6, 6.10 and 6.11).

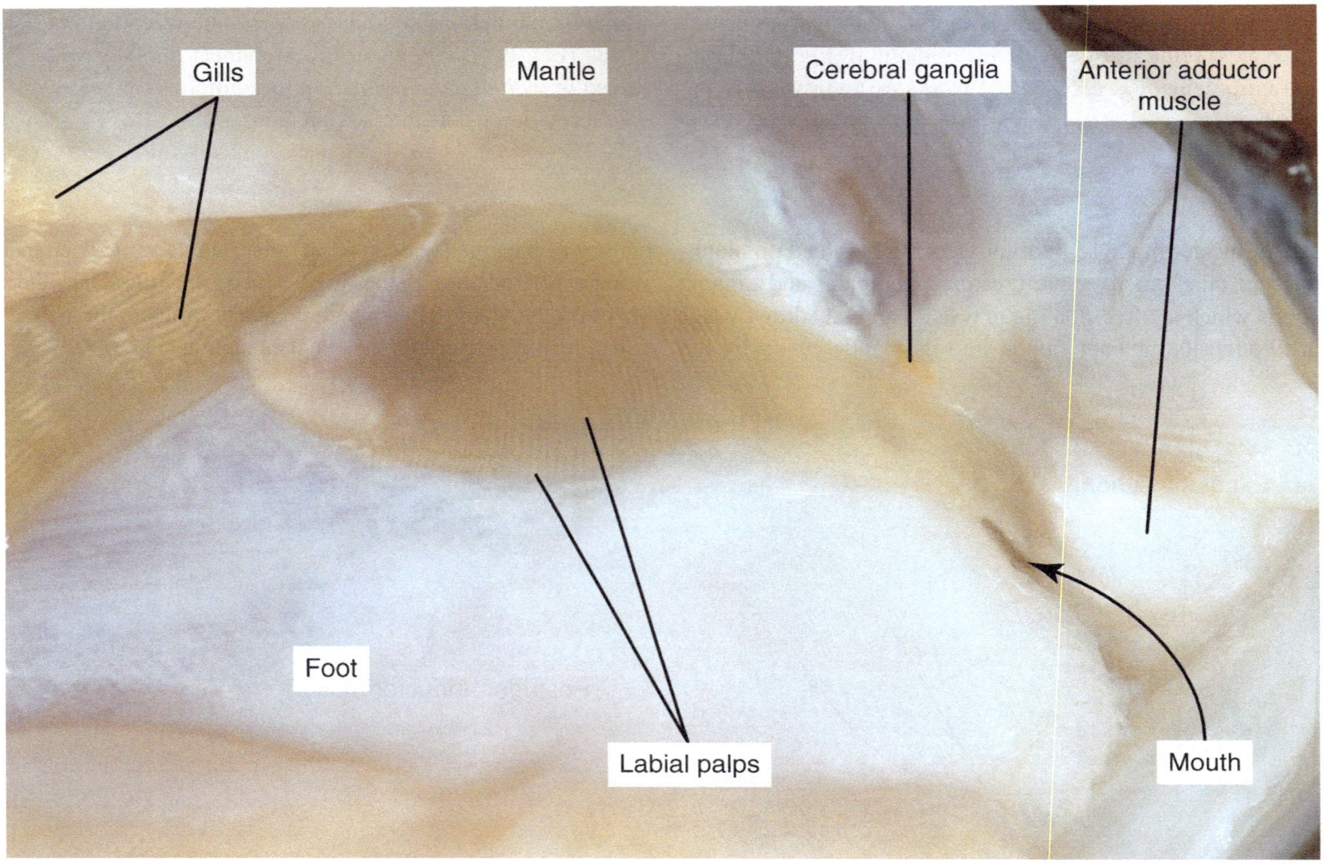

Fig. 6.10 The right pair of labial palps leading food into the slitlike mouth of the mussel

These flap-like structures surround and guide food into the slitlike *mouth* of the mussel. The palps secrete a great amount of mucus and are ciliated to guide food particles trapped in the mucus towards the mouth (Fig. 6.11).

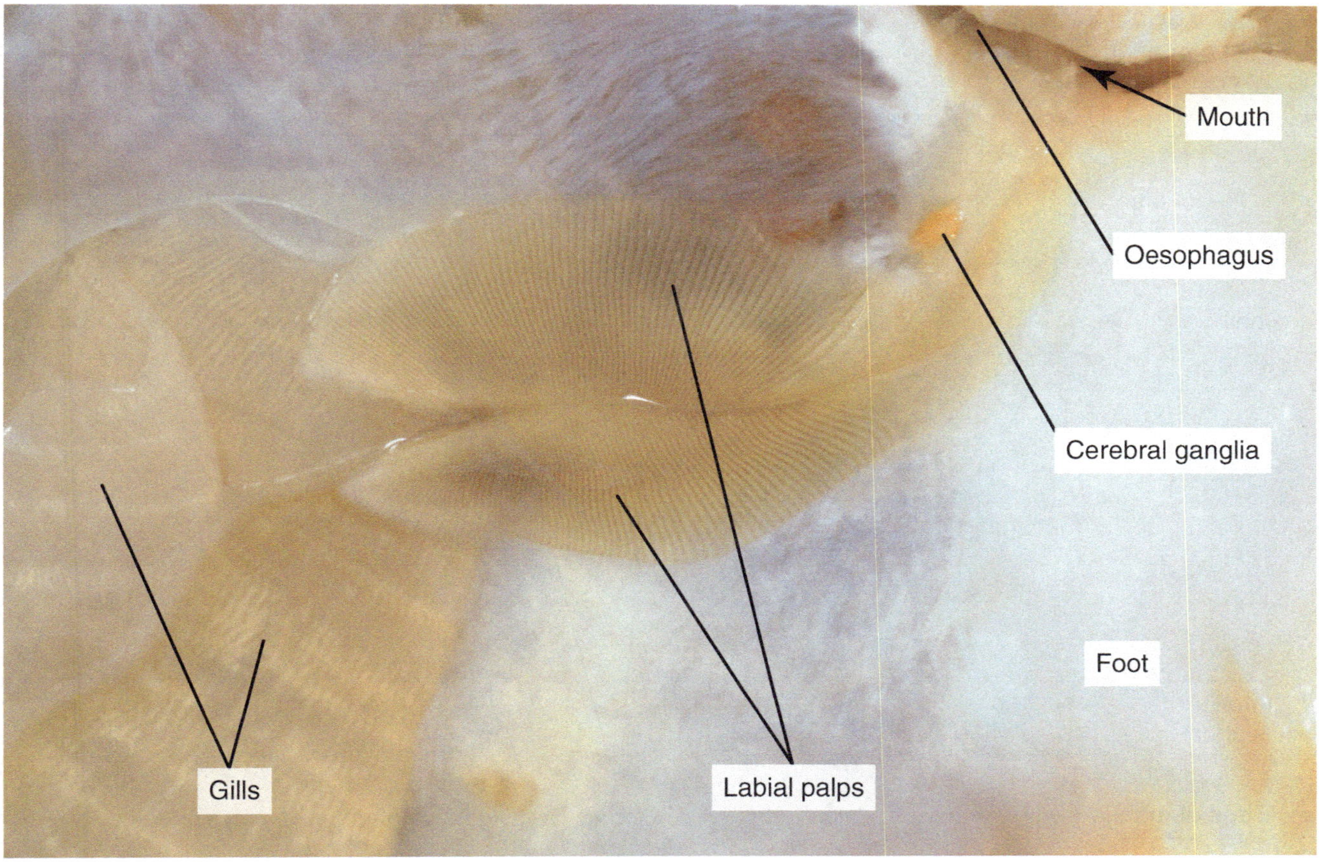

Fig. 6.11 The right pair of labial palps is a flap-like, ciliated structure and guides food particles towards the mouth. Note the cerebral ganglia and save it during further dissection

To reveal the parts of the alimentary canal, cut through the surface tissue on one side of the visceral mass and foot and strip it away. Hold the scalpel horizontally and be careful not to cut deeply (Fig. 6.12).

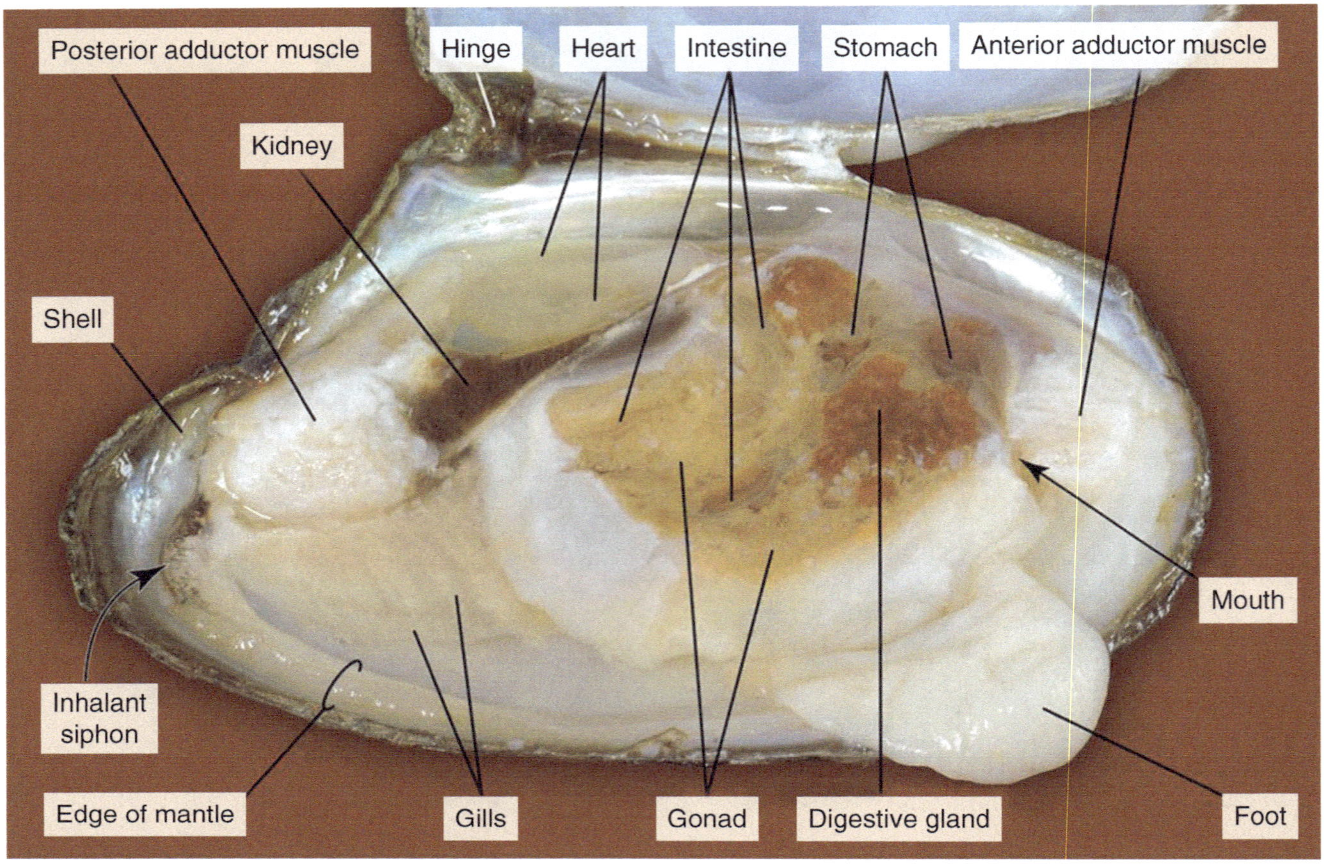

Fig. 6.12 The completely dissected mussel

The mouth leads into a short *oesophagus* that widens into the *stomach*, surrounded by the greenish-brown *digestive gland* (hepatopancreas) (Figs. 6.12, 6.13 and 6.14).

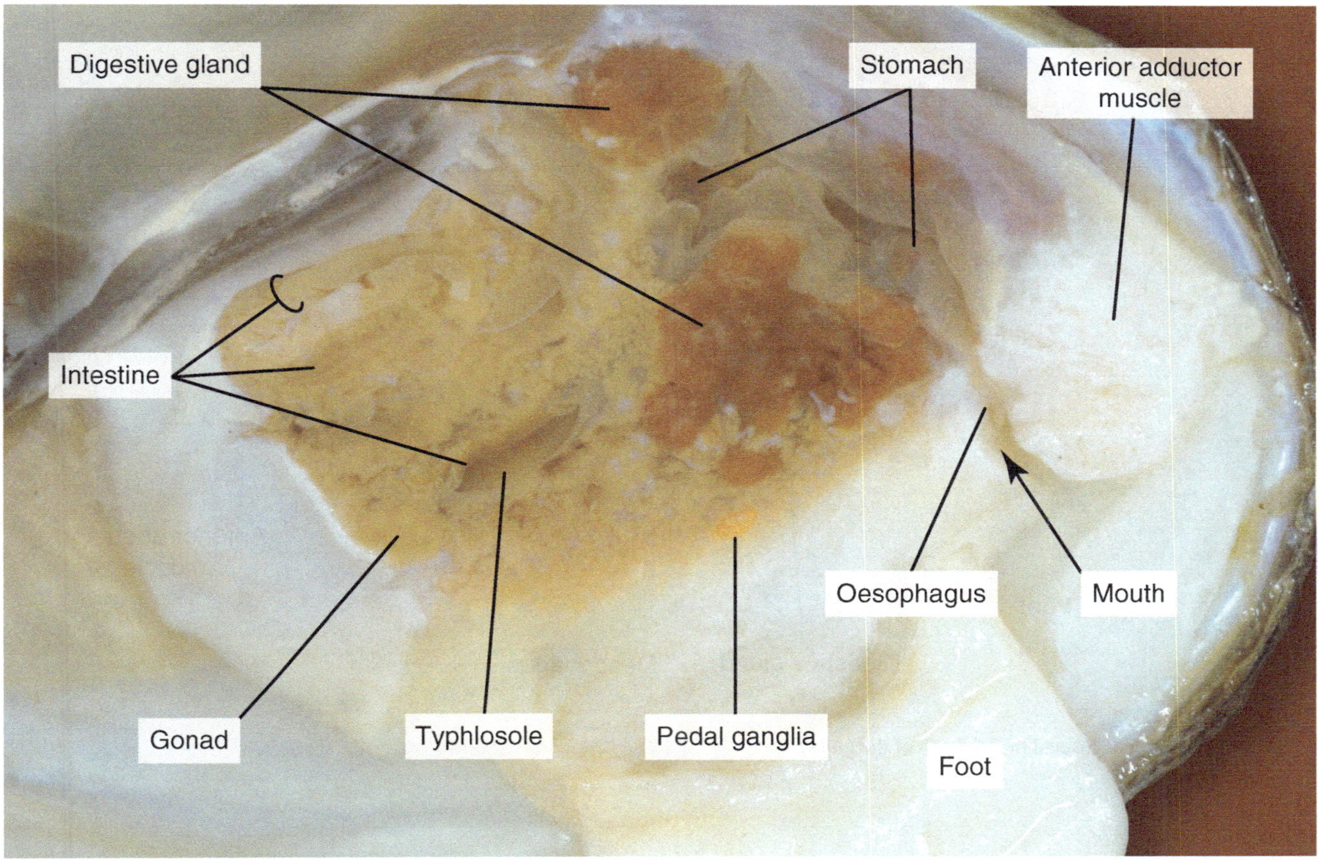

Fig. 6.13 The anterior part of the alimentary canal and surrounding organs

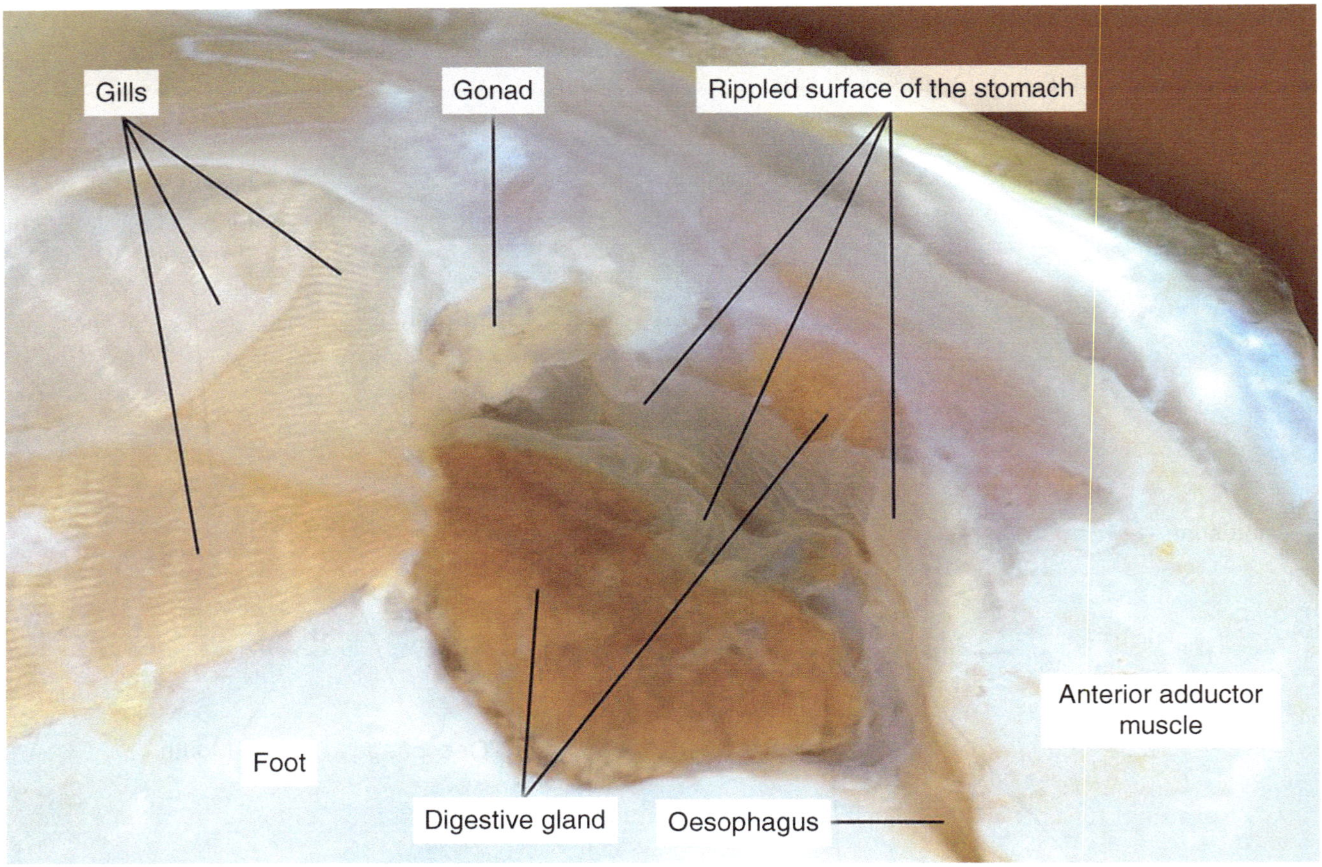

Fig. 6.14 The oesophagus and the stomach of the mussel

In freshly collected mussels, a solid gelatinous rod, the *crystalline style*, may be found, projecting into the stomach. It is composed of mucoproteins and digestive enzymes (chiefly amylase and cellulase). As the style rotates by ciliary action against a chitinous gastric shield, the tip is continually worn away, with release of small quantity of enzymes. The rotation also mixes the stomach contents and pulls in more mucus strings of food. The crystalline style disappears within a few days after mussels are collected. The constant motion of the style propels food particles into a sorting region at the rear of the stomach, which distributes smaller particles into the digestive gland, and heavier ones into the intestine. Digestion is mostly intracellular and it is carried out in the wall of the *canals* of the digestive gland (Fig. 6.15).

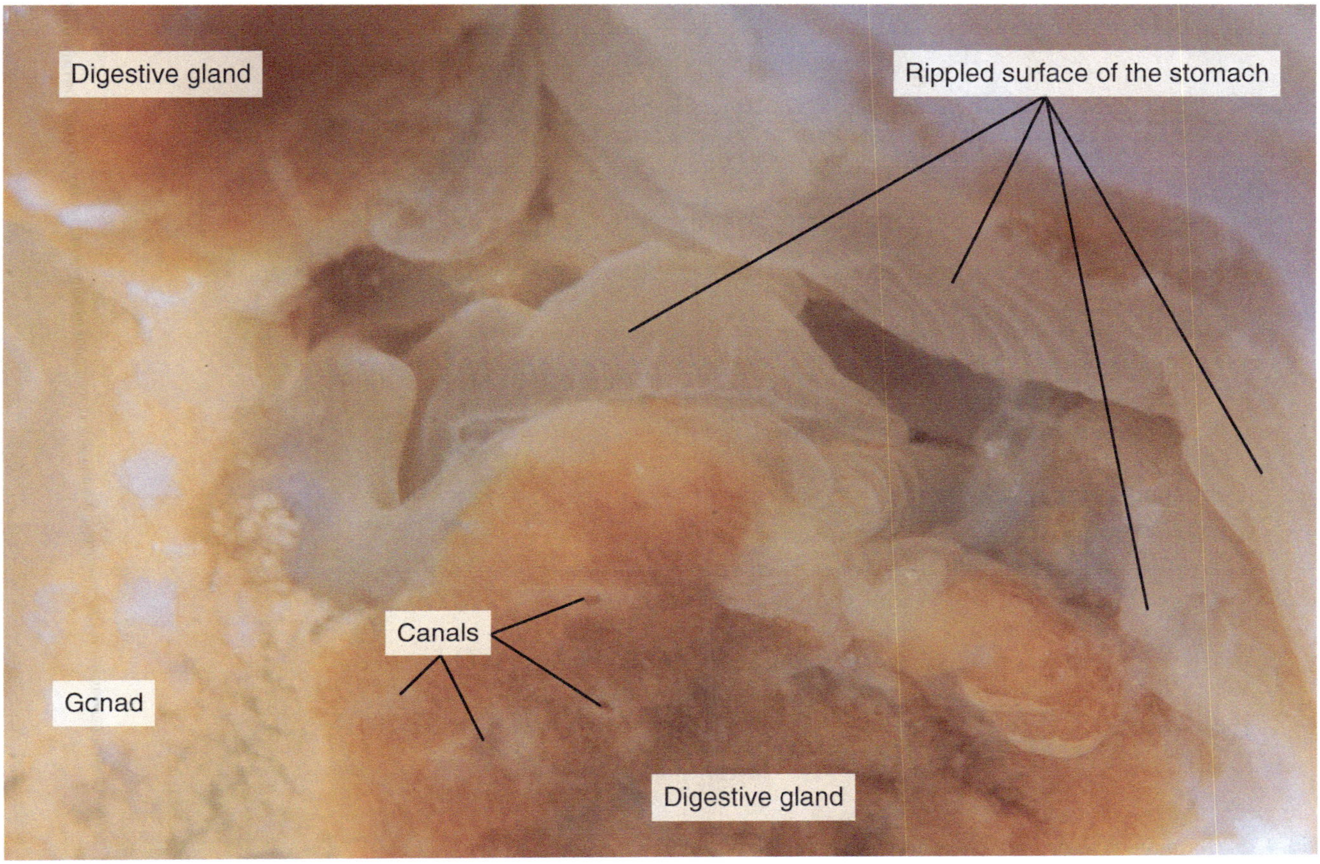

Fig. 6.15 A close up of the rippled ciliated surface of the stomach where smaller food particles are distributed into the canals of the digestive gland

The stomach narrows into the long, coiled *intestine*, which can be seen looping back and forth through the visceral mass. The *typhlosole*, a ridge-like structure projecting into the lumen of the intestine, increases the surface area for more efficient absorption of digested nutrients (Fig. 6.13). Surrounding the intestine is the yellowish or light brown tissue of the *gonad*. Follow the intestine through the mussel. The intestine connects to the *rectum*, which passes through the ventricle. Trace the rectum as it passes dorsal to the posterior adductor muscle. Find the *anus* just behind the posterior adductor muscle, which empties faeces into the exhalant siphon (Fig. 6.16).

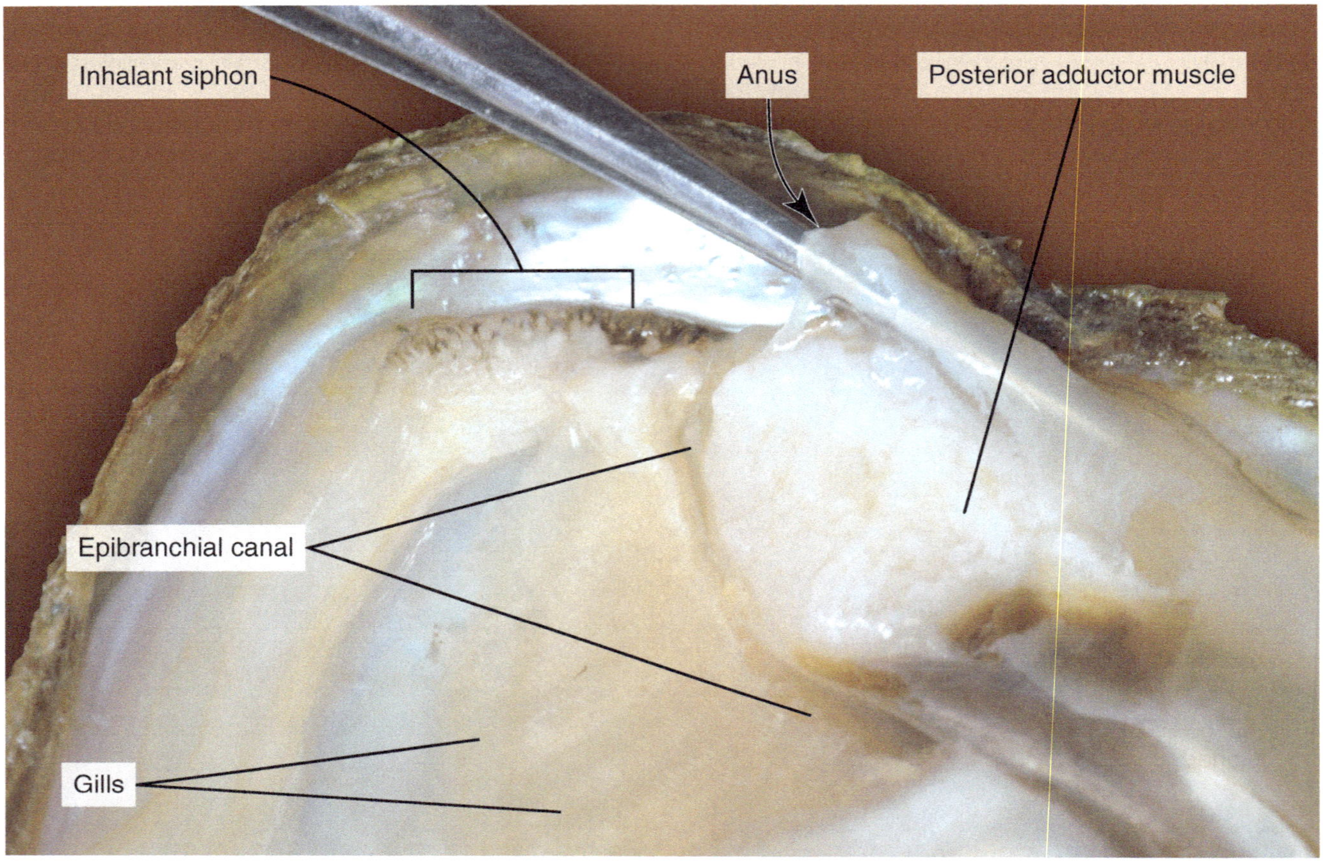

Fig. 6.16 The position of the anus (probe in the end part of the rectum)

Near the dorsal midline, just below the hinge, is the thin-walled *pericardial sac*, within which lies the *heart*. The three-chambered heart is composed of a single *ventricle* and a pair of *auricles*. The paired auricles are fan shaped and very thin walled. They can be visualised by pulling the pericardium with a pair of tweezers, as their wall is stretched (Fig. 6.17).

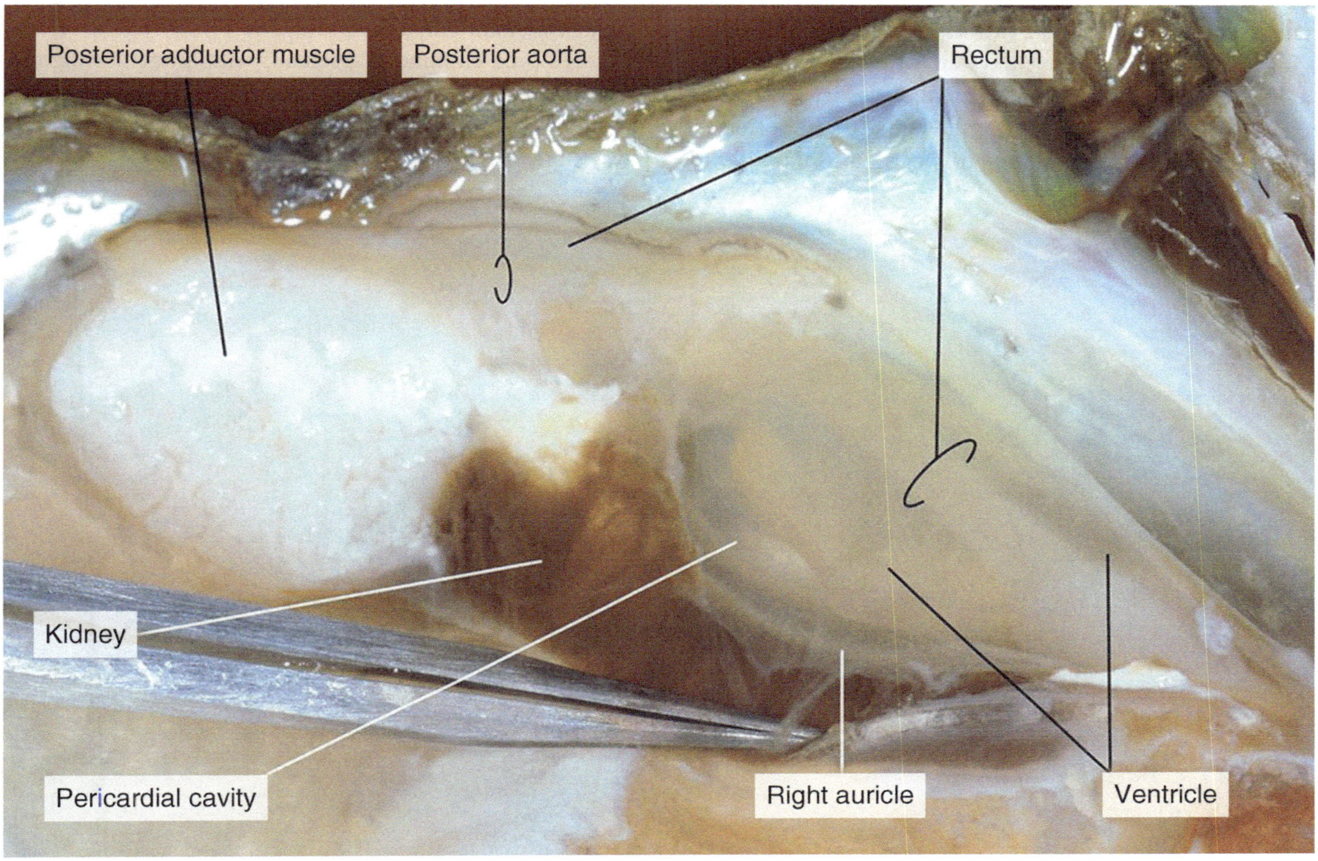

Fig. 6.17 The heart and surrounding organs. Auricles can be visualised by pulling the pericardium with a pair of tweezers, as their wall is stretched

Note that the ventricle surrounds the rectum. The *pericardial space* around the heart is a remnant of the coelomic cavity, which is greatly reduced in molluscs. Two aortae leave the ventricle: the *anterior aorta* passes to the visceral mass and intestine, and the *posterior aorta* runs along the ventral side of the rectum to the mantle (Figs. 6.17 and 6.18). Mussels have an open system of circulation with no capillaries among the tissues. From the ventricle, the aortae carry haemolymph to sinuses in the body tissues. From the visceral organs, the haemolymph is carried to the gills for gaseous exchange and then back to the auricles and ventricle. Haemolymph from the mantle, also rich in oxygen, returns directly to the auricles. The haemolymph of mussels is colourless but contains haemocyanin for oxygen transport.

Inject a small amount of toluidine blue solution or any other dye (ink, carmine suspension) with a fine needle into the ventricle. This will flow into and reveal parts of the heart and the two aortae. Press the dye slowly and carefully out of the syringe (Fig. 6.18).

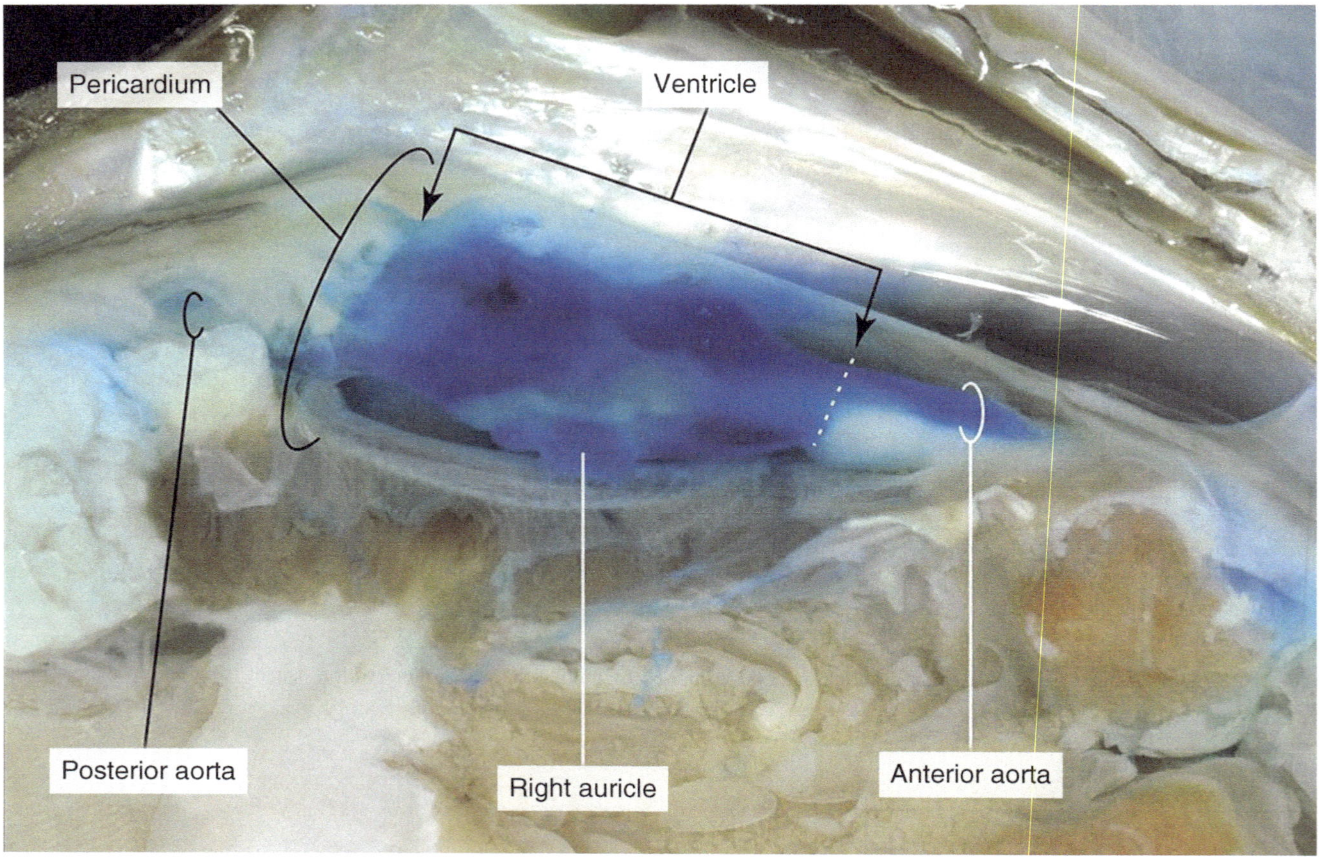

Fig. 6.18 The heart and the two aortae after injection of toluidine blue solution

A pair of dark *kidneys* (modified metanephridia) lies under the floor of the pericardium (Figs. 6.12 and 6.17). The kidneys get the ultrafiltrate from the pericardium, with which they connect. The excreted waste is discharged into the epibranchial canal and carried away with the exhalant current.

Dissection of the *nervous system* is difficult and sometimes inefficient. The nervous system of the mussel is highly centralized; only three pairs of bright orange-yellow ganglia are connected to each other by nerves. The *cerebropleural ganglia* (or cerebral ganglia) are found one on each side of the oesophagus close to the posterior surface of the anterior adductor muscle (Figs. 6.10, 6.11 and 6.19).

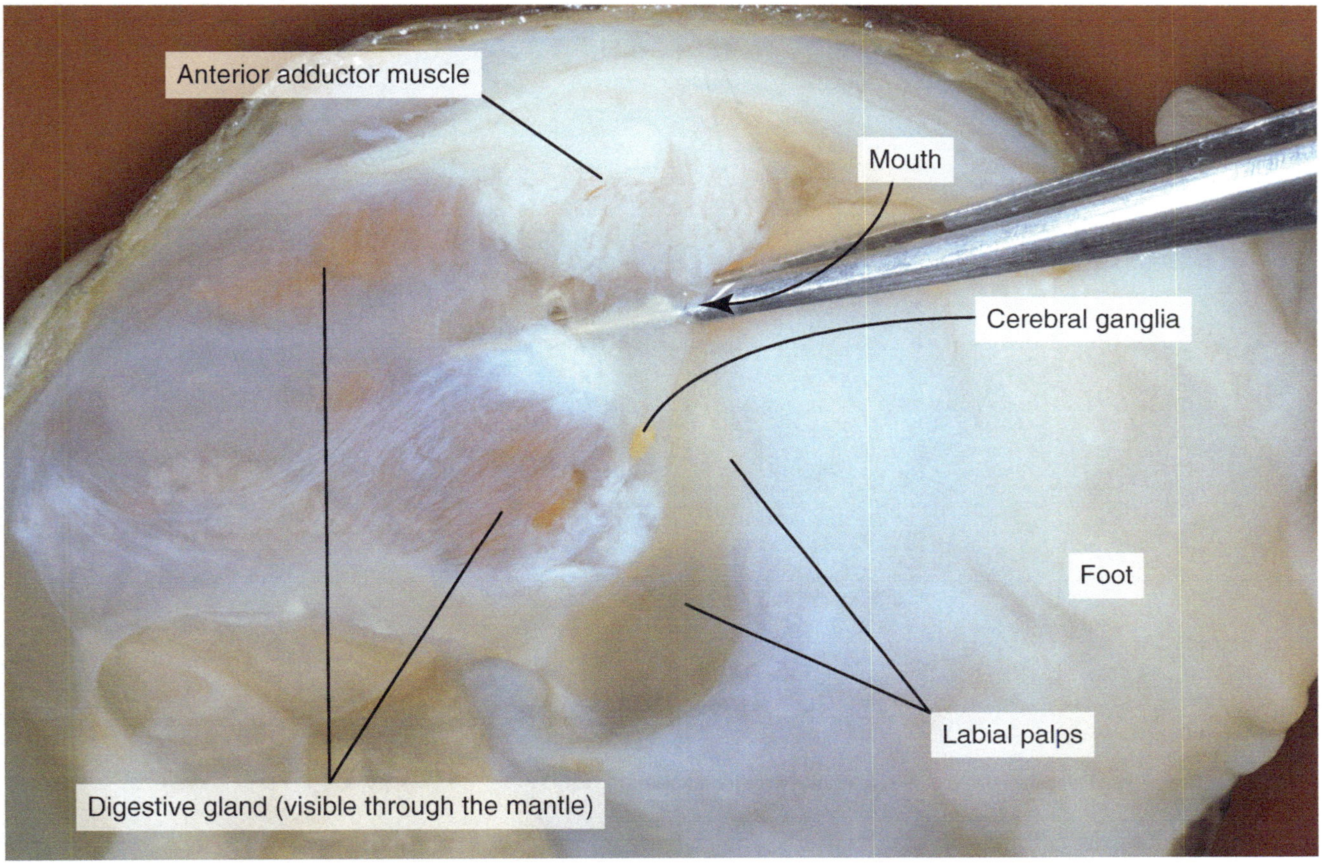

Fig. 6.19 The cerebral ganglia are found next to the oesophagus close to the posterior surface of the anterior adductor muscle

The *pedal ganglia* are fused and located in the anterior part of the foot (Fig. 6.13). The *visceral ganglia* are also fused and found against the ventral side of the posterior adductor muscle and the visceral nerve connective running through the kidney (Fig. 6.20). If this connective does not show, make no attempt to find it, as you have already removed it accidentally.

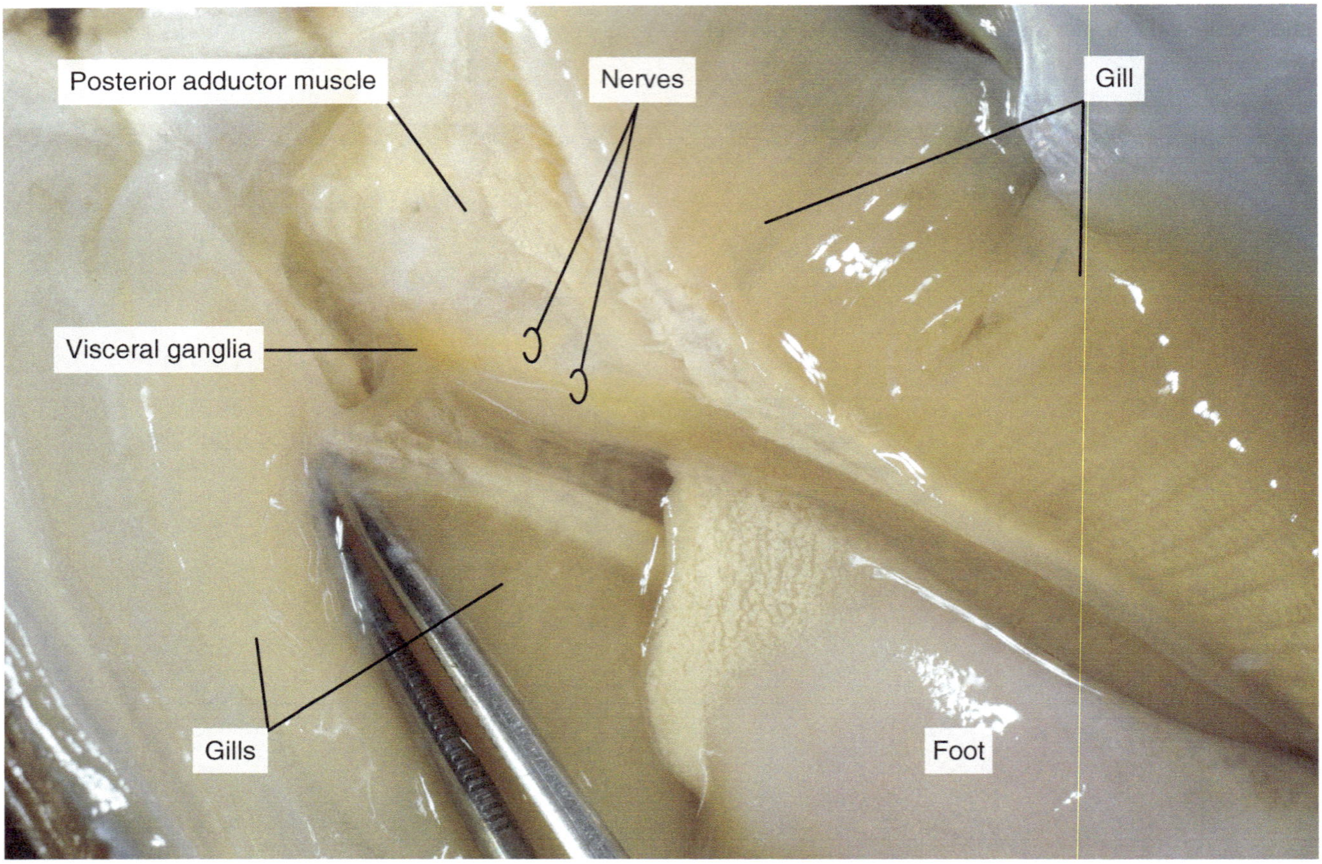

Fig. 6.20 Visceral ganglia against the ventral side of the posterior adductor muscle

Locate the spongy, yellowish or light brown reproductive organs, the *gonads* (Figs. 6.12 and 6.13). Freshwater mussels reproduce sexually. They are gonochoristic, with separate male and female individuals. The gonads discharge their products into the epibranchial canal. Sperm is released by the male directly into the surrounding water and enters the female via the inhalant siphon and fertilize eggs in the epibranchial canal. After fertilization, the zygotes settle into the water tubes of the outer gill, where they develop into a tiny bivalved larval form called a *glochidium* (plural glochidia). Here they are constantly flushed with oxygen-rich water. Later, when the glochidia are released from the female mussel, they temporarily parasitise fish, attaching themselves with hooks on their valves to the fish's fins or gills. They grow, break free from the host and drop to the bottom of the water to metamorphose and begin an independent life.

Dissection of a Crayfish (*Astacus astacus*)

7

- *Availability:* The crayfish is found in freshwater streams and ponds all over the world. There are about 300 species worldwide (*Cambarus*, *Procambarus*, *Astacus* and *Orconectes*). They are omnivorous, feeding on fish, tadpoles, worms, insects and plants. We can purchase fresh crayfishes from fishing companies supplying restaurants. Alternatively the dissection can be performed on preserved material.
- *Anaesthesia:* Put a wad infiltrated with diethyl-ether in a jar. Place in the crayfish and close the jar immediately; the crayfish can be dissected within 15 min, when it doesn't move upon shaking the jar. Wash the specimen thoroughly with water.

Before commencing the dissection, study the external anatomy of the crayfish. Like all crustaceans, a crayfish has a fairly hard *exoskeleton* that covers its body. It is a cuticle secreted by the epidermis and hardened with an organic substance called *chitin*, with addition of mineral salts, such as *calcium carbonate*. The cuticle must be shed or moulted several times while the crayfish is growing up, each time being replaced by a new soft exoskeleton that soon hardens.

P. Lőw et al., *Atlas of Animal Anatomy and Histology*, DOI 10.1007/978-3-319-25172-1_7

Its body is divided into two main parts, the *cephalothorax* and the *abdomen* (Fig. 7.1).

The cephalothorax consists of 13 segments, of which the cephalic (or head) region has five and the thoracic region eight segments. The part of the exoskeleton that covers the cephalothorax is called the *carapace*. A transverse *cervical groove* marks the head-thorax fusion line. On the thoracic region, locate the prominent *suture* or indentation on the cephalothorax that defines a central cardiac area (which covers the heart) separate from the broad lateral areas (which cover the gills). Lift up the edge of the carapace (branchiostegit) to disclose the gill chamber and the feathery gills. The *abdomen* is located behind the cephalothorax and consists of six clearly divided segments (Fig. 7.1).

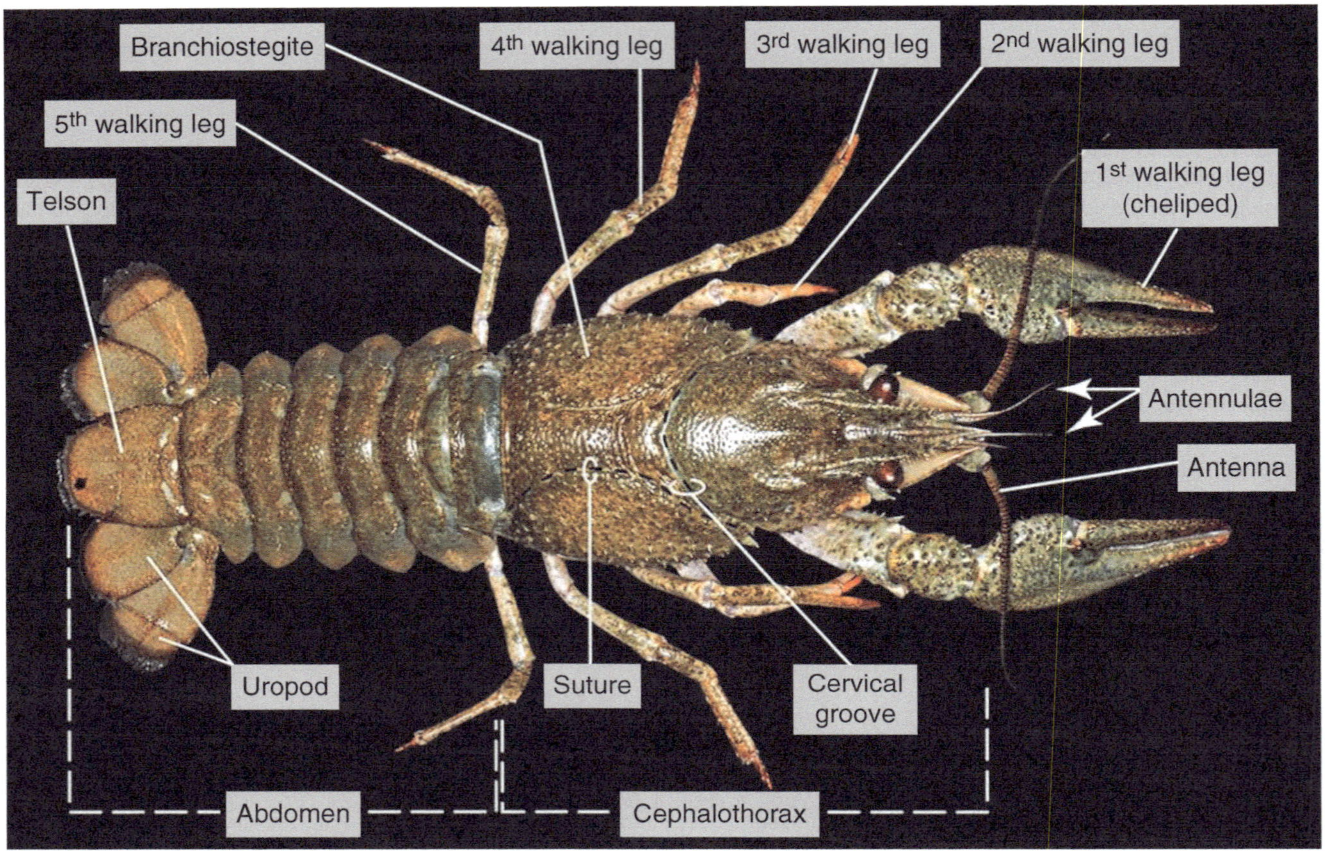

Fig. 7.1 Dorsal view of a crayfish

Find the *rostrum* at the head of the animal, which is the pointed extension of the carapace. Beneath the rostrum locate the two stalked *eyes* (Figs. 7.2, 7.3 and 7.4).

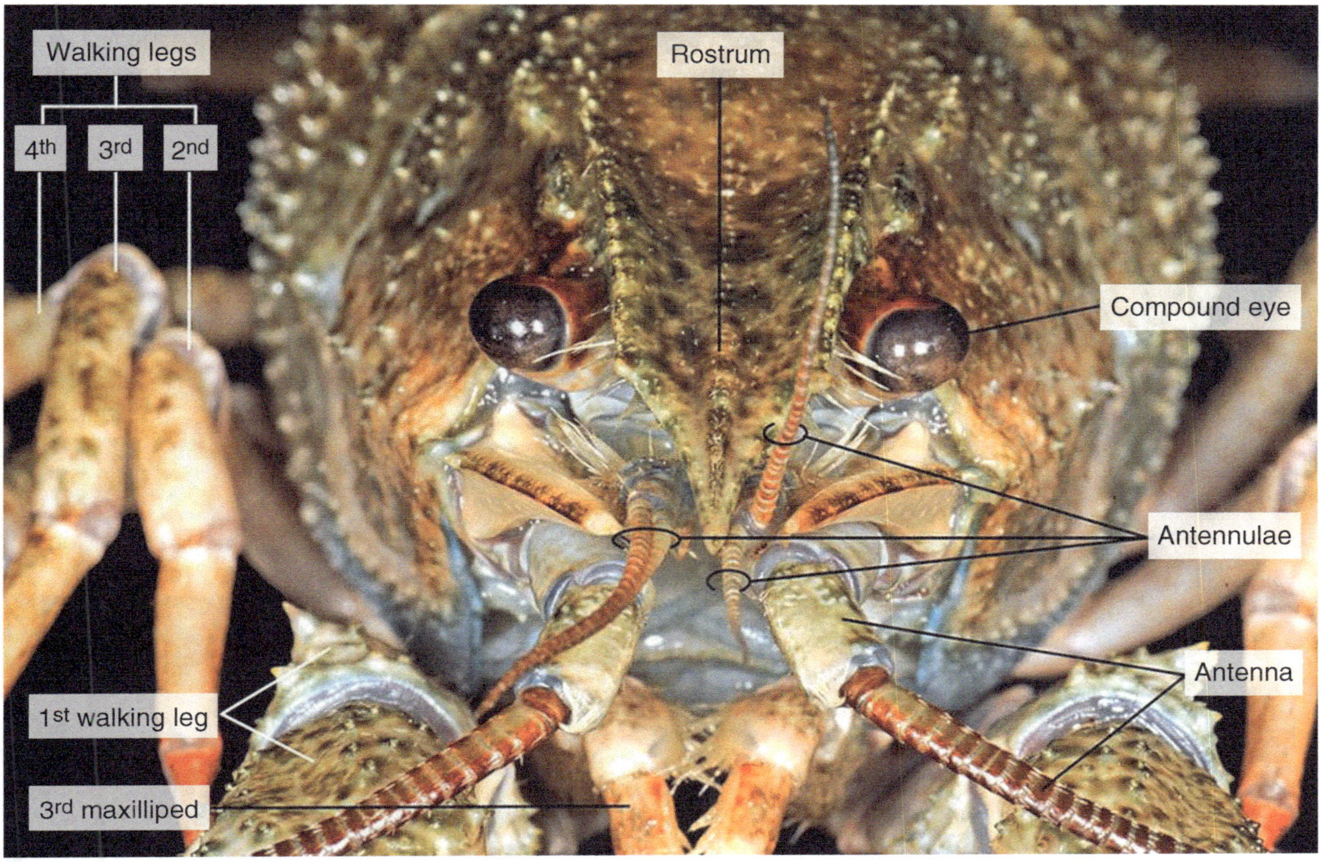

Fig. 7.2 Front view of the head region of a crayfish

Each segment of both the cephalothorax and the abdomen contains a pair of appendages. Arthropods have probably descended from an annelid-like ancestor whose appendages were all *biramous* (two-branched) and unspecialised, much like the swimmerets of the crayfish. The part of the appendage attached to the body is called the *protopod*. The medial branch is the *endopod*, and the lateral branch is the *exopod*. On some appendages even an *epipod* is attached to the proximal external part of protopod. During the evolution of the arthropods, the appendages became specialised for different functions: walking, grasping, food handling and so on. But because they are all derived from parts that were essentially alike, they are considered *homologous*.

The head (or cephalic) region has five pairs of appendages, the two pairs of antennae and three pairs of small mouthparts. The *antennules* are organs of balance, touch and taste. Long *antennae* are organs for touch, taste and smell. The endopod of each antenna is a very long, many-jointed filament (Figs. 7.2, 7.3 and 7.4).

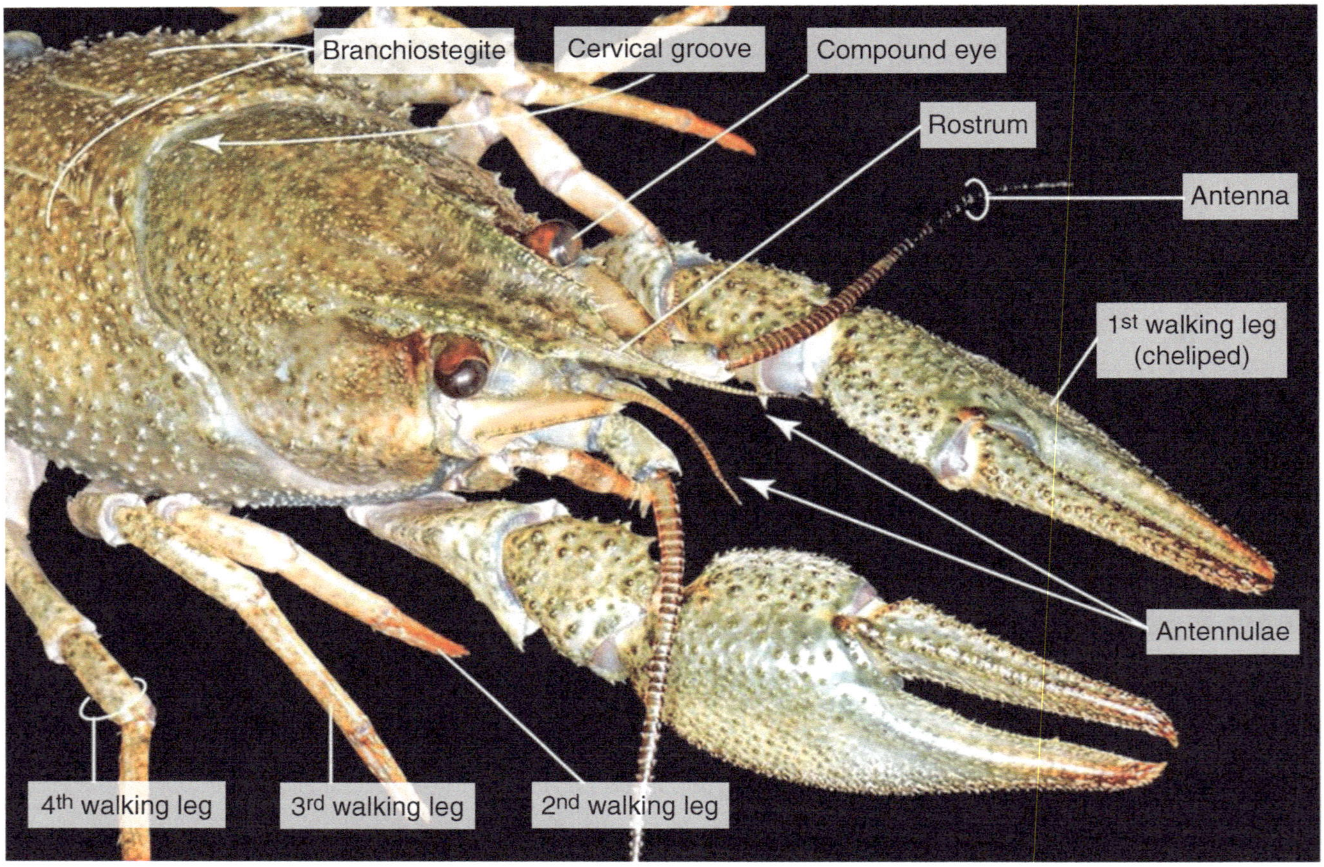

Fig. 7.3 Structures and appendages on the cephalothorax of a crayfish

The exopod of antenna is a broad, sharp, movable projection near the base (antennal scale) (Fig. 7.4).

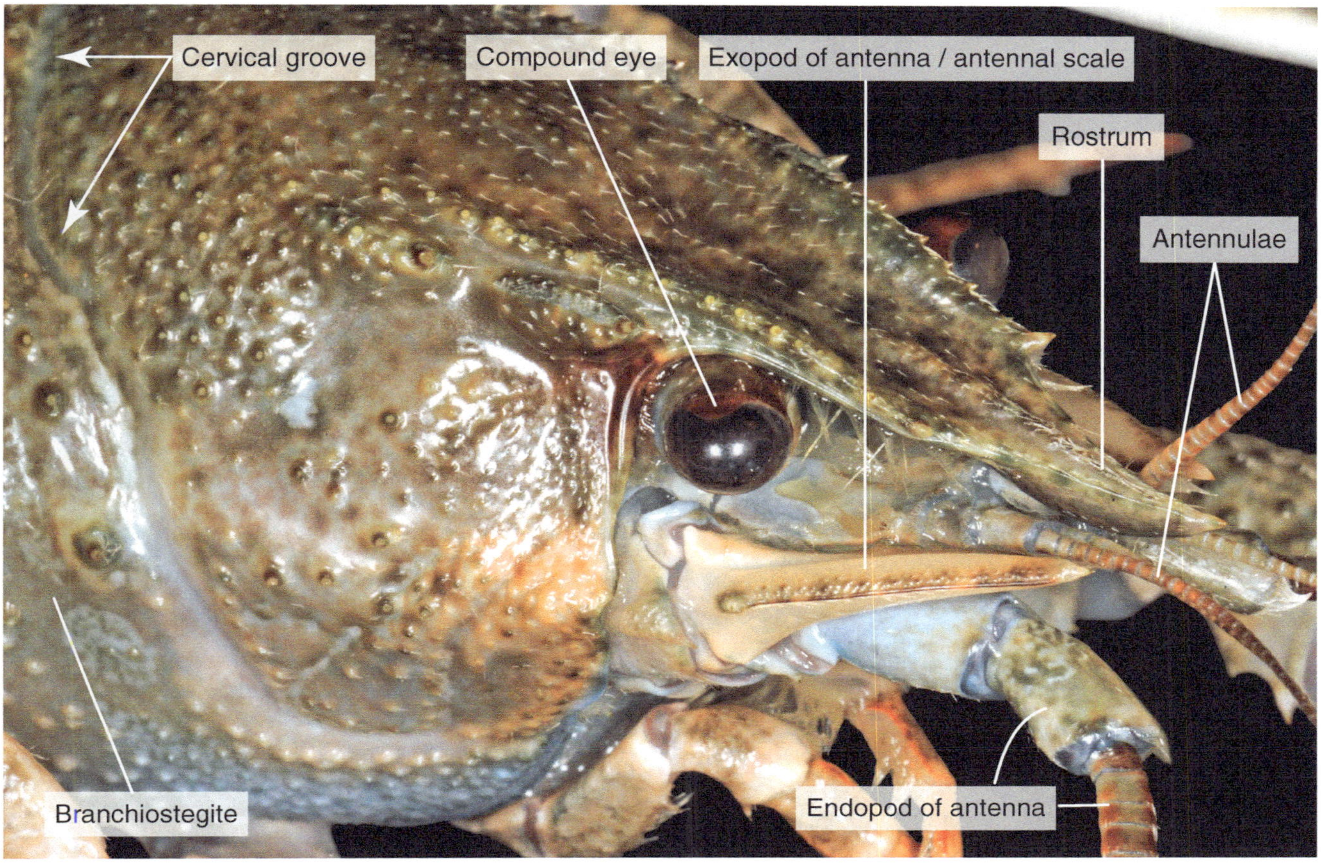

Fig. 7.4 Side view of the head region of a crayfish

On the ventral side of its broad protopod is the excretory pore of the green gland (Fig. 7.5).

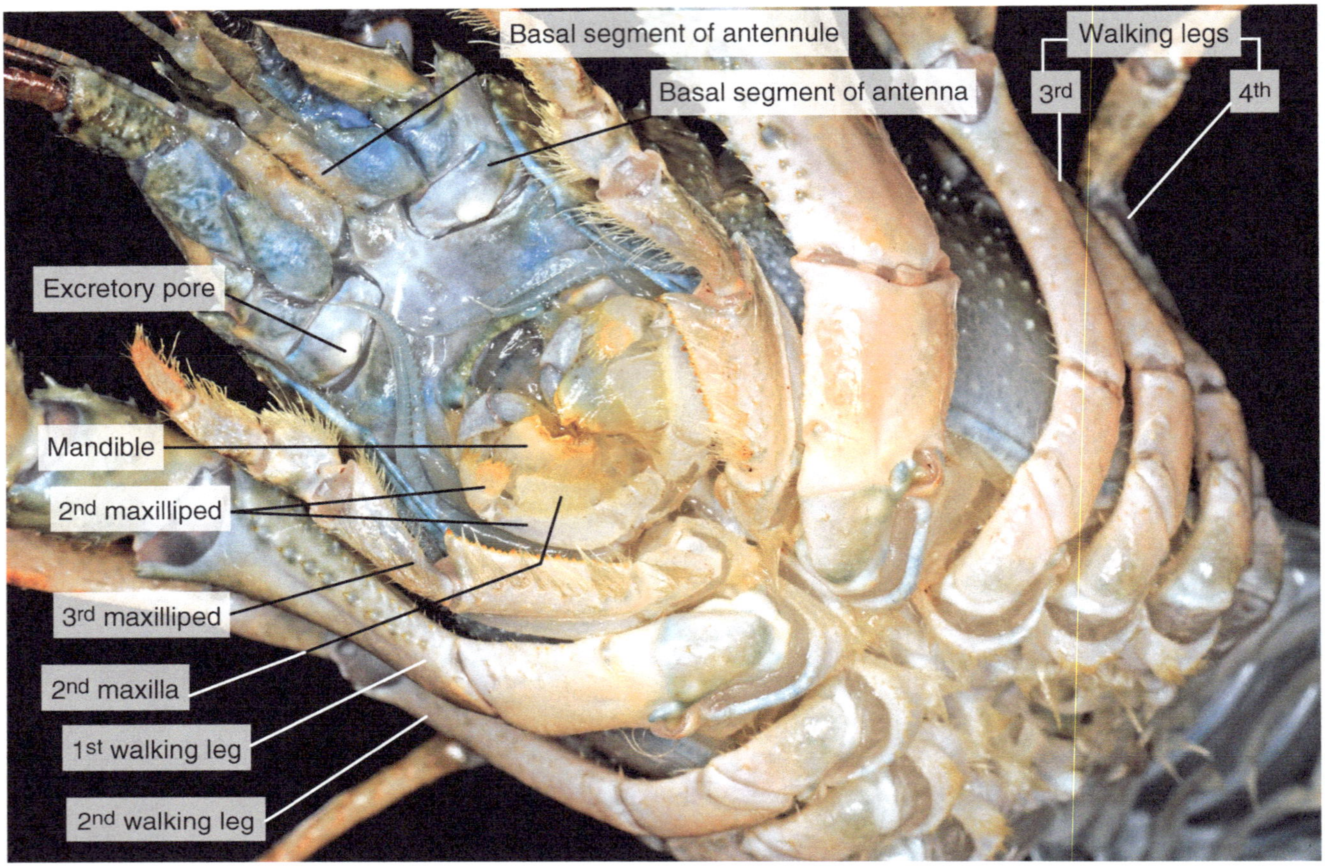

Fig. 7.5 Ventral view of the head region of a crayfish

The *mandibles*, or jaws, crush food by moving from side to side. Locate the mouth between the mandibles. Two pairs of *maxillae* hold solid food, tear it and pass it to the mouth. The second pair of maxillae bears a long *scaphognathite* (bailer) composed of exopod and epipod, the movement of which helps to draw water over the gills. Of the eight pairs of appendages on the cephalothorax, the first three are *maxillipeds*, which hold food during eating (Fig. 7.6).

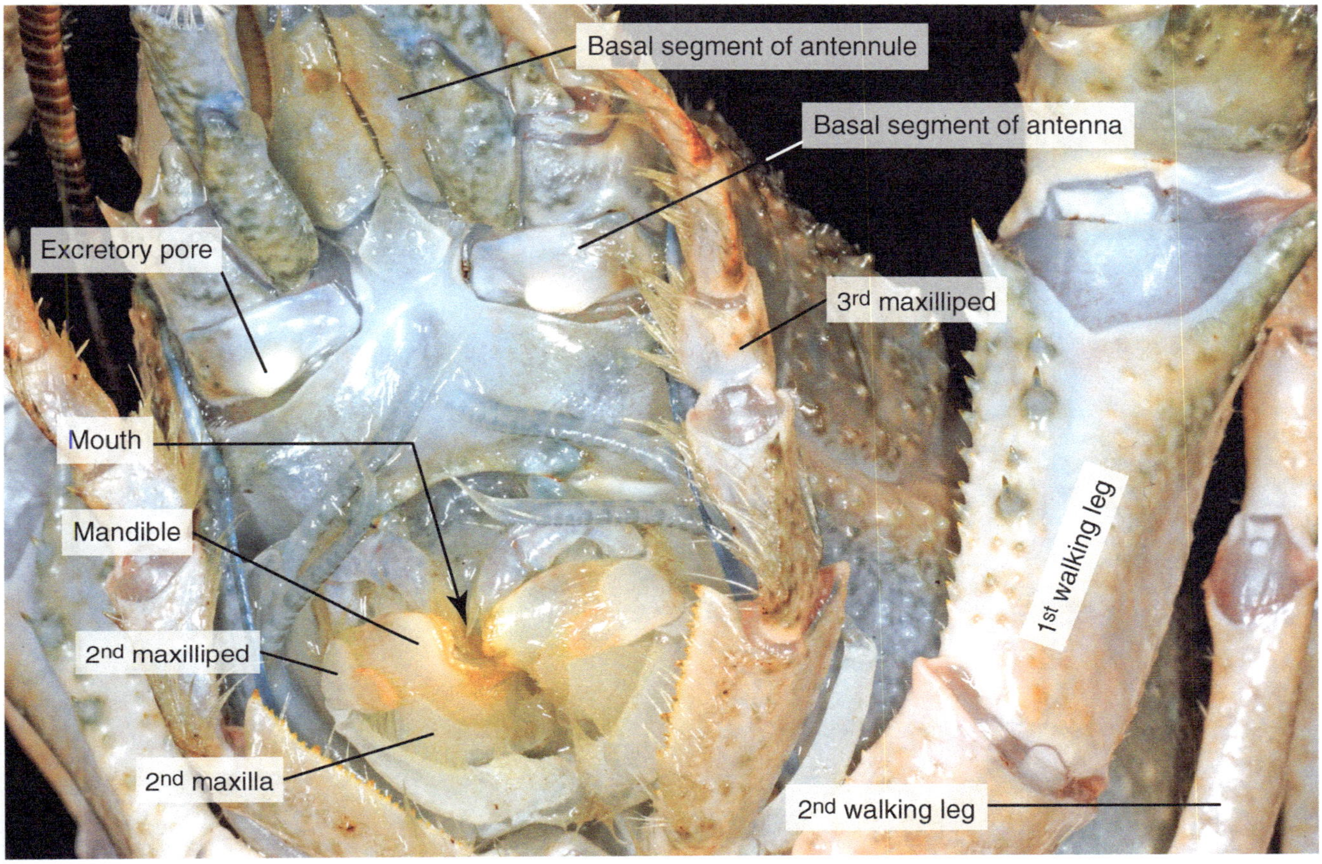

Fig. 7.6 Appendages surrounding the mouth of a crayfish

Next observe the largest prominent pair of appendages, the *chelipeds* (chela=claw) which are the large claws that the crayfish uses for defence and to capture prey (Fig. 7.7). The wide part of the claw is the "palm" (propodite) which ends in a narrow, fixed "finger". There is another narrow projection attached to it with a joint, the movable "finger" (dactylopodite).

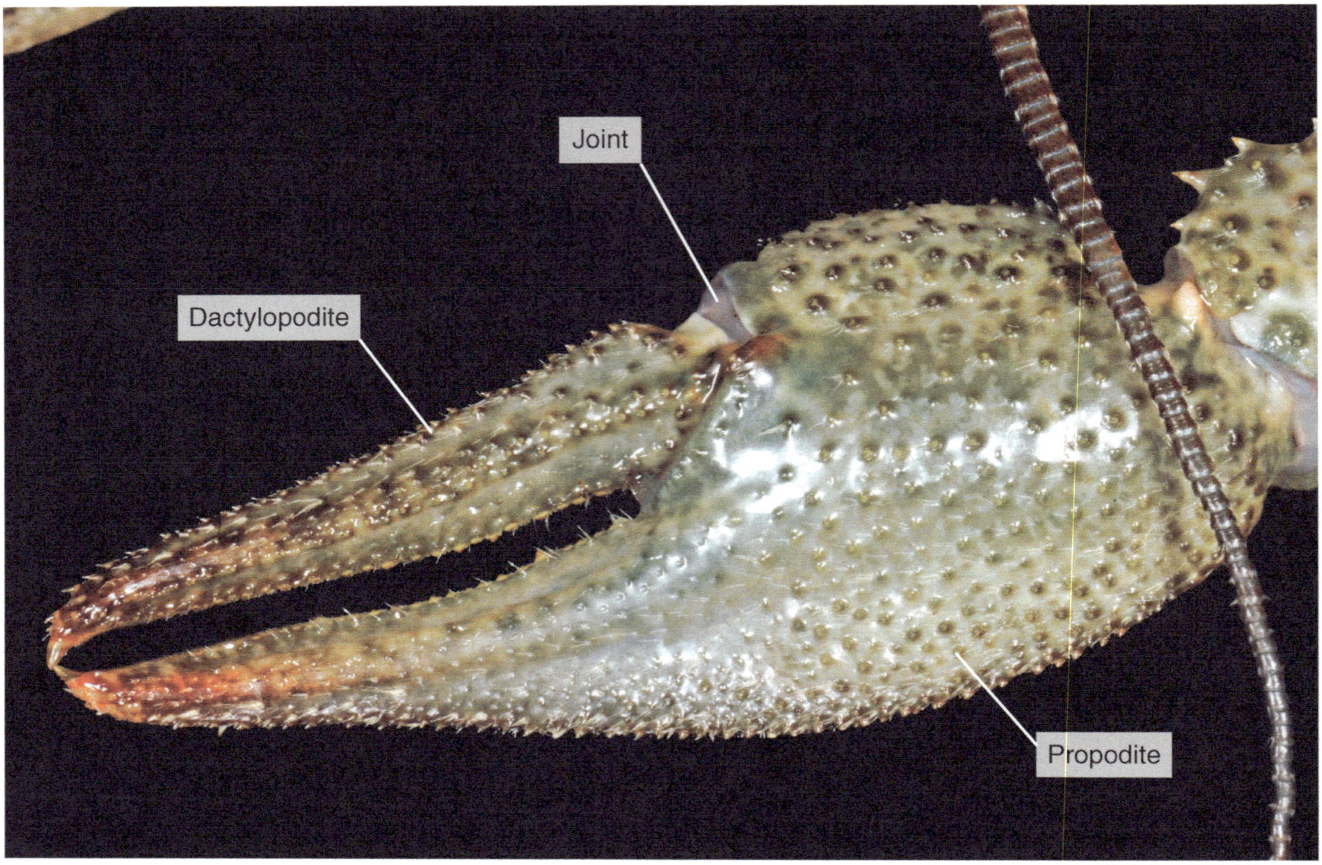

Fig. 7.7 The first walking leg, the cheliped ends in an enlarged claw

Each of the four remaining segments of the thorax contains a pair of *walking legs* (Figs. 7.1 and 7.5). In the abdomen, each body segment (somite) is enclosed in four articulated exoskeletal plates (sclerites) that form a complete ring around the segment: dorsal *tergite*, ventral *sternite* and laterally two *pleurites*. The tergite and pleurites are fused together to form a hard arch of exoskeleton (Fig. 7.8). The first five abdominal segments each have a pair of *swimmerets* (pleopods), which create water currents and function in reproduction (Figs. 7.9 and 7.10).

The sixth segment of the abdomen contains a modified pair of *uropods* (Fig. 7.8). In the middle of the uropods is a structure called the *telson*, which bears the anus on its ventral side (Fig. 7.32). The uropods and telson together make up the *tail fan*. The crayfish can also move backward if it forces water forward with its tail fan.

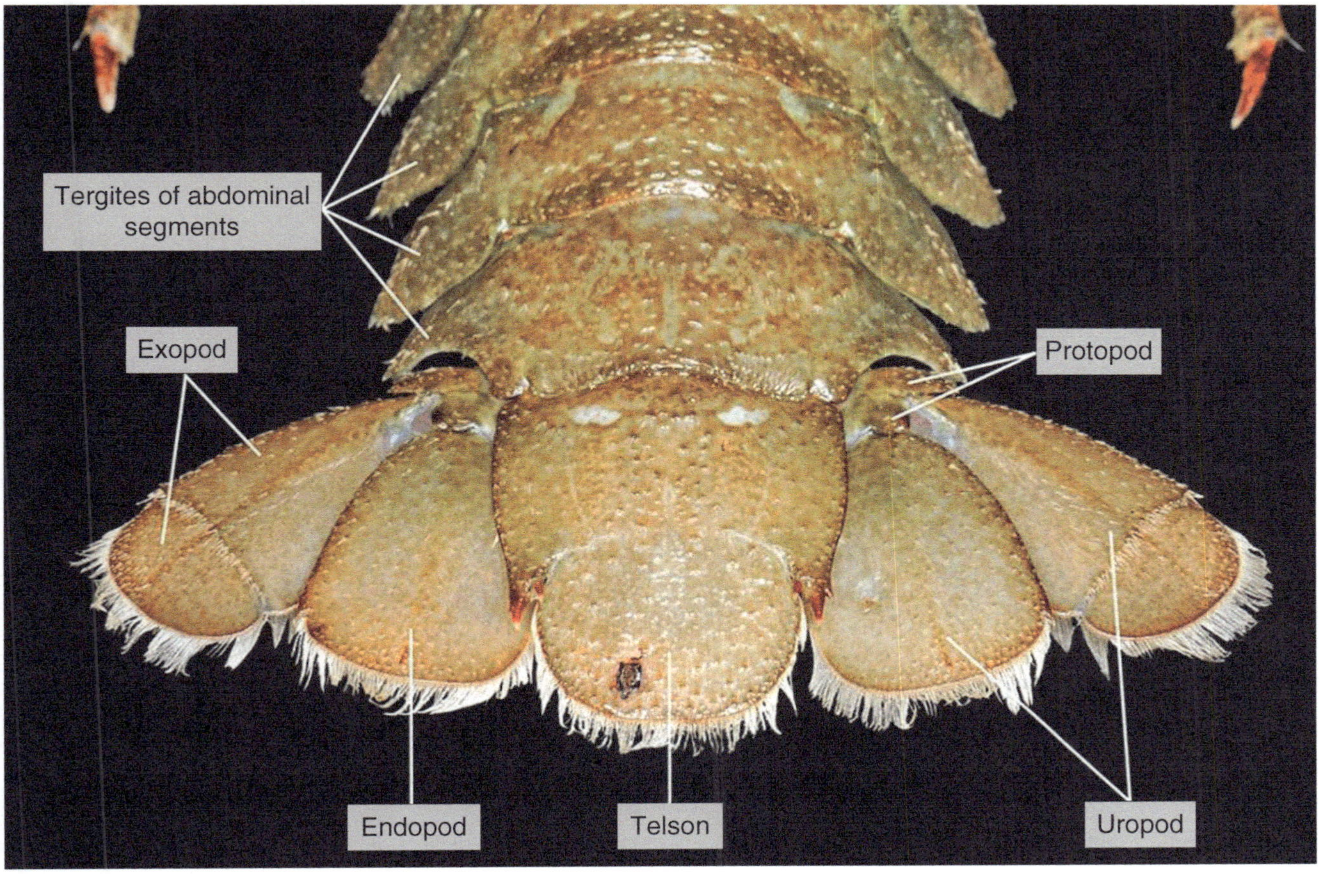

Fig. 7.8 The dorsal view of uropods and telson, which together make up the tail fan

To determine the sex of your specimen, find the base segment of each pair of walking legs where the legs attach to the body. Use a magnifying glass to study the inside surface of them. In a *female* the genital pores of the oviducts (crescent-shaped slits) are located at the base segments of the third pair of walking legs (Fig. 7.9).

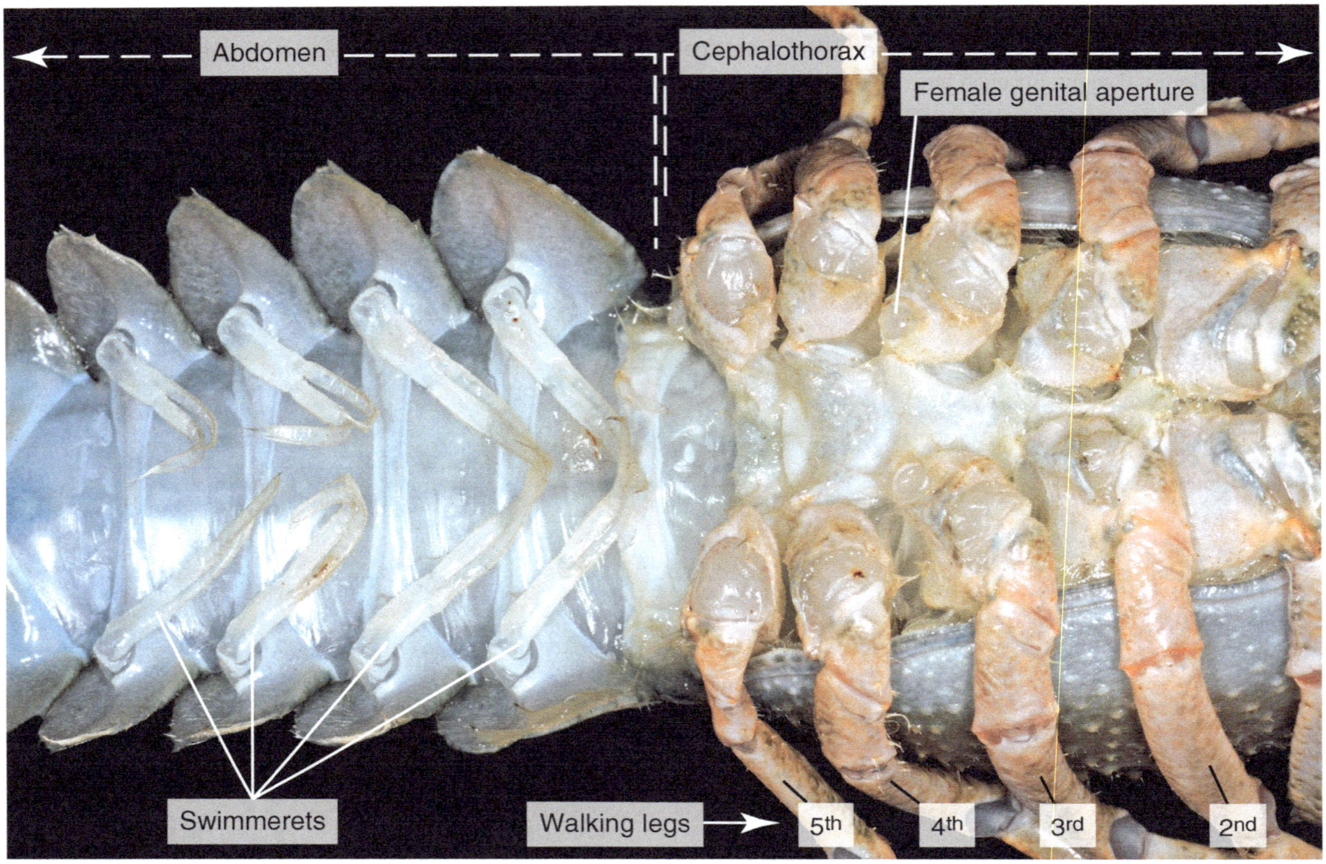

Fig. 7.9 Ventral view of the walking legs and swimmerets of a female crayfish

In a *male* the genital openings of the sperm ducts are on the base segment of the fifth pair of walking legs (Fig. 7.10). The first two pairs of swimmerets in males are modified copulatory appendages; they transfer sperm to female (Fig. 7.10). If there is a possibility, try to examine a crayfish of both sexes.

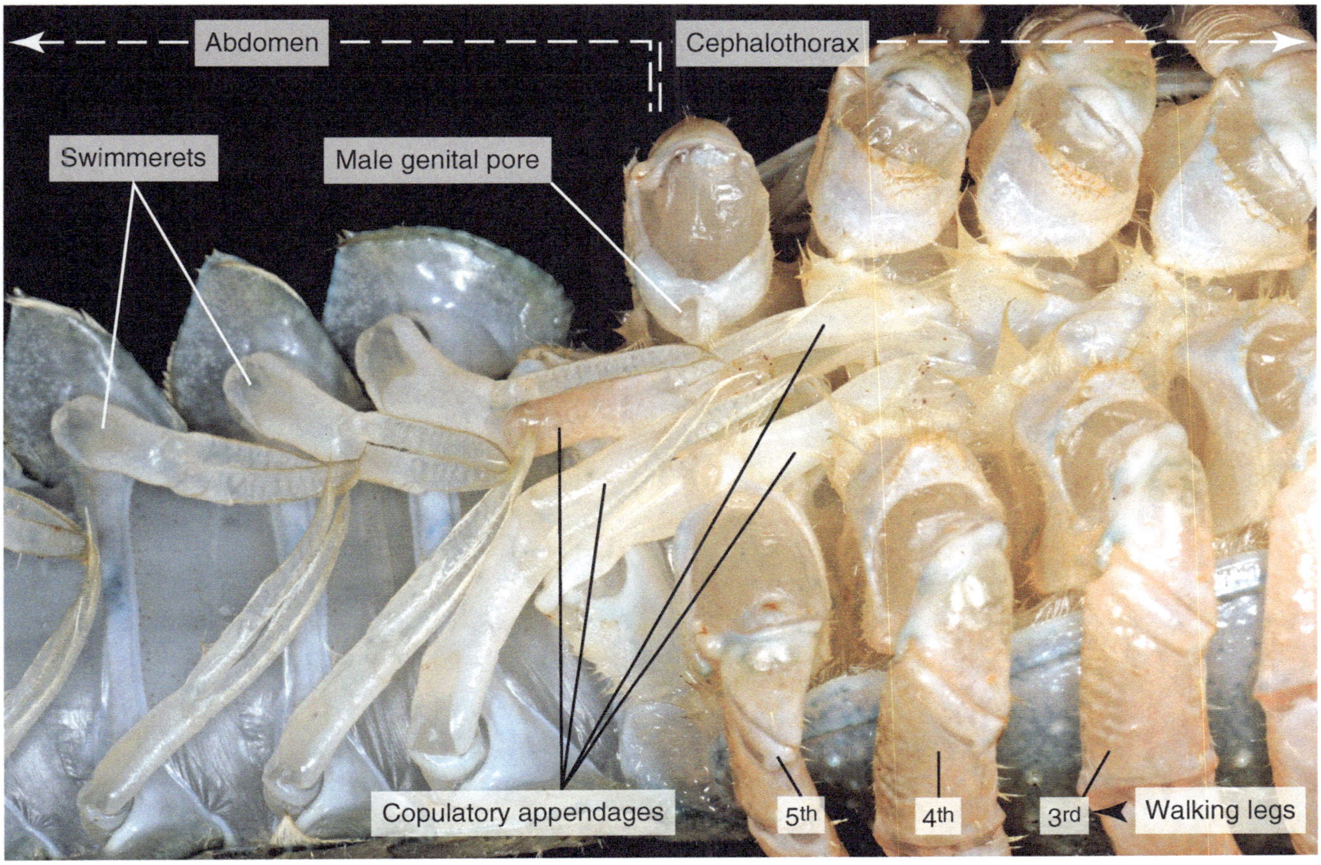

Fig. 7.10 Ventral view of the walking legs, swimmerets and modified copulatory appendages of a male crayfish

Place the specimen on its side. Using scissors cut along the side of the crayfish, as illustrated by the dashed cutline 1 on Fig. 7.11. Be careful to cut on the lateral side of the suture in order to save the integrity of the body wall.

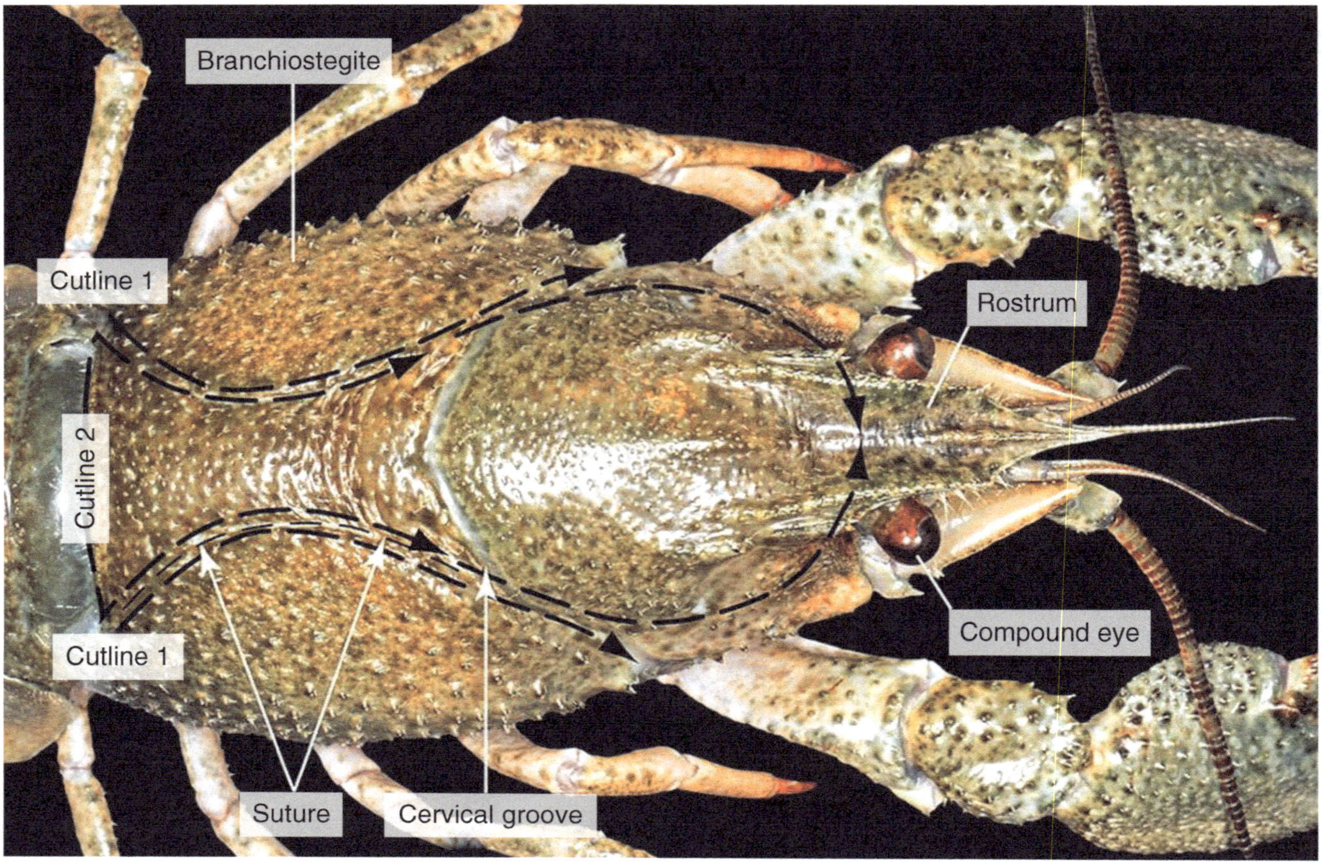

Fig. 7.11 Dashed lines indicating cuts for removal of the dorsal carapace

Use forceps to carefully remove part of the carapace (the branchiostegit), exposing the underlying feathery *gills*, the organs of the respiratory system in the branchial chamber (Fig. 7.12).

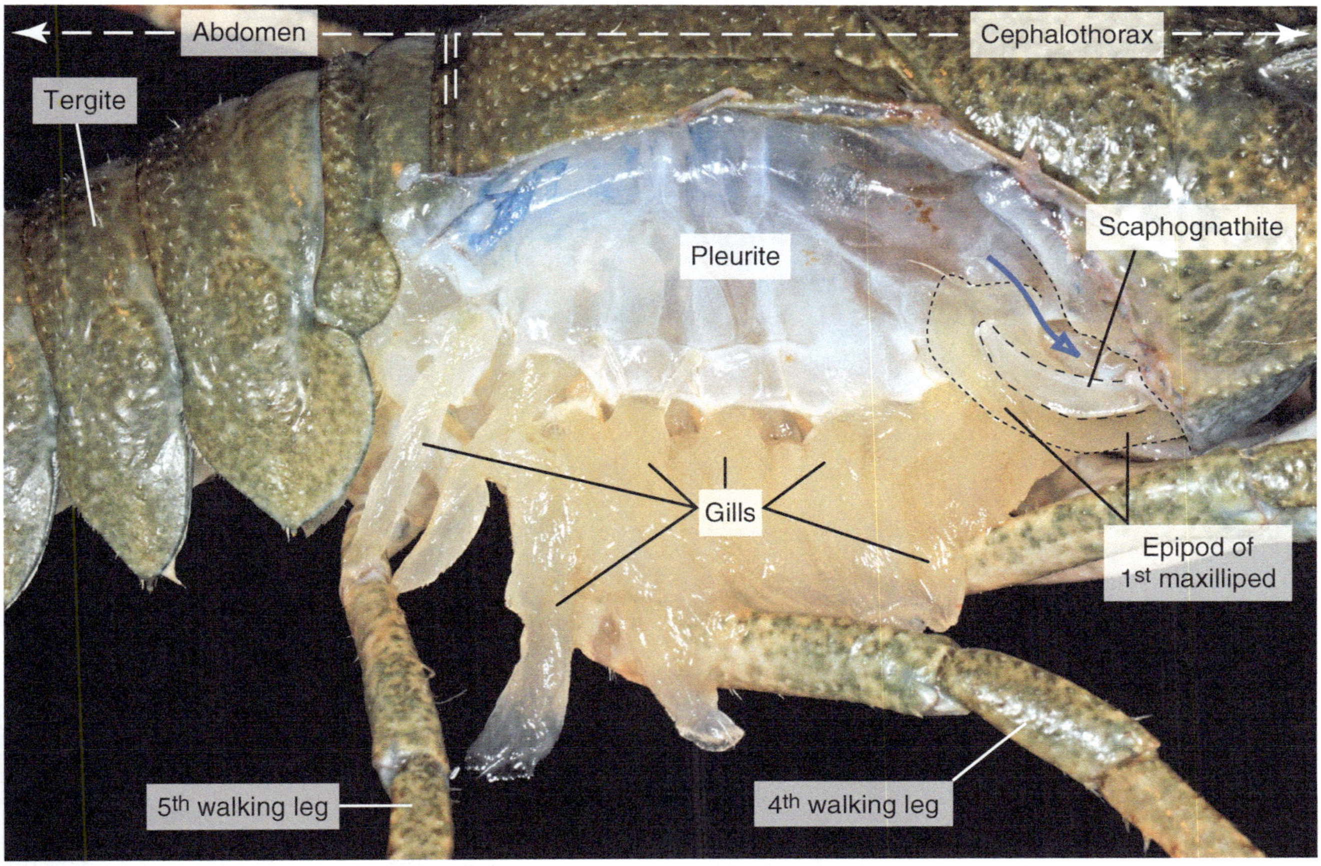

Fig. 7.12 Side view of the branchial chamber of a crayfish after the removal of the branchiostegit. Gills are folded out onto the feet. *Blue arrow* indicates the direction of respiratory water current leaving the branchial chamber

Water enters the branchial chamber by the free ventral edge of the branchiostegit and is drawn forward over the gills by the action of the *gill bailer* (scaphognathite) of the second maxilla (arrows on Figs. 7.13 and 7.14). A cuticular fold forms a narrow constriction to enhance the pumping effect of the scaphognathite.

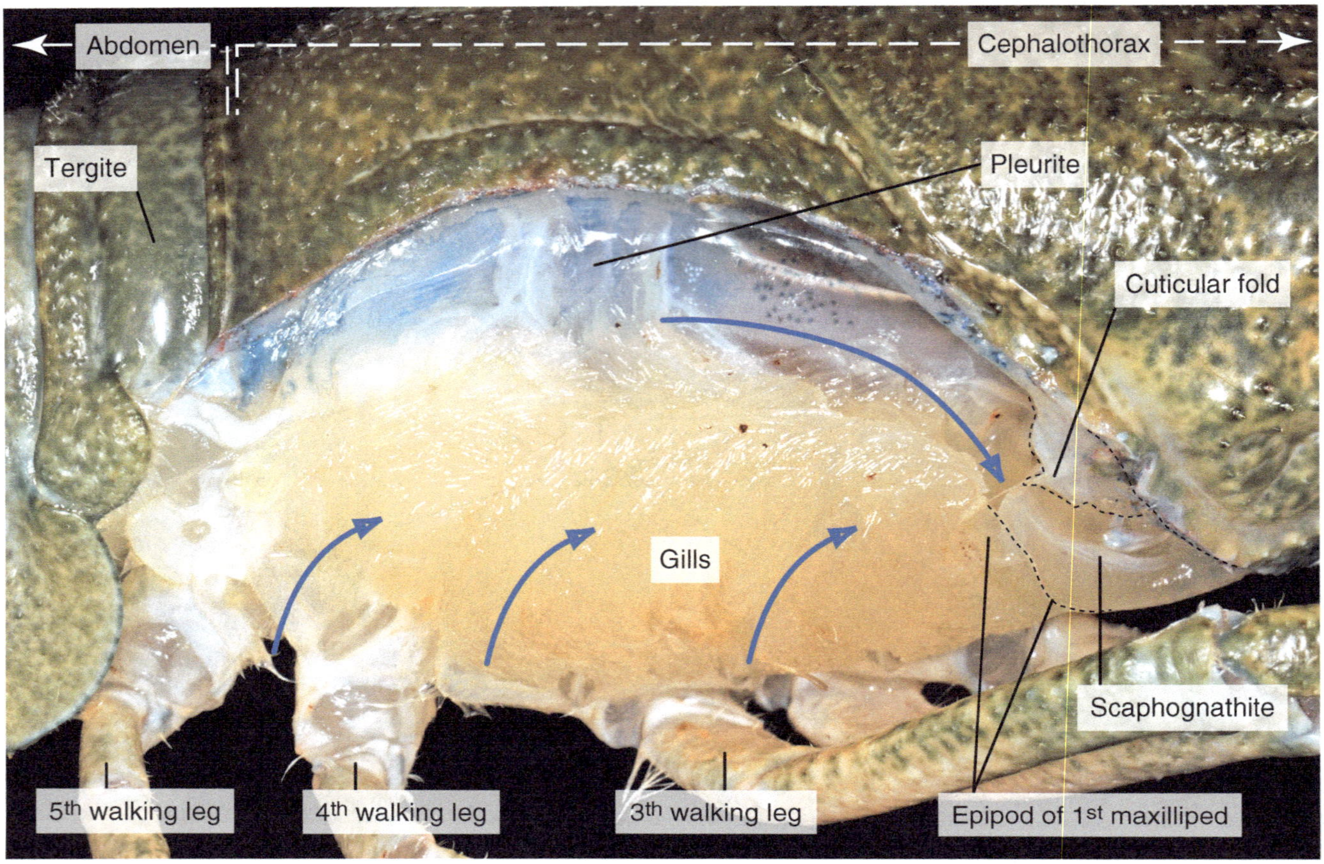

Fig. 7.13 *Blue arrows* show direction of water circulation in the branchial chamber by the action of the gill bailer (scaphognathite). Note gills are in their original position

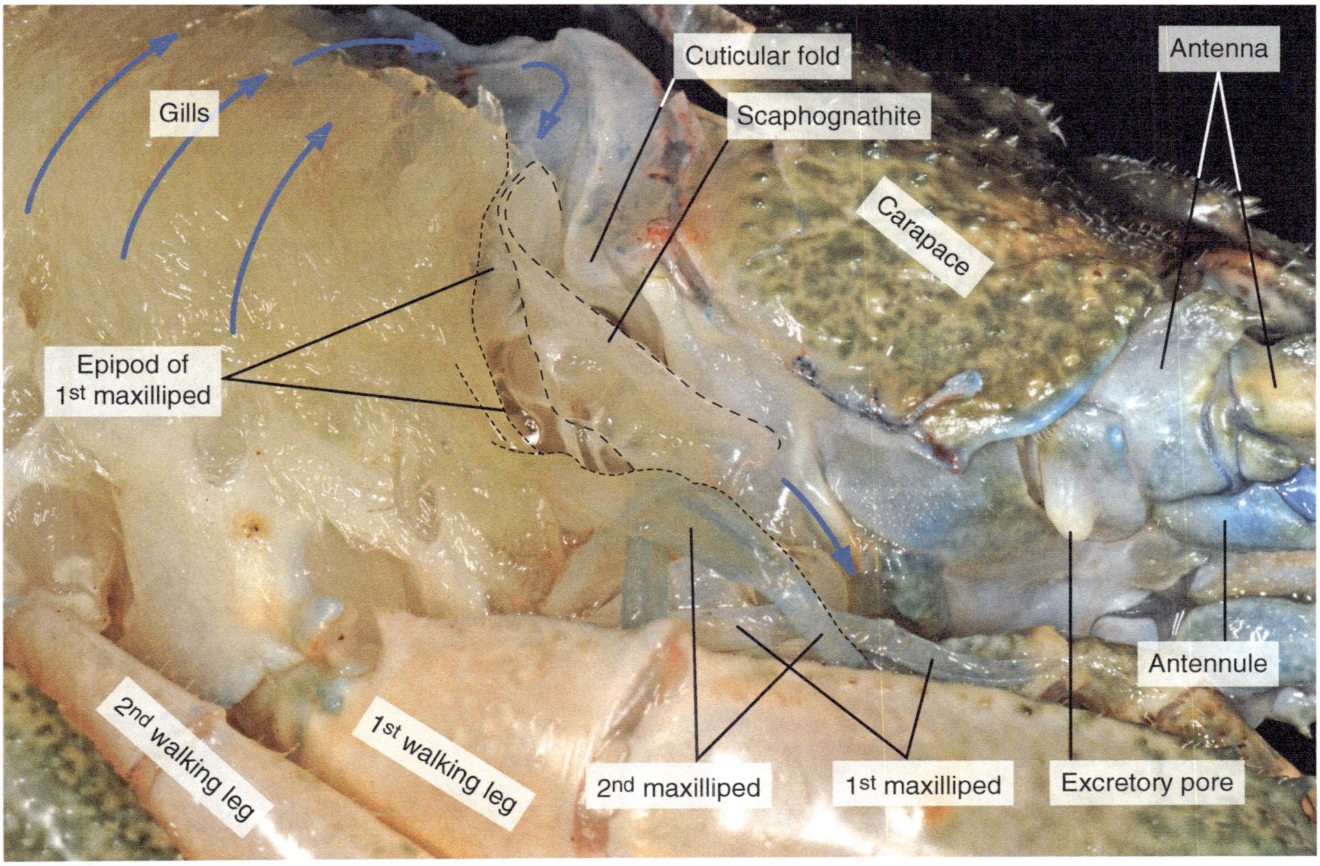

Fig. 7.14 Position of the gill bailer (scaphognathite) and the direction of water circulation (*blue arrows*) by its action

A constant circulation of haemolymph in the gills releases carbon dioxide and picks up oxygen. Separate the gills carefully, laying aside the *foot gills* (podobranch) and, another row, the *joint gills* (arthrobranch) which are attached to membranes that hold the appendages to the body. Remove a set of gills, place them under a stereomicroscope, cover them with water and examine them (Fig. 7.15). Note the three types of surface enlargements on the gills: flat folds, pointed fingerlike projections and hairs. These structures are all formed by the hypodermis and covered with a thin cuticule (Fig. 7.15).

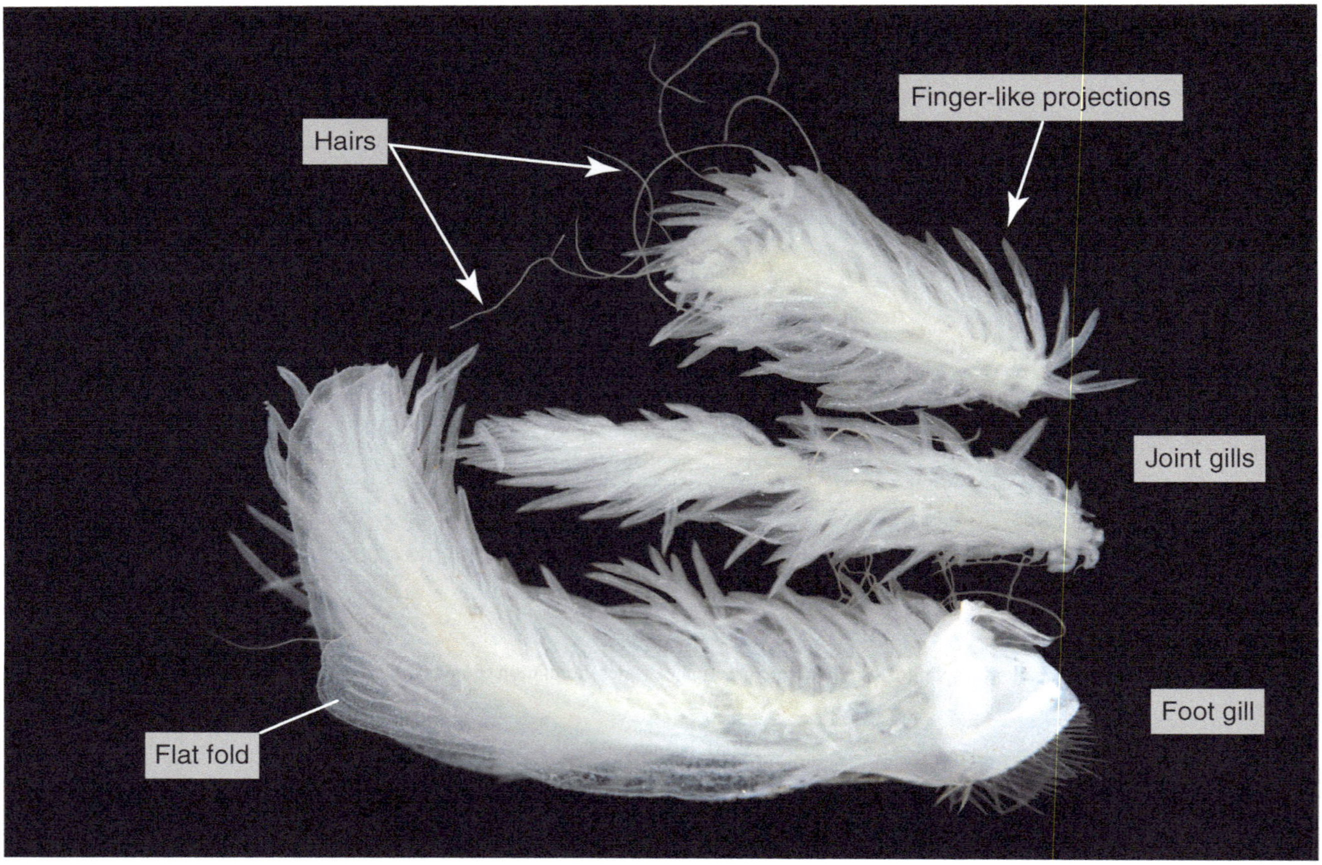

Fig. 7.15 Separated foot gills and joint gills of a crayfish

Next study the internal anatomy of a crayfish. Never let your specimen dry out during dissection. Wet it occasionally and if you keep a break, cover it with a dampened paper towel. Hold the crayfish in the left hand and use scissors to carefully cut through the back of the remaining carapace along the cut-line 2 indicated on Fig. 7.11. Insert the point of the scissors on one side under the posterior edge of the carapace, and cut forward to the cephalic region. Cut through the thoracic region just internal to the sutures that separate the thoracic portion of the carapace into three regions. Do the same on the other side, thus loosening a dorsal strip. Use forceps to carefully lift up this centre portion of the carapace (Fig. 7.16).

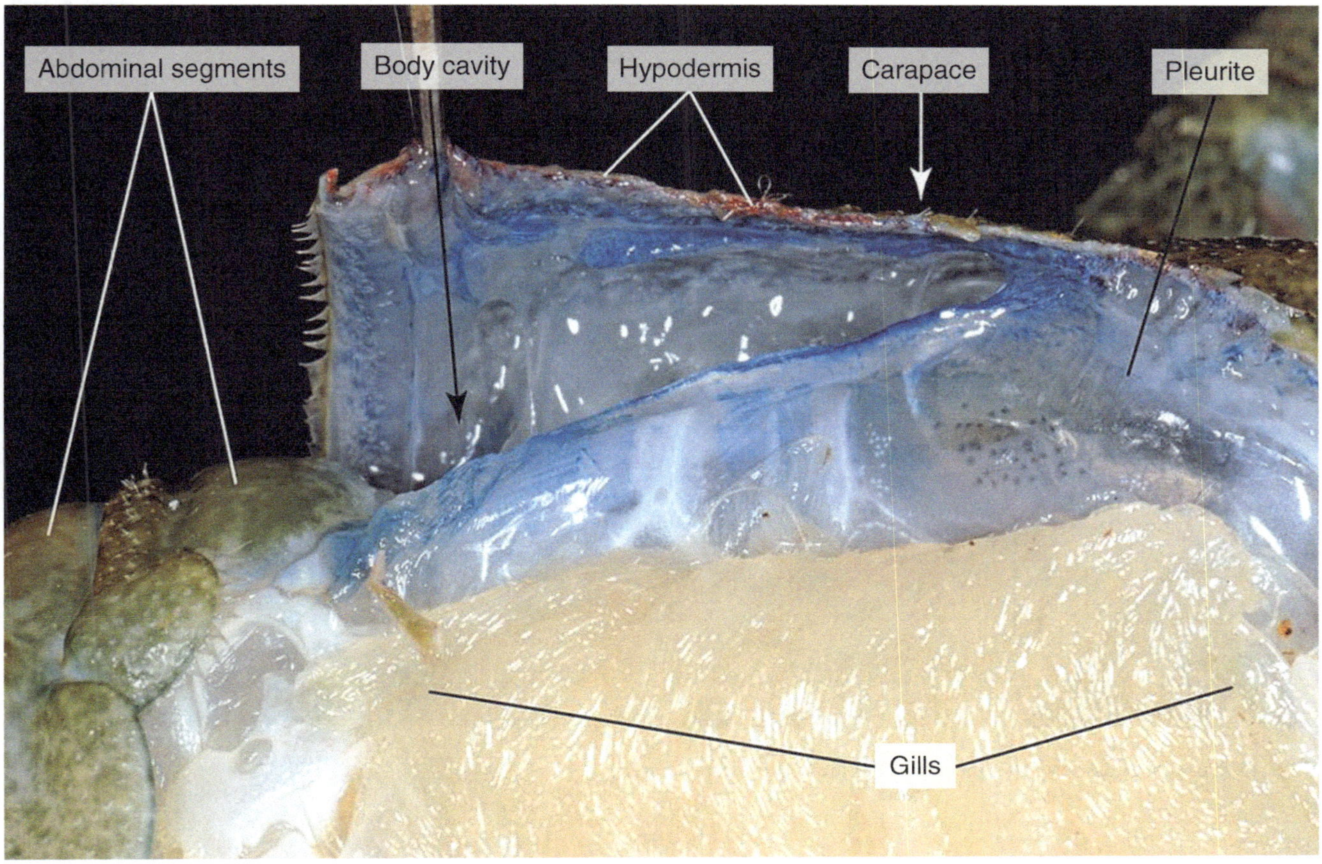

Fig. 7.16 Careful removal of the centre portion of the carapace with forceps

Do not to pull the carapace away too quickly, only a little at a time, being careful not to remove the underlying hypodermis and muscles, which cling to the epidermis. Such action would disturb or tear the underlying heart as well (Fig. 7.17).

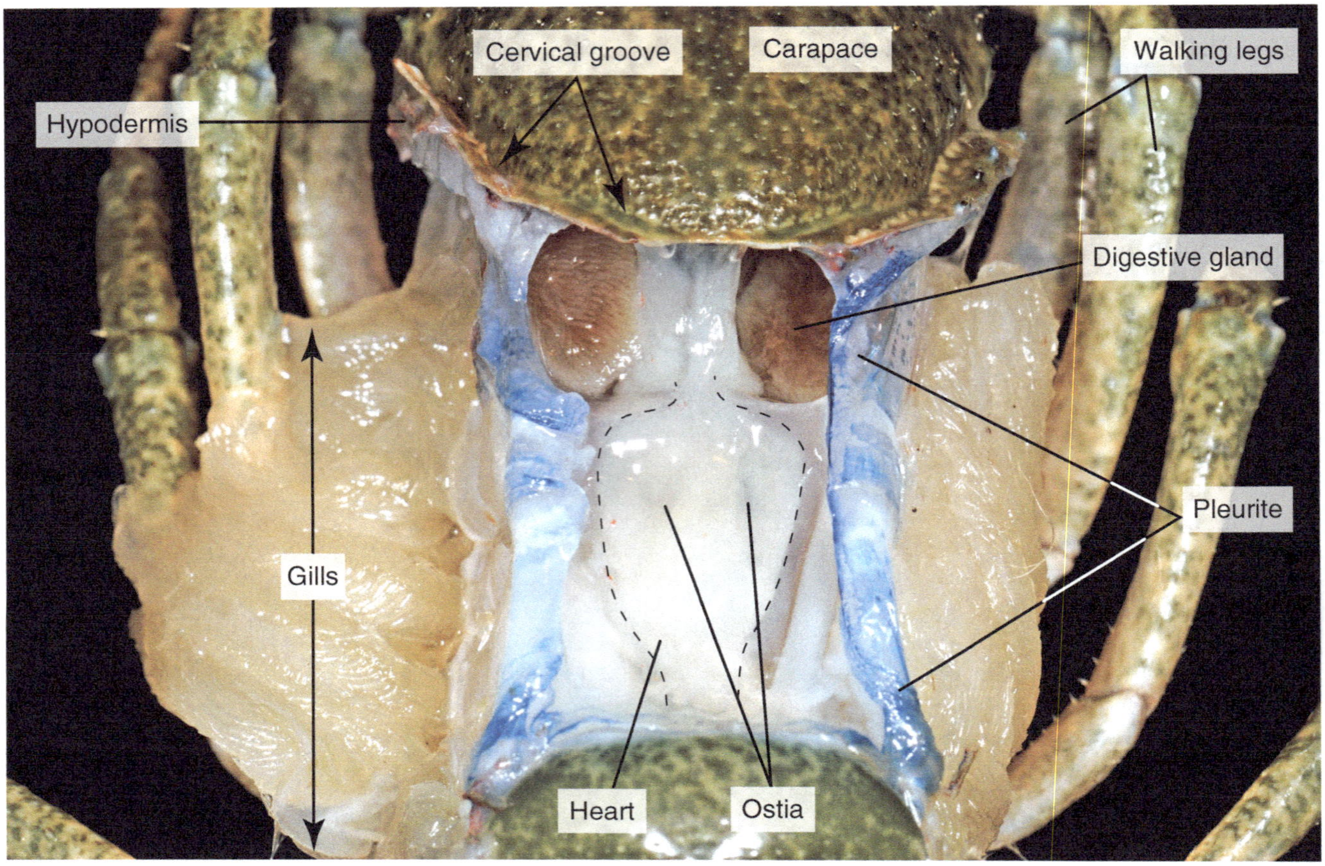

Fig. 7.17 Position of the heart after the removal of the centre part of carapace

Observe the heart and the gastric mill as exposed by the removal of the carapace (Fig. 7.18). The thin tissue covering the viscera is the hypodermis, which secretes the exoskeleton. Remove the remnants of the hypodermis very carefully to expose the viscera.

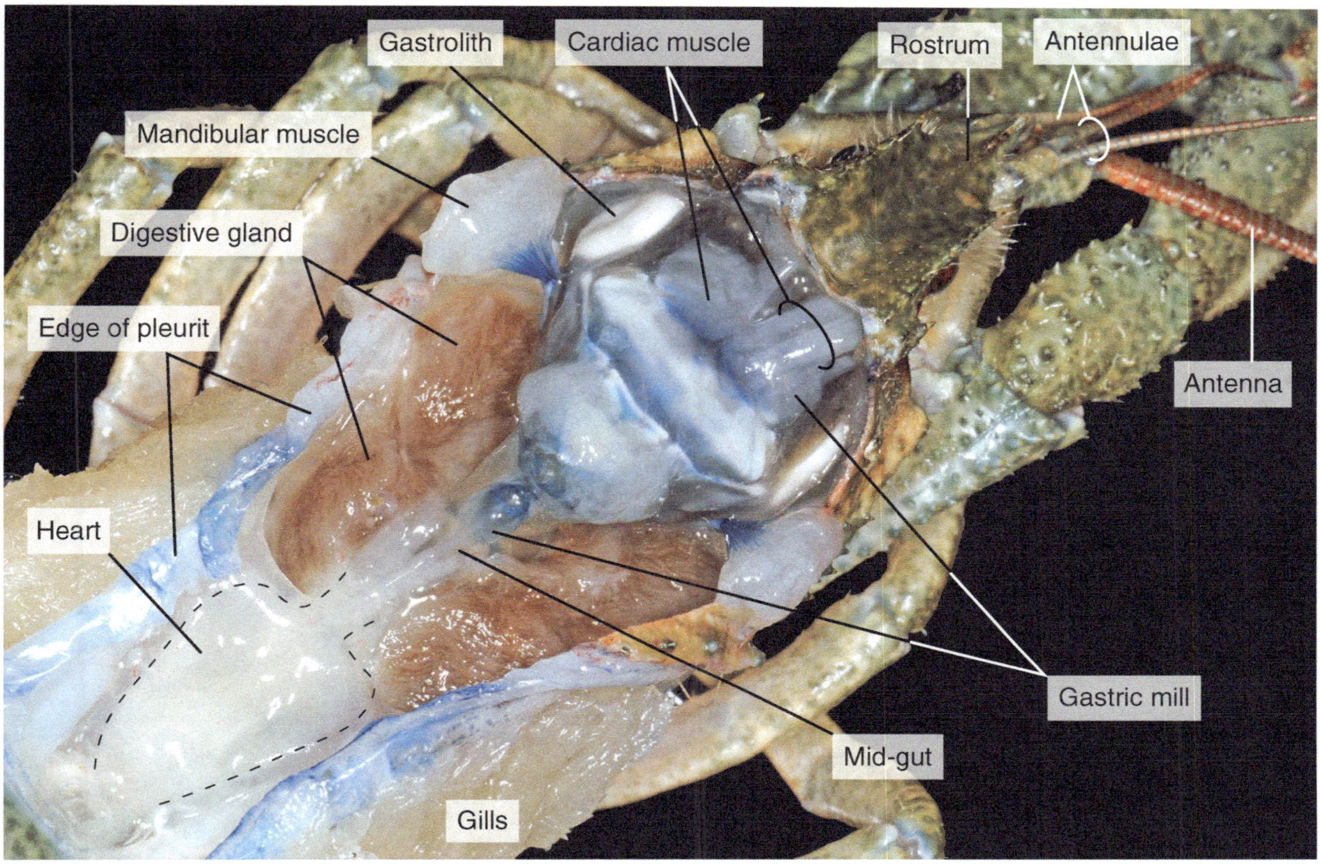

Fig. 7.18 The heart, digestive gland and the gastric mill as exposed by the removal of the carapace

Cut along the dashed lines (cutline 1 and 2) indicated on Fig. 7.19 to remove terga of the abdomen. Extensor muscles should come away with the terga.

Fig. 7.19 Dashed lines indicating the cutlines to remove terga of the abdomen

Locate and identify the organs in the opened crayfish (Fig. 7.20).

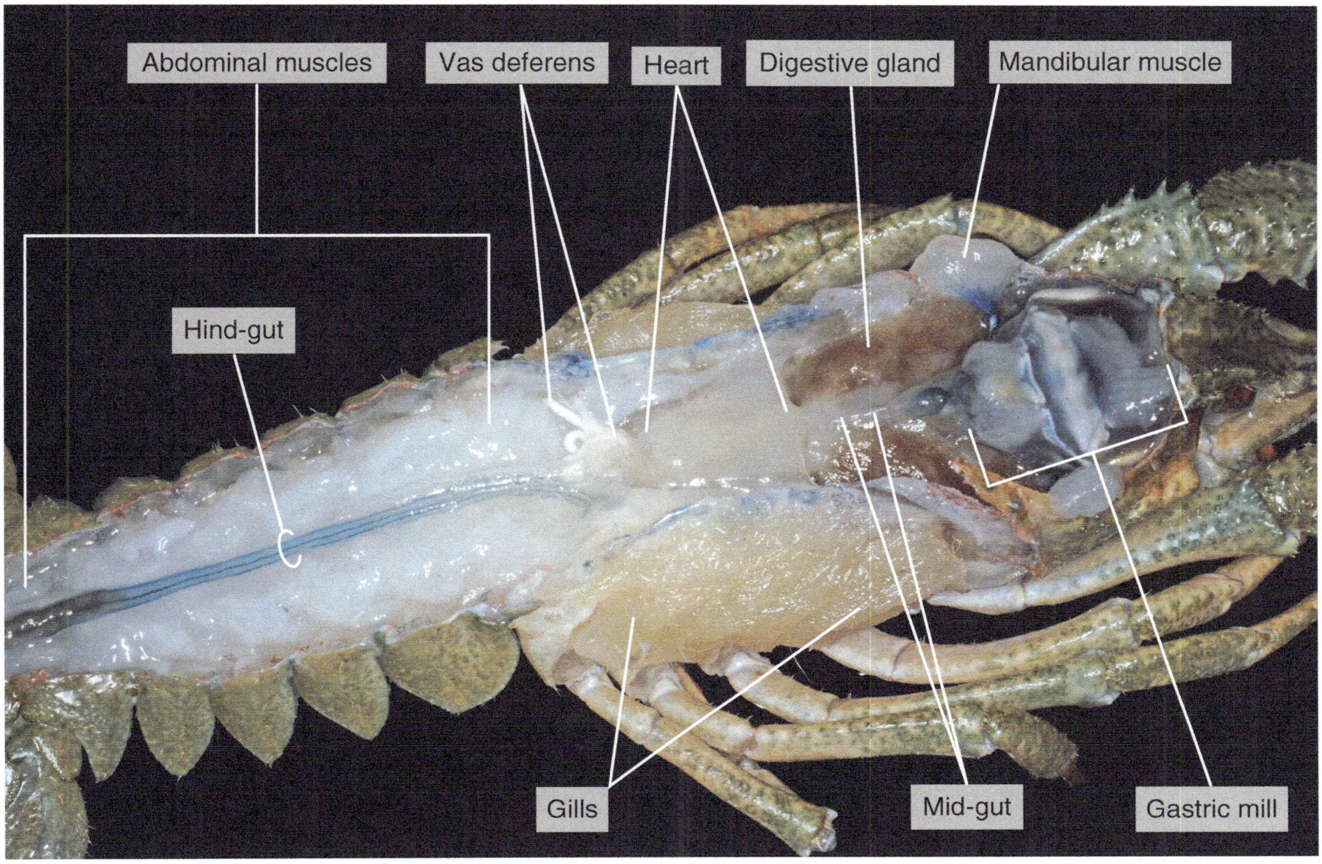

Fig. 7.20 Internal organs of the crayfish after removal of the carapace and abdominal terga

Observe first the organs of the circulatory system. The small, angular *heart* is located just posterior to the stomach. The heart lies in a cavity called *pericardial sinus*, which is enclosed in a membrane, the *pericardium* (Fig. 7.21).

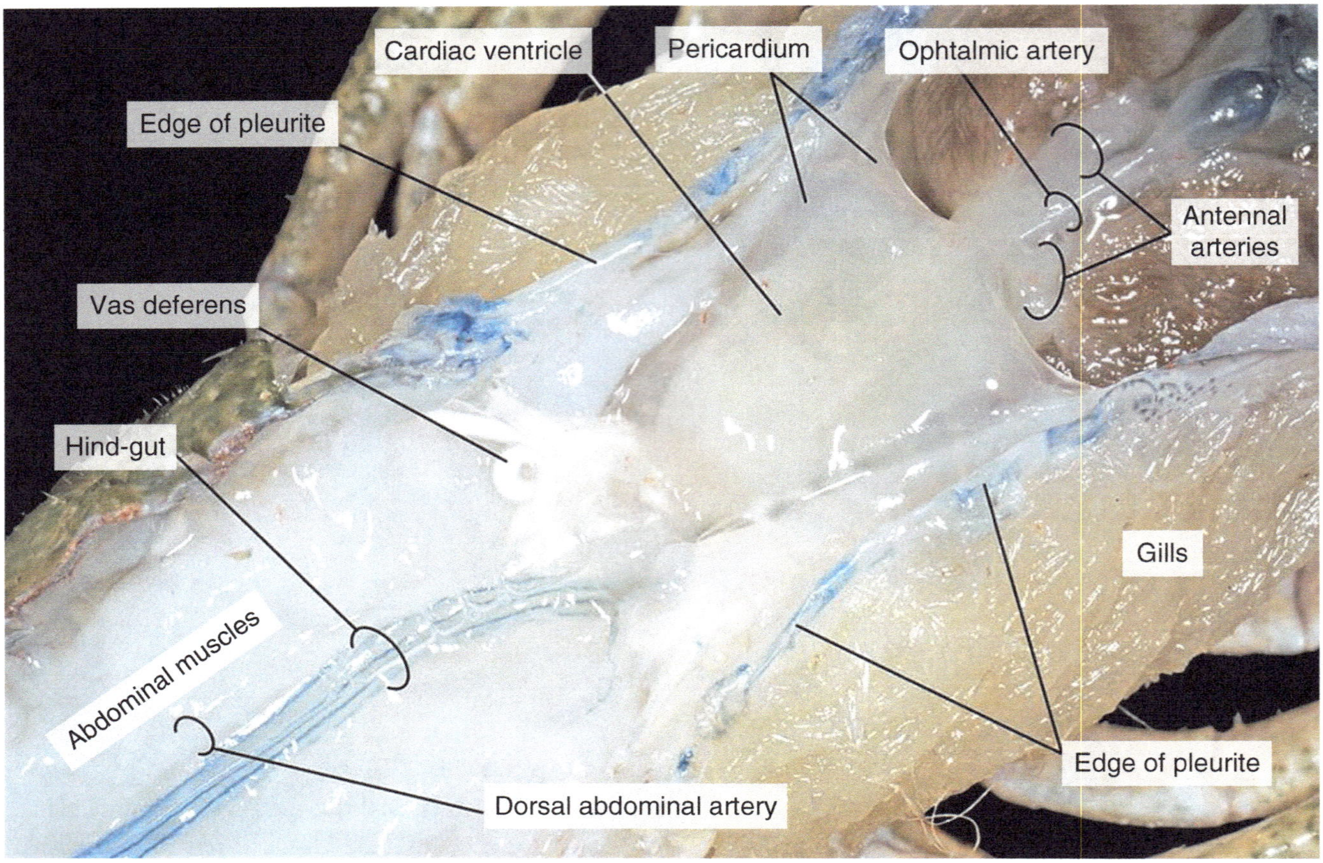

Fig. 7.21 The heart and the main arteries

The crayfish has an open circulatory system in which the haemolymph leaves the heart in arteries but flows into sinuses, or spaces, in tissues. The haemolymph flows over the gills before returning to the pericardial sinus. Haemolymph enters the heart through three slitlike openings, the *ostia*, which open to receive the haemolymph from the pericardial sinus and then close when the heart contracts to pump the haemolymph through the arteries (Fig. 7.22). There are altogether seven arteries leaving the heart, for example, the anterior ophthalmic artery, the two antennal arteries and the posterior dorsal abdominal artery (Fig. 7.21).

If desired the heart and principal blood vessels may be injected with toluidine blue or any other dye (ink or carmine suspension). Insert a fine hypodermic needle cautiously in the lumen of the cardiac ventricle and press the dye slowly and carefully out of the syringe which flows into and reveals parts of the heart and the main vessels (Fig. 7.22).

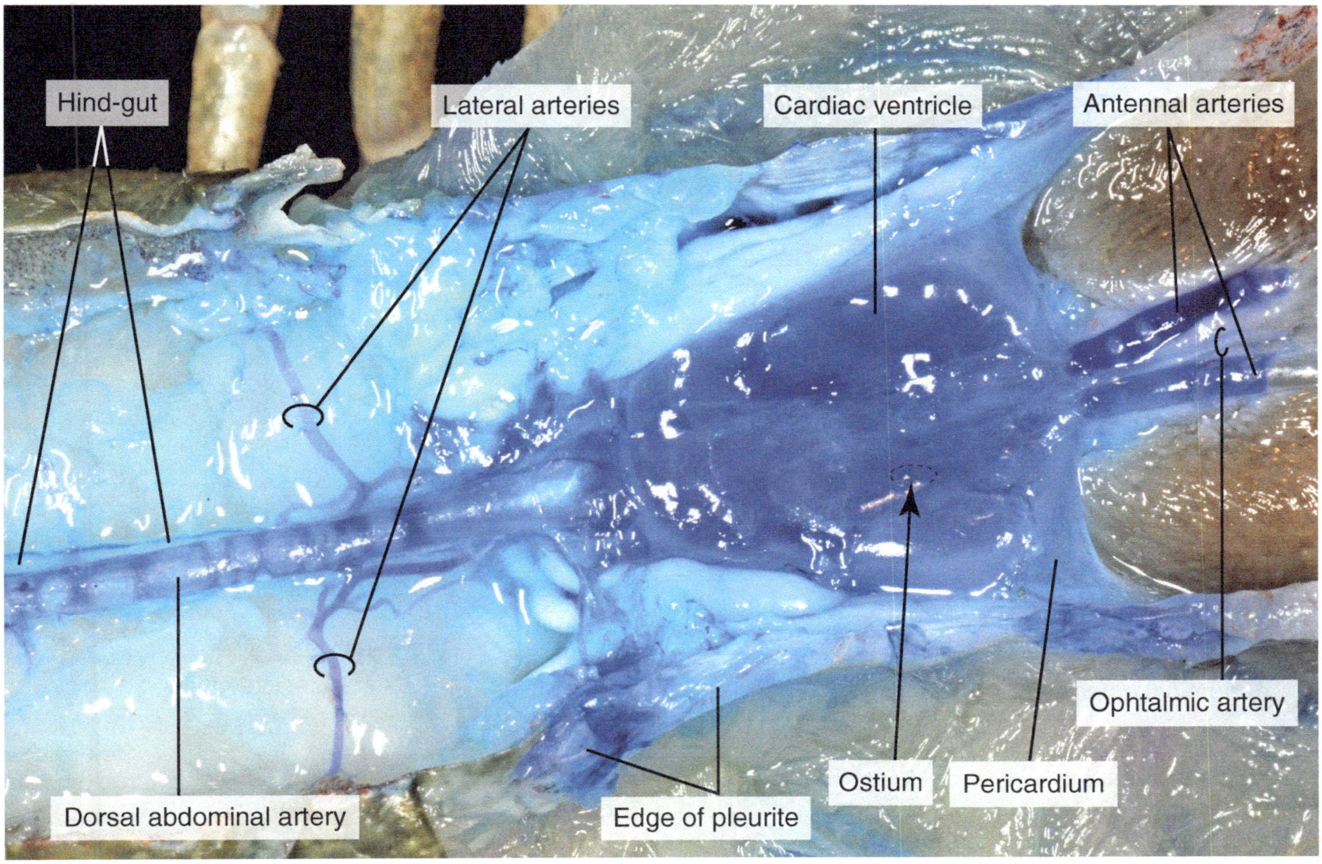

Fig. 7.22 The heart and the main arteries injected with toluidine blue

To study the reproductive system, it is necessary to remove the heart and distinguish the gonads from the much more conspicuous digestive gland. That is easier in the female because the ovaries are yellow and the eggs can be seen. The posterior ends of the two *ovaries* are fused so that they form a Y-shaped structure mostly between the right and left digestive glands (Fig. 7.23). At the junction of the stem and arms of the Y, the *oviduct*, a delicate, band-like tube, leaves each side of the ovary and extends ventrally to open on the base of the third walking leg (Fig. 7.9).

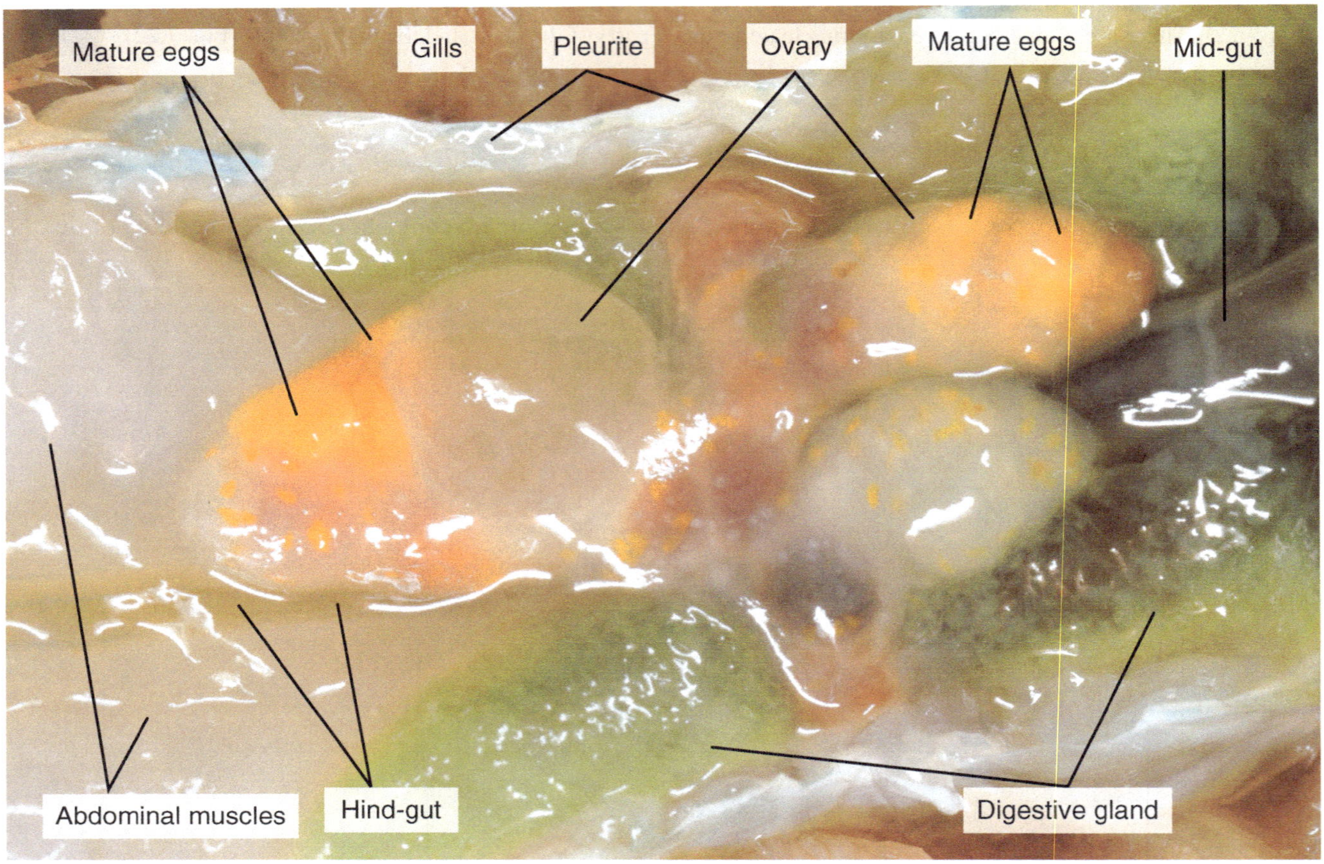

Fig. 7.23 Internal reproductive organs of a female crayfish

In the male, the *testes* are similar in form but are smaller than the ovary (Fig. 7.24). The *vas deferens* on each side is a convoluted tube extending over the surface of the digestive gland to pass ventrally and open on the base of the fifth walking leg (Fig. 7.10).

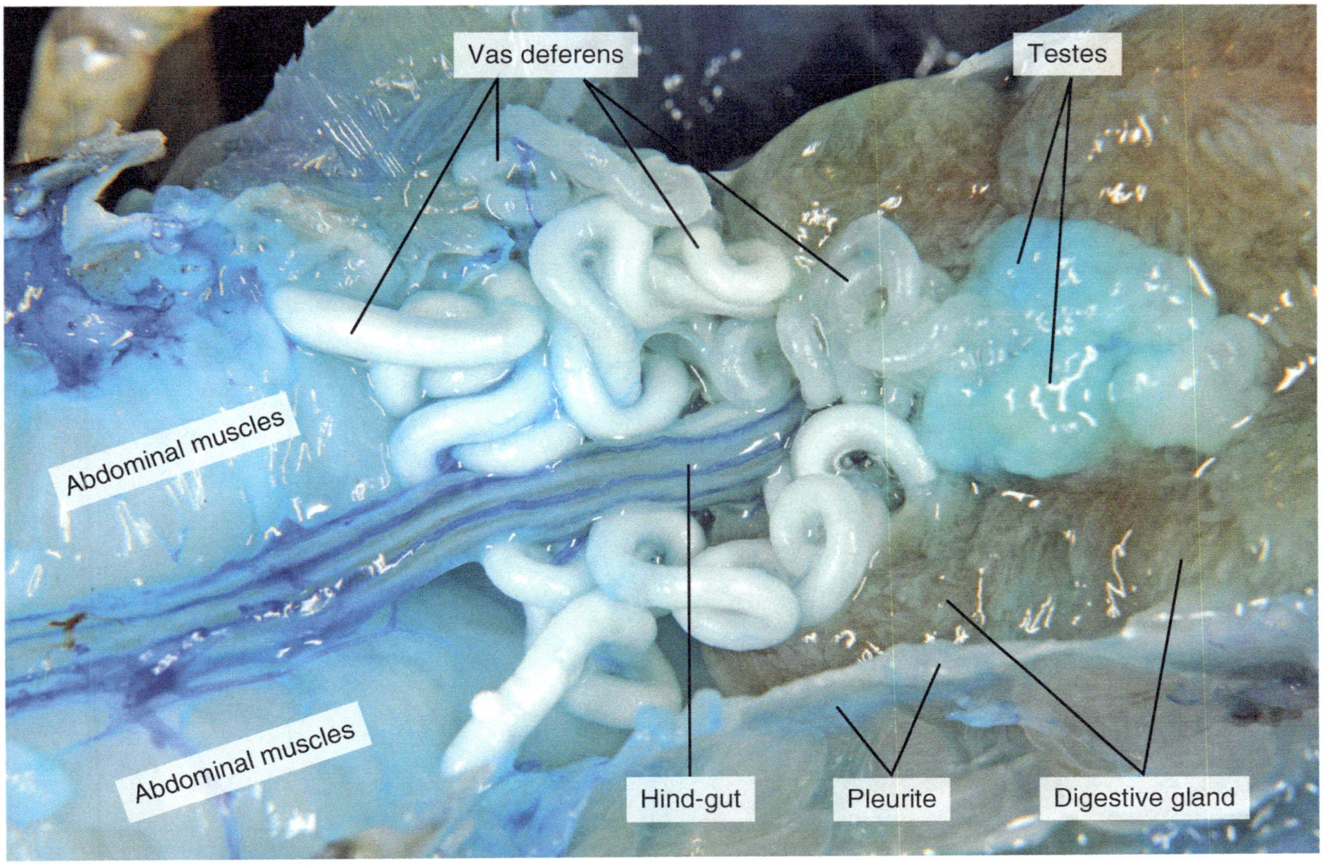

Fig. 7.24 Internal reproductive organs of a male crayfish (The figure is derived from the specimen injected with toluidine blue, which accounts for the blue background colour)

Locate and identify the organs of the digestive system. Locate the maxillae that pass the pieces of food into the mouth. The food travels down the short oesophagus into the stomach. The large stomach or *gastric mill* lies in the head region, anterior to the heart (Figs. 7.18, 7.20 and 7.25).

On each side of the stomach and heart are the large cream-coloured lobes of the *digestive gland* (hepatopancreas), which produces digestive enzymes, is the chief site of absorption of nutrients, and serves for storage of food reserves (Figs. 7.18 and 7.25). They extend full length of the thorax. This gland is the largest organ in the body. Undigested material passes into the hind-gut and is eliminated through the anus (Figs. 7.20, 7.31 and 7.32).

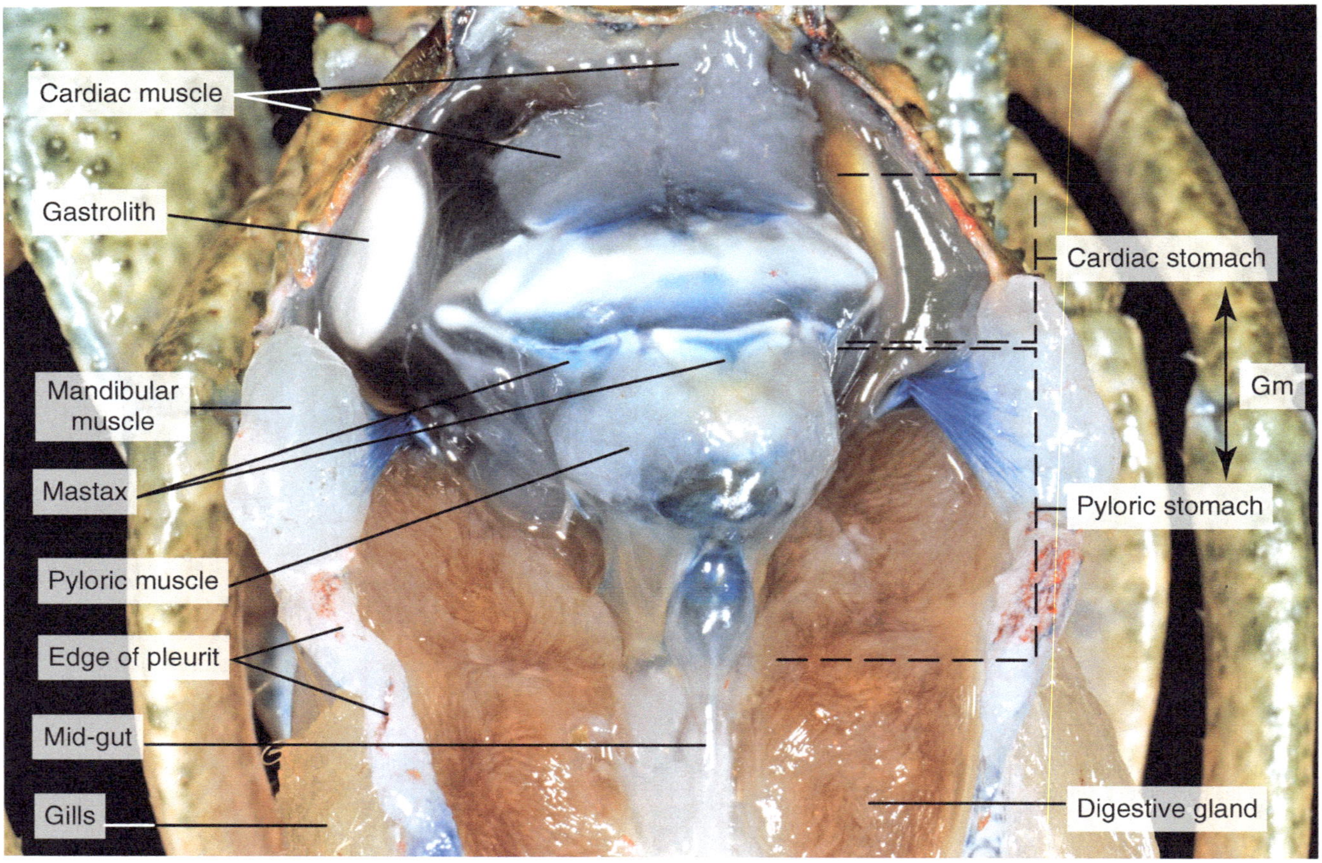

Fig. 7.25 End part of the abdomen of a crayfish in dorsal view. *Gm* gastric mill

Digestive gland of crayfish is a compound gland with long, blind-ended hepatopancreatic tubules (Fig. 7.26).

Columnar epithelium rests on a thin basal membrane, which forms the boundary of the haemolymph spaces (H) among tubules. This epithelium contains several cell types. There is a number of *proliferating stem cells* (SC) on the basal membrane to give rise to differentiated cells. *B cells* (blister-like cells, BC) are conspicuous: they have a large central vacuole or a few phagocytic vacuoles beneath the apical surface. They phagocytose nutrients from the hepatopancreatic tubular lumen, digest them intracellularly and secrete the digested products into the lumen. *R cells* (resorptive cells, RC) and *F cells* (fibrillar cells, FC) are cylindrical. Their luminal surface is covered with *microvilli* (MV). There is a clear difference between them in the position and appearance of their nuclei. R cells have central nuclei with finely dispersed chromatin, whereas F cells' nuclei are in a lower position and their chromatin is condensed. R cells have basophilic cytoplasm; they store glycogen and lipid droplets to provide energy during starvation, moulting and reproduction. F cells synthesise digestive enzymes. *Myoepithelial cells* (MYOs) can be found on the basal membrane with flattened cell nuclei.

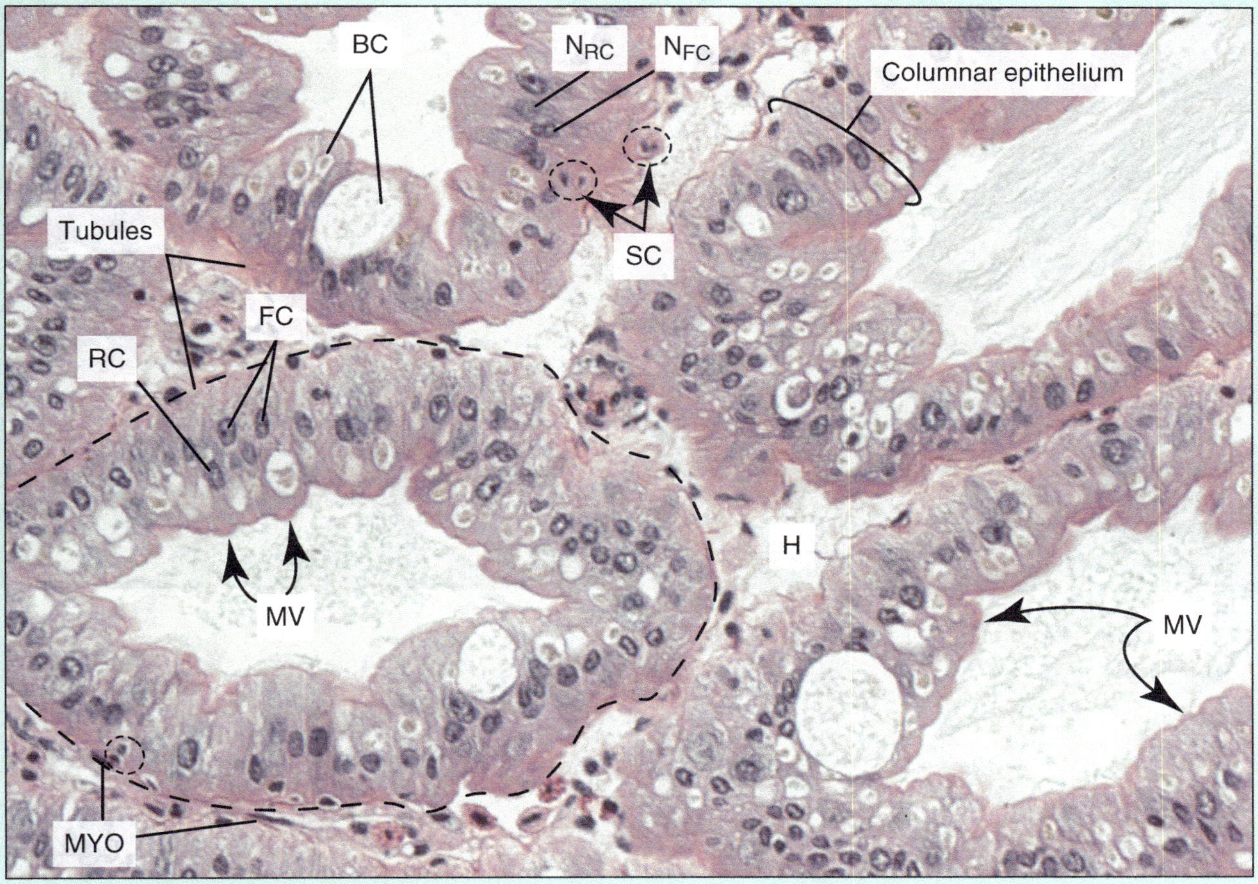

Fig. 7.26 Histological cross section of the tubules in the digestive gland of the crayfish (HE). *Dashed circle* cross section of a tubule, *dotted circles* mitotic activity, *BC* B cells, *FC* F cells, *H* haemolymph sinus, *MV* microvilli, *MYO* myoepithelial cells, N_{FC} nucleus of F cell, N_{RC} nucleus of R cell, *RC* R cells, *SC* proliferating stem cells

To examine the *gastric mill*, lay aside the mandibular muscles. Cut through the hind-gut close to the posterior end of the gastric mill. Cut through the cardiac muscles, and the oesophagus just above the circum-oesophageal connective, being careful not to damage the connective (Fig. 7.33). Remove the gastric mill together with the digestive glands (Fig. 7.27).

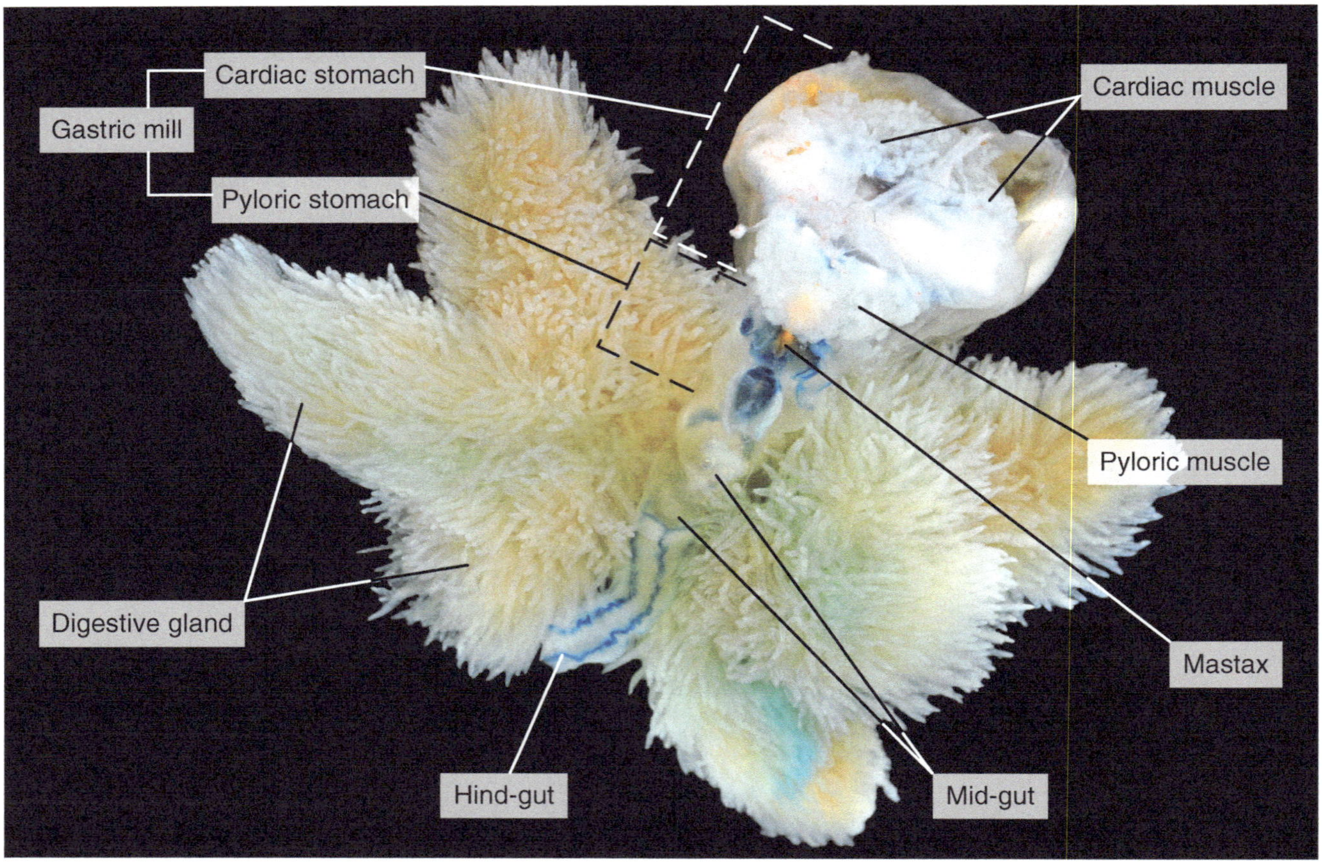

Fig. 7.27 The gastric mill, the mid-gut and the digestive glands under water cover

Cut through the dorsal wall of the gastric mill from the oesophagus to the mid-gut and the anterior wall until the mill can be opened almost flat. Wash out the contents, if necessary. Pin it out, cover it with water and examine under a stereomicroscope (Fig. 7.28). The cardiac stomach contains the *mastax*, which consists of a set of three chitinous teeth (ossicle), one dorso-medial (median tooth) and two lateral (zygocardiac ossicle), that are used for grinding food (Fig. 7.28). They are operated by gastric muscles (cardiac and pyloric muscles).

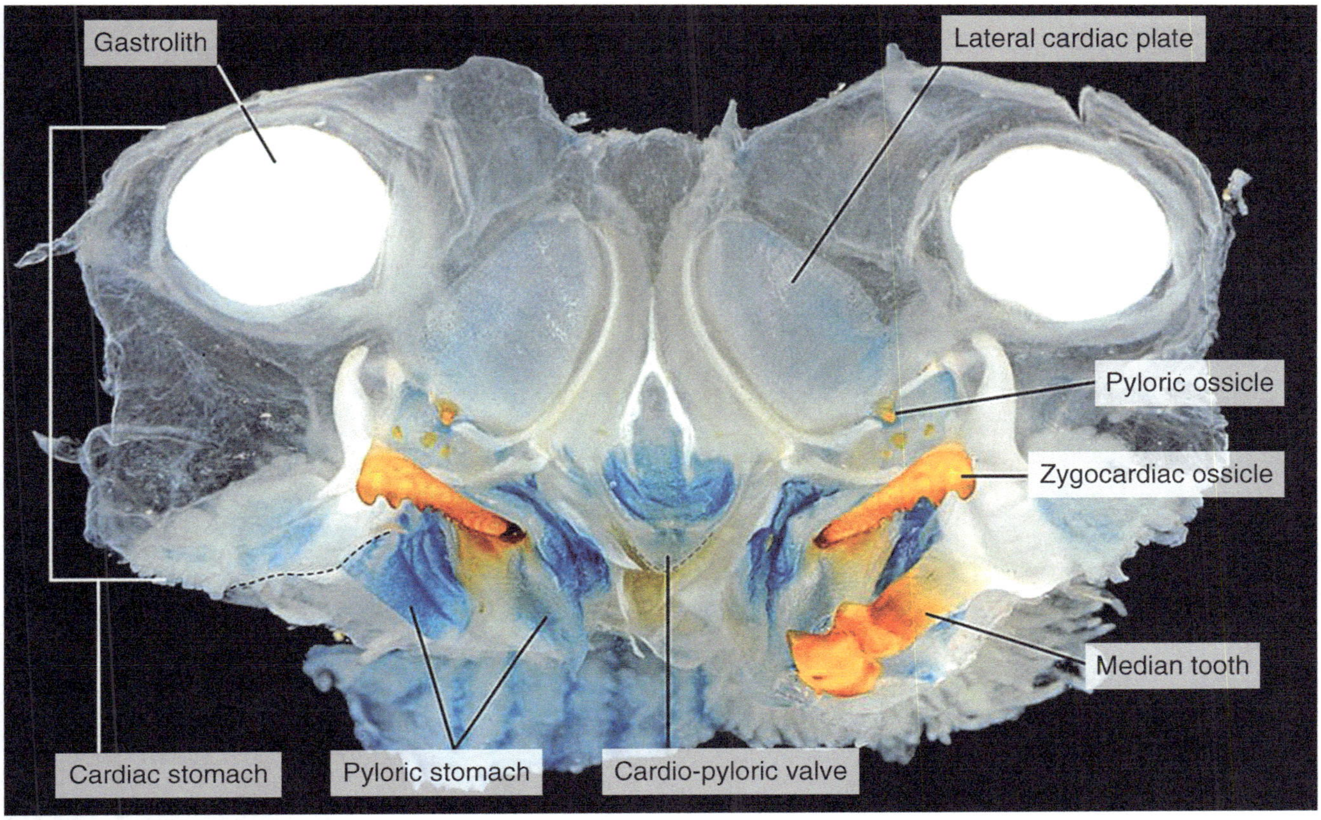

Fig. 7.28 The cardiac part of the gastric mill cut on the dorsal side with the mastax under water cover

In the cardiac stomach, the food is ground up and partially digested by enzymes from the hepatopancreas before it is filtered through the *cardio-pyloric valve* into the smaller pyloric stomach in liquid form. Rows of setae and folds of the stomach lining strain the finest particles and pass them from the pyloric stomach into the mid-gut and from there to the hepatopancreas where digestion is completed and absorption occurs. Coarse food particles and indigestible material get directly into the hind-gut through the *funnel* (Fig. 7.29).

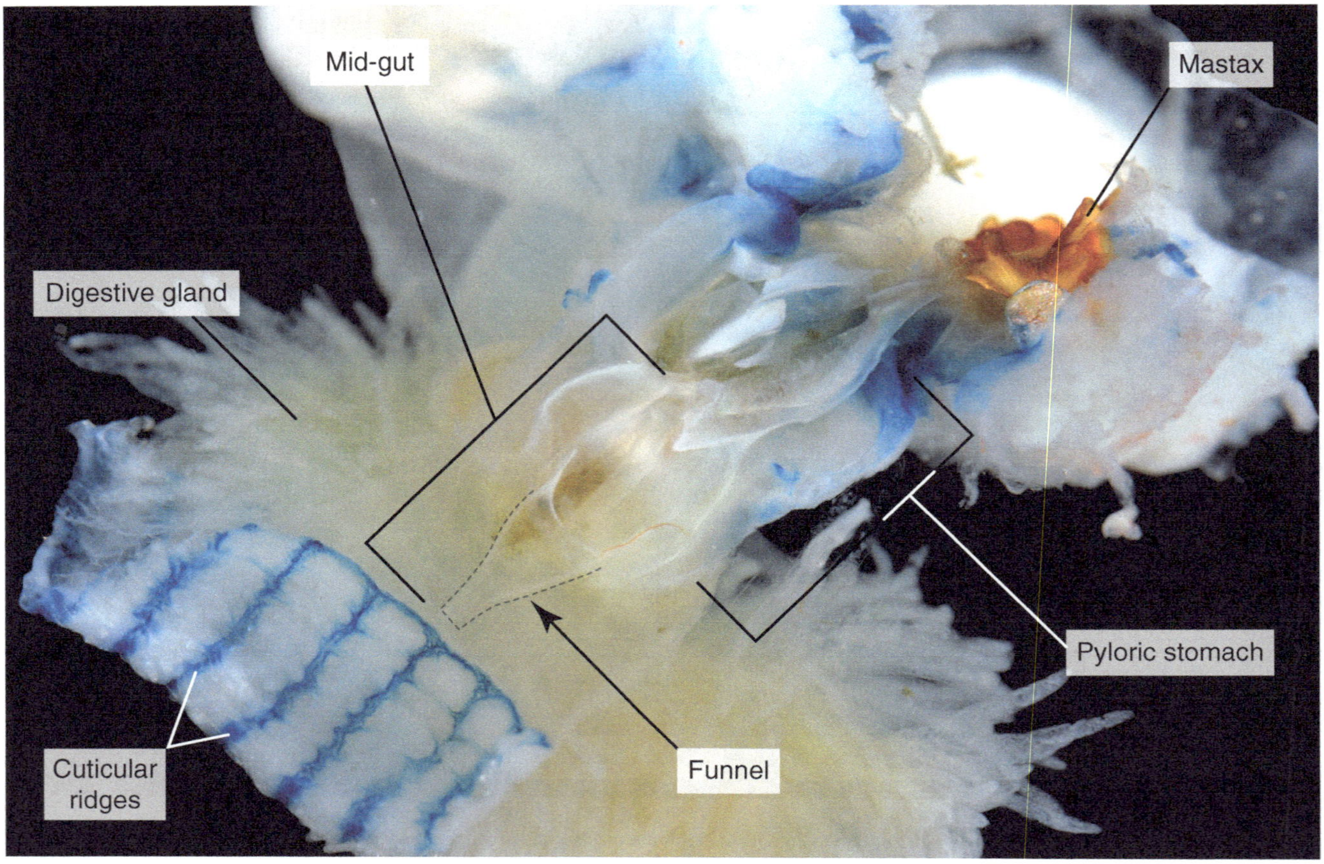

Fig. 7.29 Mid-gut is bridged over by the funnel transporting coarse food particles directly into the hind-gut under water cover

In the anterior part of the cardiac stomach, there are two lateral pouches, which synthesise the *gastroliths*, masses of calcareous crystals (Figs. 7.25 and 7.28). These calcium-storing structures first appear in hatchlings, and then they are formed prior to each moulting and completely resolved after ecdysis. They are recovered from the old exoskeleton by the haemolymph and used in the production of the new exoskeleton. Gastroliths are a specific adaptation to freshwater, where calcium is less available than in seawater.

The *mid-gut* is really short and inconspicuous, leaves the pyloric chamber and immediately joins into the hind-gut (Fig. 7.30).

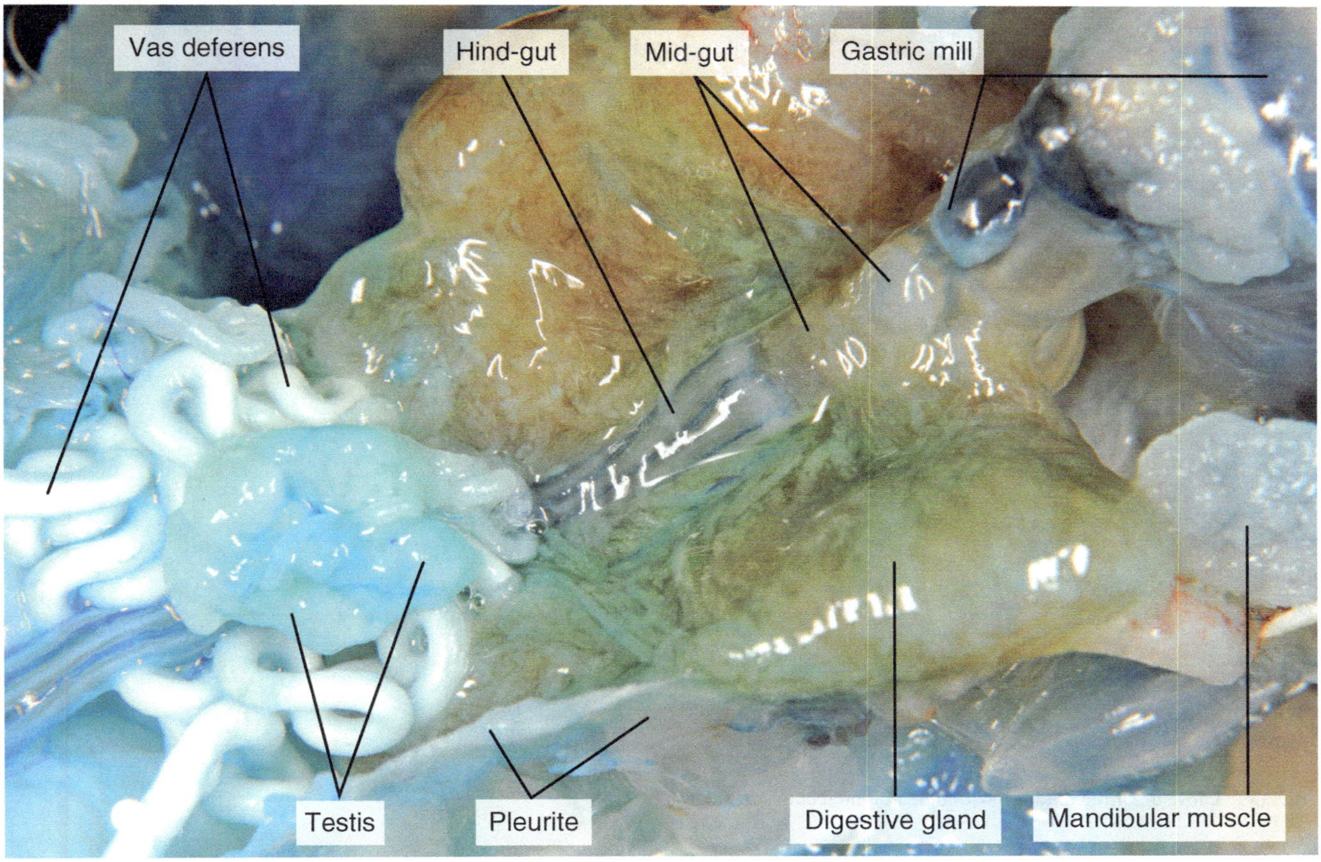

Fig. 7.30 The mid-gut of a crayfish is very short. (The figure is derived from the specimen injected with toluidine blue, which accounts for the blue background colour)

It is totally bridged over by the *funnel* leading bigger food particles directly into the hind-gut (Fig. 7.29). Fine canals of the hepatopancreas join to the mid-gut. The *hind-gut* is long and straight and easy to identify because of the six parallel blue cuticular ridges running along its length (Figs. 7.20 and 7.31). It ends at the *anus* on the telson (Fig. 7.32).

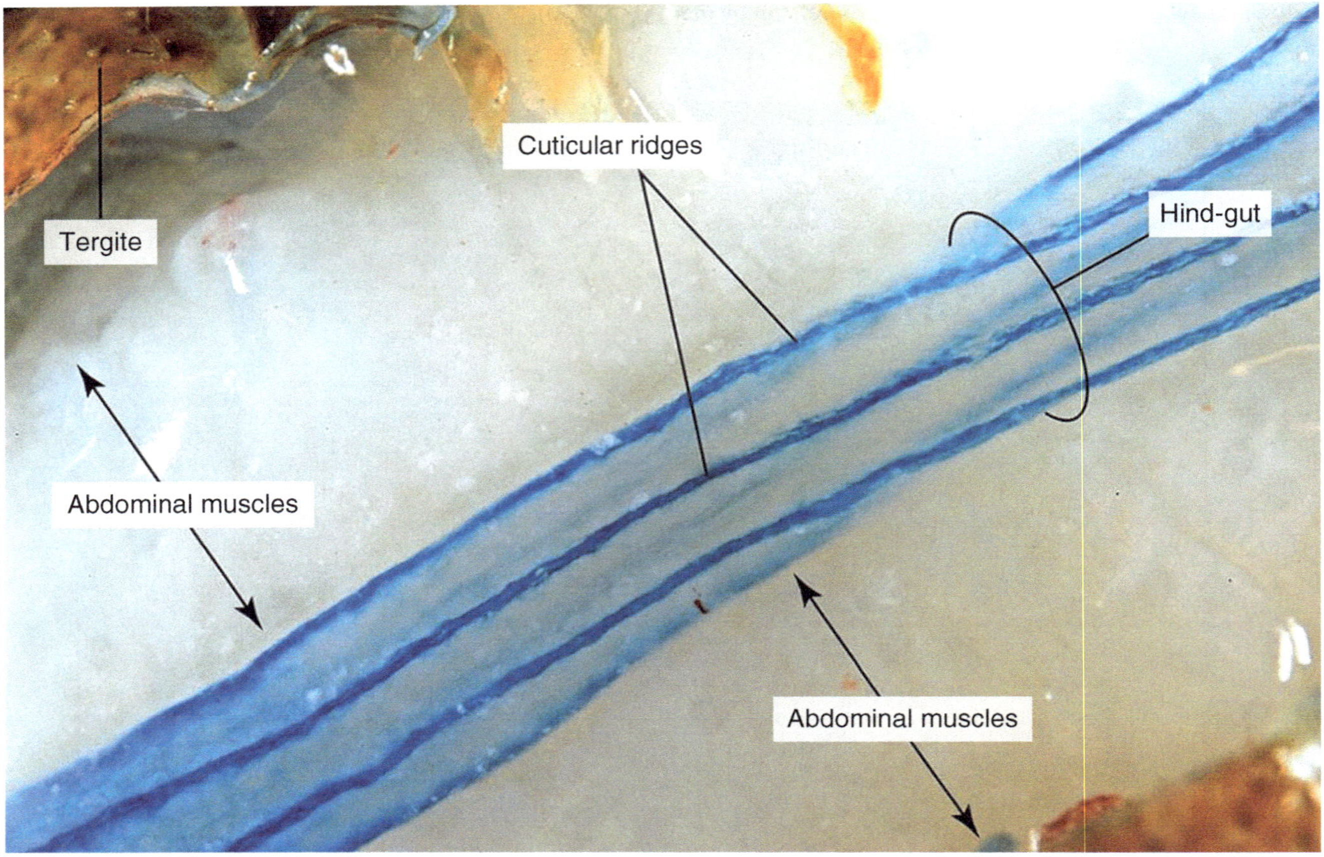

Fig. 7.31 Close up of the hind-gut with blue-coloured cuticular ridges

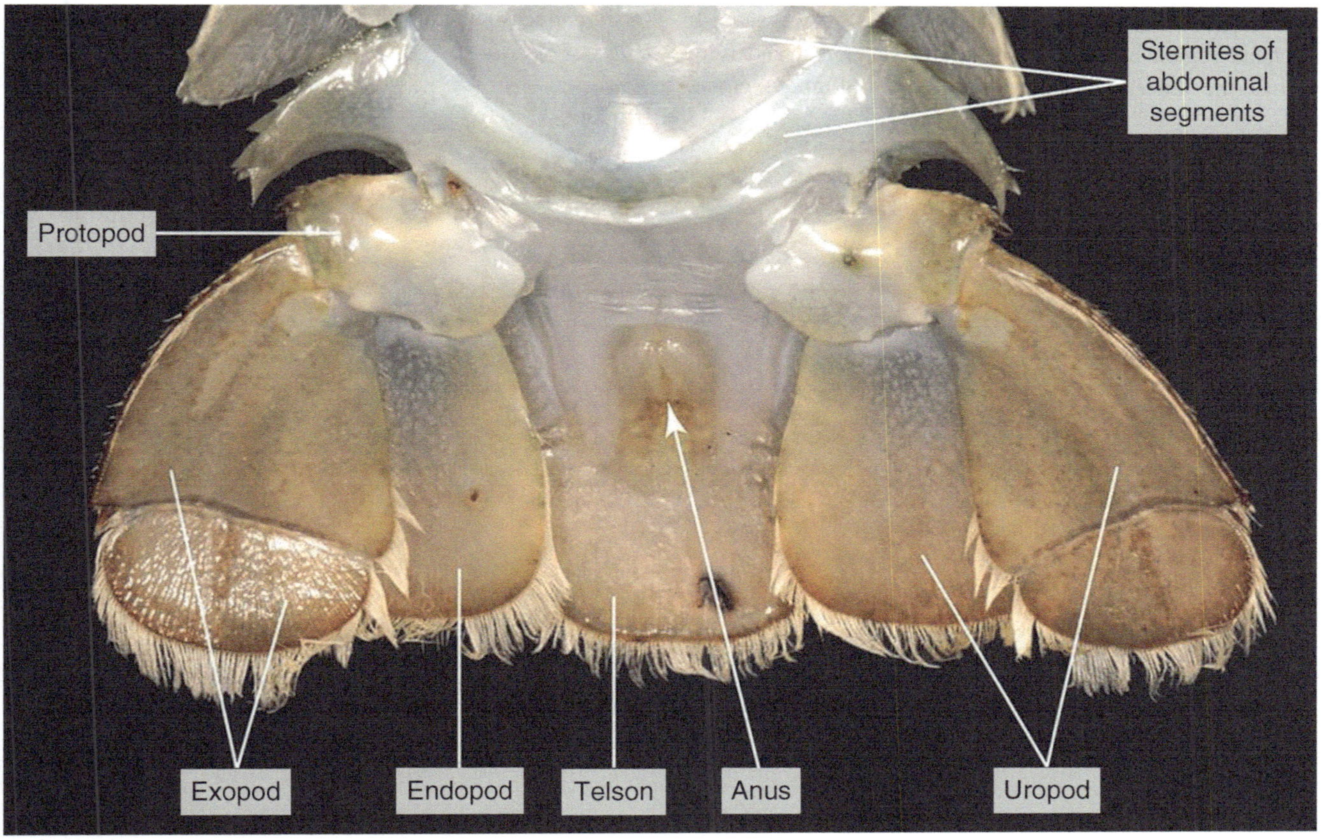

Fig. 7.32 Ventral view of the tail fan of a crayfish

Locate and identify the organs of the excretory system. In the head region anterior to the digestive glands and lying against the anterior body wall is a pair of *green glands* (also called antennal glands or modified metanephridia) (Fig. 7.33).

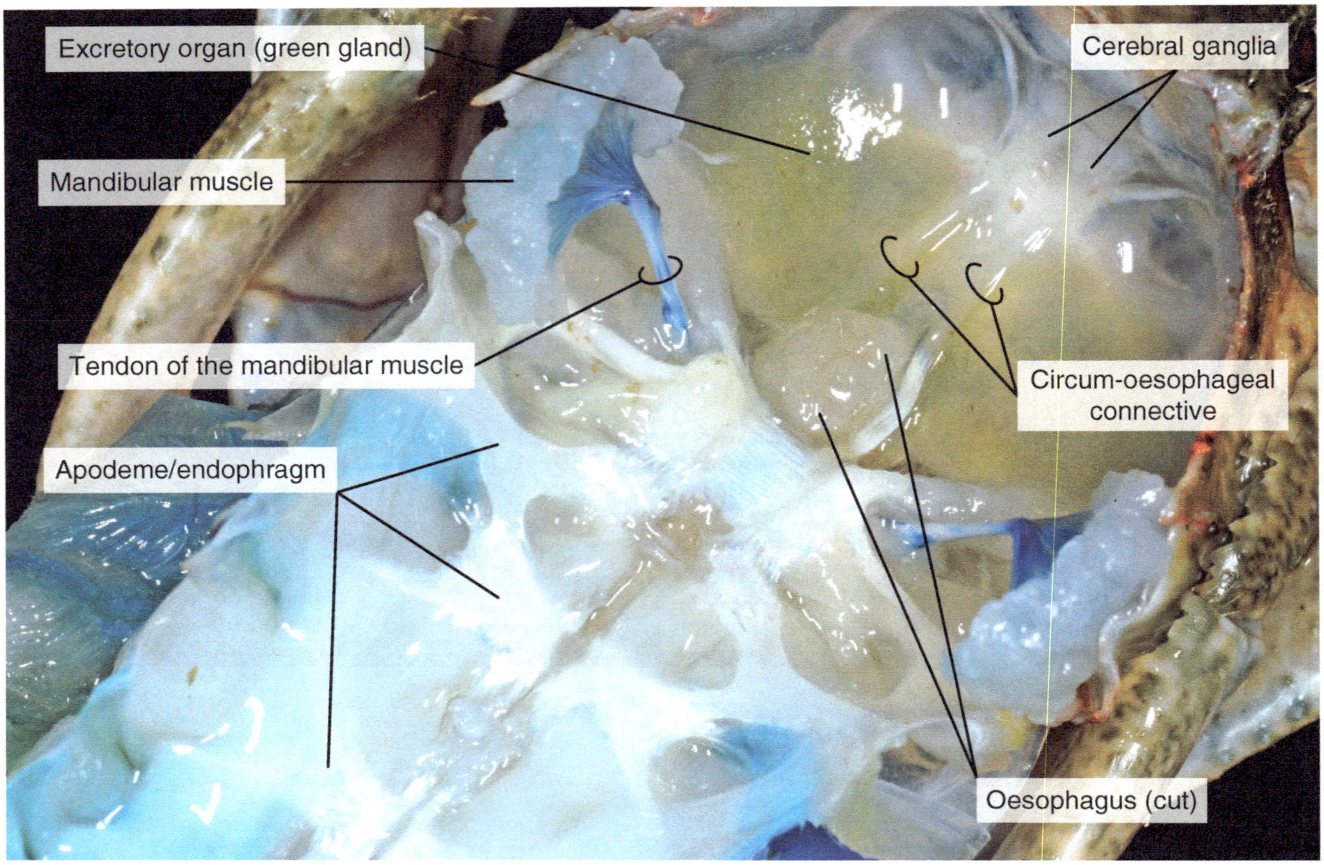

Fig. 7.33 The green glands (excretory organs) and the apodeme after the removal of viscera

They are round and cushion shaped. The haemolymph is filtered into the end sac by hydrostatic pressure in the haemocoel. As the filtrate passes through the excretory tubule, reabsorption of salts and water occurs, leaving the urine to be excreted. A duct from the bladder empties through a renal pore at the base of each antenna (Figs. 7.5 and 7.6). The role of the green glands is largely the regulation of the ionic and osmotic composition of the body fluids. In the freshwater crayfish, the urine is copious and hypotonic. Excretion of nitrogenous wastes (mostly ammonia) occurs by diffusion in the gills and across thin areas of the cuticle.

Clear away all of the viscera. Trim the base of the rostrum so that the brain is fully exposed. The brain is a pair of *supra-oesophageal ganglia* that lie against the anterior body wall between the green glands (Fig. 7.33). Note the *circum-oesophageal connective*, the two large and long nerves that lead from the brain, around the oesophagus, and join the *sub-oesophageal ganglia* (Fig. 7.34). In the arthropods there is some fusion of ganglia. The brain is formed by the fusion of three pairs of head ganglia, and the sub-oesophageal ganglion is formed by the fusion of at least five pairs. Chip away the calcified plates, the apodemes and connective tissue that conceal the ventral nerve cord in the thorax on either side of the midline, and follow the cord posteriorly (Fig. 7.34). Locate a ganglion, one of the enlargements of the ventral nerve cord. By removing the big flexor muscles in the abdomen, trace the cord for the length of the body. Observe the abdominal part of the nerve cord (Fig. 7.34).

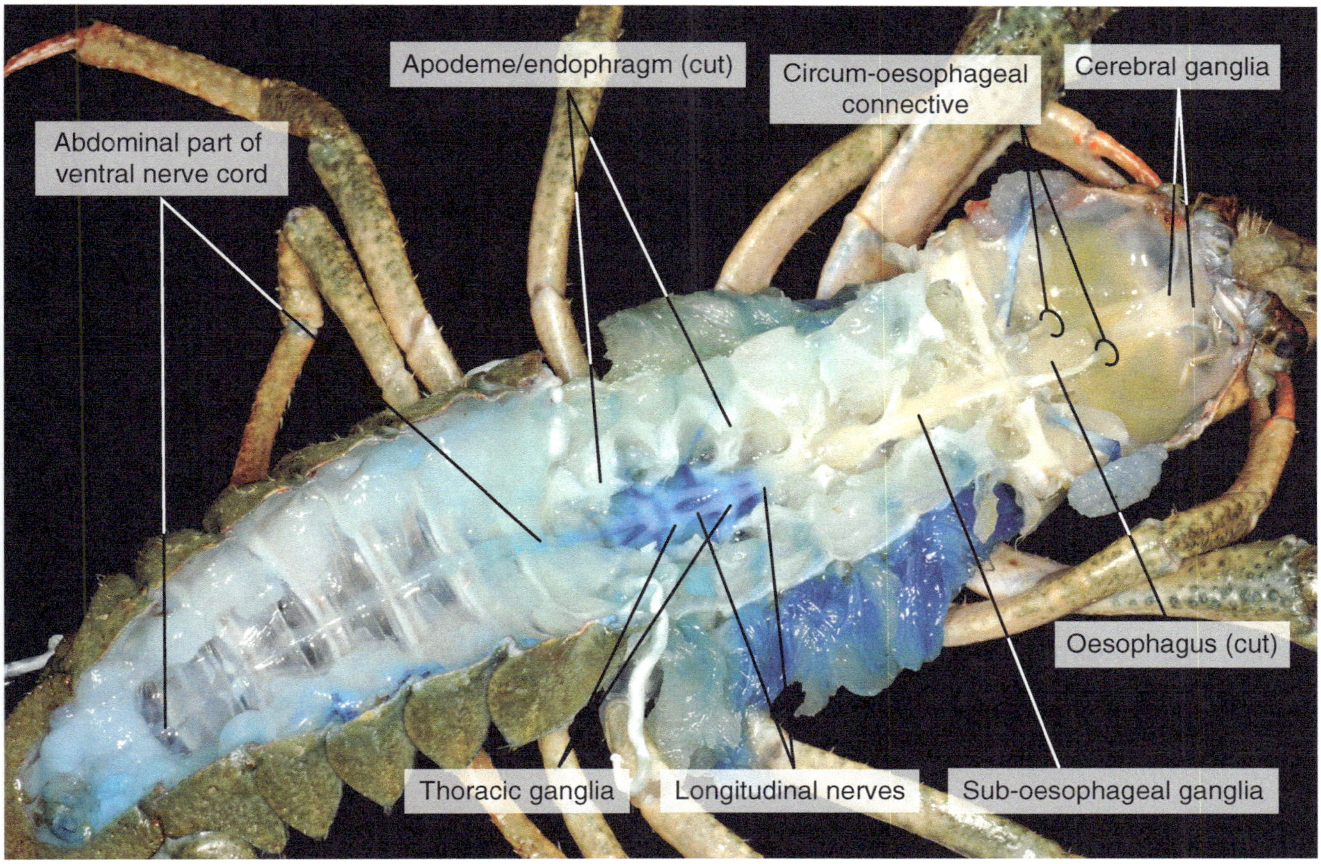

Fig. 7.34 The nervous system (ventral nerve cord) of a crayfish. (The figure is derived from the specimen injected with toluidine blue, which accounts for the blue background colour)

The crayfish has many sense organs: tactile hairs over many parts of the body, antennules, antennae, statocysts and compound eyes. The stalked *compound eyes* projecting from beneath each side of the rostrum are composed of functional units, the *ommatidia* (Figs. 7.2, 7.3 and 7.36). With a sharp scalpel, cut off the tip of one eye, mount in a drop of water and examine with a microscope. Note the many facets. They are square shaped for the first sight, but after a careful inspection their hexagonal nature is revealed; nevertheless, two opposite sides of the hexagon are very short indeed (Fig. 7.35). Each facet is the external surface of an ommatidium.

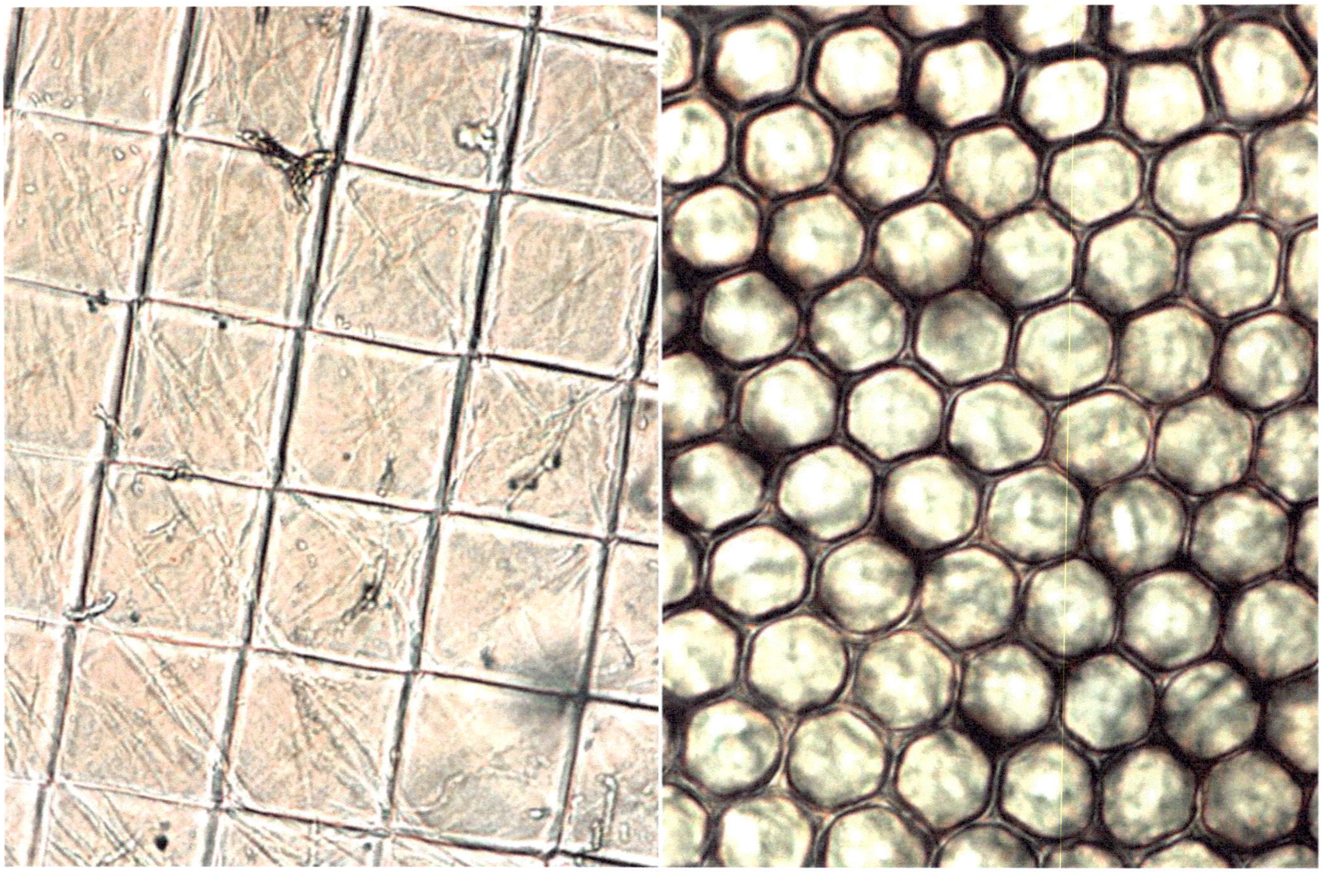

Fig. 7.35 The square (unevenly hexagonal) facets of the compound eye of the crayfish (*left*) compared with the symmetrically hexagonal facets of the insects on equal magnification (400x)

A balance organ (statocyst) is located on the dorsal side of the basal segment (protopod) of each antennule. These are small invaginations with sensory hair cells on their surface. The sensory pits are covered with dense chitinous bristles (Fig. 7.36). The pressure of sand grains, taken up from the environment, against sensory hairs in the statocyst gives the crayfish a sense of equilibrium.

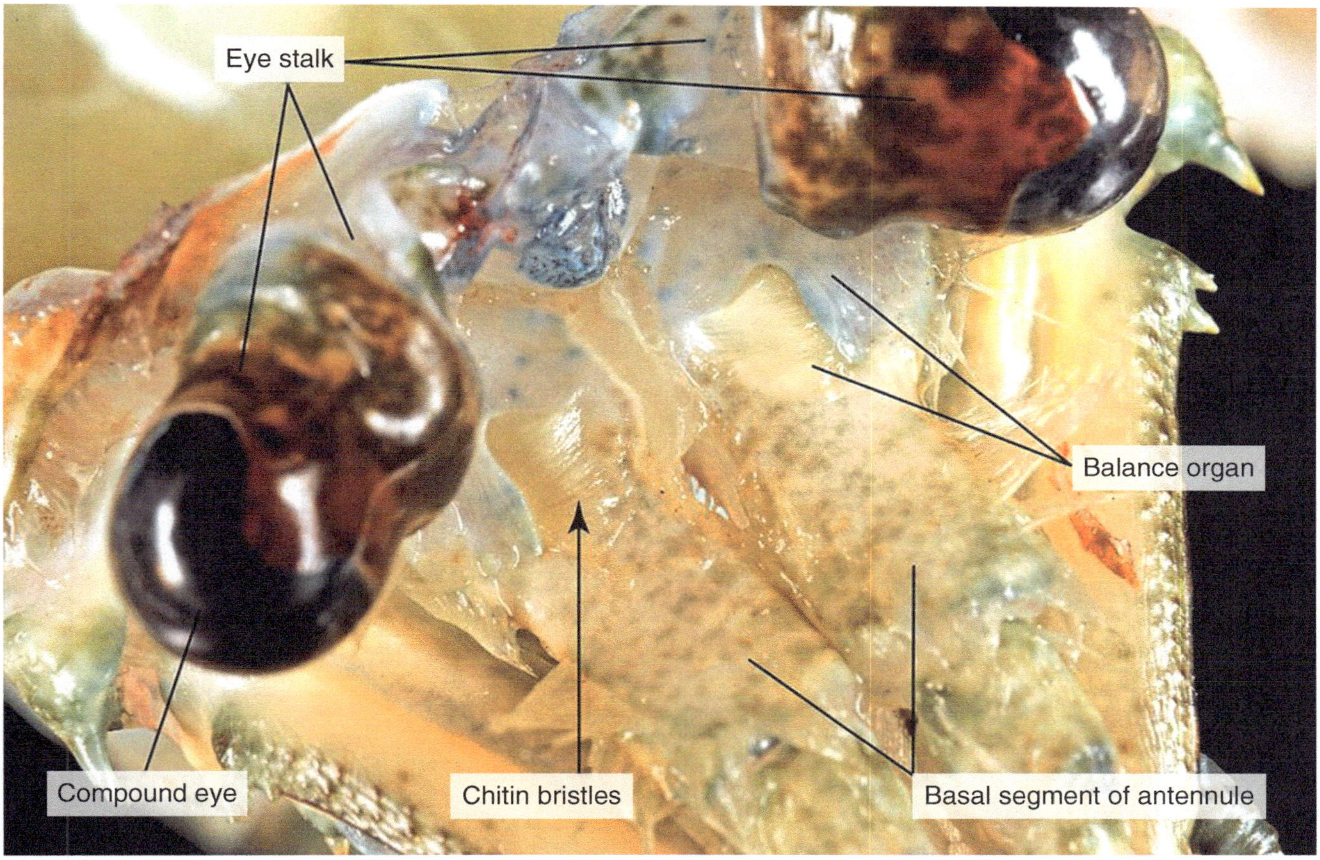

Fig. 7.36 The balance organs (statocysts) on the dorsal side of the basal segments of antennules

8 Dissection of a Cockroach (*Blaberus* sp.)

- *Availability:* Insects are the most extensive group of animals in the world. Approximately 800,000 species of insects have been recorded, and probably as many more remain to be discovered. Cockroaches are insects of the order Blattodea, of which about 30 species out of 4600 are associated with human habitats. They are terrestrial, cosmopolitan, nocturnal insects, found on warm, damp, dark places. Cockroaches are generally omnivorous scavengers; they feed on all sorts of organic debris. We can purchase big tropical species suitable for dissection from pet shops or zoos.
- *Anaesthesia:* Put some cotton wool infiltrated with ethyl acetate in a killing jar. Place in the cockroach and close the jar immediately; the cockroach can be dissected within 5 min, when it does not move upon shaking the jar. The use of ethyl acetate is preferable to ether or chloroform as it leaves the insect relaxed.

The body of the cockroach is elongated and segmented. It is dark or reddish brown in colour. The chitinous exoskeleton is secreted by the underlying hypodermis. It is made up of thick and hard plates, which are bounded by sutures of soft cuticle. It protects the body from loss of water and provides rigidity and surface for attachment of body muscles (Fig. 8.1).

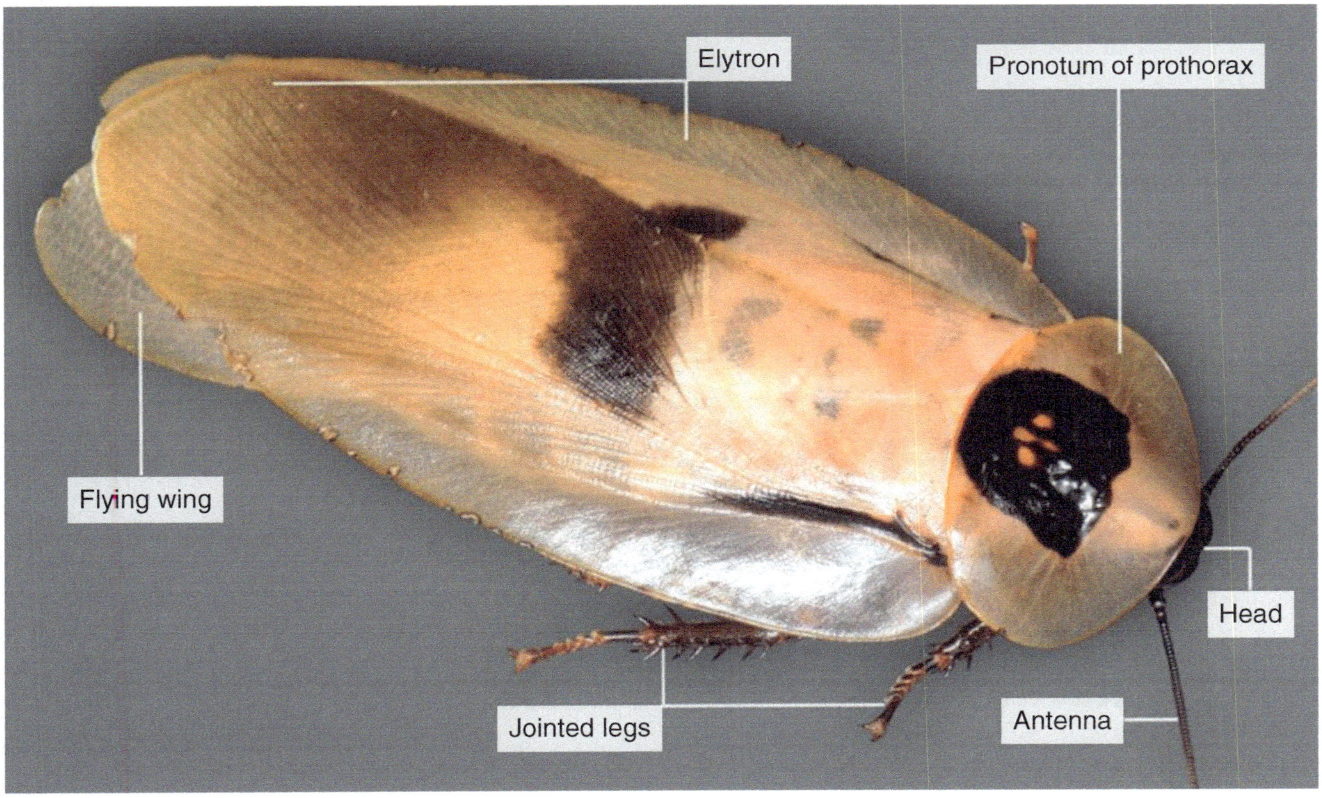

Fig. 8.1 Dorsal view of the cockroach

P. Lőw et al., *Atlas of Animal Anatomy and Histology*, DOI 10.1007/978-3-319-25172-1_8

The body is divisible into *head*, *thorax* and *abdomen* (Fig. 8.2). The basic 18 segments of the insect are functionally organised into the three body regions: the head (5 segments), the thorax (3 segments), and the abdomen (10 segments). The adjacent segments are joined by thin, soft and flexible arthrodial membrane.

Fig. 8.2 Ventral view of the cockroach

The entire dorsal plate is called the *tergum* (or notum) and any one specific segment is a *tergite*. The ventral body surface is the *sternum*; one specific segment is a *sternite*. The lateral body surface is the *pleuron*, and *pleurite* for one segment. A pair of small holes, called *spiracles*, is located on the lateral side of the body in thoracic and abdominal segments to allow air to enter the tracheae. Find and identify all of these parts of the cuticle (Fig. 8.3).

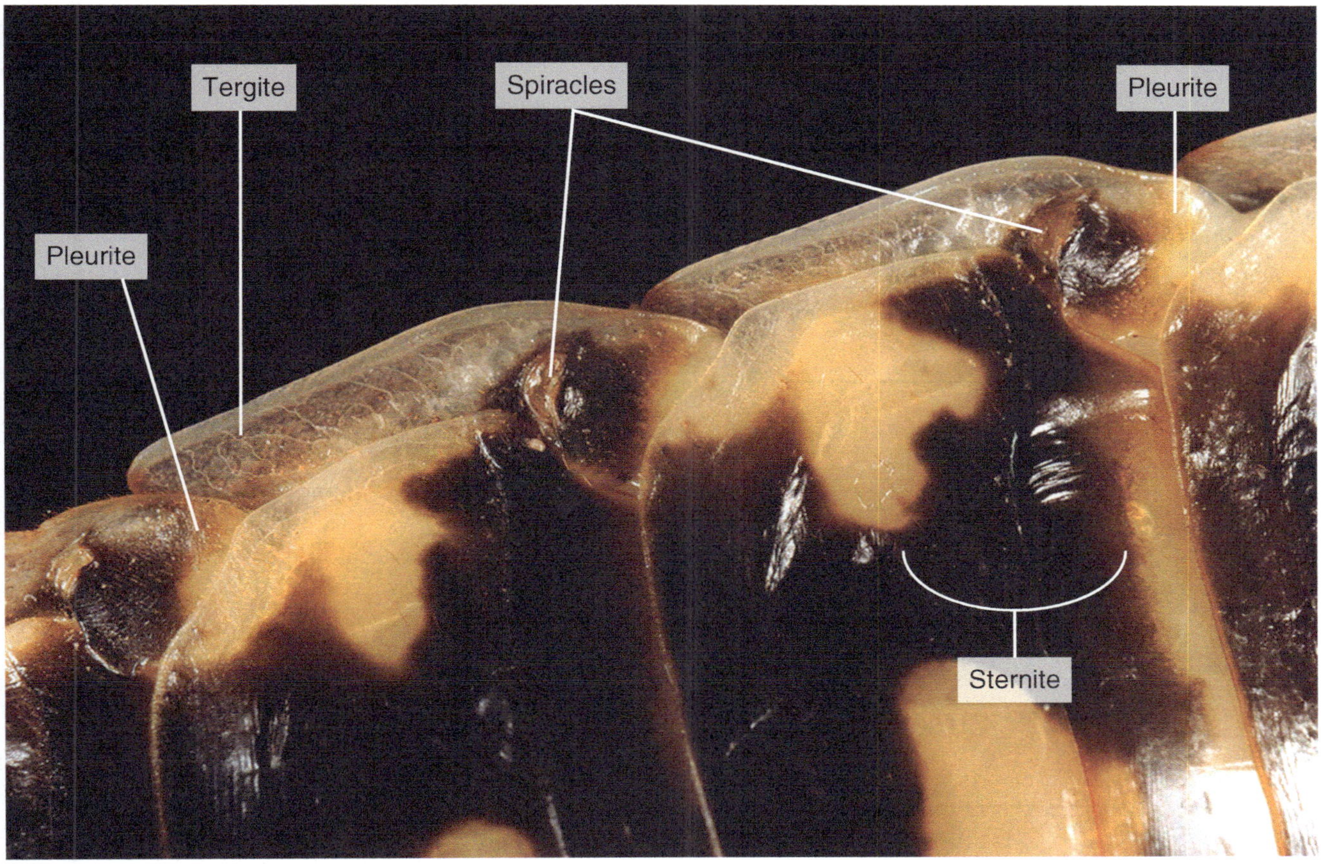

Fig. 8.3 The abdominal spiracles are located laterally on the anterior-dorsal corner of the pleura of the first eight abdominal segments

The three thoracic segments are the *prothorax*, *mesothorax* and *metathorax*. The locomotory organs of the cockroach (three pairs of jointed appendages and two pairs of wings) are on the thorax. The leathery forewings are mesothoracic and are called *elytra* or wing covers. These are dark, stiff and opaque. They cover the hindwings and are protective in function. The *hindwings* are large, thin, membranous and transparent. They are kept folded below the elytra and are used for flying. Examine the cockroach in dorsal view with two pairs of elytra and wings extended (Fig. 8.4).

Fig. 8.4 Dorsal view of the cockroach with two pairs of elytra and wings extended

Cut a little square out of the thinnest inner edge of either flying wing, put it on a slide and cover it with a coverslip dry and examine in a microscope (Fig. 8.5).

The thin membrane of the wings is made up by two layers of integument closely apposed. Where the two layers remain separate, the *veins* are formed. Here, the lower cuticle may be thicker and more heavily sclerotized to provide strength and rigidity to the wing. As the cavities of the veins are connected with the haemocoel, they form *haemolymph sinuses*, so haemolymph can flow into the wings (Fig. 8.5). Within each of the major veins there is a nerve and a trachea in addition to haemolymph. Two types of *hair* may occur on the wings: microtrichia, which are small and irregularly scattered, and macrotrichia, which are larger, socketed and may be restricted to veins (Fig. 8.5).

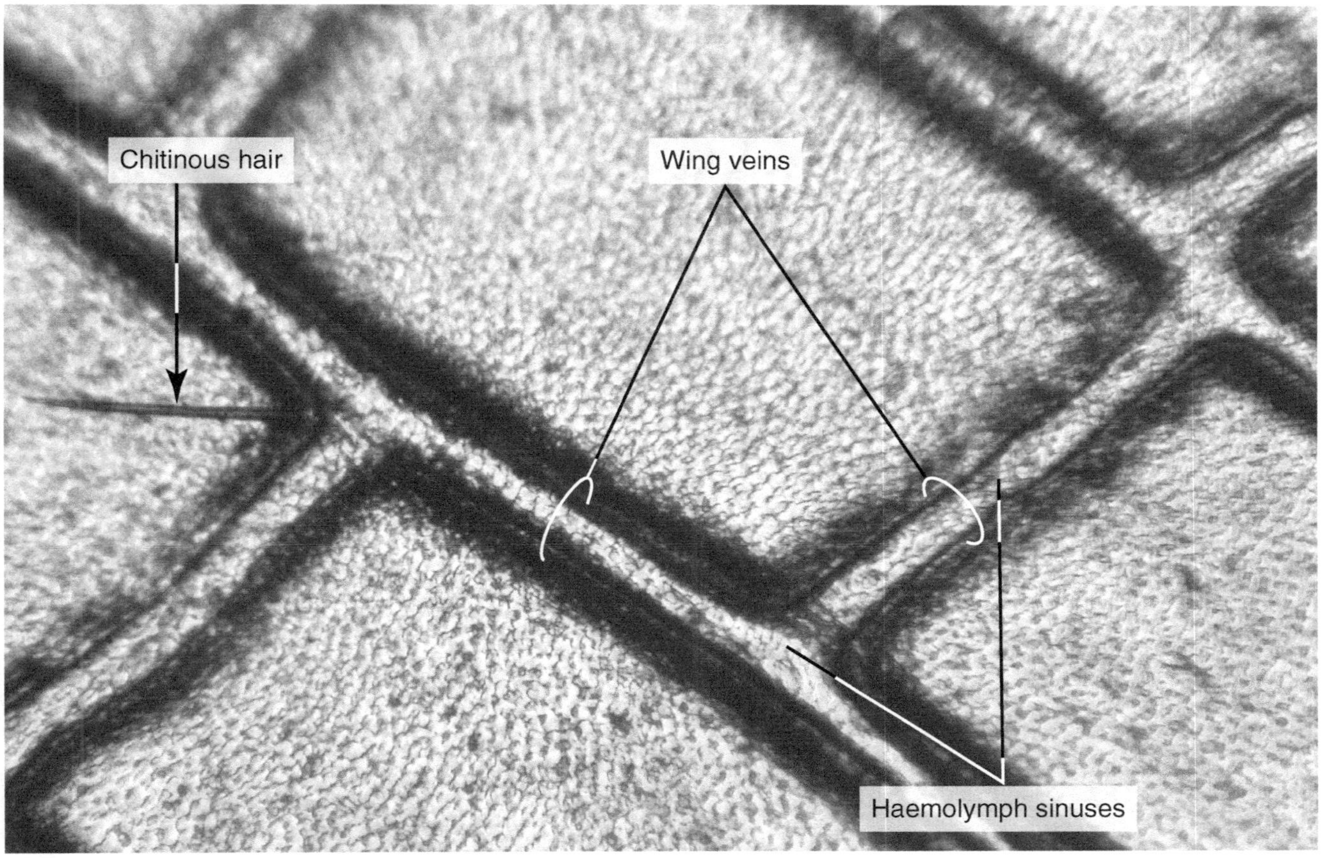

Fig. 8.5 System of veins in the flying wing of the cockroach. Note the haemolymph sinuses within the veins and a socketed, large chitinous hair connected to the vein

A cockroach's thorax bears three pairs of legs as well. Each of the three pairs of legs is named after the region of the thorax to which it attaches: the prothoracic legs are closest to the cockroach's head. These are the shortest legs, and they act like brakes when the roach runs. The middle legs are the mesothoracic legs. They move back and forth to either speed the cockroach up or slow it down. The very long metathoracic legs are the cockroach's back legs, and they move the cockroach forward (Fig. 8.6).

These three pairs of legs are different in lengths and functions, but they have the same parts and move the same way. The upper portion of the leg, called the *coxa*, attaches the leg to the thorax. The small *trochanter* acts like a hinge and lets the cockroach bend its leg. The large *femur* and slender, spiny *tibia* are the two longest elements of the leg. The most distal part of the leg is the five-jointed *tarsus* with two hook-like *claws* at its end and *pulvilli* (singular pulvillus) on its segments (Fig. 8.6). The latter two components of the tarsus help cockroaches to climb walls and glass and walk upside down on ceilings.

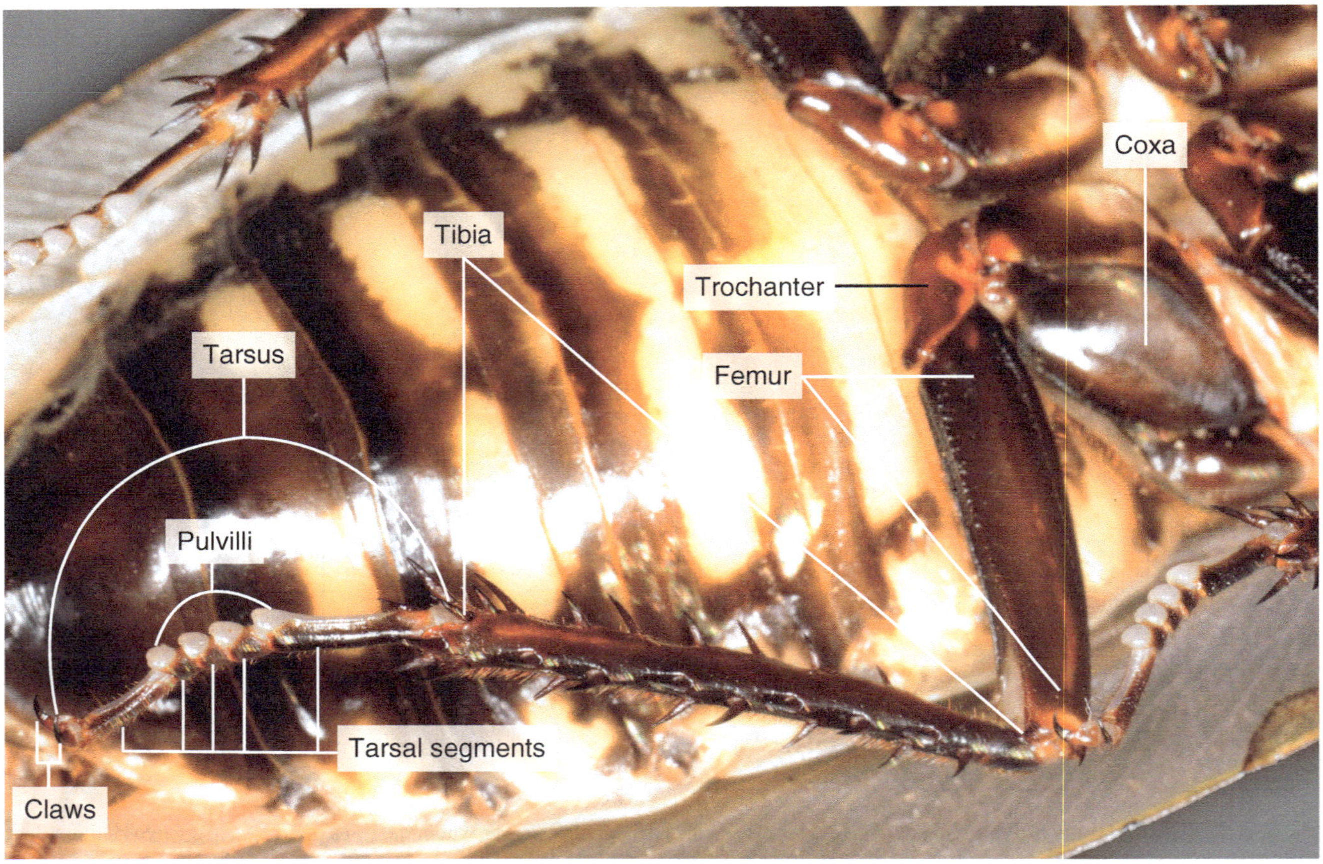

Fig. 8.6 Ventral view of the cockroach with a metathoracic leg

In cockroach, sexes are separate, so it is dioecious. Determine the gender of your specimen. You can distinguish female and male cockroaches from one another by the differences in their abdominal tips (Figs. 8.7 and 8.8). Four small projections (2 *cerci* (singular cercus) and 2 *styli* (singular stylus)) are visible on the terminal abdominal tergite (epiproct) of the male (Fig. 8.7), whereas only two projections (the cerci) are visible in the female (Fig. 8.8). Cerci and styli are sense organs.

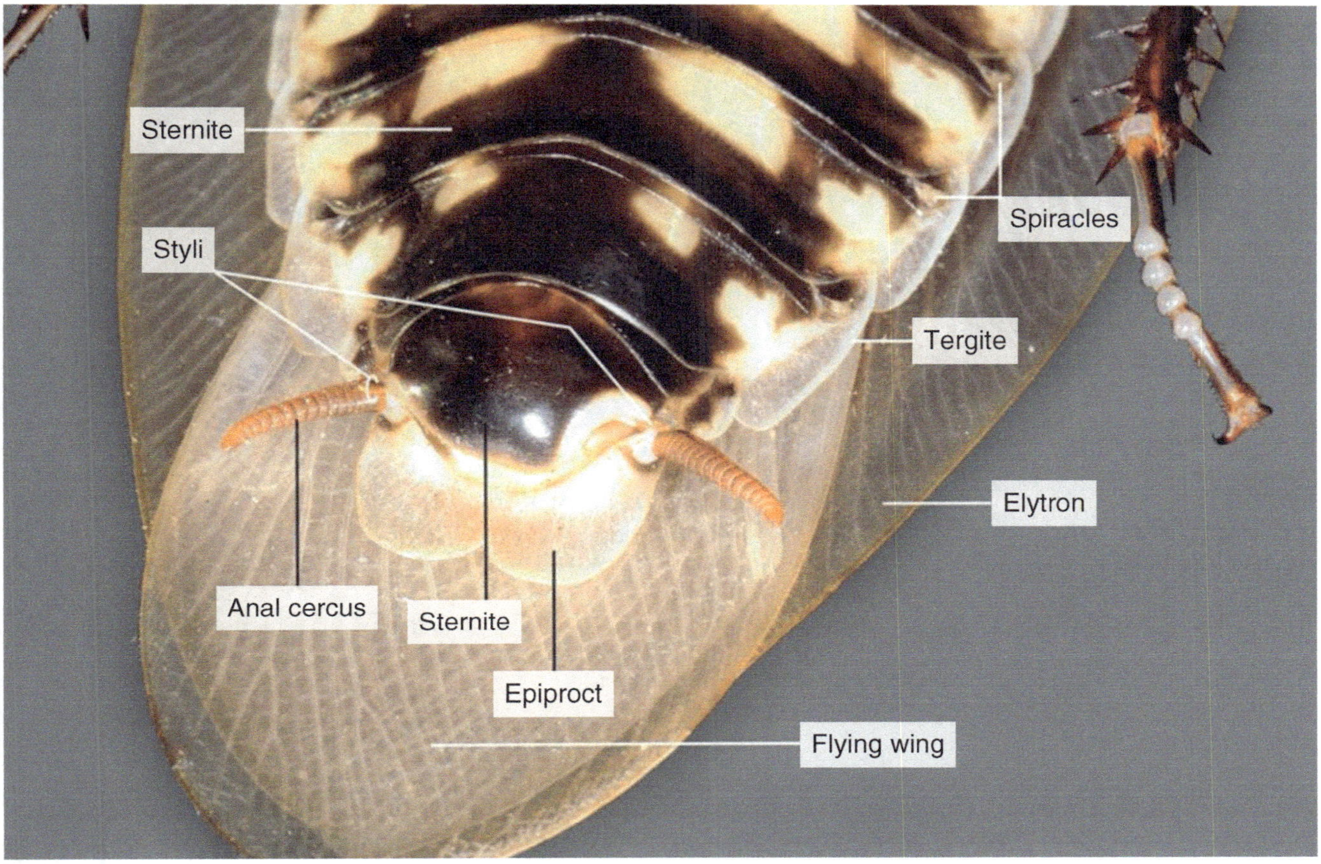

Fig. 8.7 Ventral view of the posterior abdominal segments of a male cockroach

Fig. 8.8 Ventral view of the posterior abdominal segments of a female cockroach

Feeding and sensory organs are on the *head*. Notice the *compound eyes*, the *antennae* and two *ocelli* (fenestrae), one dorsal to the base of each antenna (Fig. 8.9).

All the head segments are fused into a single head capsule, and mouthparts represent modified appendages. The head consists of a dorsal *vertex*, the cheeks, or *genae*; the front of the face, or *frons*; the *clypeus* below the frons; and the movable upper lip, or *labrum*, with a row of bristles on its free margin. Ventrally, an opening called mouth is present on the head that is surrounded by the mouth parts consisting of a pair of mandibles, maxillae and the labium (Fig. 8.9).

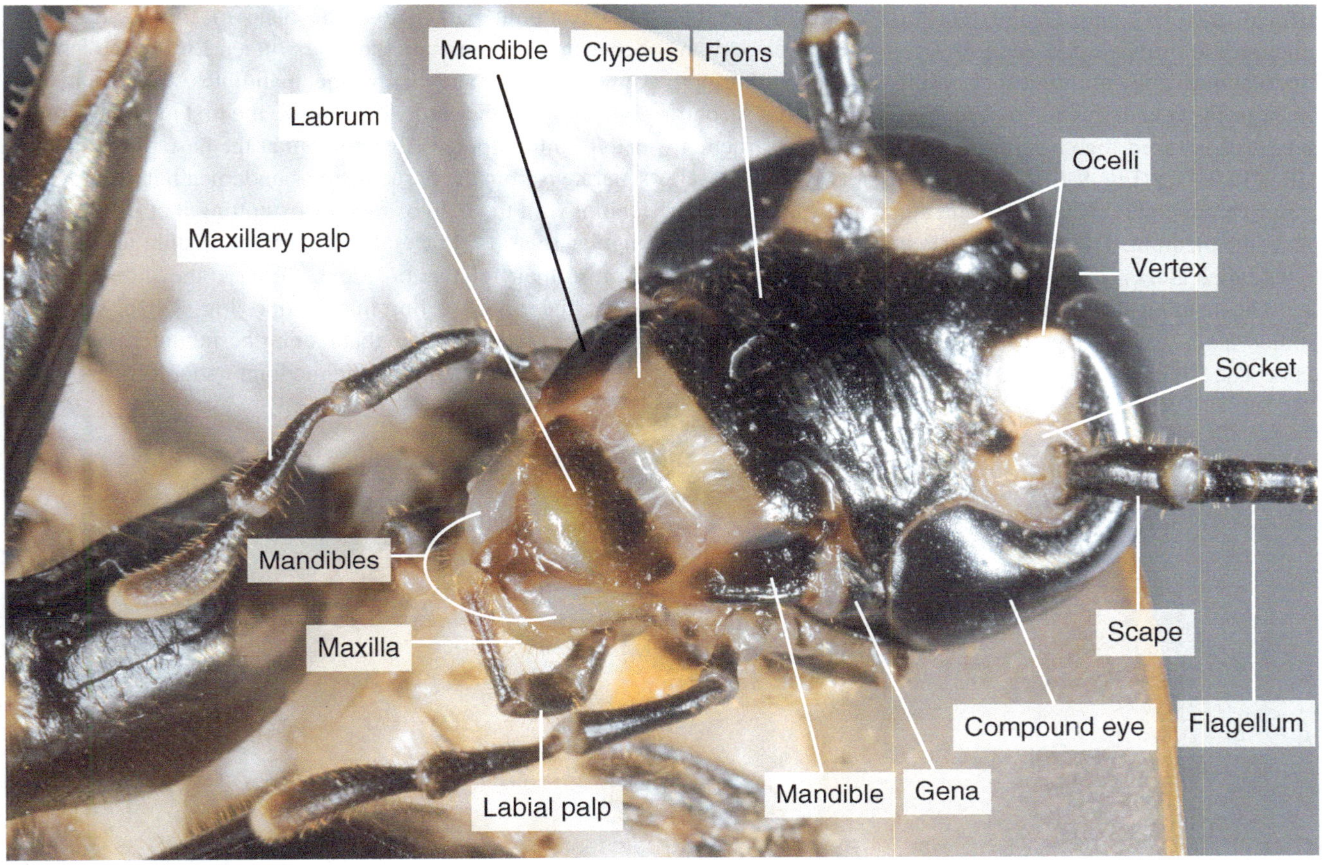

Fig. 8.9 Details of the head capsule of a cockroach and feeding and sensory organs on the head

Mandibles are a pair of hard, triangular, chitinous structure present below the labrum on each side. Each mandible is provided with three strong pointed teeth: a condyle, a set of abductor muscles and a set of adductor muscles (Fig. 8.10). Mandibles are jaws used for cutting and masticating the food. First *maxillae* are mouth parts of cockroach situated behind the mandible (Fig. 8.10). Each one consists of two basal segments: cardo and stipes. Stipes bear a five segmented maxillary palp having olfactory bristles. From the inner side of stipes arise two lobes, an outer galea and an inner lacinia. They are used for holding the food and bringing it to the mandibles for mastication. *Labium* (or second maxillae) is fused together forming a single large structure called lower lip (Fig. 8.10). It is made up of two broad basal parts: a broad lower plate, the submentum, and an oval upper plate, the mentum. The mentum bears in front a pair of inner lobes called glossae and a pair of outer lobes called paraglossae. The mentum also bears on the lateral sides a pair of three jointed labial palps; they bear tactile and gustatory sensory hairs. The labium works like a tray, prevents the loss of food material from the mandibles and pushes the masticated food material back in the mouth (Fig. 8.10).

Hypopharynx is a small, cylindrical organ, sandwiched between first maxillae and covered by labrum and labium on dorsal and ventral sides, respectively. It bears several sensory setae on its free end and the opening of common salivary duct upon its basal part.

Use the forceps and teasing needles and carefully remove all the mouth parts. Hold the head of the cockroach in between the thumb and the index finger. Lift up the labrum. Cut the membrane below the mandibles, take them out. Catch hold of the cardo and remove the first maxilla with the help of a fine forceps. Insert a pin at the base of the labium, separate it from the tissue that lies underneath it. Remove the labium with the help of forceps by cutting it at its base with angular scissors. Place all the parts on a slide and arrange them in their relative positions. Observe the slide thus prepared under a dissecting microscope (Fig. 8.10). After drying for a week, the mouthparts can be mounted in Canada balsam or synthetic mounting medium under a coverslip.

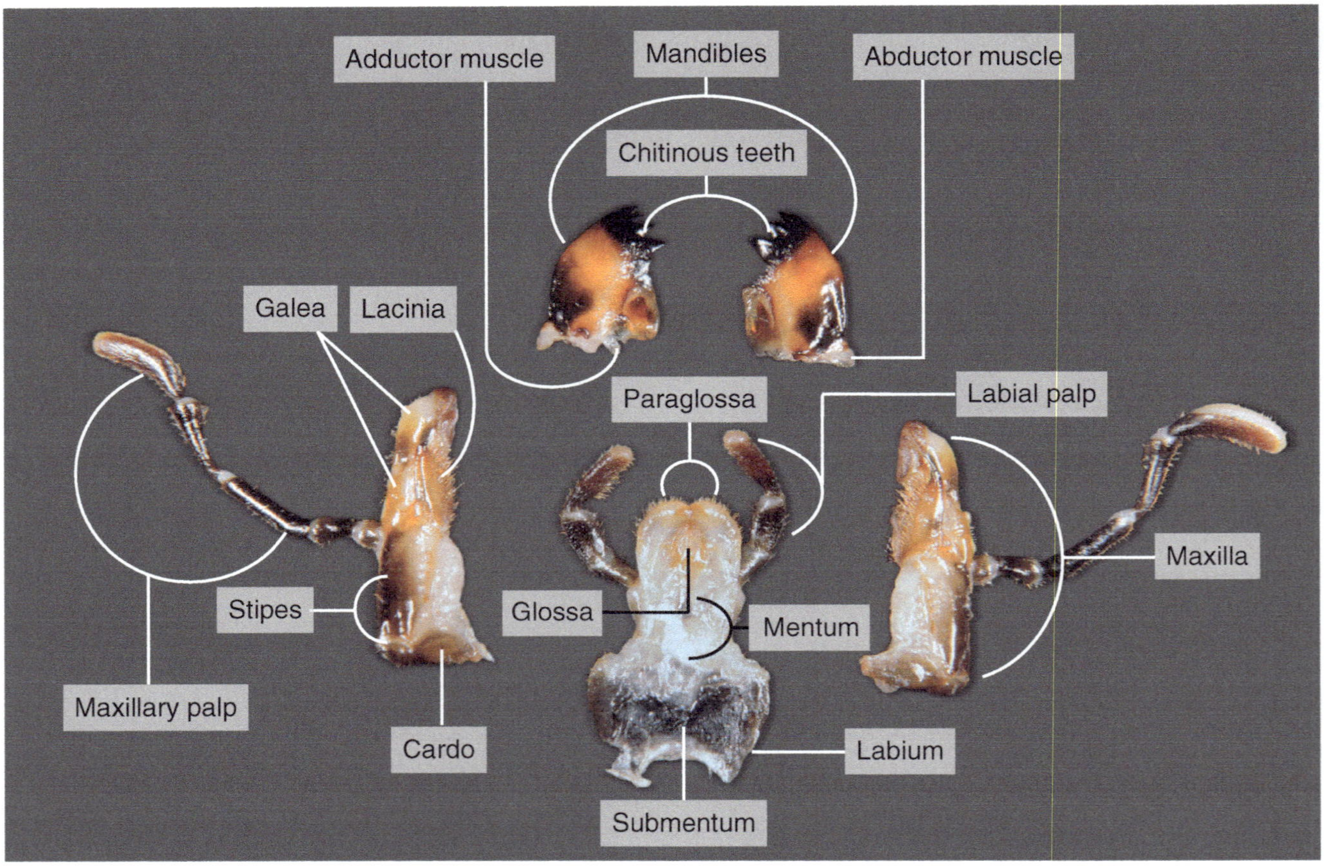

Fig. 8.10 Isolated mouthparts of a cockroach

Cut off the elytra and the wings close to their bases. Cut the *pronotum* away from the neck (cervix) (Fig. 8.11). Under the pronotum slender, longitudinal cervical *sclerites* strengthen the otherwise soft integument of the neck (Figs. 8.15 and 8.28).

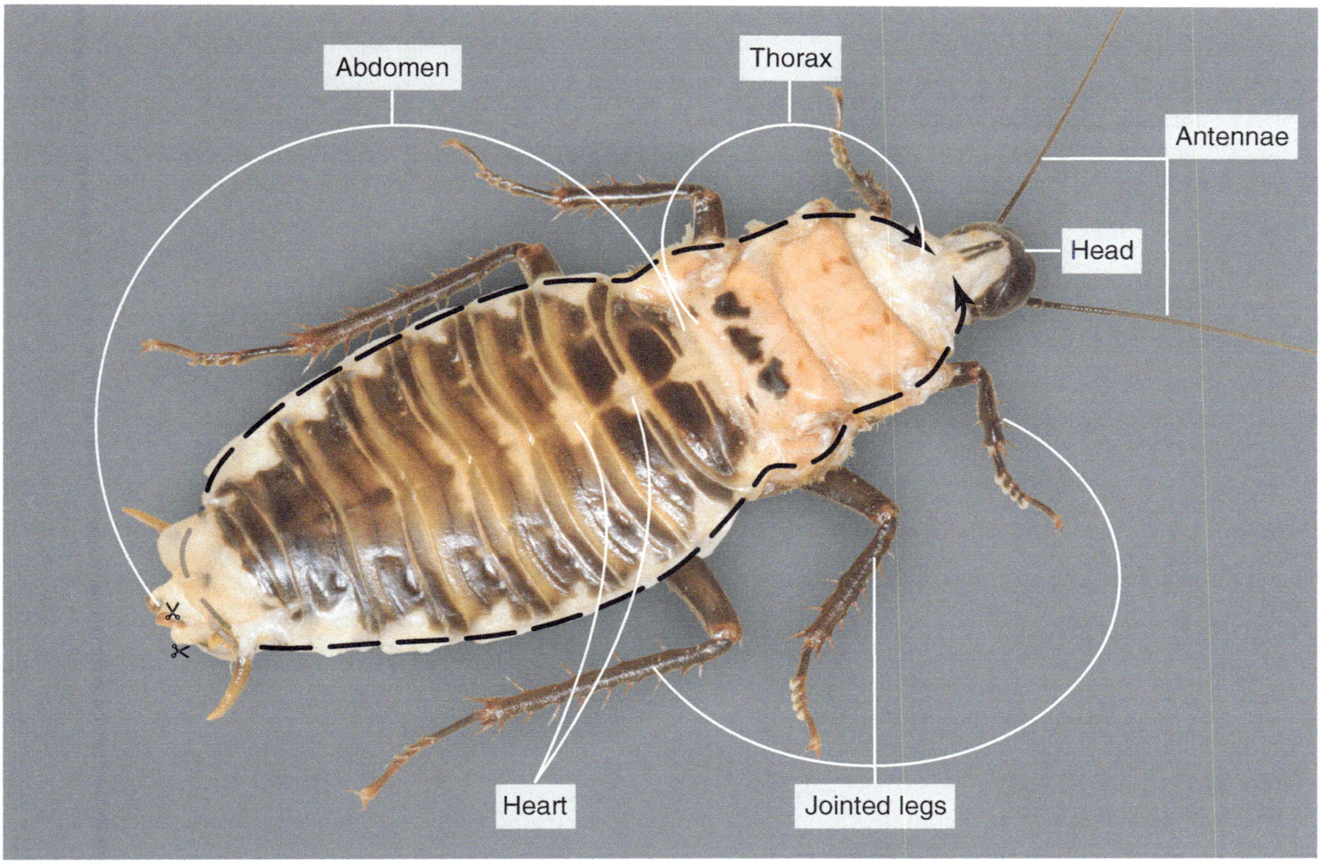

Fig. 8.11 A cockroach prepared for dissection. *Dashed lines* indicate the cut line where abdominal and thoracic tergites should be detached along the lateral side of the insect

Cut up one side of the abdominal and thoracic tergites longitudinally near the lateral margins of the insect, being careful not to probe too deeply with the scissors as indicated by the dashed lines in Fig. 8.11. Keep to line of the body wall. Lift the epiproct, the tenth abdominal tergite with forceps carefully leaving the anus and hind-gut in the ventral body part. Holding the tergum with forceps by the epiproct, carefully remove it from the posterior end forwards (Fig. 8.12).

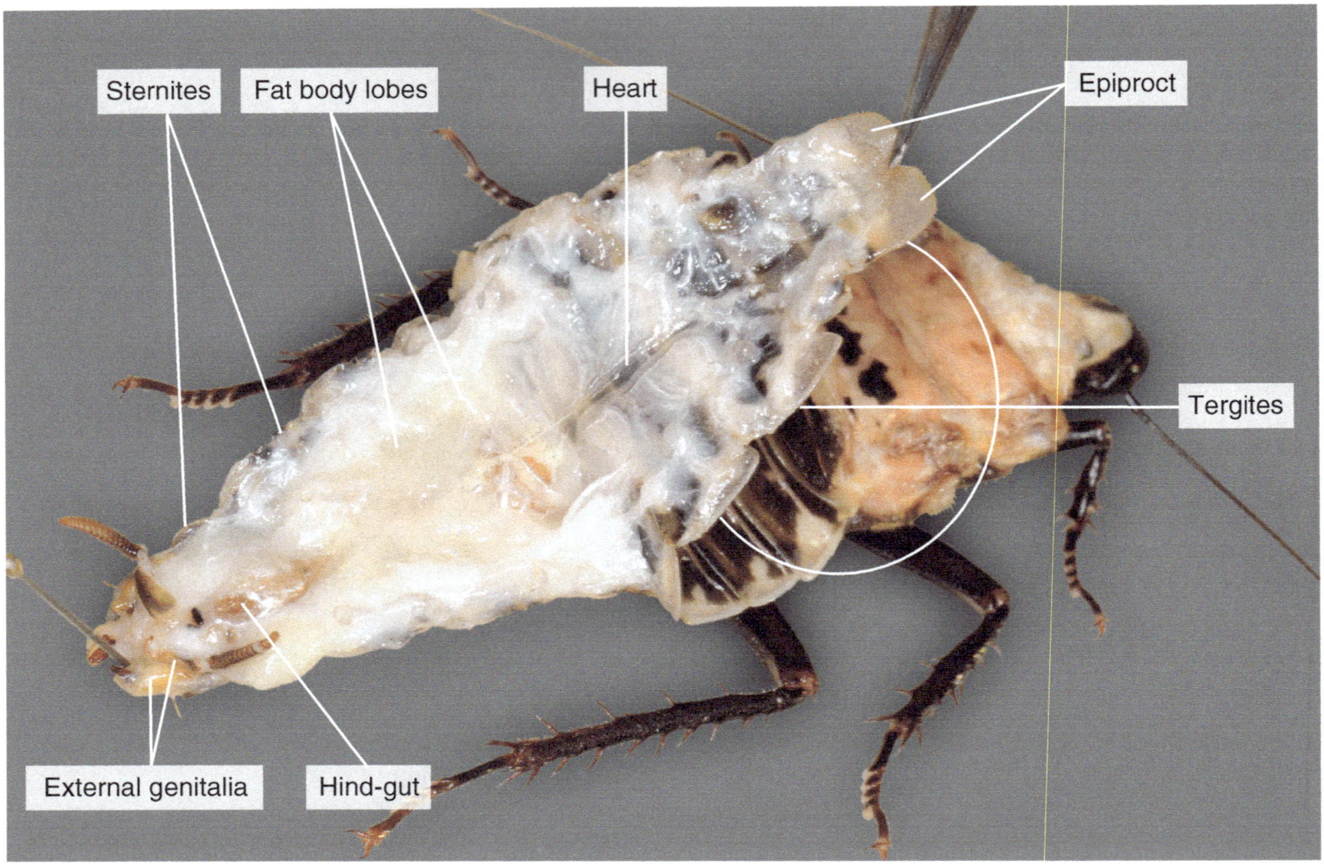

Fig. 8.12 The removal of the tergum by holding it through the epiproct and cutting the dorsoventral muscles

Use a scalpel to cut the dorsoventral muscles and loosen connective tissue where necessary. The thoracic tergites are more tightly fastened by dorsoventral muscles, which have to be cut as well. Remove the tergum, lay it onto its dorsal side and place two pins to fix it. A transparent median dorsal vessel (heart) should now be visible along the midline of the body. It is flanked laterally by silvery-white tracheae of the respiratory system (Fig. 8.13).

Insects have an *open circulatory system*, so they have *haemolymph*, instead of blood. Body cavity contains the haemolymph, which bathes viscera in it therefore known as *haemocoel*. The haemocoel is subdivided by two membranous horizontal partitions into three wide and flattened sinuses, the dorsal *pericardial sinus* containing the heart, the middle *perivisceral sinus* containing the gut and most of the viscera, and the ventral *perineural sinus* or sternal sinus containing the nerve cord. The partition between pericardial and perivisceral sinuses is called *dorsal diaphragm*, and between perivisceral and perineural sinuses is called *ventral diaphragm*. The sinuses intercommunicate by pores in the respective diaphragms.

The circulatory system consists of a tubular *heart*, a dorsal vessel called *anterior aorta* and a system of haemolymph spaces, the lacunae or *sinuses*. The heart is a longitudinal middorsal tube extending the length of the body in the pericardial sinus and resting on the dorsal diaphragm. Together with the dorsal diaphragm, the heart is attached along the inner side of the tergum and visible also from the outside through the integument (Figs. 8.11 and 8.13). It is closed at the rear end of the insect. At the front end of the heart, the aorta opens into the head and the body cavity that surrounds all the organs. The heart has segmental swellings or *heart chambers*, paired segmental openings or *ostia*. The ostia and most of the swellings are inconspicuous. A pair of fan-like, triangular *alary muscles* in each segment is attached by their broad bases to the wall of a chamber and also connects it, by their pointed tips with the tergite of the segment (Fig. 8.13).

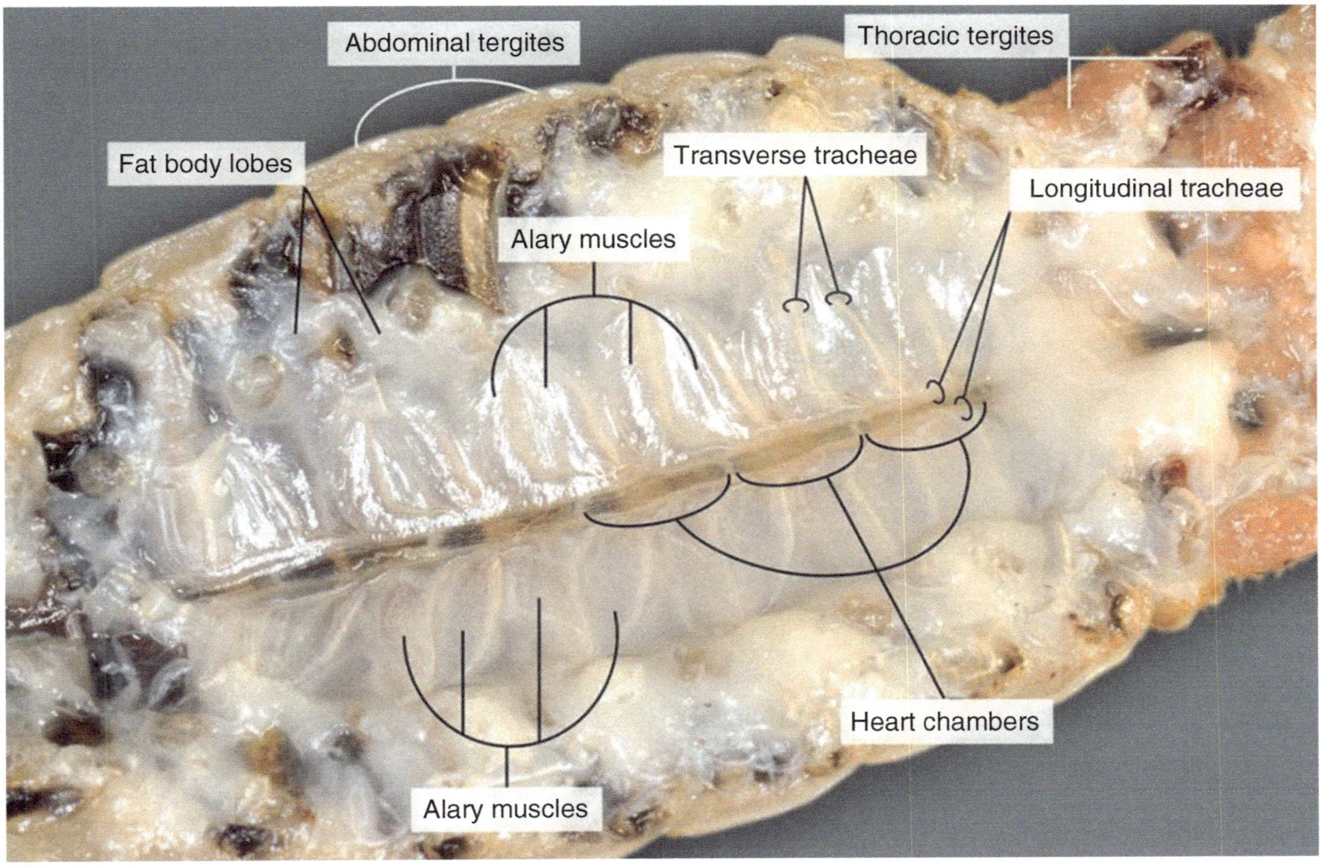

Fig. 8.13 The heart of a cockroach is a chambered median dorsal vessel attached to the inner side of the tergum

During diastole, contractions of radiating alary muscles and relaxation of heart wall muscles cause the heart to dilate and draw haemolymph into its lumen through the ostia from the pericardial sinus. Then alary muscles relax and circular muscles in the heart wall contract (systole), while valve-like ostia close, preventing back flow of haemolymph into the pericardial sinus. Haemolymph moves through the heart towards the head by continual peristaltic contractions of the heart wall. The contractions begin at the posterior chamber of the heart and continue forward, pushing the haemolymph into the head sinus through the terminal opening of anterior aorta. From the head sinus, the haemolymph flows backward into the thorax and abdomen. While flowing backwards from head sinus, the haemolymph remains in the ventral part, so it fills into the perineural sinus. From the perineural sinus, the haemolymph flows into the perivisceral sinus through the pores of ventral diaphragm in abdominal region. Then from perivisceral sinus, it flows into pericardial sinus through the pores of dorsal diaphragm. The pumping force that propels the haemolymph is provided by the pulsations of the alary muscles. The respiratory movements of abdomen and contraction of alary muscles increase this force.

Pin the ventral side of the cockroach down in the dissecting dish with thin but rigid pins (the best are the kind used by entomologists to pin their specimens). The pins should anchor your specimen securely and should be placed at an angle so as not to obscure your vision (Fig. 8.14).

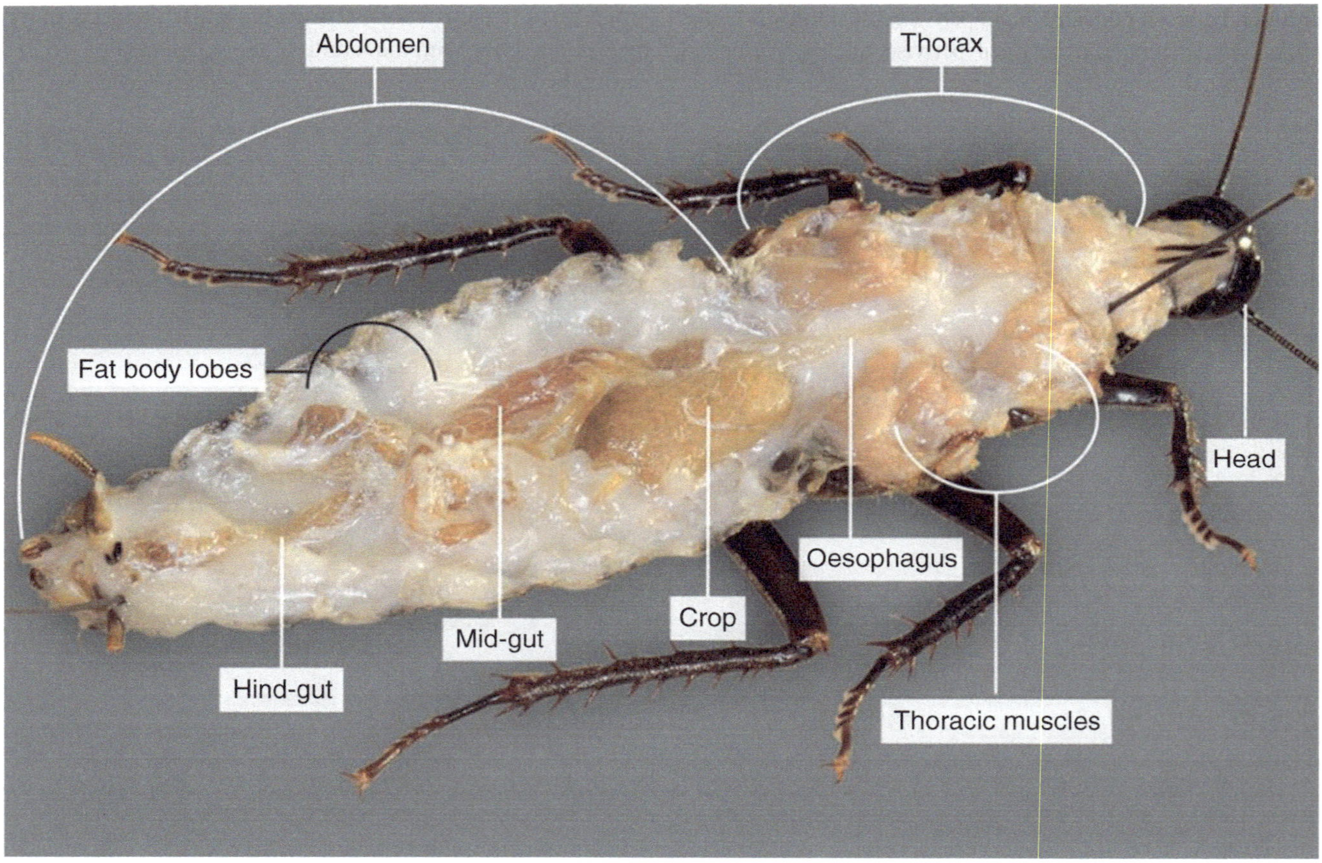

Fig. 8.14 Dorsal view of the internal organs of a cockroach before water cover

All this can be done in the air, but the following manipulations give better results if they are made under water. Cover the dissection completely with water (Fig. 8.15).

Insects store fat, protein and excretory products in their *fat body* lobes (Fig. 8.15). The chalky white fat body lobes that surround the abdominal organs should be removed. With an angled forceps, grip a piece of fat body (make sure it is not another organ), tear it away and wipe it into a paper towel. Repeat this until the grease is worked out.

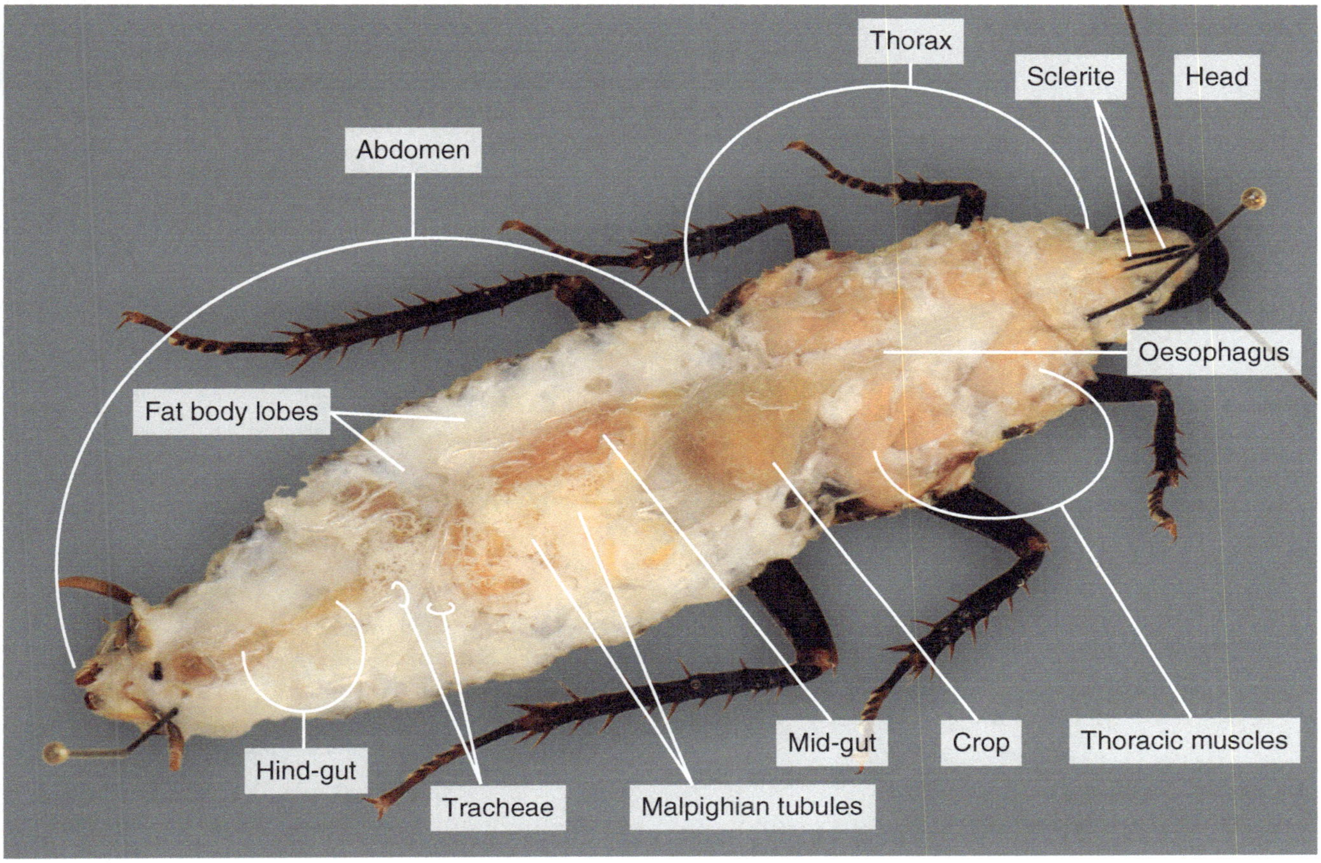

Fig. 8.15 The internal organs of a cockroach covered with water

The digestive system is held in place by connective tissues and aerated by glistening tracheae. As you remove the fat body to better expose the gut, be careful not to accidentally remove other organs, especially the more translucent reproductive organs that lie alongside the gut. The alimentary canal is easily isolated and it can be set free from its ties. If you want to separate it more or less completely, the gizzard must be gently moved with the tweezers until the oesophagus is seen and then the rest of the digestive tract can be liberated with the teasing needle to the level of the anus (Fig. 8.16).

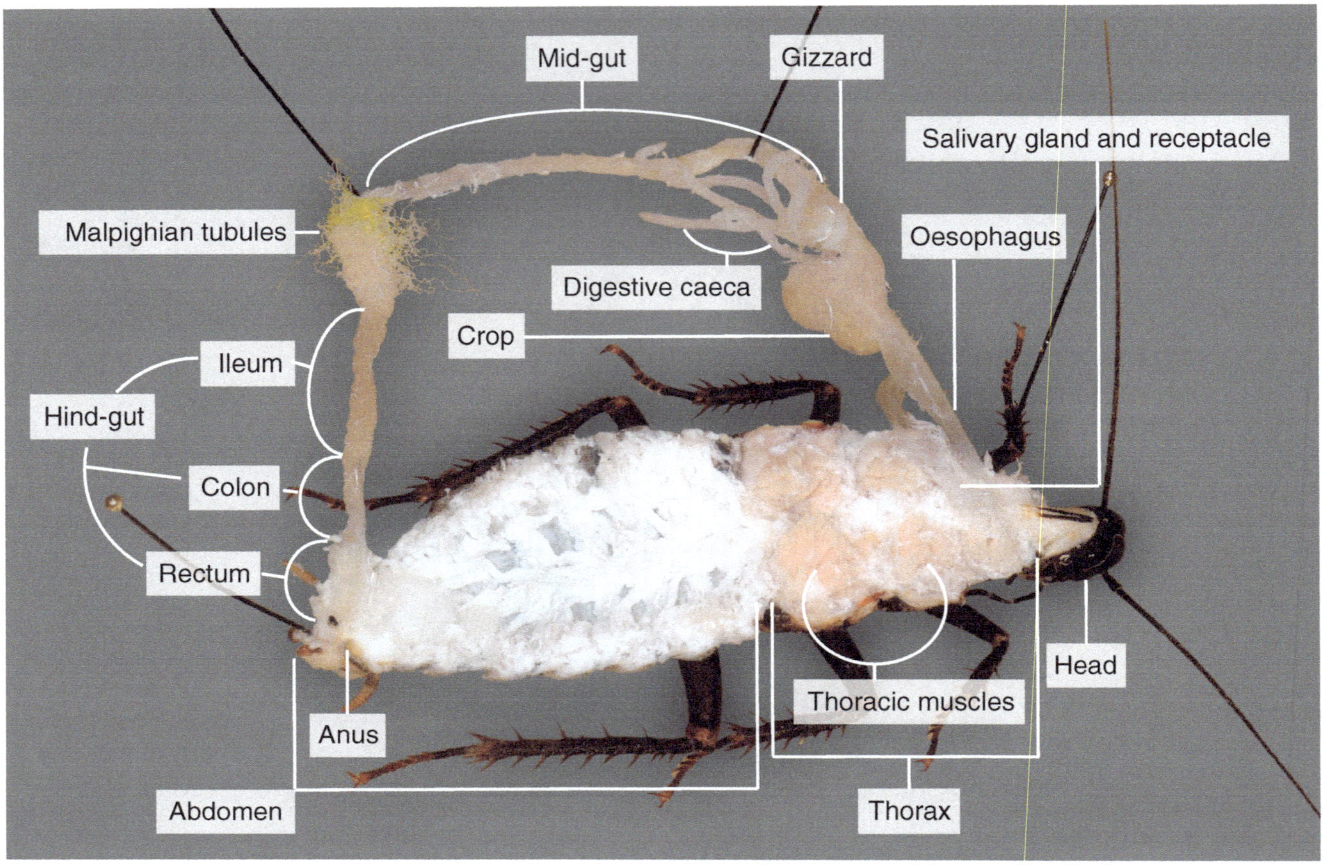

Fig. 8.16 The isolated alimentary canal of a cockroach

Comb through the tissues just behind the head until you see a pair of translucent *salivary glands* and *salivary receptacles* (or reservoirs) lying along each side of the gut (Fig. 8.16). The glands (and associated ducts) form branching, tree-like networks, whereas the receptacles are thin, membranous bladders. The receptacles of either side have a common receptacular duct which opens into the common salivary duct. This common salivary duct opens behind the hypopharynx at the labium. Take a small sample of the salivary gland with a fine forceps, place it on a slide and cover it under a drop of water with a coverslip, then examine it under a microscope (Fig. 8.17).

The salivary glands of the cockroach consist of several lobes of secretory acini and an extensive duct system. The acini are grape-like structures. Tracheae accompany the salivary gland ducts and branch into small tracheoles (Fig. 8.17).

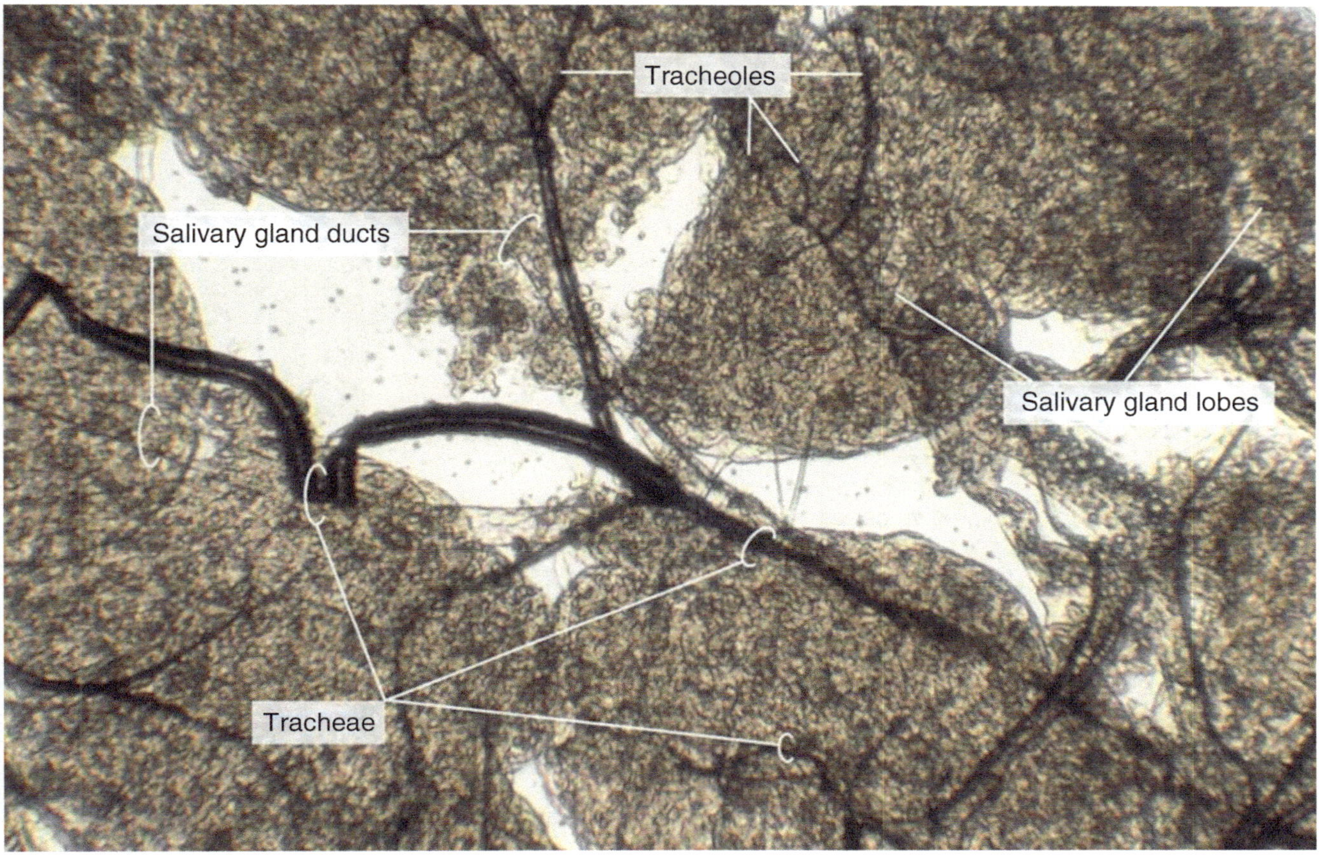

Fig. 8.17 Lobes of the salivary gland of a cockroach as seen in a microscope

The alimentary canal is long and somewhat coiled divisible into three main parts, namely, foregut, mid-gut and hind-gut. Gently pull the foregut away from the body cavity, sever some of the tracheae and membranes that hold it in place, and pin it at an angle to the body (Fig. 8.16). *Foregut* (stomodeum) is differentiated into four parts: pharynx, oesophagus, crop and gizzard. The *oesophagus* is just a short, narrow tube. It opens into a large, brown *crop* that may fill much of the space in the abdomen. The *gizzard* (proventriculus) forms a distinct conical bulge just behind the crop (Fig. 8.18). This region of the foregut has cuticularized walls with heavily sclerotized, black or golden brown teeth.

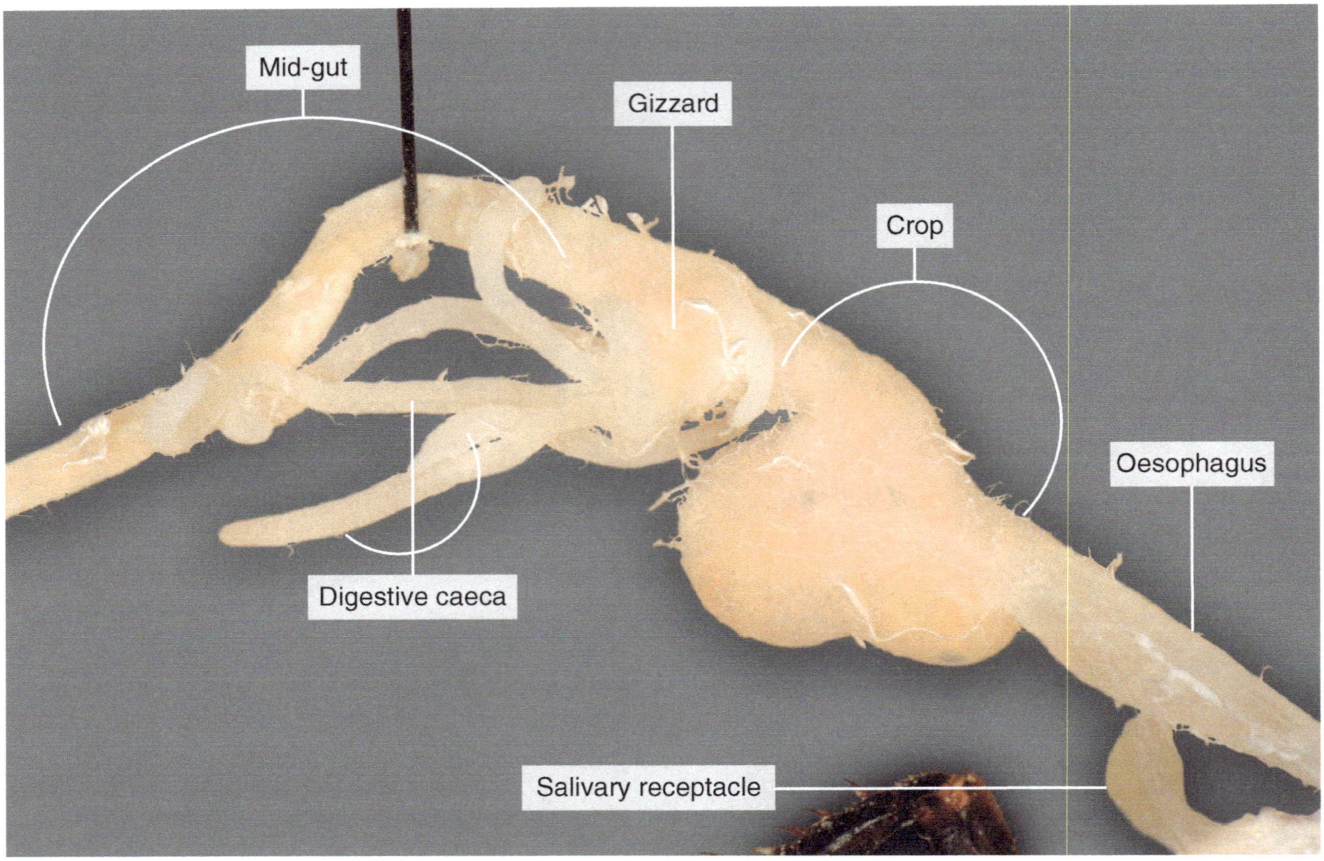

Fig. 8.18 Parts of the foregut (oesophagus, crop and gizzard) and the mid-gut with digestive caeca

Cut out a section from the alimentary canal from the posterior part of the crop to the anterior part of the mid-gut. Use fine scissors to cut open the wall of this tubular structure along its length. Inside the gizzard, there are 12 plates in a circle in two alternating sets adjacent to the crop (Fig. 8.19).

Six are *toothed plates*, each with a small, dark tooth, and six are *ridged plates*, each with several parallel ridges. The *teeth* continue the mechanical breakdown of food particles initiated by the mandibles and maxillae. Six additional ridges are arranged in a second whorl closer to the mid-gut. These form *soft cushions* (pulvilli), bearing short fine setae which act as a filter to exclude large particles from the mid-gut. When food particles are crushed fine enough, *stomodeal valve* opens and they pass into the mid-gut.

For the most part, the *mid-gut* (mesodeum) is just a simple tube, but it is the site of most digestion and absorption of nutrients. The front end of the mid-gut is marked by the *digestive caeca* (singular caecum), 6–8 blind, glandular, fingerlike processes that produce an assortment of enzymes and other secretory products (Figs. 8.18 and 8.19).

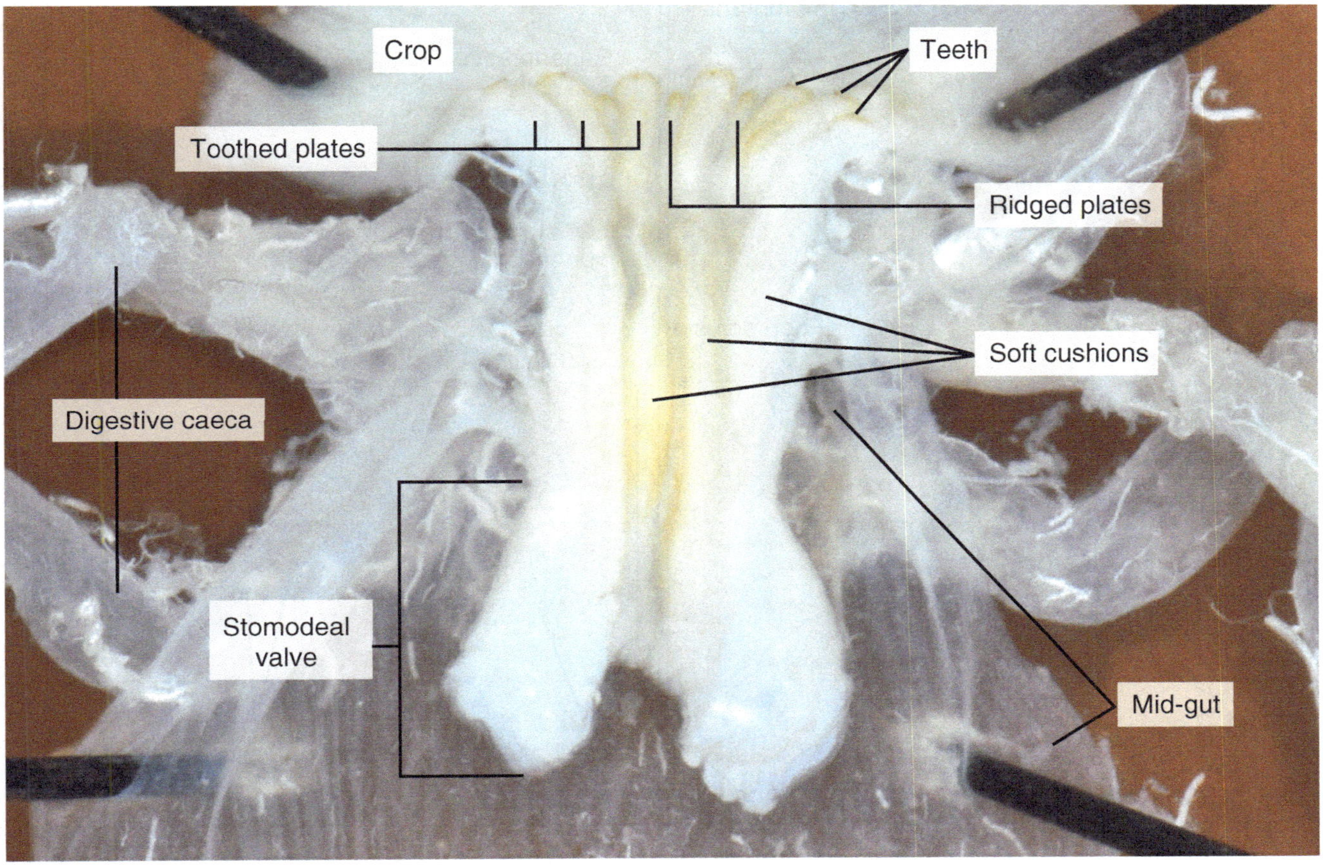

Fig. 8.19 The longitudinally cut gizzard with the chitinous teeth and ridged plates

Mid-gut of cockroach has a thin wall which is formed mainly by an epithelial layer (Fig. 8.20, left). It has a characteristic pattern caused by repetitive occurrence of *regenerative cell* groups. Stem cells divide in the centre of these groups and give rise to cylindrical secretory cells that reach the surface of the epithelium. They produce digestive enzymes and absorb nutrients. The food bolus is bounded by a *peritrophic membrane* formed by fibrillary network of chitinous material. It is permeable for enzymes and nutrients, but stop coarse food particles to cause injury of the epithelium. There are very thin, continuous circular (CM) and reticular longitudinal muscle (LM) layers in the outer wall of the mid-gut (Fig. 8.20, left).

Proximal portion of mid-gut has blunt-ended *digestive caeca*. Their histological composition is similar to that of the mid-gut, but they do not contain food bolus and a peritrophic membrane (Fig. 8.20, right). Some tracheae can be attached to them (Fig. 8.20, right).

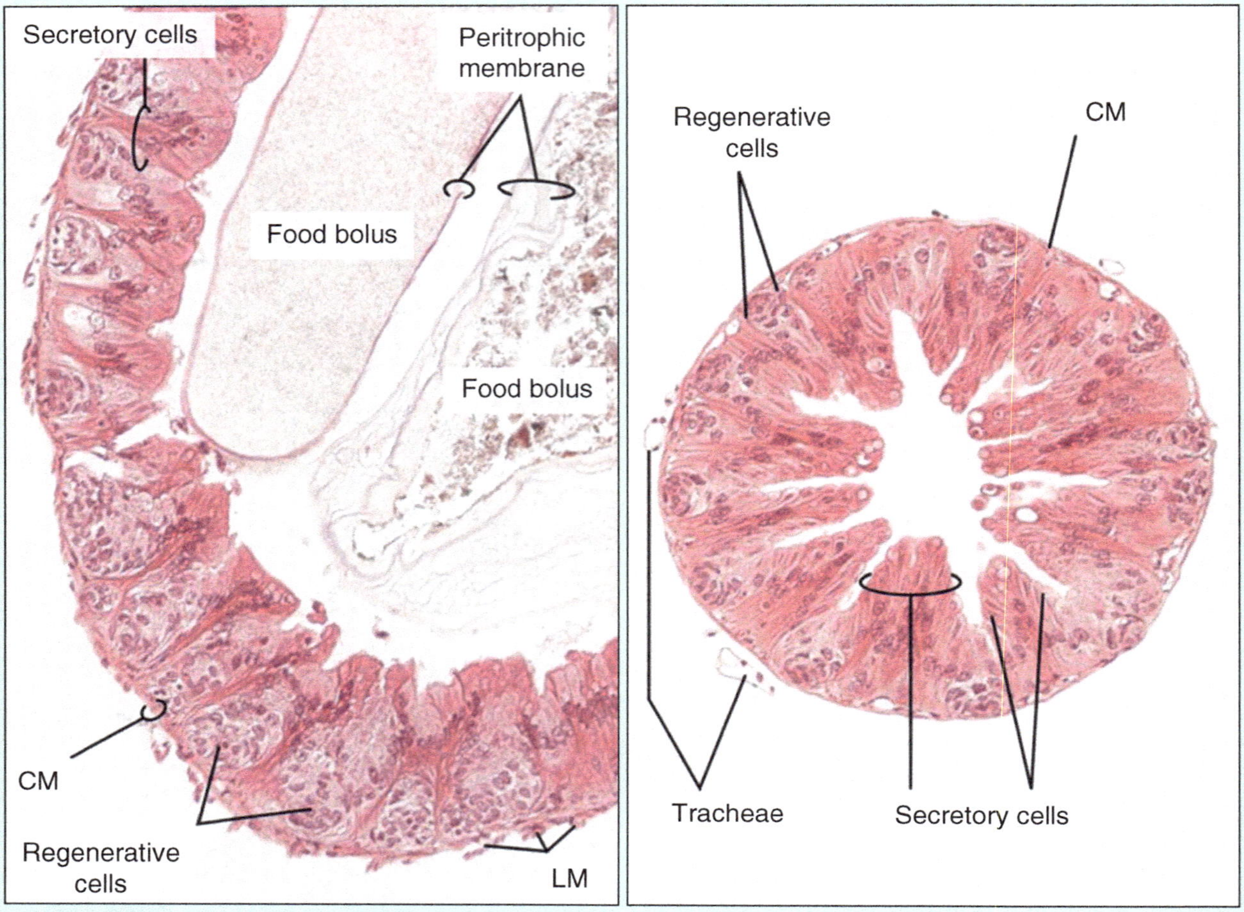

Fig. 8.20 Histological structure of mid-gut and a digestive caecum of cockroach (HE). Note that because of a fold on the peritrophic membrane, there are virtually two separate membranes in the section. *CM* circular muscle layer, *LM* longitudinal muscle layer

The *hind-gut* (proctodeum) is morphologically subdivided into an ileum, a colon and a rectum. The anterior-most *ileum* is large, dark and sculptured. The *colon* is shorter, lighter in colour and thinner. The *rectum* is just a small bulge tucked under the last abdominal tergite, but if you look carefully, you can find the six opaque rectal papillae (pads) that are instrumental in removing most of the water from the faecal pellet (Fig. 8.16).

The respiratory system of insects consists of a network of *tracheae* that open through ten pairs of small holes called *spiracles* present on the lateral side of the body in each of the cockroach's segments, excluding the head (Figs. 8.3, 8.15, 8.22 and 8.23). Air enters the cockroach's body through the spiracles. The opening of the spiracles is regulated by sphincters. Tracheae appear silvery white because of the contained air. The tracheae branch into smaller tubes, called *tracheoles* (Fig. 8.21). The sphincters open when the CO_2 level in the insect rises to a high level; then the CO_2 diffuses out of the tracheae to the outside and fresh O_2 diffuses in. Unlike in vertebrates that depend on blood for transporting O_2 and CO_2, the tracheal system brings the air directly to cells. The tracheal tubes branch continually like a tree until their finest divisions, the tracheoles. The tracheoles are associated with each cell, allowing gaseous O_2 to dissolve in the cytoplasm lying across the fine cuticle lining of the tracheole. CO_2 diffuses out of the cell into the tracheole and leaves the insect's body through the tracheae and spiracles. In the cockroach and other large species, the body musculature may contract rhythmically to forcibly move air out and in the spiracles; this may be considered a form of breathing.

Cut out a small sample of the light brown, thoracic flight muscles with a fine pair of scissors, place it on a slide and cover it under a drop of water with a coverslip. Gently squash the preparation between your thumb and index finger, and then examine it under a microscope (Fig. 8.21).

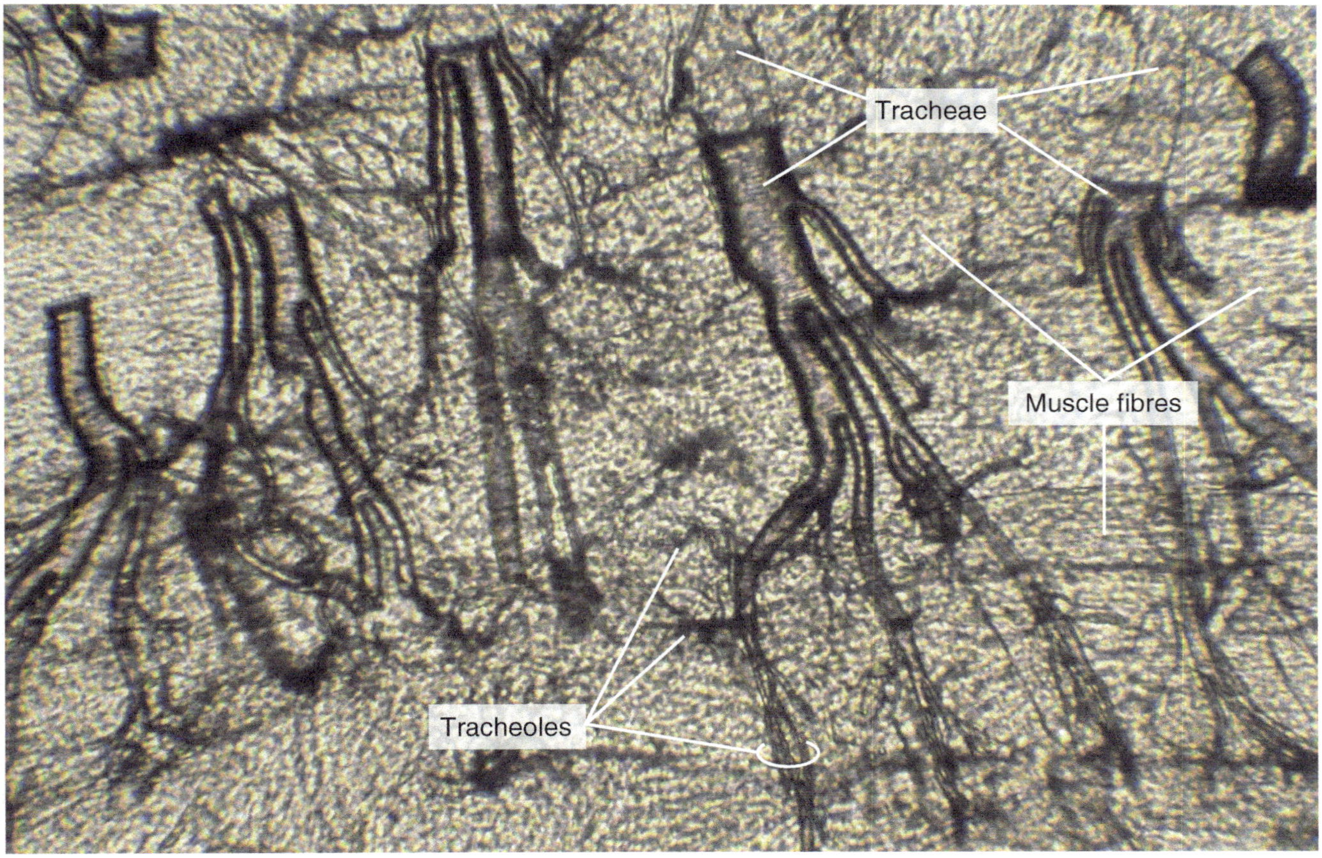

Fig. 8.21 Details of the tracheae and tracheoles supplying the flight muscle of a cockroach as seen in a microscope (tracheae are partly filled up with water and invisible; they are contrasted where they contain air)

Tracheoles, the small tubes that form the terminal endings of the tracheal system, range from 1 to 0.1 μm in diameter. Tracheoles are formed within single tracheolar cells. These tracheolar cells have many branching processes, some of which contain an air-filled channel (the tracheole) that connects to the air-filled lumen of the trachea. Tracheole walls are capable of transporting oxygen at high rates by diffusion because they are extremely thin (usually <0.1 μm) and have a very large surface area-to-volume ratio. Thus, the tracheoles are the major site of gas exchange between the tissues and the tracheal system.

Tracheoles are particularly dense in metabolically active tissues such as flight muscle (Fig. 8.21). Most tracheoles occur outside of the cells in the insect's body, but sometimes in histological sections, they appear to be within cells, particularly in flight muscle. The high tracheolar densities and penetration of flight muscle cells by tracheoles allow flying insects to achieve oxygen consumption rates that are among the highest in the animal kingdom.

Throughout the abdominal cavity, there is a multitude of very thin, spaghettilike structures that are yellow in colour. These are the *Malpighian tubules* (Figs. 8.15 and 8.22).

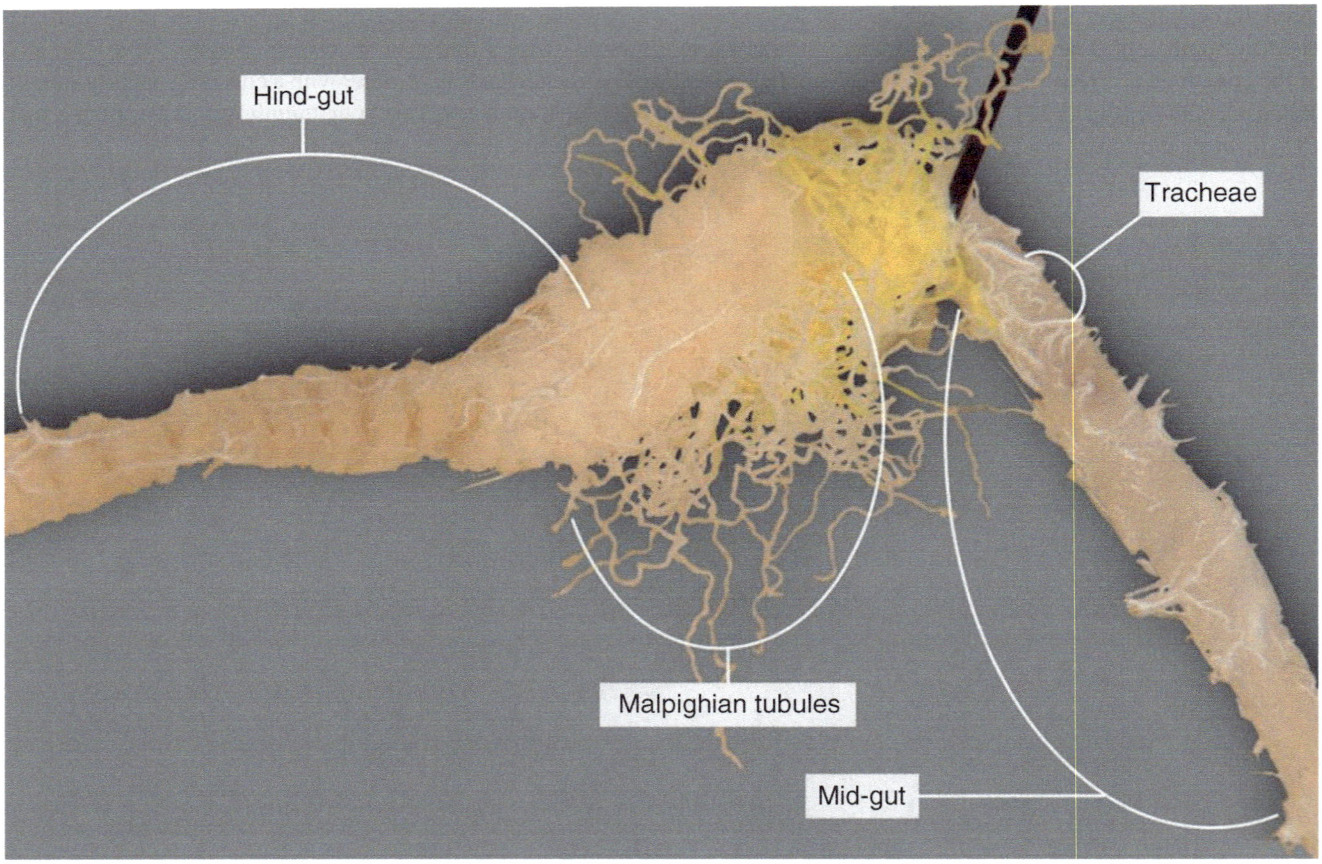

Fig. 8.22 The Malpighian tubules are connected to the digestive tract at the border of the mid-gut and hind-gut

They absorb nitrogenous waste products from the haemolymph and convert them into uric acid. Therefore, cockroach is called uricotelic. These organs also regulate the balance of water and salts in the cockroach's body. The contents of the Malpighian tubules are emptied into the digestive tract at the front end of the hind-gut. In the rectum, the last part of the hind-gut, water is reabsorbed from the faeces and urine. The remaining wastes leave the body through the anus, which is the exit of the digestive and the excretory systems in insects. In addition, the fat body and *nephrocytes* (groups of special excretory cells around the pericardium and under the oesophagus) also help in excretion.

As it was stated earlier, cockroaches are dioecious, so sexes are separate. The male cockroach's internal reproductive organs include a pair of *testes* (Fig. 8.23). Sperm is produced in each testis. Ducts (*vas deferens*) lead from the testes out of the male cockroach's body to deliver sperm to the female cockroach. The mushroom-shaped *accessory gland* is a white ball of tubules near the base of the last abdominal tergite (Fig. 8.23).

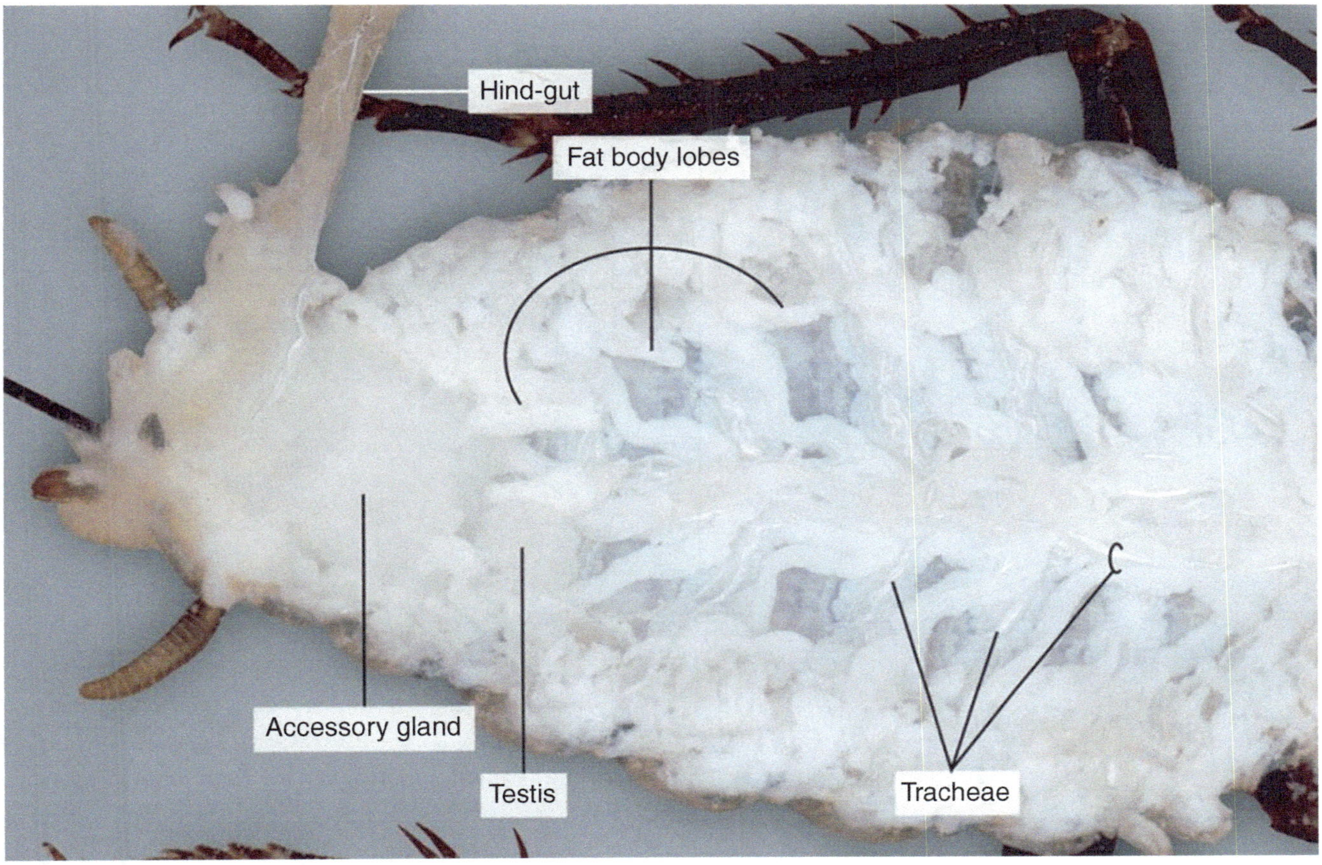

Fig. 8.23 Reproductive organs of a male cockroach (only one testis is visible)

Accessory gland consists of two types of tubules, the long slender tubules are the utriculi majores of peripheral tubules, and short tubules, the utriculi breviores, making up of the major part of the gland. Small seminal vesicles are also found associated with accessory gland. These accessory glands surround and obscure the testes and seminal vesicles. An ejaculatory duct emerges posteriorly from the accessory glands. All sperms of a seminal vesicle are glued together by the phallic gland into a large bundle called spermatophore. External genitalia are quite complex (Fig. 8.12). The opening of the ejaculatory duct, the gonopore, is on the dorsal surface of the ventral phallomere.

The female cockroach's internal reproductive organs consist of a pair of *ovaries* in which its eggs are formed and the oviduct that leads from the ovaries to the outside of the insect. Next to the duct is a pouch in which the male's sperms are stored. The sperms fertilise the eggs. Each ovary contains eight *ovarioles* and each ovariole contains a linear array of about a dozen eggs with the most mature egg nearest the base (Fig. 8.24).

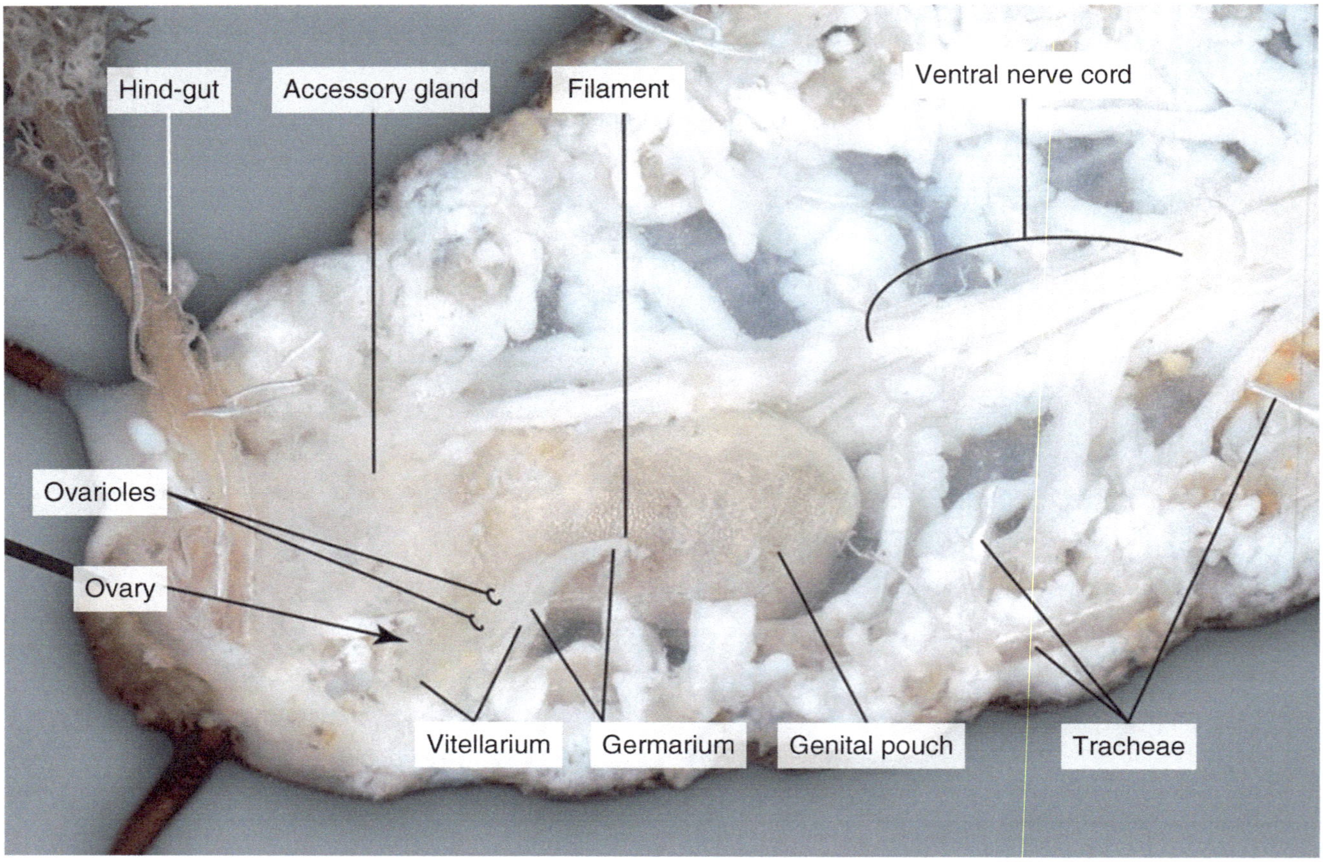

Fig. 8.24 Reproductive organs of a female cockroach (only a part of one ovary is visible)

An ovariole is a connective tissue tube beginning with a terminal *filament*. Ovarioles can be divided into two regions: an anterior germarium and a posterior vitellarium. The *germarium* contains the oogonia; it is the site of differentiation of oogonia into primary oocytes. Yolk deposition occurs in the *vitellarium* as the maturing egg cell proceeds posteriorly. A translucent tube, the *oviduct*, emerges from the base of each ovary. The two lateral oviducts merge to form a common oviduct, but this junction as well as the bursa copulatrix (or vagina) is often hard to see because they lie hidden under a part of the ventral nerve cord. The vagina opens into a genital chamber. The mass of female *accessory glands* is easily mistaken for the fat body, but these glands are more spaghettilike in shape and pure white in colour (Fig. 8.24). If you look carefully, you may be able to find the spermatheca buried amid the accessory glands. It is small, club shaped, with a dark centre that encases a remnant of the male's spermatophore. The *genital pouch* (gynatrium) is divisible into a genital chamber in front and oothecal chamber behind (Figs. 8.24 and 8.25).

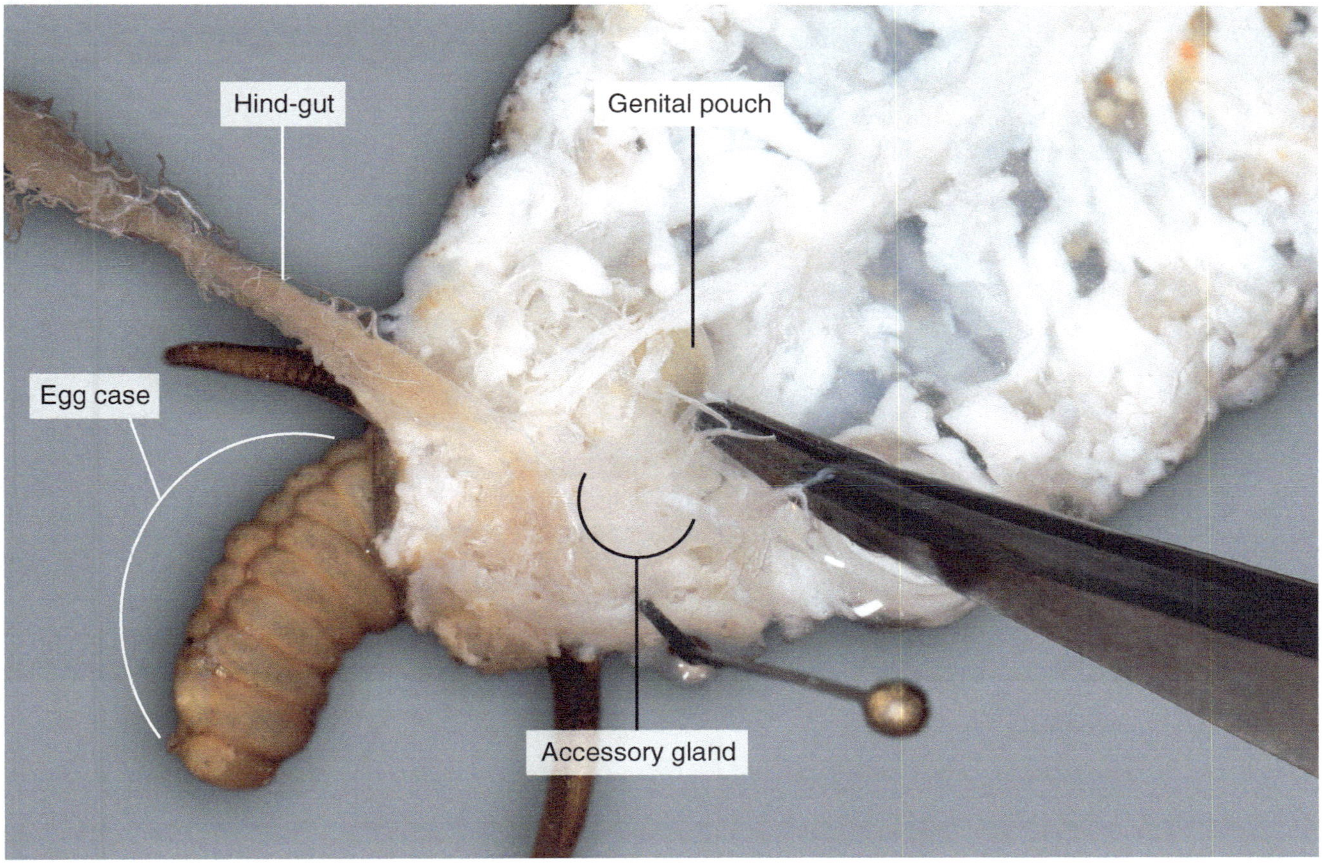

Fig. 8.25 Egg case is half way pressed out of the genital pouch of a female cockroach

A pair of collateral glands also opens into the genital chamber. *Egg case* (ootheca) of cockroach contains 32 fertilised eggs (Fig. 8.26).

The egg case is formed of a protein secreted by collateral glands. It protects the developing fertilised eggs. The developing larva gets oxygen through the opening of the egg shell as it is otherwise impermeable for gases (Fig. 8.26). This opening is the site of hatching as well. The egg case can be seen protruding from a female's rear end before it is deposited on the ground outside the female's body (Fig. 8.25).

Fig. 8.26 Egg case of a female cockroach

Fat body (FB) is a lobular tissue, which occupies most of the body cavity (Fig. 8.27, left). Its main cell type is the adipocyte, which stores energy by forming glycogen and lipid droplets in its cytoplasm. Several organs are embedded among fat body lobes, such as the Malpighian tubules and the reproductive organs. *Malpighian tubules* (MTs) are appendages of the gut at the boundary of mid-gut and hind-gut. They are formed by a simple cuboidal epithelium lying on a thin basement membrane (Fig. 8.27, left and small micrograph in the middle). Well-developed basal labyrinth and brush border are characteristic to tubular cells which indicate their excretory function. Emerging data prove that tubules work not only like a kidney but also similar to the vertebrate liver, and they represent an autonomous immune system as well.

Tracheal system is a complex network of tubules making up the respiratory system of insects. The system develops as an invagination of the epidermis, so the whole network is lined with a cuticular layer. Cuticle forms spiral reinforcement (*taenidia*) on the surface to prevent the tubules from collapse (Fig. 8.27, left). The narrowest elements of tracheal system (*tracheoles*) reach every organ and tissue and provide respiratory gases for gaseous exchange of the cells (Figs. 8.20, right; 8.21 and 8.27, left).

Cockroach's *testis* is a drop-like organ composed of tube-like follicles running parallel to each other. Free spermatogonia are found in their proximal ends. Developing spermatocytes form cell groups in which sister cells are interconnected by cytoplasmic bridges – this connection is responsible for synchronous development as a clone. Every cell group is enclosed by enveloping cells to form a cyst (Fig. 8.27, right). Spermatids are gathered in bundles while they elongate, and their heads are oriented towards an enveloping cell of the cyst. Morphologically mature spermatozoa degrade their cytoplasmic bridges, escape from the cyst and leave the testis.

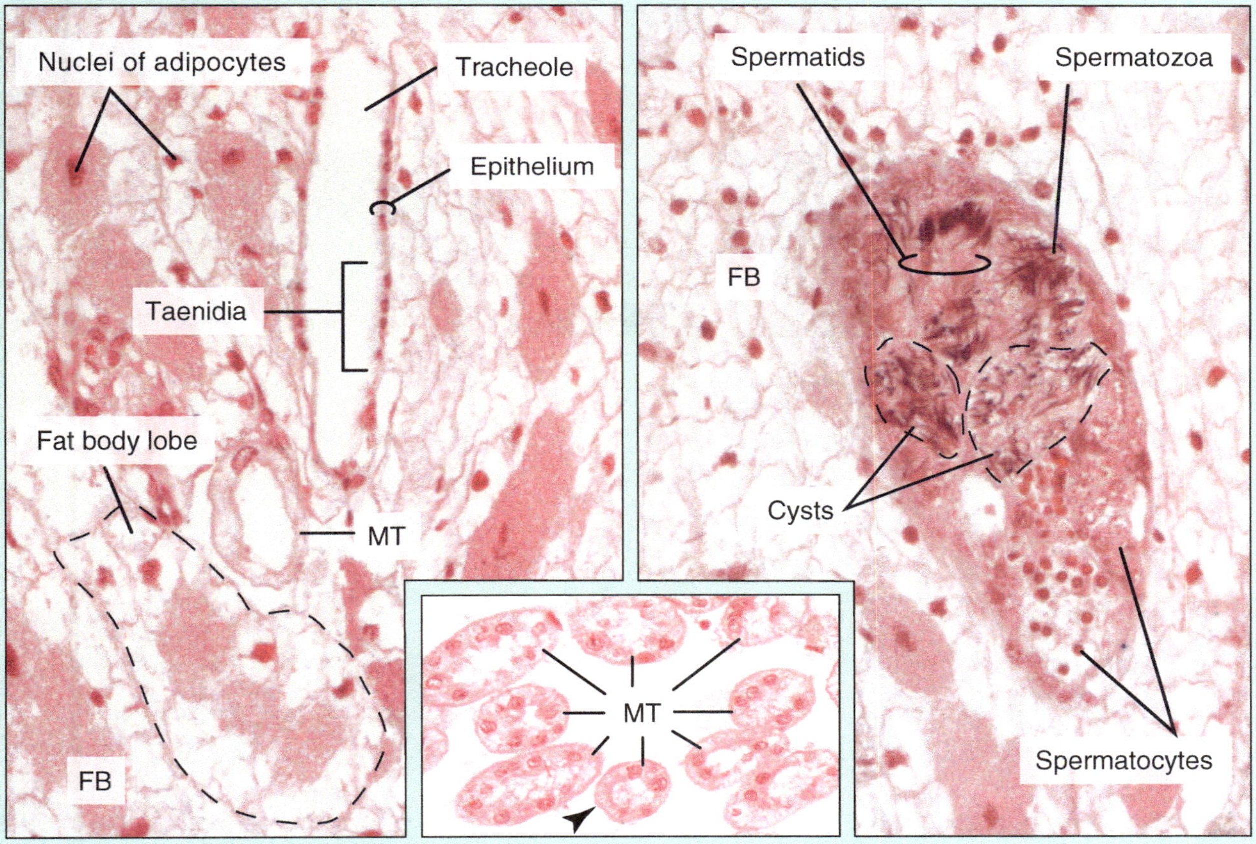

Fig. 8.27 Histological appearance of three internal organs of cockroach: fat body lobes, Malpighian tubules and testis (HE). *Arrowhead* well-developed basal labyrinth in the epithelium of Malpighian tubule, *FB* fat body, *MT* Malpighian tubules

When all the digestive and reproductive organs have been removed from the abdomen, you will be able to find the most visible part of the nervous system of the cockroach, the *ventral nerve cord* lying along the length of the ventral side (Figs. 8.28 and 8.29).

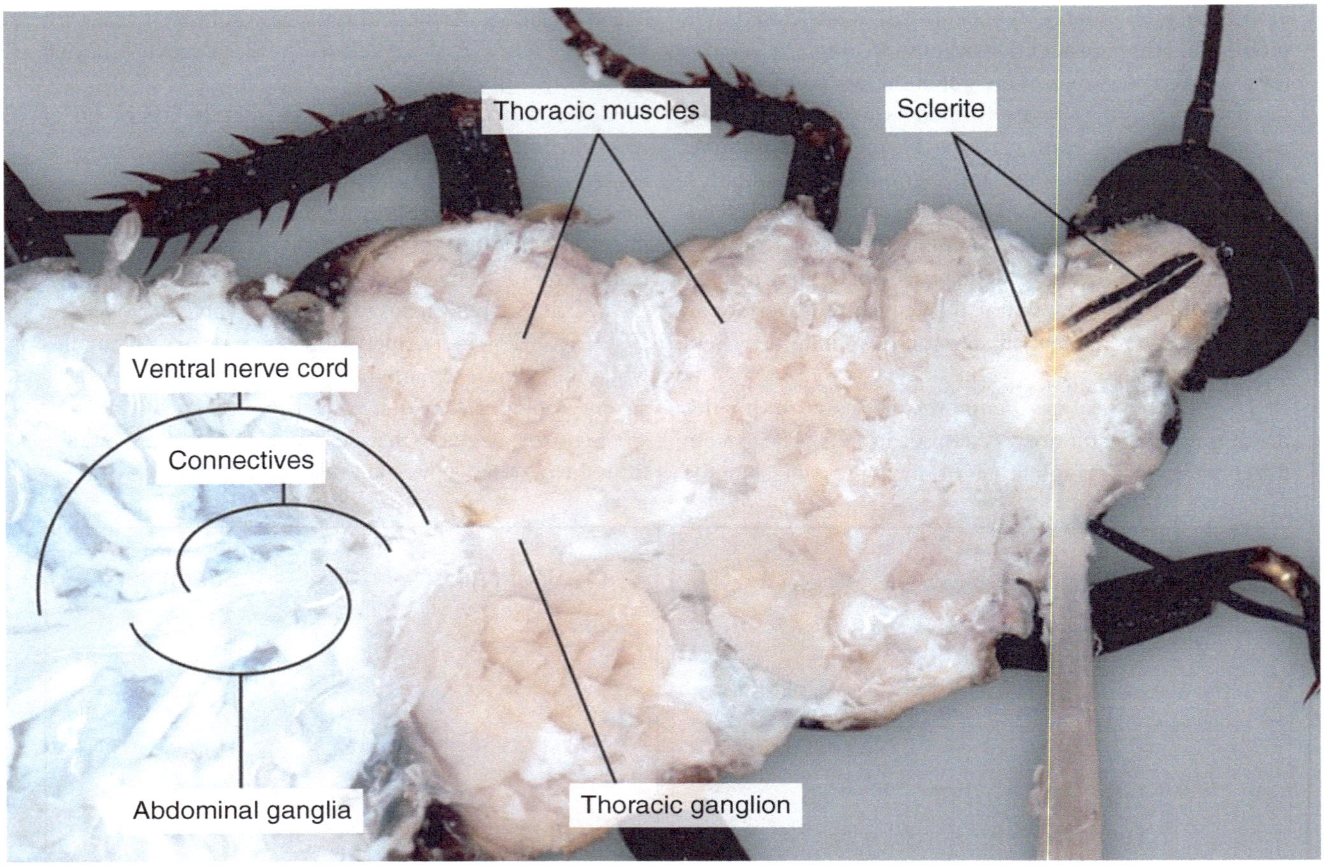

Fig. 8.28 The anterior part of the ventral nerve cord of a cockroach

The nerve cord is shiny and white and consists of two parallel strands, running close together alongside the ventral body wall. The nerve strands join together at each body segment to form lumps called *ganglia*. The ganglia are joined to one another longitudinally by intersegmental *connectives* (Figs. 8.28 and 8.29). Use the point of an insect pin to separate the two parallel nerves within each connective. Small peripheral nerves radiate out laterally from each ganglion to innervate adjacent parts of the body, the muscles and the sense organs.

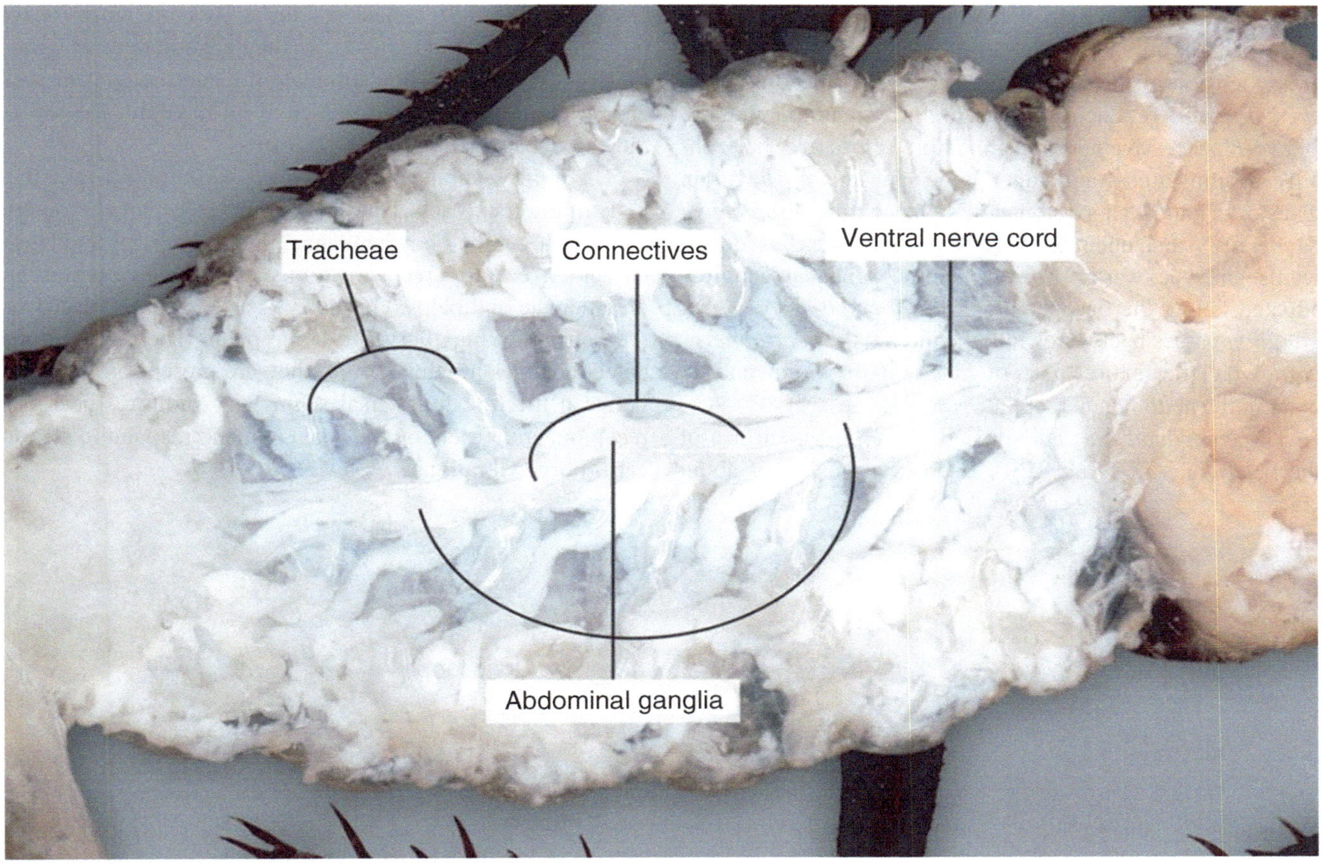

Fig. 8.29 The posterior part of the ventral nerve cord of a cockroach

The exposure of the brain should only be attempted by those who have considerable experience in dissection or are above average in manual dexterity. Use a fine pair of scissors to cut away the dorsal part of the head capsule and parts of the eyes. Scrape aside the blocks of muscle in order to expose the ganglia (Fig. 8.30).

The central nervous system includes the *brain* (or supra-oesophageal ganglion) which has three parts, protocerebrum, deuterocerebrum and tritocerebrum. The *optic lobes* are lateral extension of the protocerebrum towards the compound eyes (Fig. 8.30). Each consists of three neuropil masses. Between successive neuropils, the fibres cross over horizontally forming outer and inner optic chiasmata. The brain gives off a pair of short, stout cords, the *circum-oesophageal connectives* that encircle the oesophagus and pass downwards and backwards over the sub-oesophageal ganglion situated below the oesophagus. From the *sub-oesophageal ganglion* passes backwards into the thorax, the *ventral nerve cord*, which bears three pairs of ganglia in the thorax and six pairs in the abdomen (Fig. 8.29).

The *compound eyes* are situated at the dorsal surface of the head (Figs. 8.9 and 8.30). Each eye consists of about 2000 hexagonal *ommatidia* (singular ommatidium). With the help of several ommatidia, a cockroach can receive several images of an object. This kind of vision is known as mosaic vision with more sensitivity but less resolution, being common during night (hence called nocturnal vision).

In addition to the pair of compound eyes, cockroach possesses two single-lens simple eyes, known as *ocelli* (Fig. 8.9). Ocelli are generally considered to be far more sensitive to light than the compound eyes, so they are suitable for light-metering functions. They stimulate the nervous system and enable the insect to judge the length of daylight. However, they are incapable of perceiving form.

There are several kinds of *sensory hairs* on all over the body of the cockroach. Thermoreceptors (receptors of touch) are present on the body, antennae, maxillary palps and legs. Olfactory receptors receive various smells; they are present on antennae and palps. Gustatory (sense of taste) receptors are present on maxillary and labial palps. Thermoreceptors detect changes in temperature; they are present on the pads between the first four tarsal segments. Auditory (hearing) receptors are present on the anal cerci and respond to air- or earth-borne vibrations.

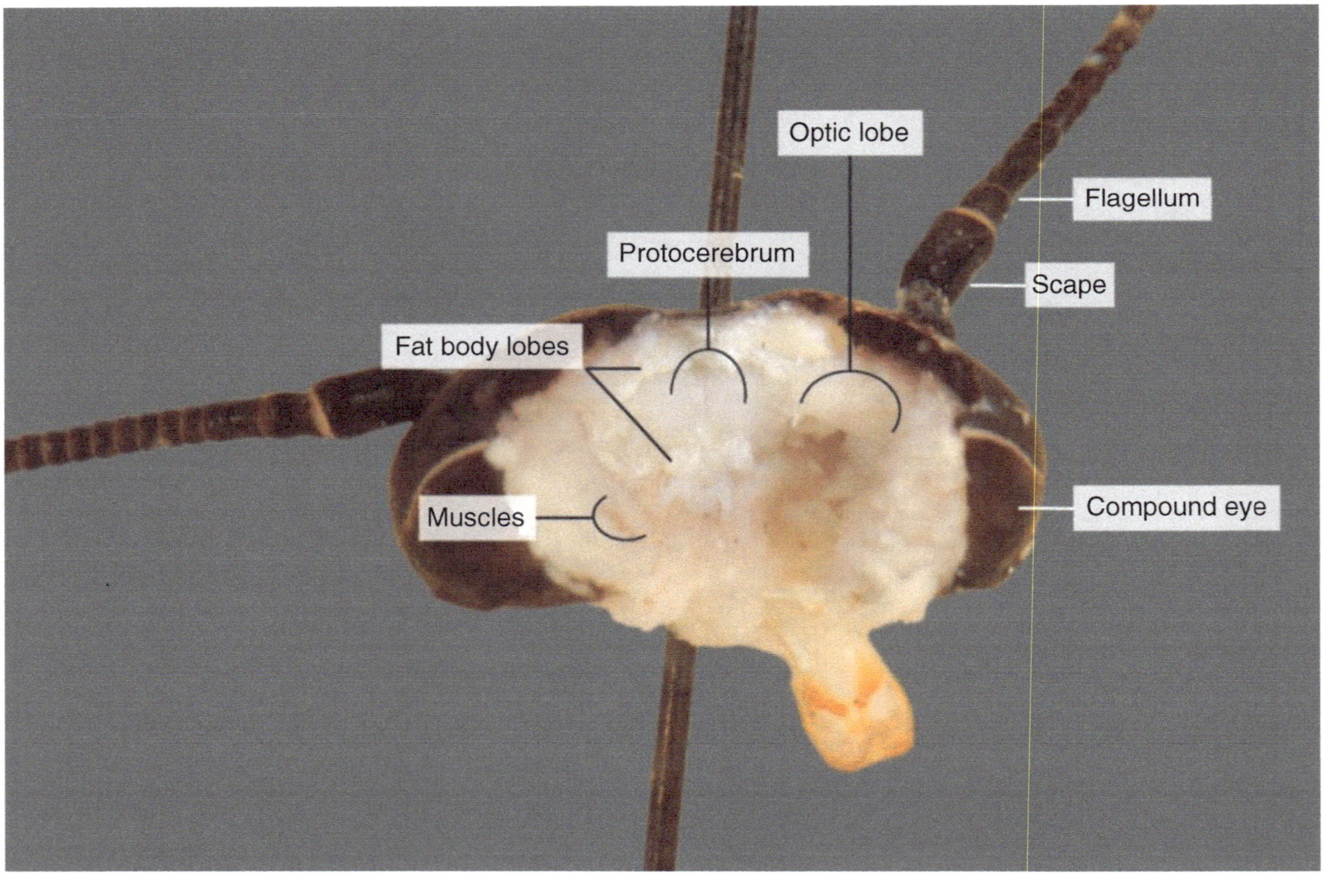

Fig. 8.30 The dorsally open head case of a cockroach with the anterior part (protocerebrum) of the brain

Central nervous system of cockroach is formed by paired cerebral ganglia, circum-oesophageal connectives, sub-oesophageal ganglia and ventral nerve cord. The *brain* is situated in the head capsule. It has own connective tissue capsule, and there are fat body lobes and haemolymph spaces around it. Cerebral ganglia have some structures very characteristic to insects, such as mushroom bodies and antennal lobes (Fig. 8.31). *Mushroom body* is formed by small neurons, often referred to as Kenyon cells. Kenyon cell bodies are situated near the surface of the brain (Fig. 8.31, top left and right). Their dendrites receive information (definitely from the antennal lobe) at a specialised dendritic zone (neuropil) called the *calyx*, found just below the mushroom body cell bodies. The axons of Kenyon cells are arranged parallel to each other in a tightly packed tract called the *peduncle*. Leaving the peduncle, these axons form three lobes (only one of them is visible in the sections). Mushroom body has been shown to be involved in a variety of complex neural functions, such as sensory processing, motor control, multimodal integration, memory and learning, sleep and decision-making. *Antennal lobe* keeps contact with the antenna bearing olfactory receptors (Fig. 8.31, bottom left). There is a correlation between the developmental stage of antennal lobes and mushroom body calyces. Both depend on the importance of the olfactory input.

The anterior midline of the brain, also known as the *pars intercerebralis*, contains the largest neurosecretory cells in the central nervous system (Fig. 8.31, top right). These cells are involved in control of reproductive diapause, cuticular tanning, sugar metabolism and diuresis.

Compound eye consists of individual receptor units (ommatidia). Optical elements of an ommatidium are the corneal lens on the surface, and the crystalline cone just beneath the lens. Layer of lenses of the compound eye is continuous with the cuticle layer of the head capsule. There is a cluster of sensory cells under the cone in every ommatidium. The cluster is surrounded by pigment cells which contain dark pigment to isolate ommatidia from one another. Receptor cells are dark as well, because of the synthesised photopigment for light detection (Fig. 8.31, bottom right).

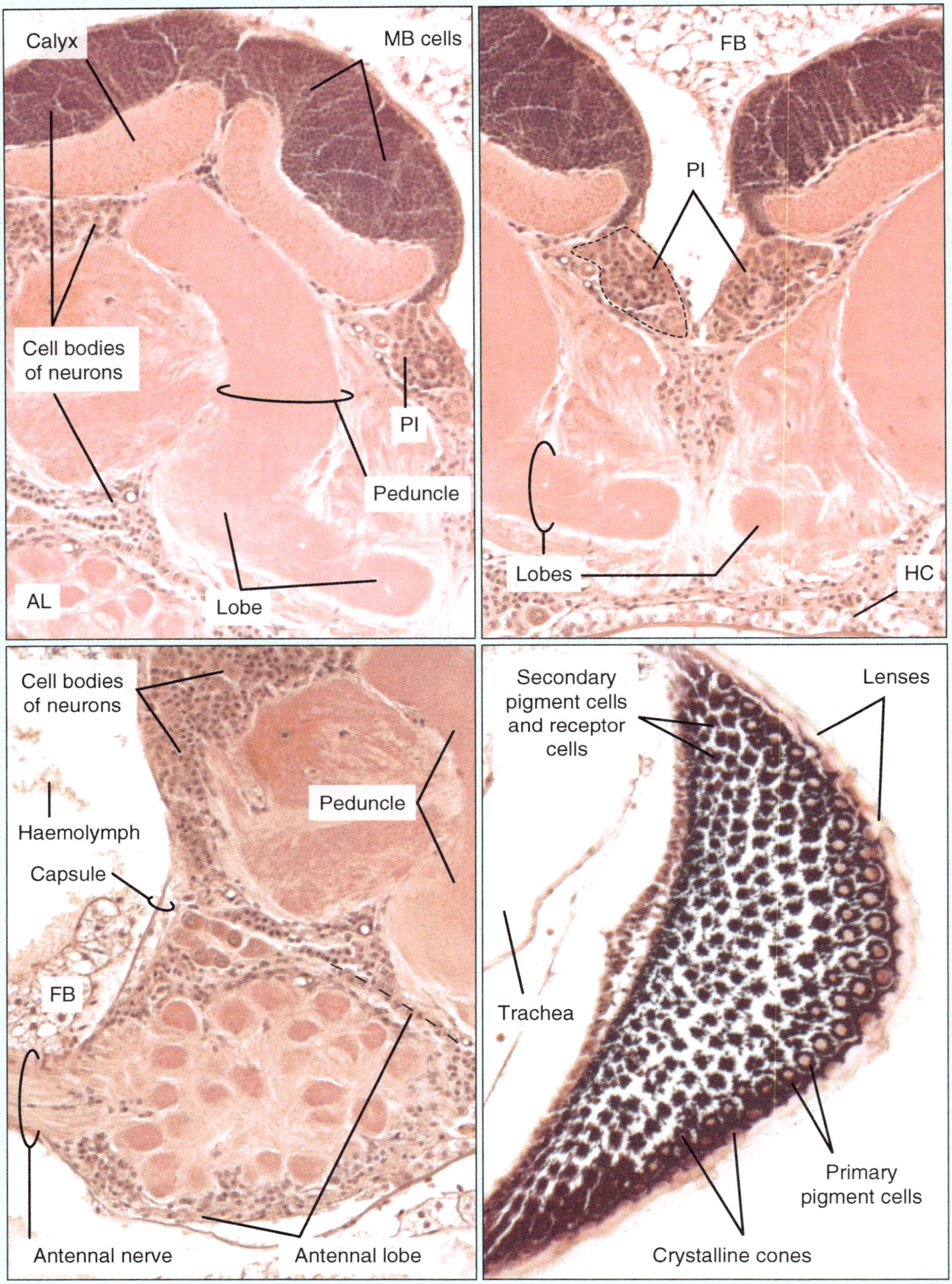

Fig. 8.31 Histological section of the brain and compound eye of a cockroach (HE). *AL* antennal lobe, *FB* fat body, *HC* haemocytes, *MB cells* mushroom body cells or Kenyon cells, *PI* pars intercerebralis (encircled by *dotted line top right*)

Part II

Vertebrates

Dissection of the Crucian (*Carassius carassius*)

9

- *Availability*: The taxon of bony fishes (*Osteichthyes*) represents the largest group of vertebrates both in number of species (more than 20,000) and in number of individuals. The crucian is a common freshwater fish widely distributed through Europe from England to Russia and it is a heavily farmed fish worldwide. Fresh specimens can be purchased from local fish markets.
- *Anaesthesia*: Hold down the fish wrapped in a towel with one hand leaving the head accessible. Wearing examination gloves, dip two cotton balls in 20 % chloral hydrate with a pair of forceps, drain them and then place one under each operculum. Hold the mouth and operculum of the fish while it becomes motionless then put it into the dissecting dish. The crucian sleeps over, becomes tranquil permanently within approximately 40–60 min. Anaesthesia is successful if the fish doesn't react when its nostrils are touched with a pair of forceps. Before starting the dissection, take out the cotton wool with a pair of forceps.

P. Lőw et al., *Atlas of Animal Anatomy and Histology*, DOI 10.1007/978-3-319-25172-1_9

Skeleton: The bony skeleton of the crucian consists of the *axial skeleton* (skull, vertebral column, ribs and medial fins) and the *appendicular skeleton* (pectoral girdle and fins, pelvic girdle and fins) (Figs. 9.1 and 9.2).

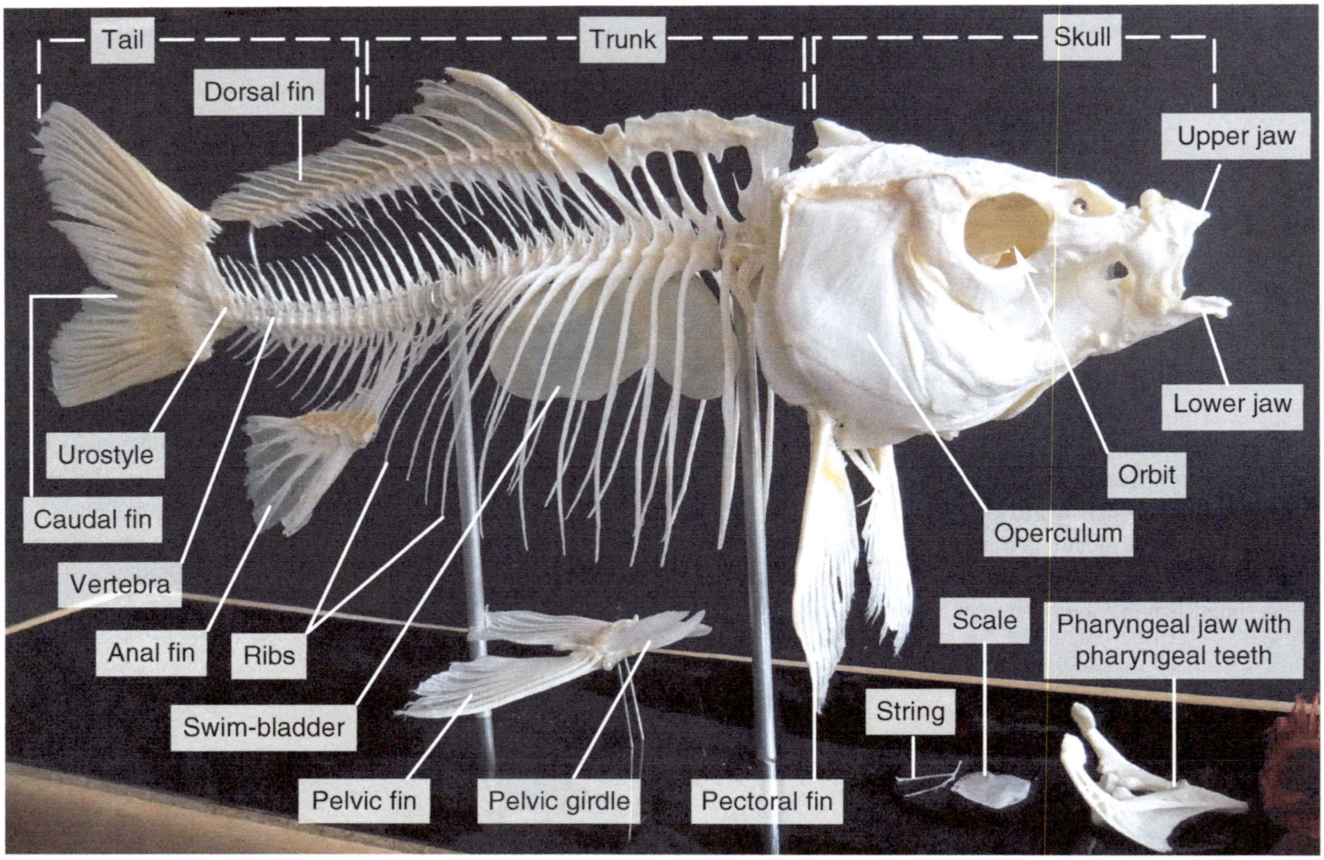

Fig. 9.1 Mounted carp skeleton. Carp is closely related to crucian, they both belong to the same subfamily (Cyprininae) of the carp family (Cyprinidae). Note pelvic girdle and pharyngeal jaw are not in their original position

In crucian, like in most fish, the vertebral column does not bear the weight of the body but functions as a flexible support for muscles. Consequently, the connections between the vertebrae need not be as strong as those seen in tetrapods. The characteristic ball-and-socket-type joints between tetrapod vertebrae are absent. The *amphicoelous* vertebrae of fish make edge-to-edge contact. Bony fish possess ventral ribs. The *pectoral girdle* (cleithrum, scapula, coracoid plate) is connected firmly to the bones at the back of the skull by the supratemporal bone. This is quite different from the situation in land vertebrates, where the head is not connected to the pectoral girdle. The *pelvic girdle* is a single triangular plate, each half of which is embedded in connective tissue with no bony connection with the vertebral column (Fig. 9.1). Movement of the tail of the fish is the main source of propulsion during locomotion.

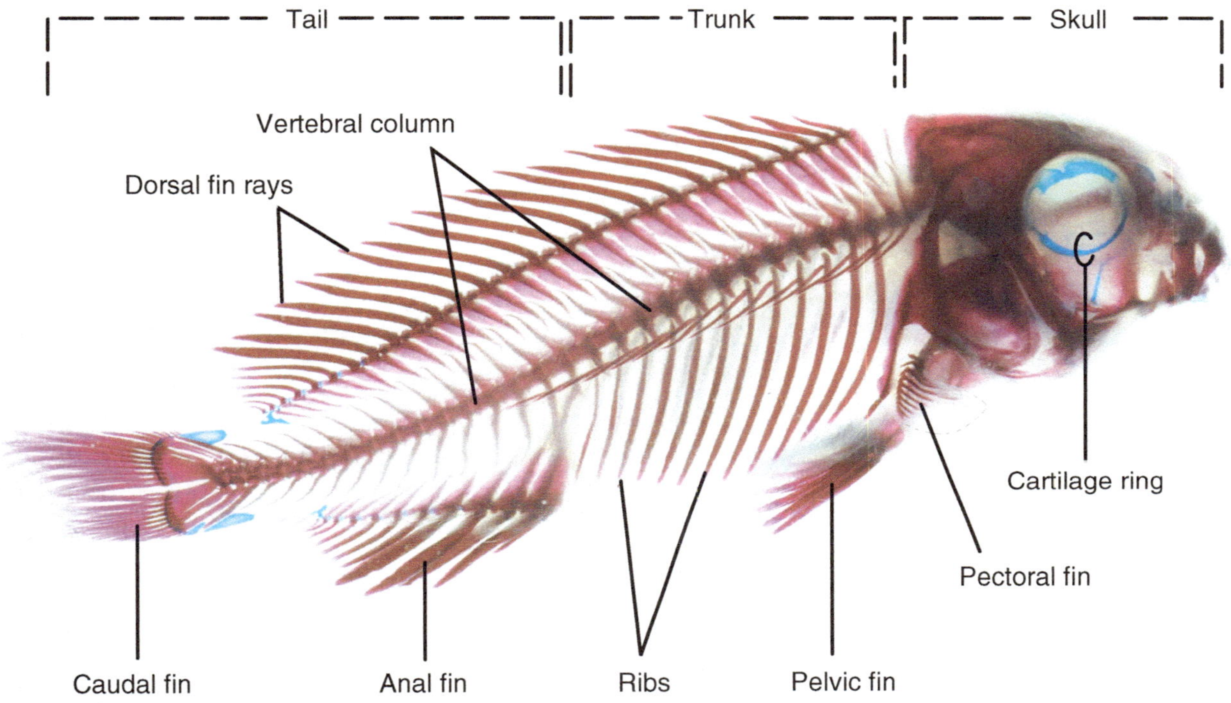

Fig. 9.2 Skeleton of a bony fish. This specimen has been skinned, eviscerated and treated chemically to clear the muscles and stained. Cartilages are blue, as they bind the dye alcian blue selectively, bones are red, because they bind alizarin red selectively. (Courtesy of György Csikós)

Wash the mucous from the lifeless fish then study its external anatomy (Fig. 9.3). The body of the crucian is torpedo shaped or fusiform which offers little resistance to movement in water. The *head* extends to the posterior edge of the operculum; the *trunk* extends to the anus. The third body part is the *tail* with a homocercal caudal fin meaning that it is externally symmetrical, but internally the upper and lower halves are slightly different because the end of the vertebral column, the urostyle, tilts upwards (Figs. 9.1 and 9.3).

Identify the median fins (anal, dorsal and caudal) and paired fins (pectoral and pelvic). Note the fin rays which support the thin membrane of each fin. These fins are used in steering and stabilising the body (Figs. 9.1, 9.2 and 9.3).

Fig. 9.3 External anatomy of the crucian

The terminal *mouth* is adapted for catching prey while swimming. The mouth of crucian is facing downwards (inferior mouth) as it is a bottom feeder. Observe the rudimentary, inflexible tongue in the bottom of the mouth cavity supported by the hyoid bone (Fig. 9.4).

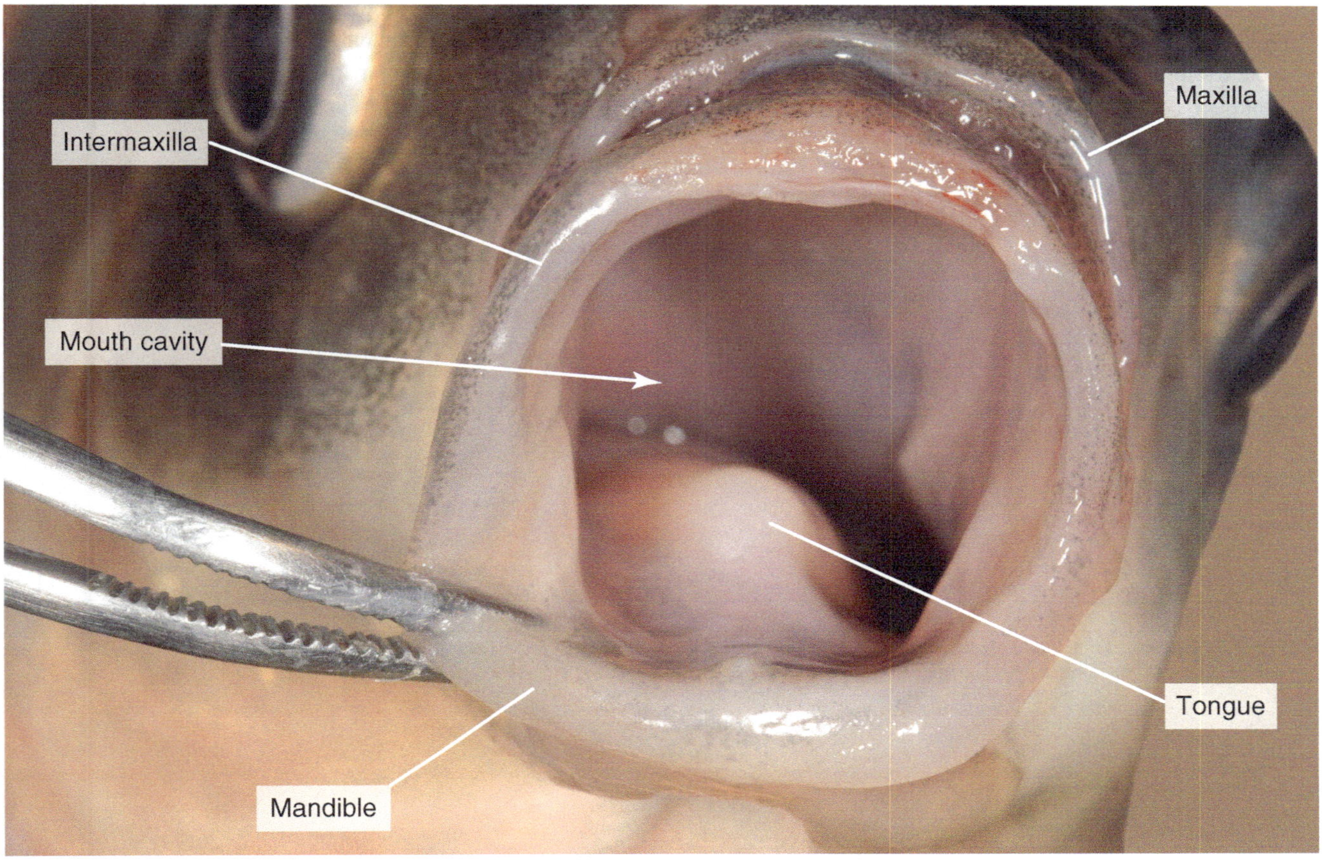

Fig. 9.4 Inflexible tongue in the bottom of the mouth cavity

The *epidermis* of fish is non-keratinized, stratified squamous epithelium containing several gland cells (Fig. 9.5). The *basal layer* of the epidermis (BL) proliferates and produces cells moving upwards and form the *prickle cell layer* (PCL) and *superficial layer* (SL). Some of them differentiate as PAS-positive *mucous cells* (MC), which secrete slippery coating onto the surface. The other type of gland cells is the PAS-negative *alarm cell* (AC); this cell has round shape and a central nucleus. The alarming material is released in case of injury. The epidermis contains barrel-shaped *taste buds* near by the mouth (Fig. 9.5, left).

The mucosa of the *oral cavity* has similar tissue composition to that of the epidermis. Its epithelium is non-keratinized, stratified squamous epithelium with mucous *goblet cells* (GC) (the alarm cells are missing). Several *taste buds* can be seen (Fig. 9.5, right).

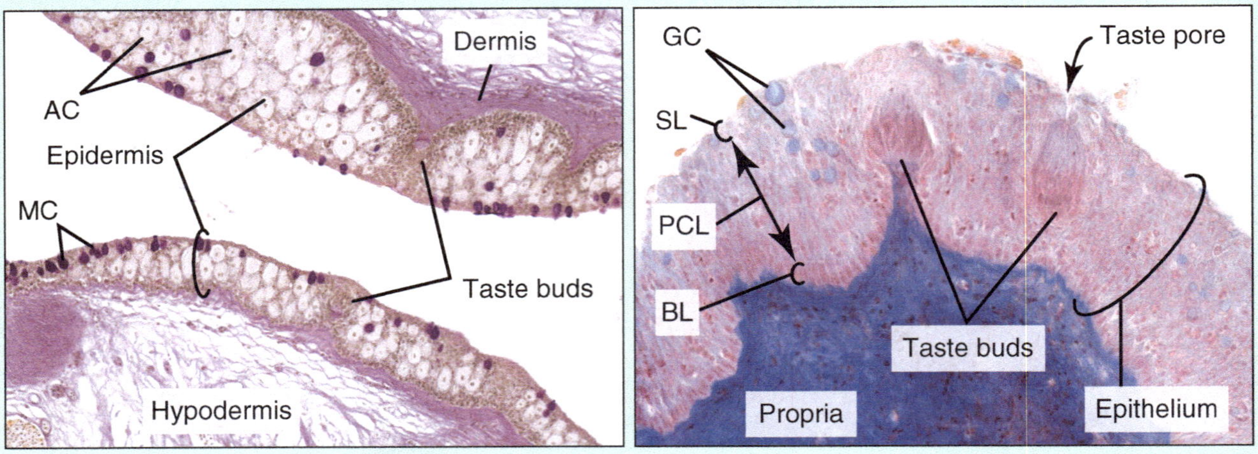

Fig. 9.5 Histological sections of the scale-free fish skin (*left*, PAS) and oral mucosa (*right*, Azan). *AC* alarm cells, *BL* basal layer, *GC* goblet cells, *MC* mucous cells, *PCL* prickle cell layer, *SL* superficial layer

Chromatophores are pigment-containing and light-reflecting cells, which are largely responsible for generating skin colour in fishes. The most common type of chromatophores, *melanophores* contain melanin that appears black or dark-brown because of its light-absorbing qualities (Fig. 9.6).

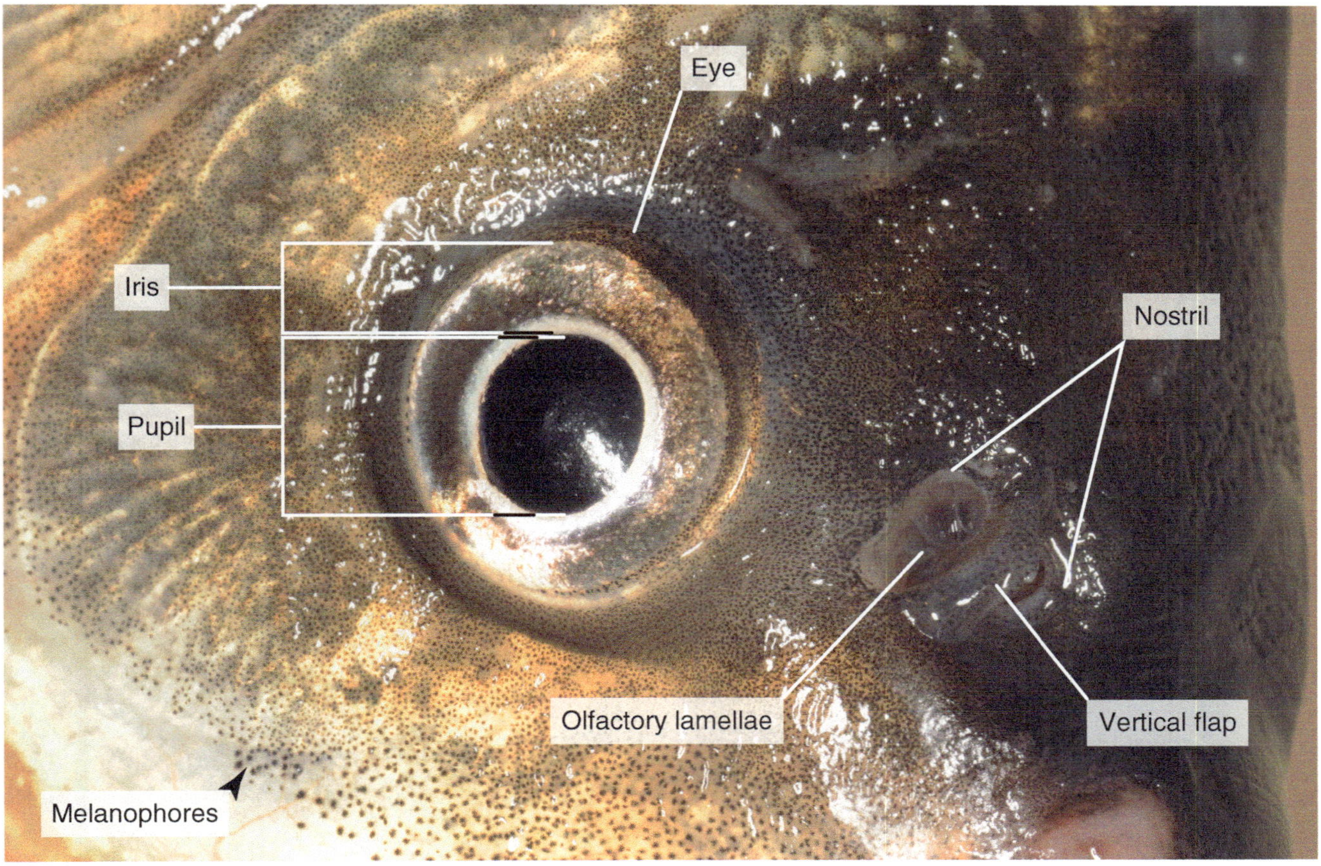

Fig. 9.6 Sensory organs on the head: eye and nostril

Most fish undergo a limited physiological colour change in response to a change in environment. This type of camouflage is known as background adaptation. When the pigment is dispersed in flat dermal melanophores throughout the cell, the skin appears dark. When the pigment is aggregated towards the centre of the cell, the skin gets lighter.

Another type of chromatophores is the silvery *iridophores* (guanophores), which participates to render a shiny appearance to the fish. They are pigment cells that reflect light using plates of crystalline guanine.

On each side in front of the eyes, a pair of *nostrils* open into an olfactory sac (Fig. 9.6). Water enters the sac through the anterior aperture, which is provided with a flap-like valve, and leaves through the posterior aperture. Lift the *operculum* and study the gills beneath (Fig. 9.7).

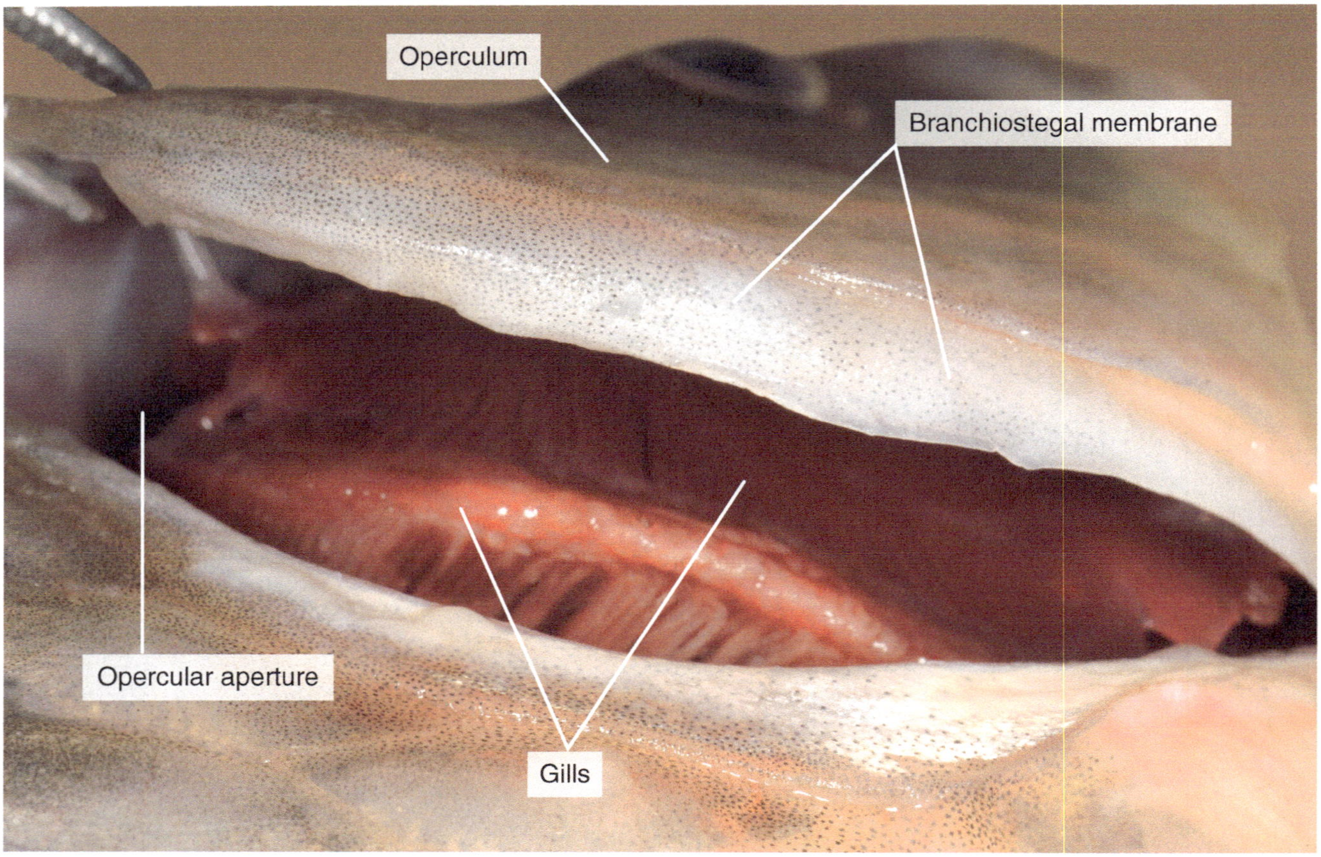

Fig. 9.7 The gills are visible under the lifted operculum

Along the posterior margin of the operculum, find the *branchiostegal membrane*. This membrane fits snugly against the body to close the opercular aperture during the first part of inhalation (Figs. 9.7 and 9.8).

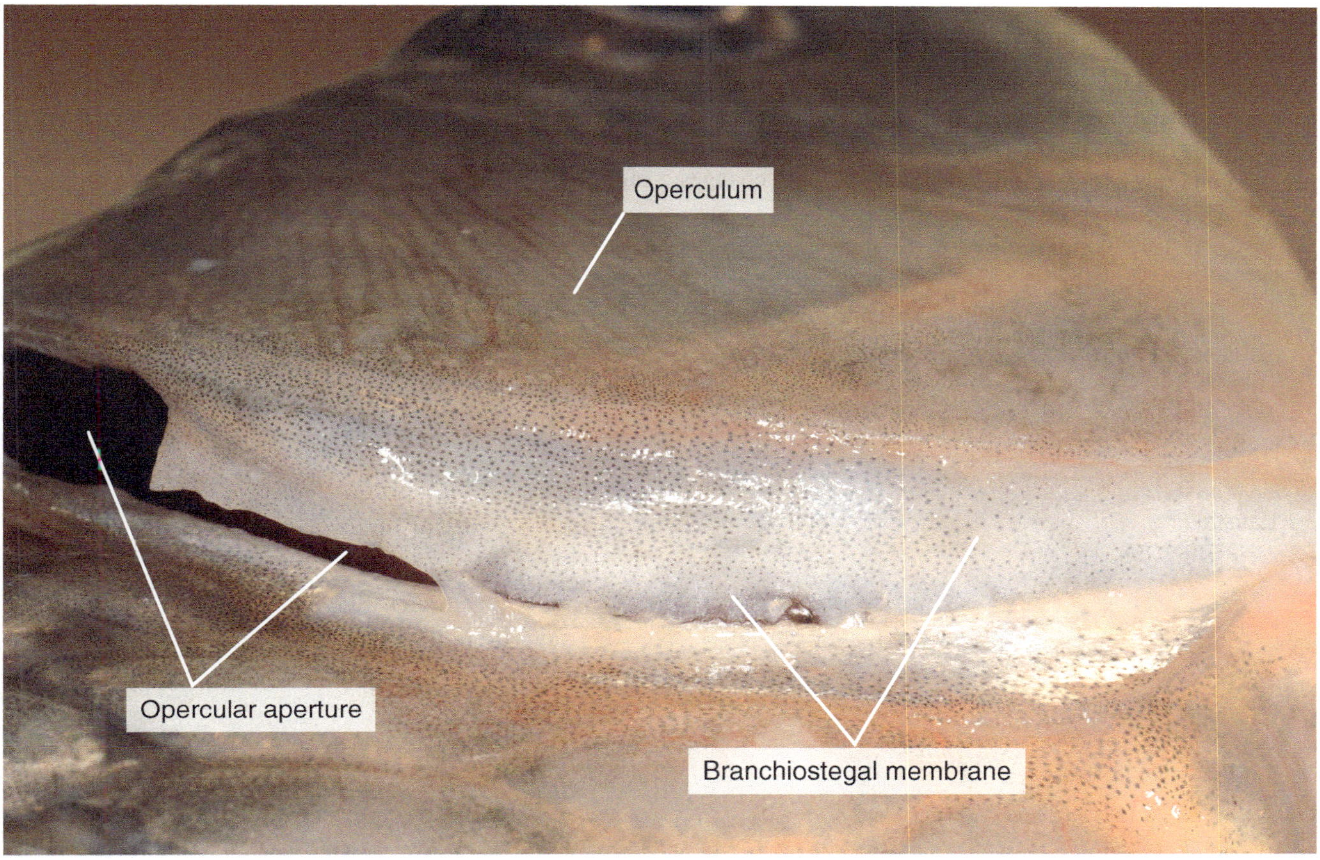

Fig. 9.8 Branchiostegal membrane closes the opercular aperture during the first part of inhalation

The *lateral line* along the side of the body is a row of external openings (small pores) connecting with a long tubular canal bearing sensory organs (Fig. 9.9).

The lateral line is made up of a series of mechanoreceptors, called *neuromasts*, arranged in an interconnected network of fluid-filled canals. These canals are placed just underneath the skin, they open to the environment through several pores and only the receptor portion of each neuromast extends into the canal. The lateral line serves to detect direction and rate of water movement and pressure changes in the surrounding water. The fish can gain a sense of its own movement, that of nearby predators or prey, and even the water displacement of stationary objects.

Note the arrangement of the *scales* (Fig. 9.9). Remove a scale with a pair of forceps from the lateral line region, cover with water in a Petri dish and examine with a dissecting microscope. These are cycloid scales. Note the fine, concentric lines of growth. The scales are covered with a very thin epidermis. In the underlying dermis, there are capillaries and pigment cells (melanocytes) (Fig. 9.9).

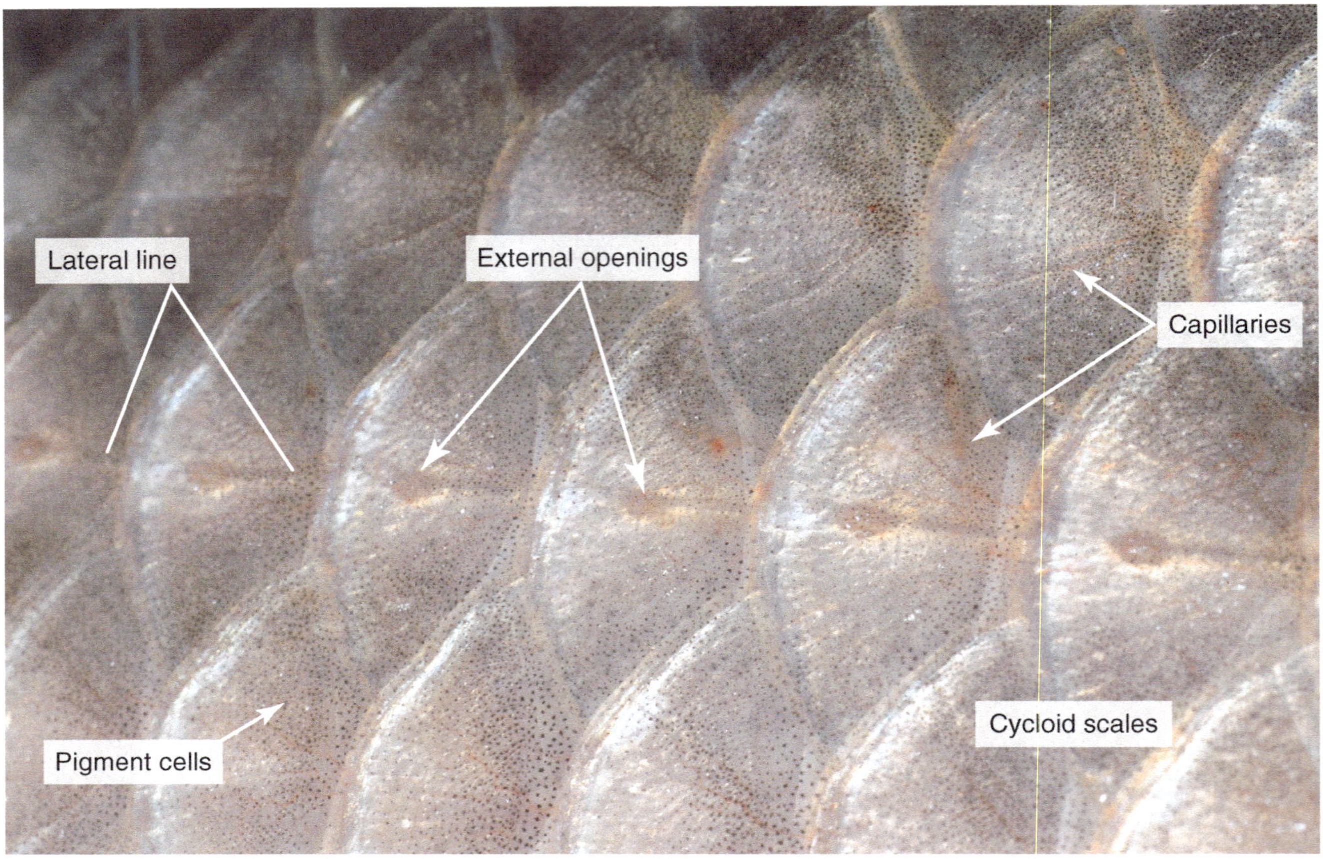

Fig. 9.9 The bony cycloid scales of crucian sit in the dermis covered with a very thin epidermis

Find the *anus* near the base of the anal fin and the small, slitlike *urogenital opening* just posterior to the anus (Fig. 9.10).

Stand the crucian on its back and hold it with one hand and with a paper towel. Starting near the anus and being careful not to damage the internal organs, cut anteriorly on the midventral line up to the *transverse septum* just posterior to the pectoral fins (Figs. 9.10, dashed arrows; and 9.11).

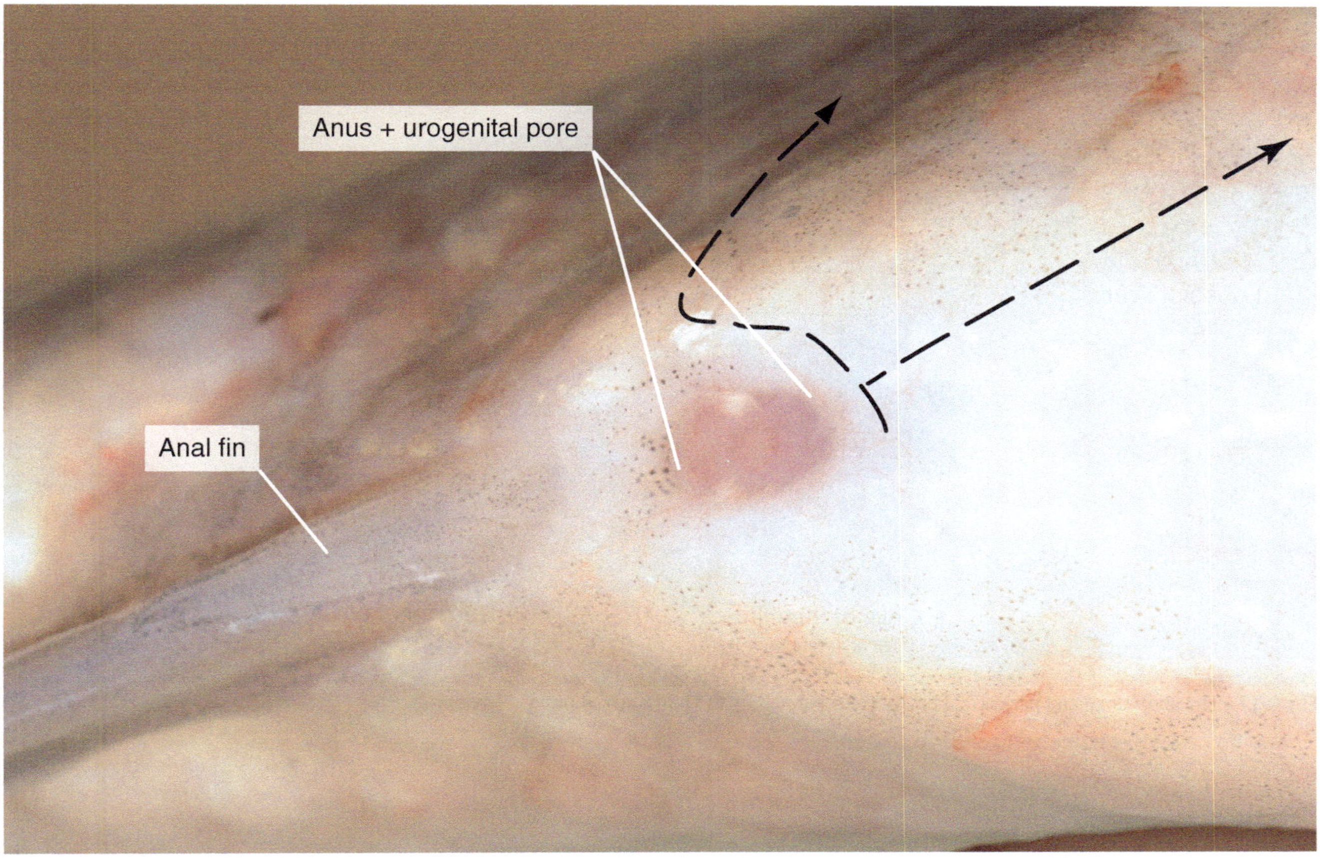

Fig. 9.10 The urogenital opening and the anus can be found near the base of the anal fin. *Dashed arrows* indicate the cutlines to open the body cavity

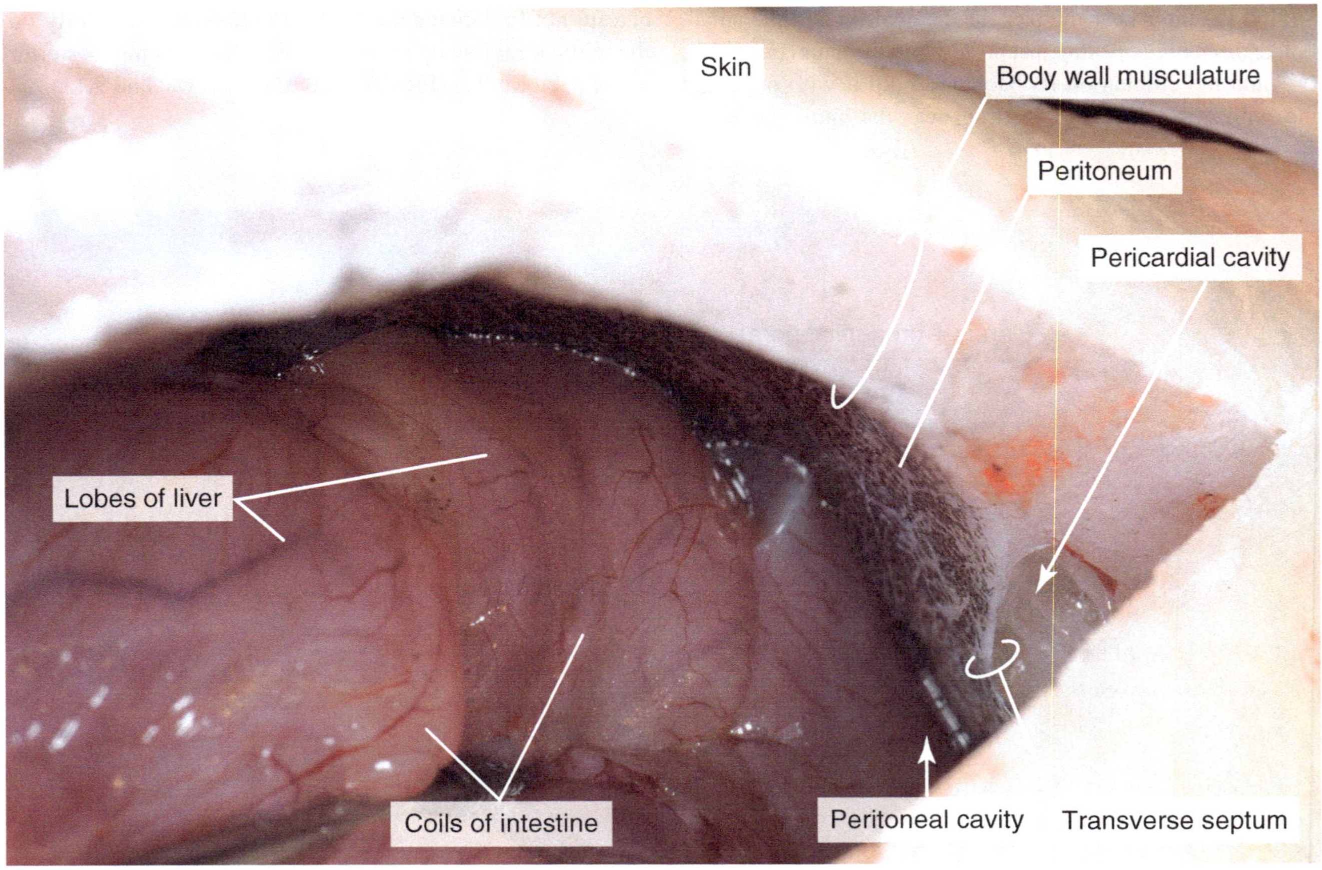

Fig. 9.11 The abdominal and pericardial cavity are separated by the transverse septum

For that, you have to cut the pelvic girdle between the pelvic fins. Now on the fish's right body wall, make a transverse cut, extending dorsally from the anal region. Make another cut dorsally parallel with the transverse septum. Then remove the right body wall by cutting the ribs between these two incisions. Be cautious to fold and not to injure the lateral lobe of the kidney between the two parts of the swim bladder (Fig. 9.12).

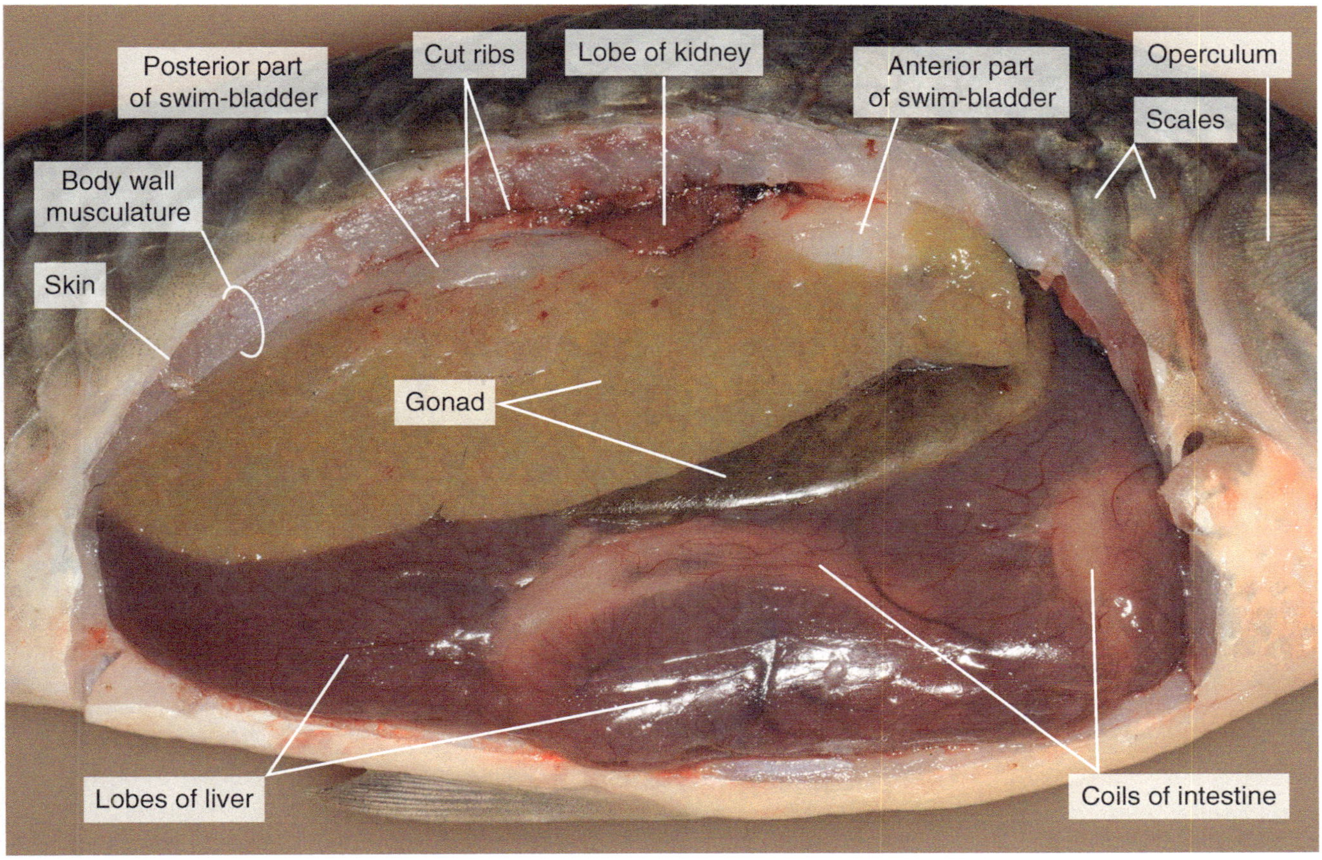

Fig. 9.12 Internal organs of the crucian after removal of the right body wall

You have now exposed the abdominal cavity. This together with the pericardial cavity, which contains the heart, makes up the *coelomic cavity*. Note the shiny lining of *peritoneum*. It contains melanophores and iridophores (Fig. 9.11). Observe the gonad, coils of the intestine, lobes of the liver, the kidney and the two parts of the swim bladder in their original position (Fig. 9.12). Now cut away the operculum from the right side exposing the *gill chamber* (branchial cavity) with the four gill arches (Fig. 9.13).

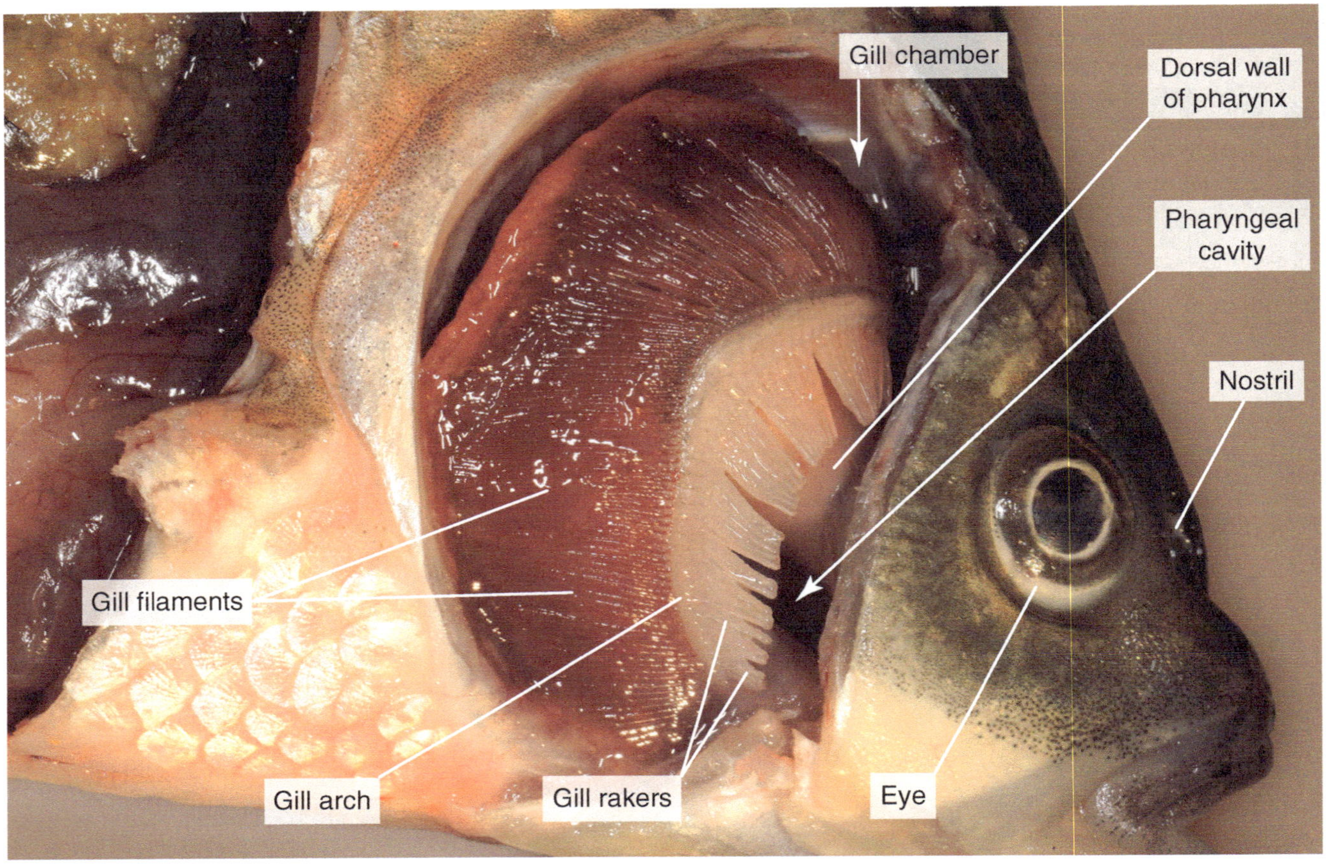

Fig. 9.13 The right gill chamber (branchial cavity) with gill arches after removal of the right operculum

Note the double row of *gill filaments* borne on the posterior side of the arch. *Gill rakers* on the opposite, anterior side of each gill arch strain out food organisms and offer some protection to the gill filaments which form any contamination passing through from the pharynx with the water (Fig. 9.14).

Cut a few filaments from an arch, place in water and examine with a hand lens or dissecting microscope. Exchange of gases takes place on the surface of gill lamellae which contain capillaries from the branchial arteries. Explore the spacious pharynx, noting the size and arrangement of the gill arches and gill slits. The oxygen-containing water is pulled through the gills by a continuous pumping action of the respiratory movements. The mouth opens, the operculum closes, the pharynx expands and the water is drawn in. Then the moth closes; the operculum elevates, this enlarges the branchial cavity; the pharynx constricts and the water is forced out over the gills through the operculum (Figs. 9.4, 9.7, 9.8 and 9.14).

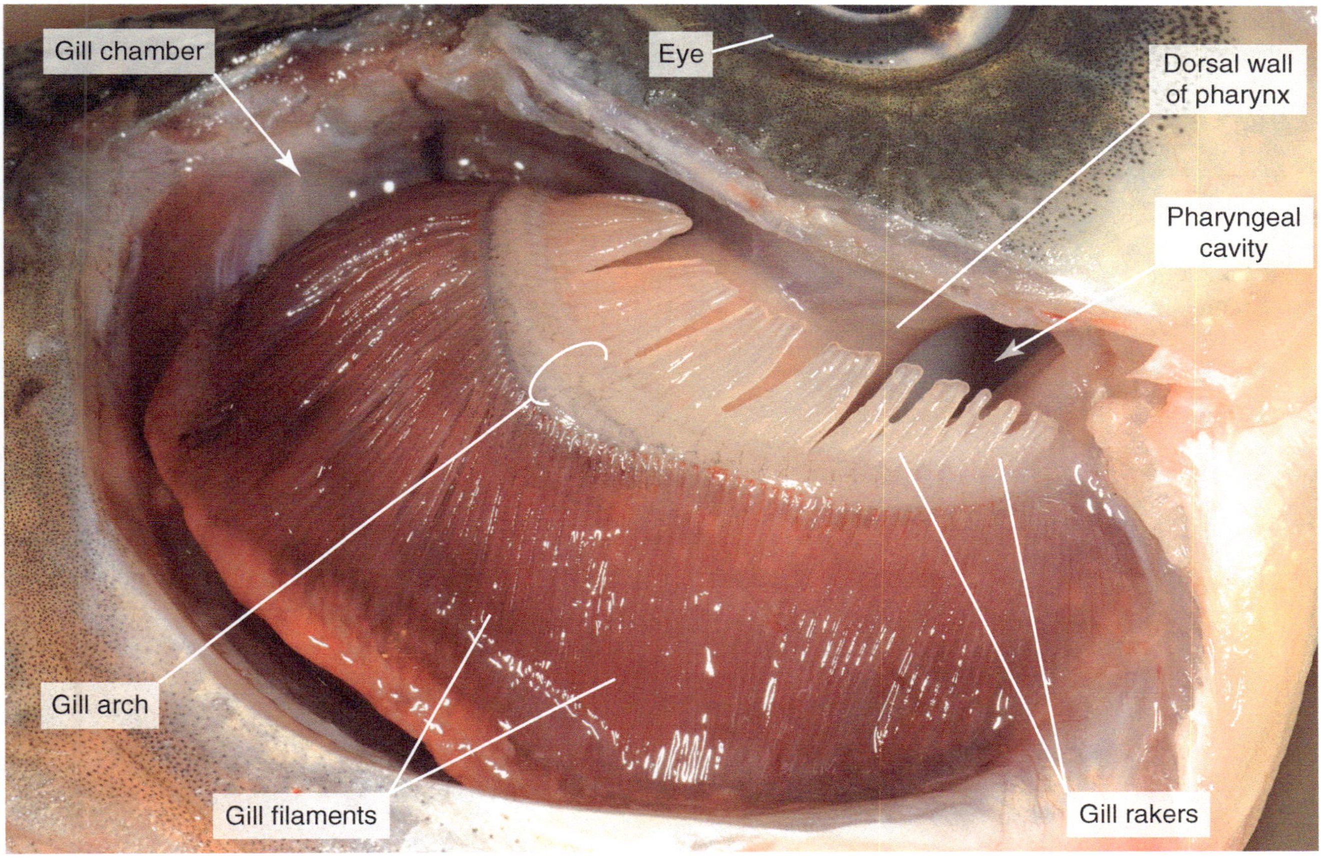

Fig. 9.14 The right gill chamber (branchial cavity) with four gill arches sticked together

The *gills* are supported by bony *gill arches*, which bear *gill filaments* (primary lamellae) and *rakers* (Fig. 9.15, top left and right).

Gill filaments line up in the gill chamber; rakers develop on the opposite (pharyngeal) side. Filaments have a central cartilaginous support (filament cartilage, FC), *afferent arterioles* (AA) and *efferent arterioles* (EA) and multilayered epithelial covering in which PAS-positive *mucous cells* (MC) are scattered (Fig. 9.15, bottom left). Secondary lamellae originate on both surfaces of primary lamellae and are oriented perpendicular to the filaments. The *respiratory epithelium* (RE) is bilayered and covers the secondary lamellae. *Lamellar lacunar system* (LL) develops in the central plane of lamellae, which connect the functional circulatory (respiratory) system (afferent and efferent arterioles) (Fig. 9.15, bottom left). These lacunae are supported by *pillar cells* (PC) – the pillars and red blood cells (RBC) show a very typical, alternating pattern in this plane (Fig. 9.15, bottom right). The *central venous sinus* (CVS) belongs to the supporting blood system of the gill: it is formed by irregular, endothelial lined spaces with different width and fewer red blood cells. The main branch of central venous sinus follows the filament-supporting cartilage in the plane of the primary lamella (Fig. 9.15, bottom right). It gives branches into every secondary lamellae, which lace the respiratory epithelium under the surface (Fig. 9.15, bottom left).

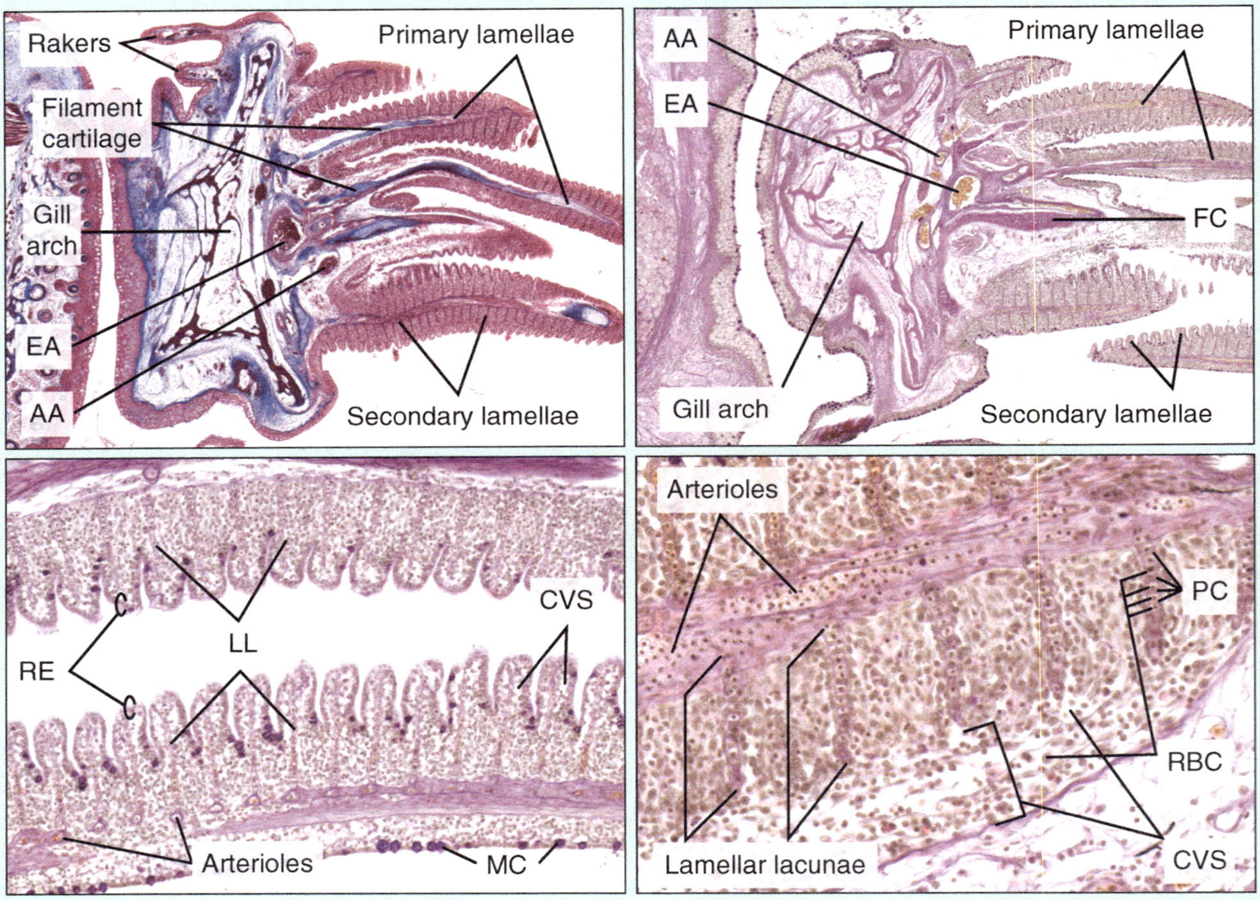

Fig. 9.15 Histological sections of fish gill (cross sections, *left* upper section is stained with Azan, others with tri-PAS). *AA* afferent arteries, *CVS* central venous sinus, *EA* efferent arteries, *FC* filament cartilage, *LL* lamellar lacunae, *MC* mucous glands, *PC* pillar cells, *RBC* red blood cells, *RE* respiratory epithelium

Extend the midventral incision to the branchial cavity and remove the pectoral fin and its suspensory girdle together with the neighbouring tissues along the dashed line on Fig. 9.16.

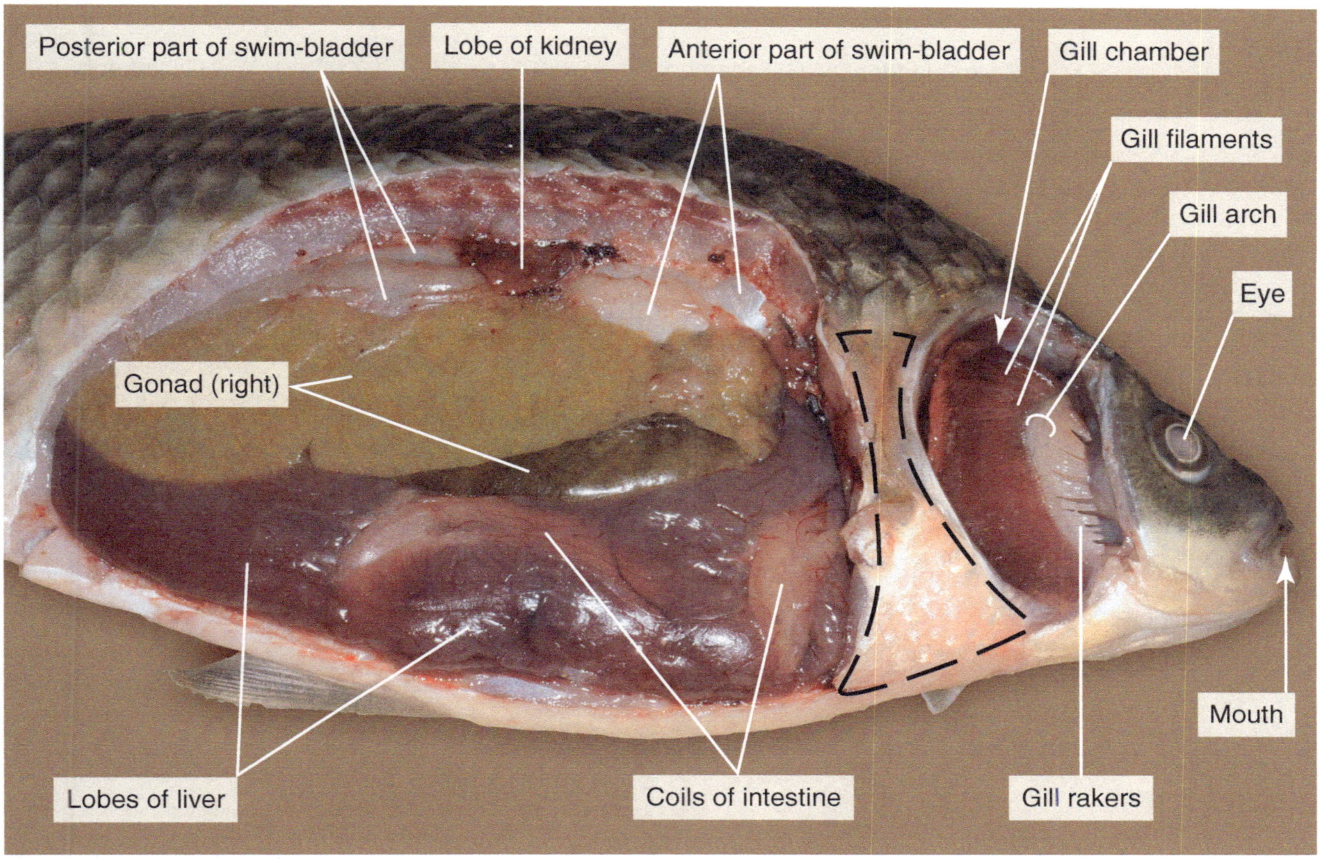

Fig. 9.16 The two windows to be connected by the removal of the pectoral fin, pectoral girdle and surrounding tissue (*dashed line*)

Be careful not to injure the underlying heart and the head kidney. The pericardial cavity is separated from the abdominal cavity by the *transverse septum*. The septum is not homologous to the diaphragm of mammals (Figs. 9.11 and 9.17).

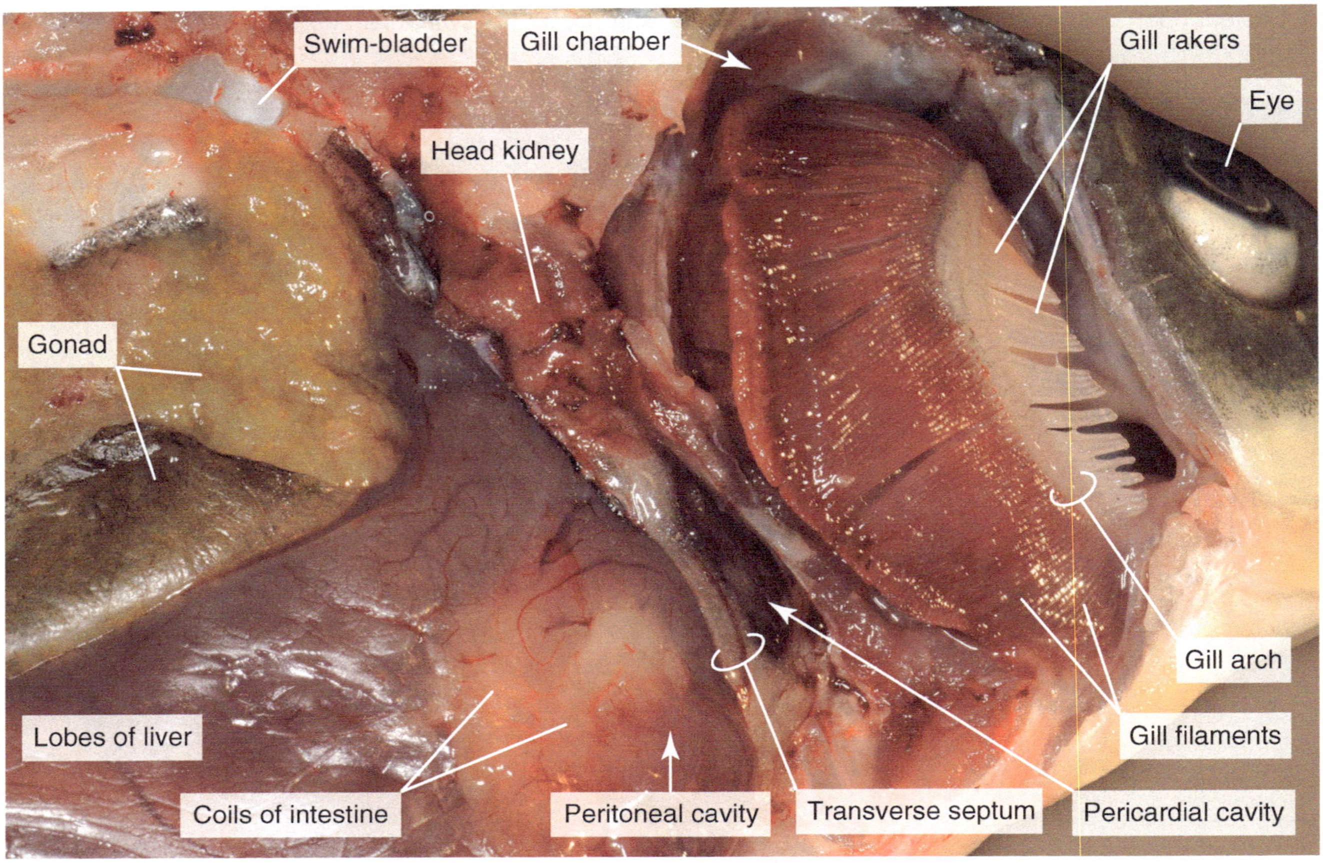

Fig. 9.17 Pericardial cavity after removal of the pectoral fin and its suspensory girdle

The *head kidney* (the rudiment of the embryonic pronephros) in fish is a basic organ forming the blood elements and immune cells (Figs. 9.17 and 9.18). The *heart* which is surrounded by the *pericardium* has two chambers, a thinner *atrium* and a muscular *ventricle* (Fig. 9.18). Blood collected from the venous system enters the *sinus venosus*, a dark red, very thin-walled sac adjoining the atrium posteriorly. Blood flows from this into the atrium and from there into the ventricle, which lies ventral to it and is the most prominent part of the heart. The ventricle pumps into a short, swollen, white *bulbus arteriosus*, which is the first part of the ventral aorta (Fig. 9.18).

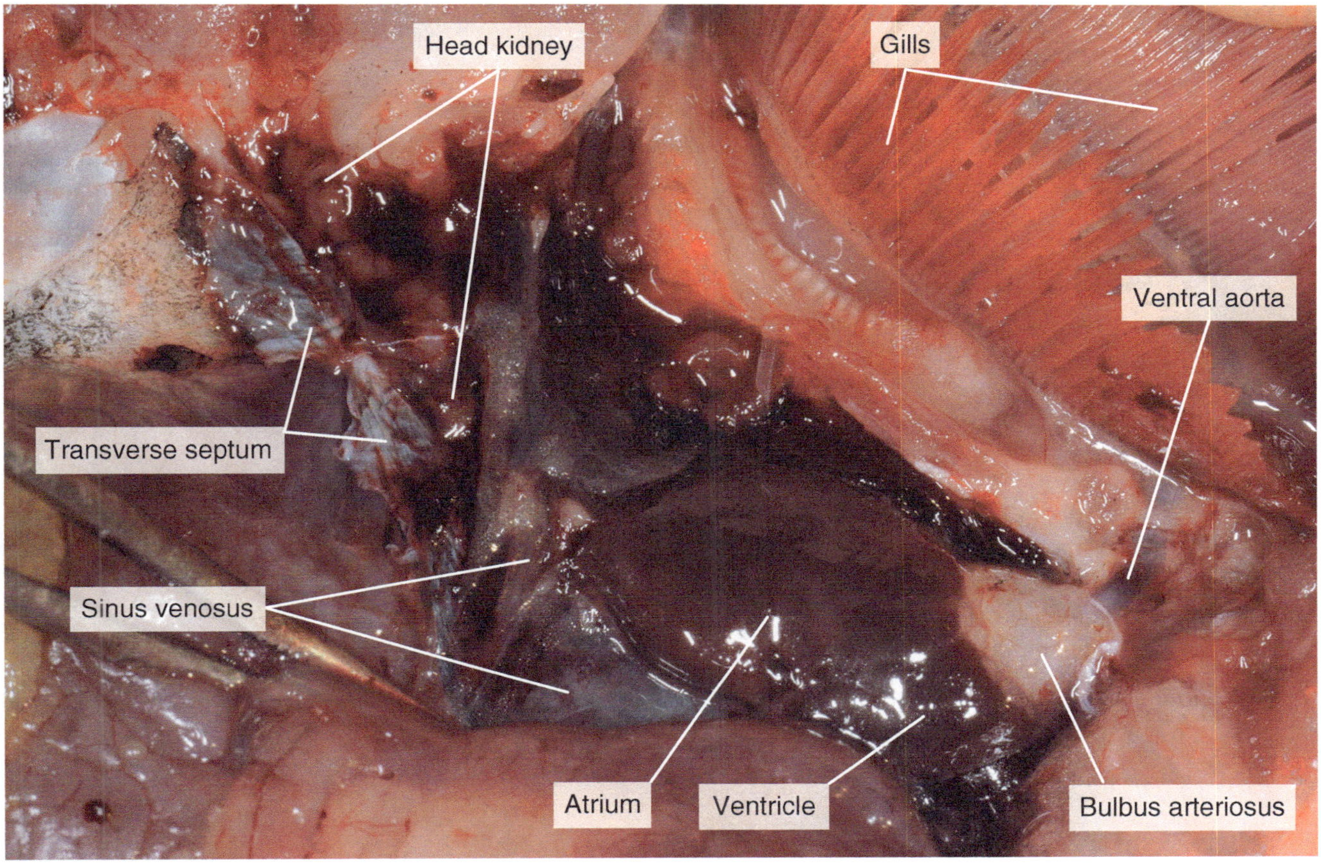

Fig. 9.18 Lateral view of the heart (parietal wall of the pericardium is removed)

If you dissect a freshly killed specimen, its heart may be still beating. This doesn't mean that the fish is alive. The control of the heart contractions is so independent from the brain in fishes that the heart continues beating until the metabolic energy stores are exhausted. Observe the movements of the heart and the order of contraction of the parts: first the sinus venosus, then the atrium and last the ventricle.

Remove the gill filaments from gill arches by a small pair of scissors. Trace with an angled forceps the ventral aorta and its branches and tear with the tip of the forceps the surrounding connective tissue (Fig. 9.19).

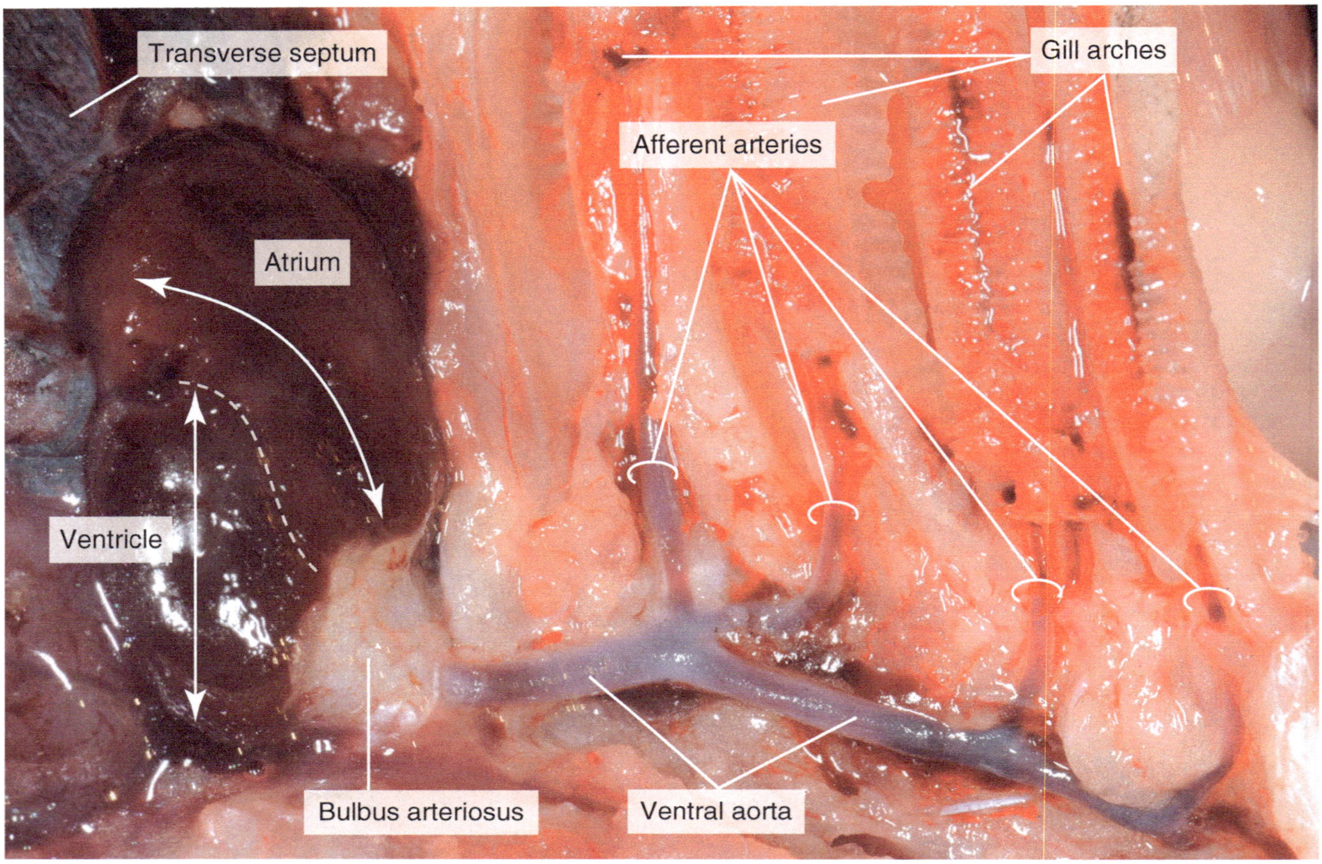

Fig. 9.19 The ventral aorta gives off four pairs of afferent arteries to the gill arches (gill filaments are removed from gill arches). *Dashed line* indicates the boundary between the ventricle and the atrium

The ventral aorta gives off four pairs of *afferent branchial arteries* to the gill arches. From the capillaries of the gill filaments, the oxygenated blood is collected by *efferent branchial arteries* and is emptied into the two roots of dorsal aorta. These roots join immediately to form the *dorsal aorta*, which you can find further posteriorly attached to the dorsal body wall (Fig. 9.29).

Fish have one blood circuit; their heart only has one atrium and one ventricle. The oxygen-depleted blood that returns from the body enters the atrium, and then the ventricle, and is then pumped out to the gills where the blood is oxygenated, and then it continues through the rest of the body.

Return to the examination of the organs of the abdominal cavity. Pull the transverse septum with a forceps anteriorly and find the *hepatic vein* returning directly into the sinus venosus (Fig. 9.20).

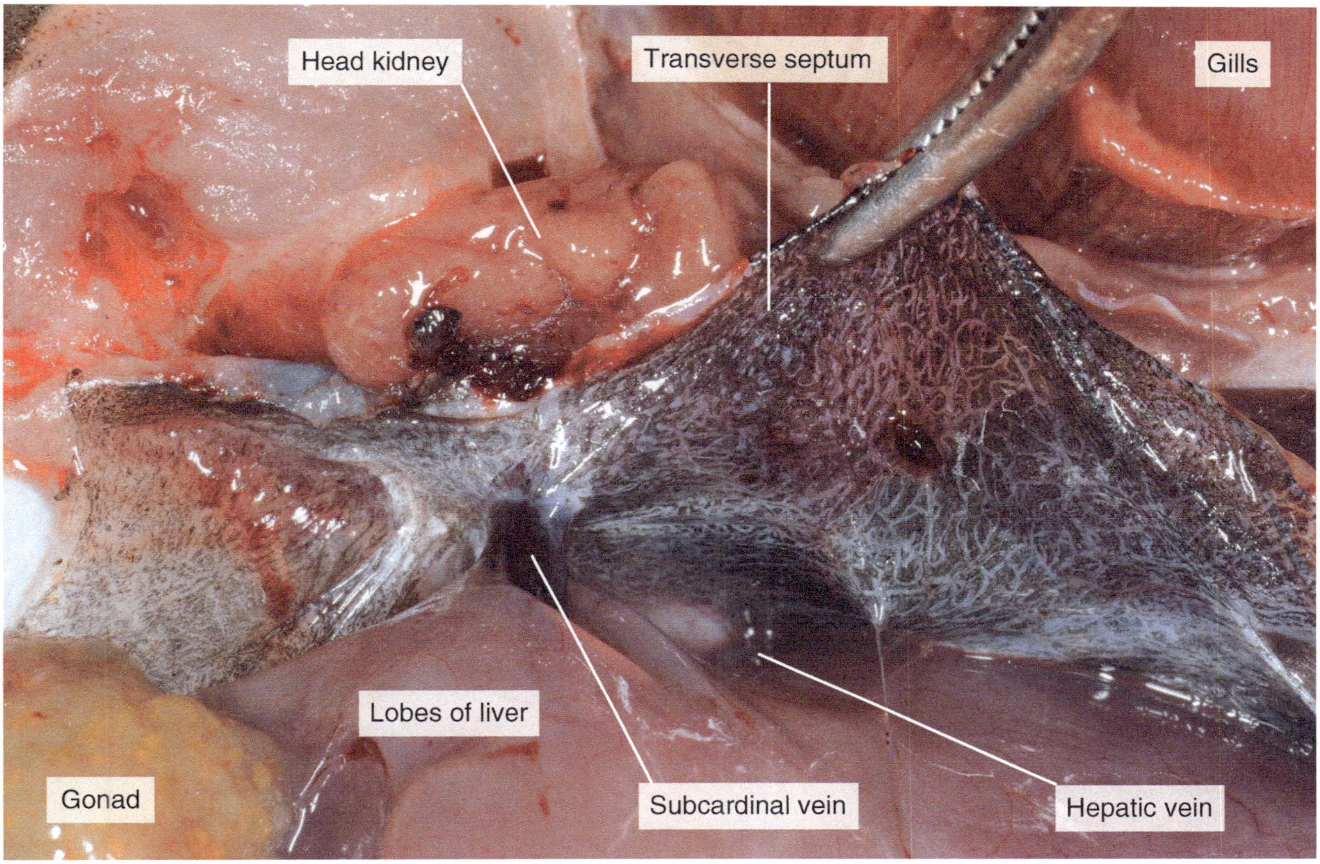

Fig. 9.20 Pericardial cavity after removal of the pectoral fin and its suspensory girdle

Sexes are separate, but it is difficult to distinguish them externally. Gonads (*ovaries* or *testes*) are paired and their size varies seasonally, being the largest during winter before spawning. They are sac-like organs with a prolongation posteriorly which serves as a short *oviduct* or *vas deferens* extending to the urogenital pore posterior to the anus. Find the *gonadal artery* (Fig. 9.21) and then separate the gonad from other organs and remove the whole gonad in one piece (Fig. 9.22).

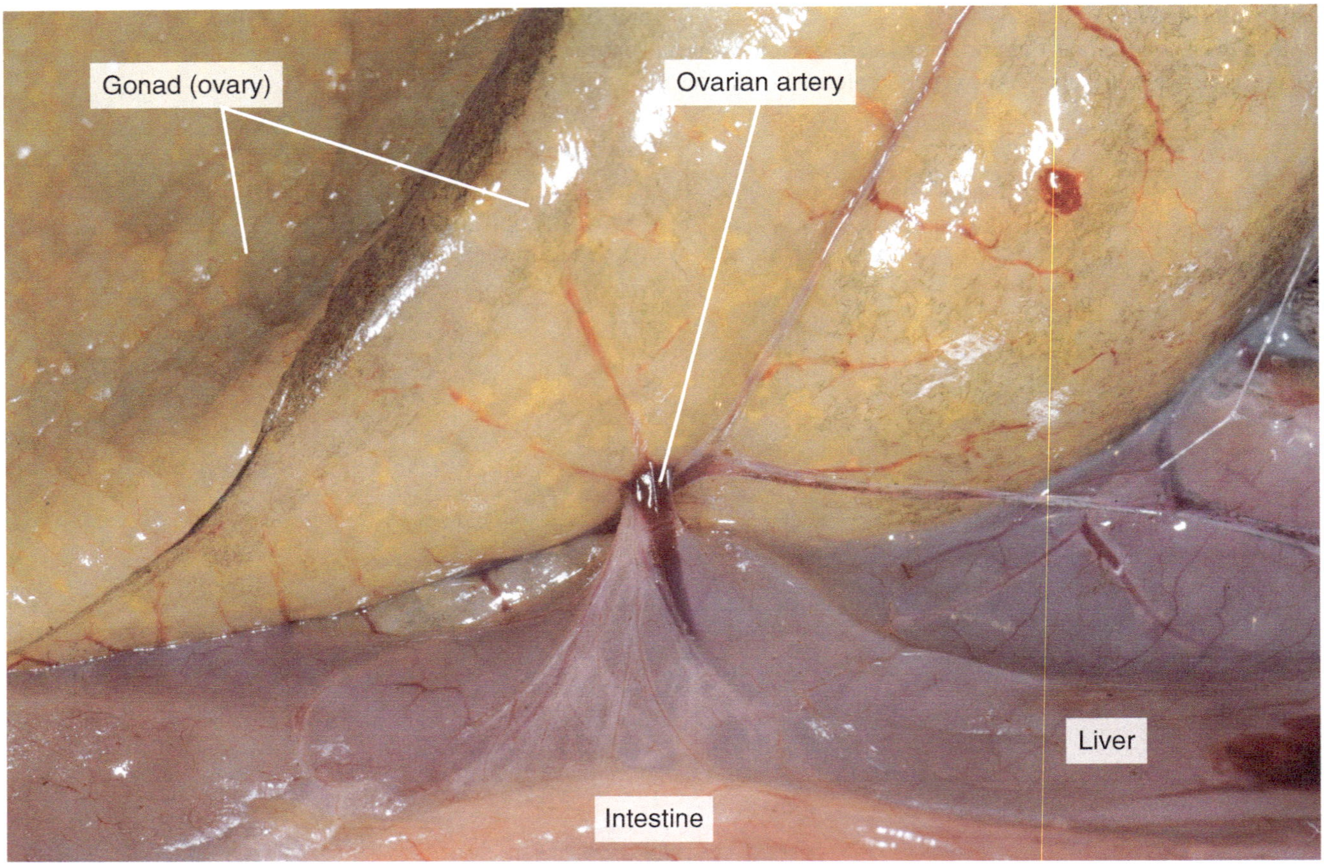

Fig. 9.21 The gonadal (ovarian) artery entering the gonad (ovary)

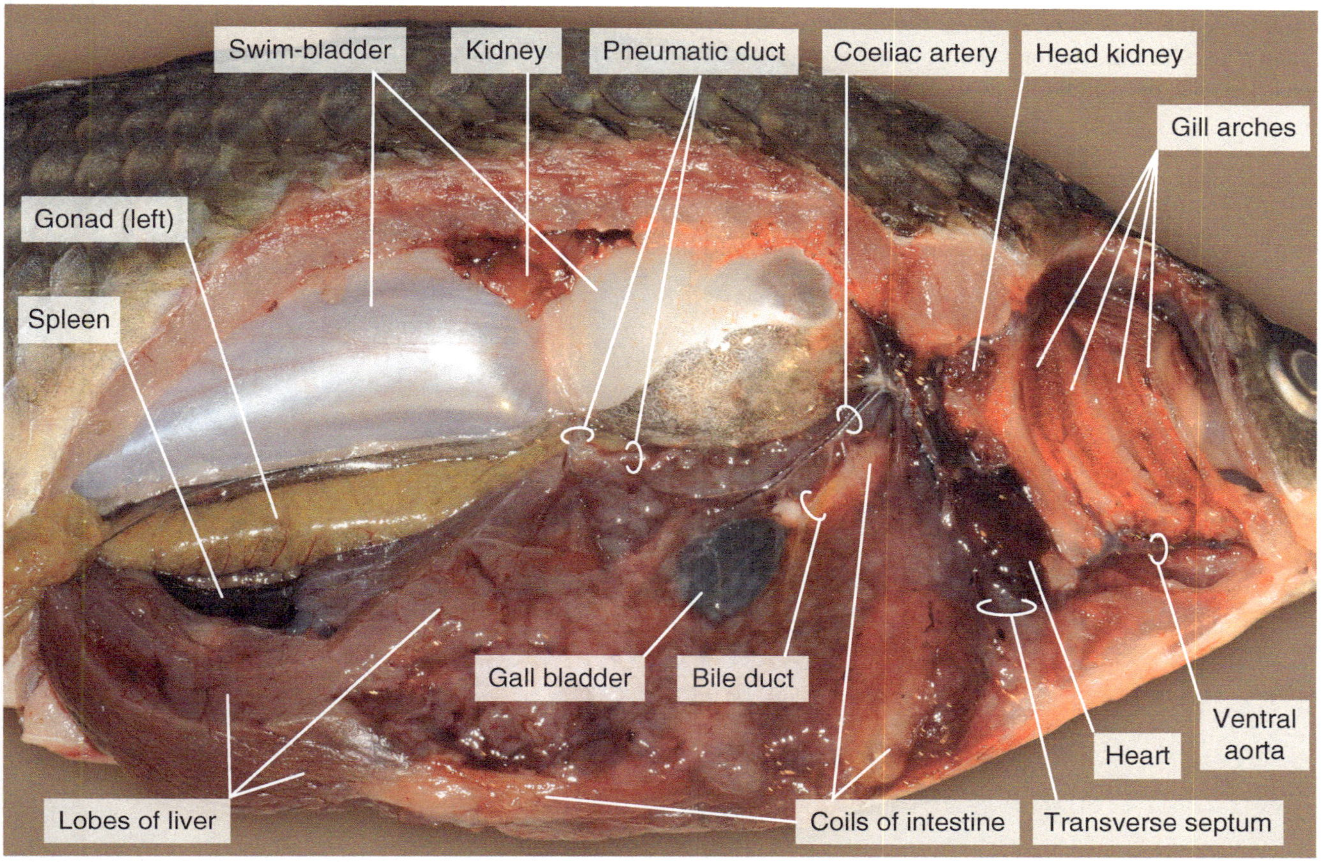

Fig. 9.22 Right-hand view of the internal organs of the crucian

Observe the *swim bladder* in the dorsal part of the body cavity (Fig. 9.22). It is a long, shiny, thin-walled sac with a constriction in the middle. In juvenile crucian, the posterior part connects with the alimentary canal by the *pneumatic duct* (Figs. 9.22 and 9.23).

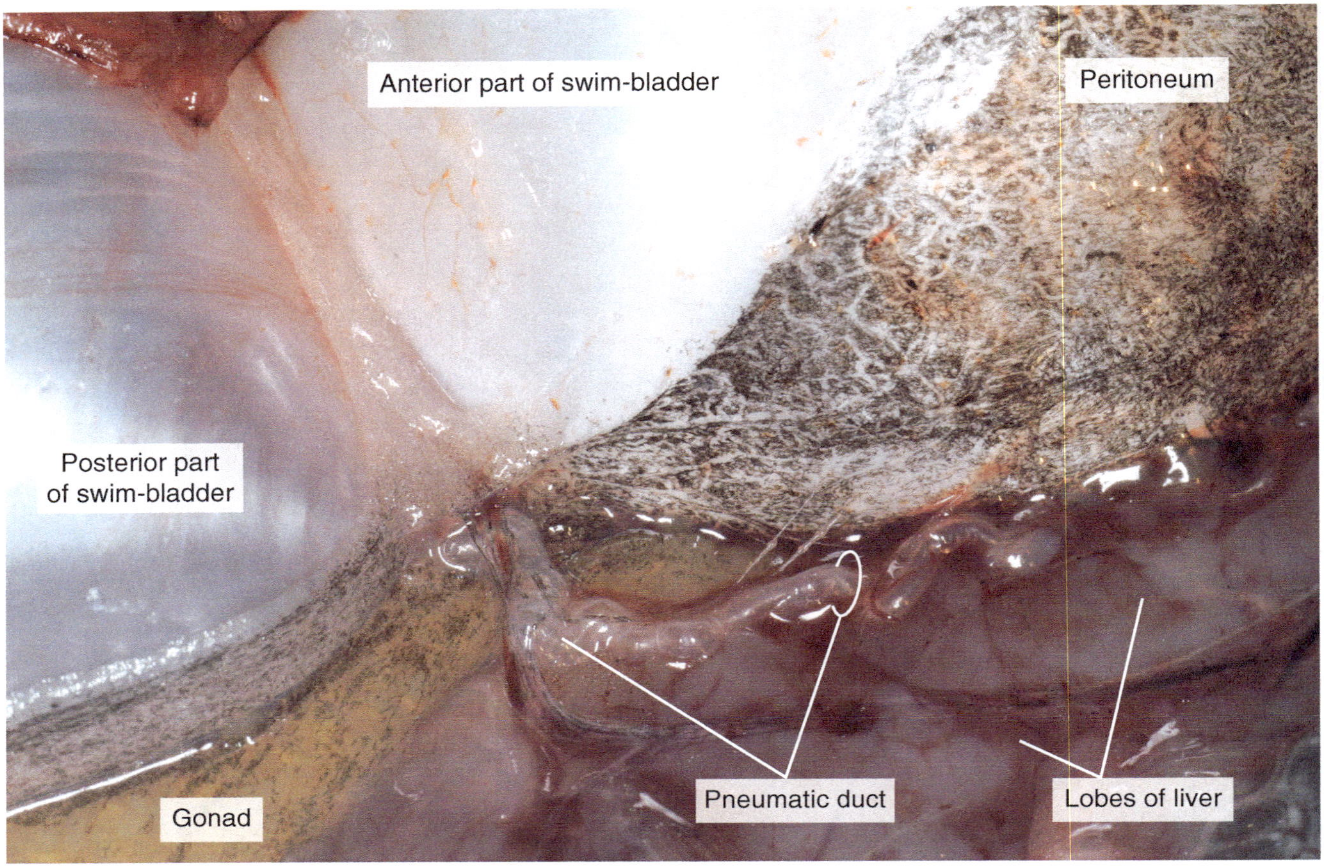

Fig. 9.23 The pneumatic duct extends between the alimentary canal and swim bladder

Once the swim bladder is filled with air through the pneumatic duct, the duct is closed. After that the air pressure inside the swim bladder is regulated by two glands in its wall. The *gas gland* or red body, a network of capillaries (rete mirabile) that secrete gases from the blood into the bladder, is situated in the anterior ventral wall of the posterior part of the swim bladder (Fig. 9.24).

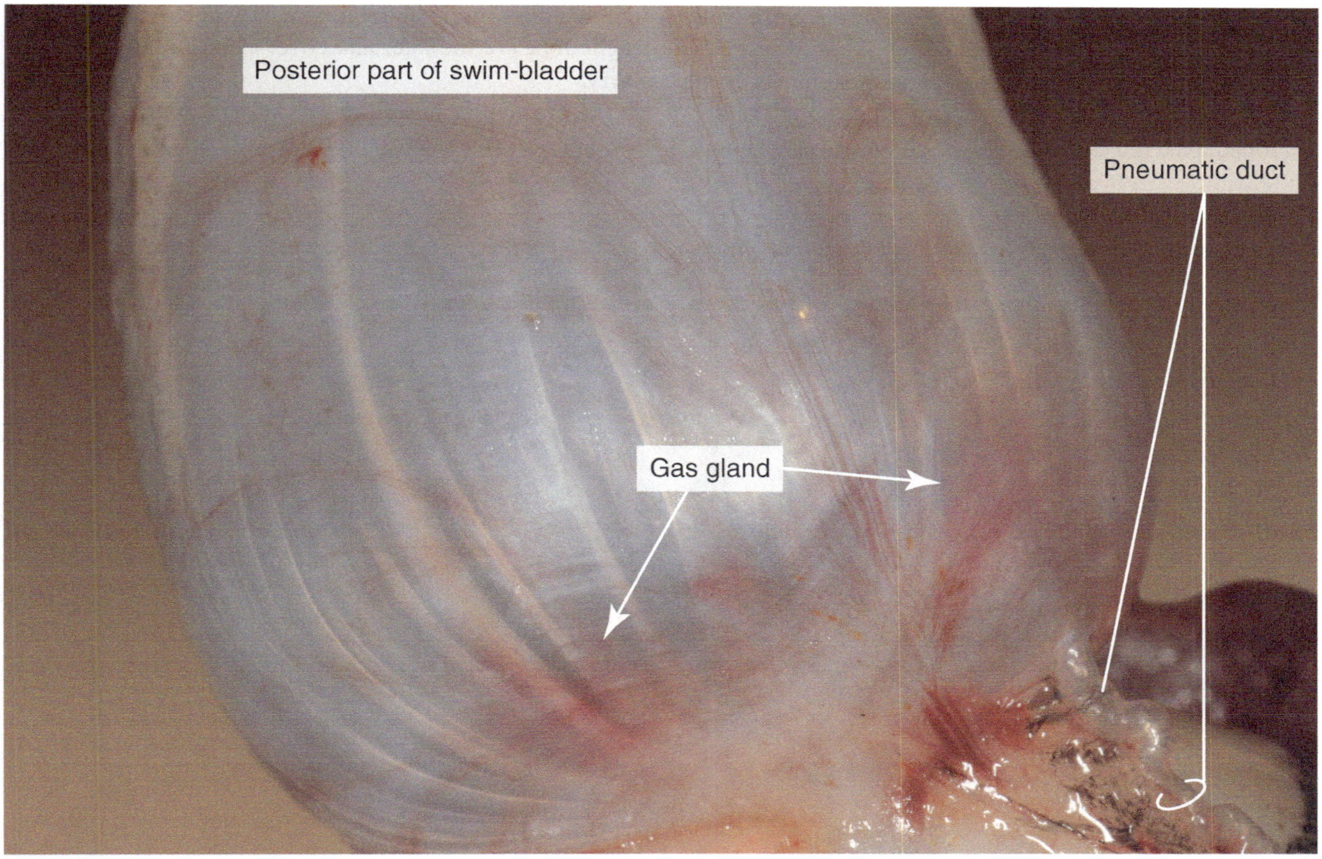

Fig. 9.24 The gas gland is situated in the anterior ventral wall of the posterior part the swim bladder

Another capillary bed, the *oval*, lies in the dorsal wall of the posterior part, but it is rather inconspicuous. Gases may be reabsorbed in this area from the swim bladder into the blood. The swim bladder is a hydrostatic organ that adjusts the buoyancy of the fish to varying depths of water. This means that the fish can remain at various levels in the water without having to expend energy.

The major visible organ in the abdominal cavity is the liver (Figs. 9.16 and 9.22). The *liver* of the crucian is large and lobed, pale red or pinkish organ. It covers other important structures. To find these, separate the lobes of the liver in the midline by your fingers (or with the blunt end of your forceps) just behind the head beginning from the anterior end of the swim bladder. Try to find the organs listed below by poking the liver as little as possible. If you fumble about in the abdominal cavity cells, break off the fragile lobes of the liver and cover everything with a pink slurry.

The crucian has a well-developed *gall bladder*; the *bile duct* joins into the duodenum (Figs. 9.22 and 9.25).

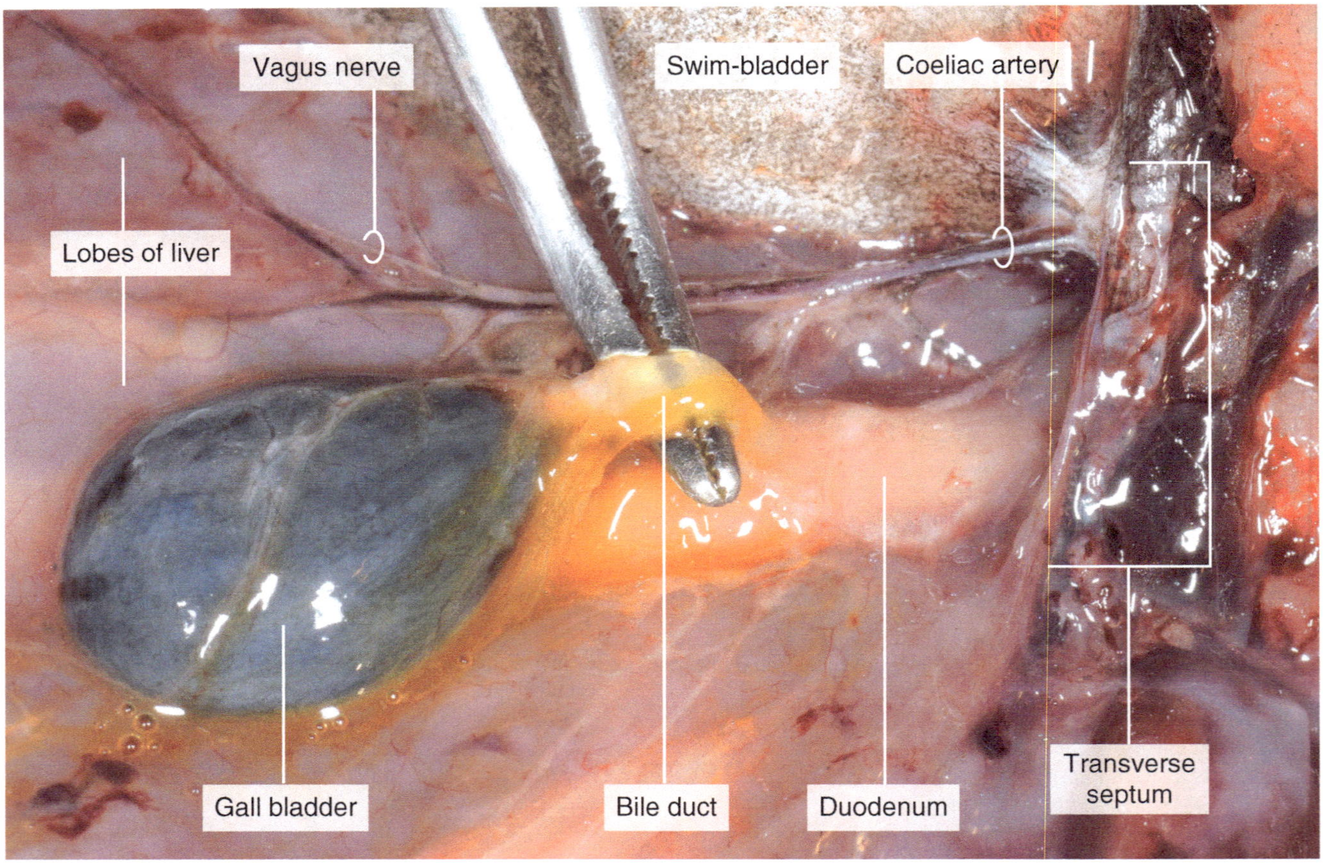

Fig. 9.25 The gall bladder and the bile duct. Yellowish colour is caused by the bile that escaped due to minor injuries of bile canaliculi

The crucian does not have discernible stomach. The gut is only a little wider after the *oesophagus*; swallowed food moves into the small *intestine* (mid-gut). The *pancreas* is not visible either, it has microscopic acini scattered among the liver lobes. The intestine is quite long followed by the *hind-gut*. The last part of the hind-gut, the *rectum*, opens to the outside in the *anus* (Fig. 9.26).

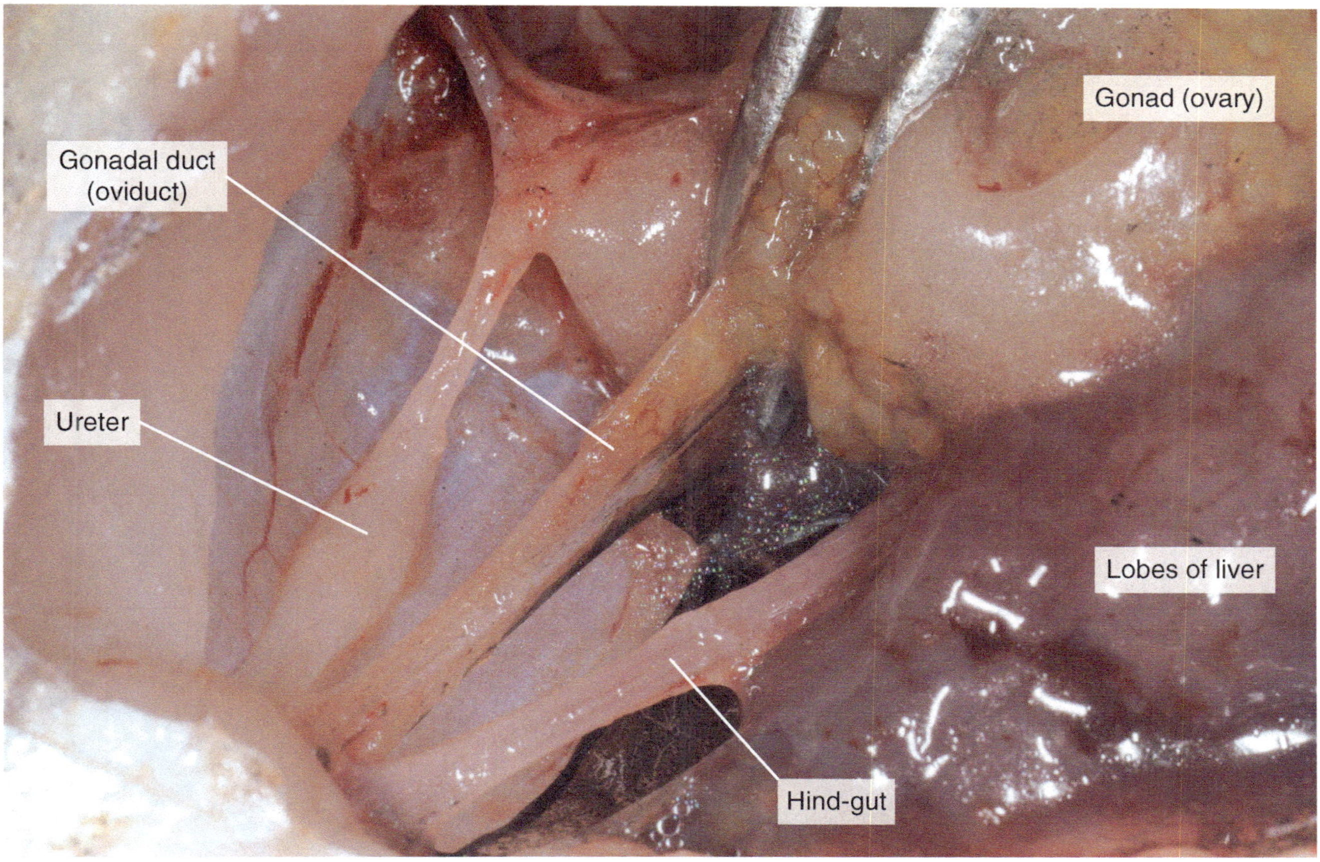

Fig. 9.26 The hind-gut running to the anus and the gonadal duct and the ureter leading to the urogenital pore

The single *coeliac artery* arises from the dorsal aorta. Branches from the coeliac artery supply intestine (intestinal), liver (hepatic), gonads (spermatic or ovarian), spleen (splenic) and swim bladder (pneumatic). The coeliac artery and its branches run together with the *vagus nerve* (Fig. 9.27).

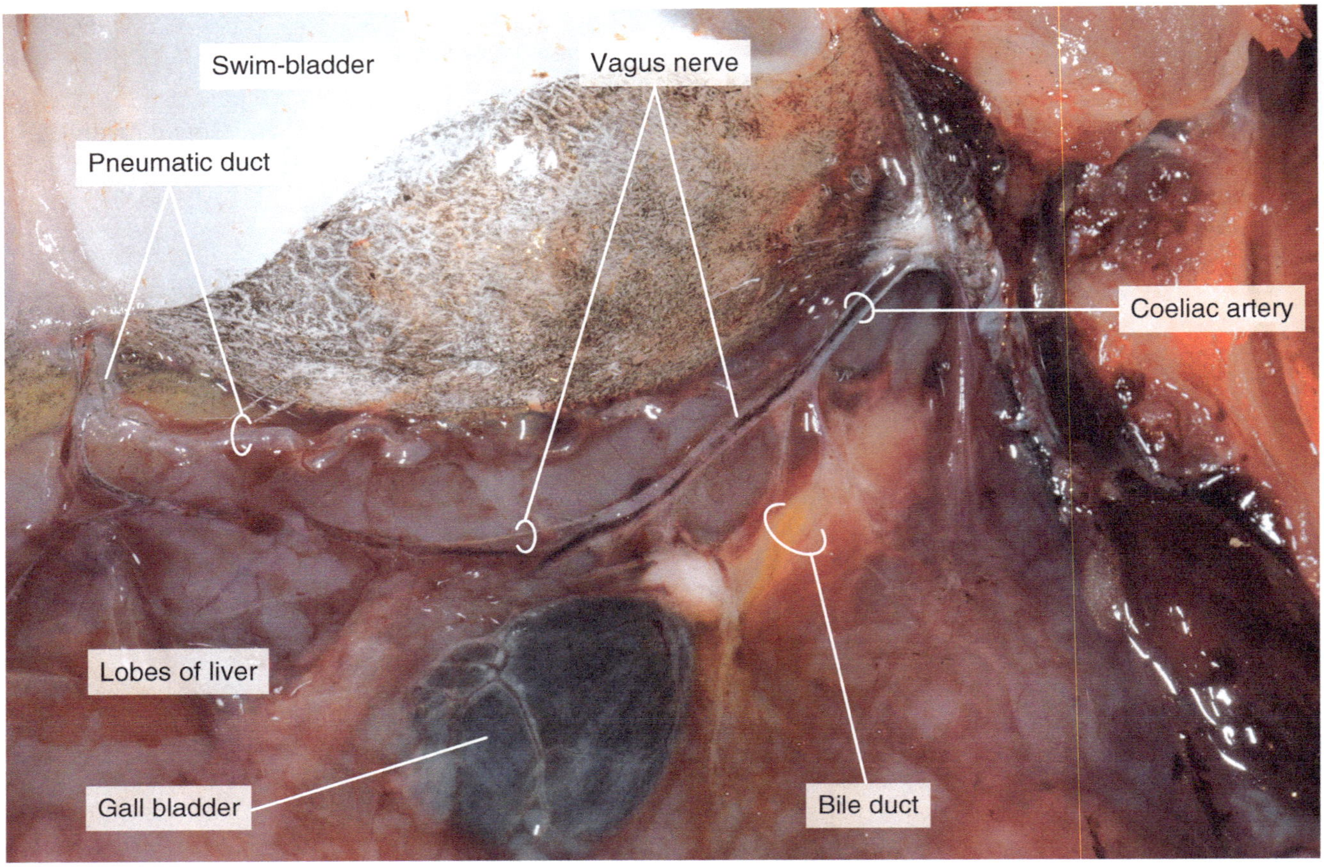

Fig. 9.27 Branches of the coeliac artery and the vagus nerve

The *spleen* is a dark, slender organ lying among the gonads and the lobes of the liver (Fig. 9.28).

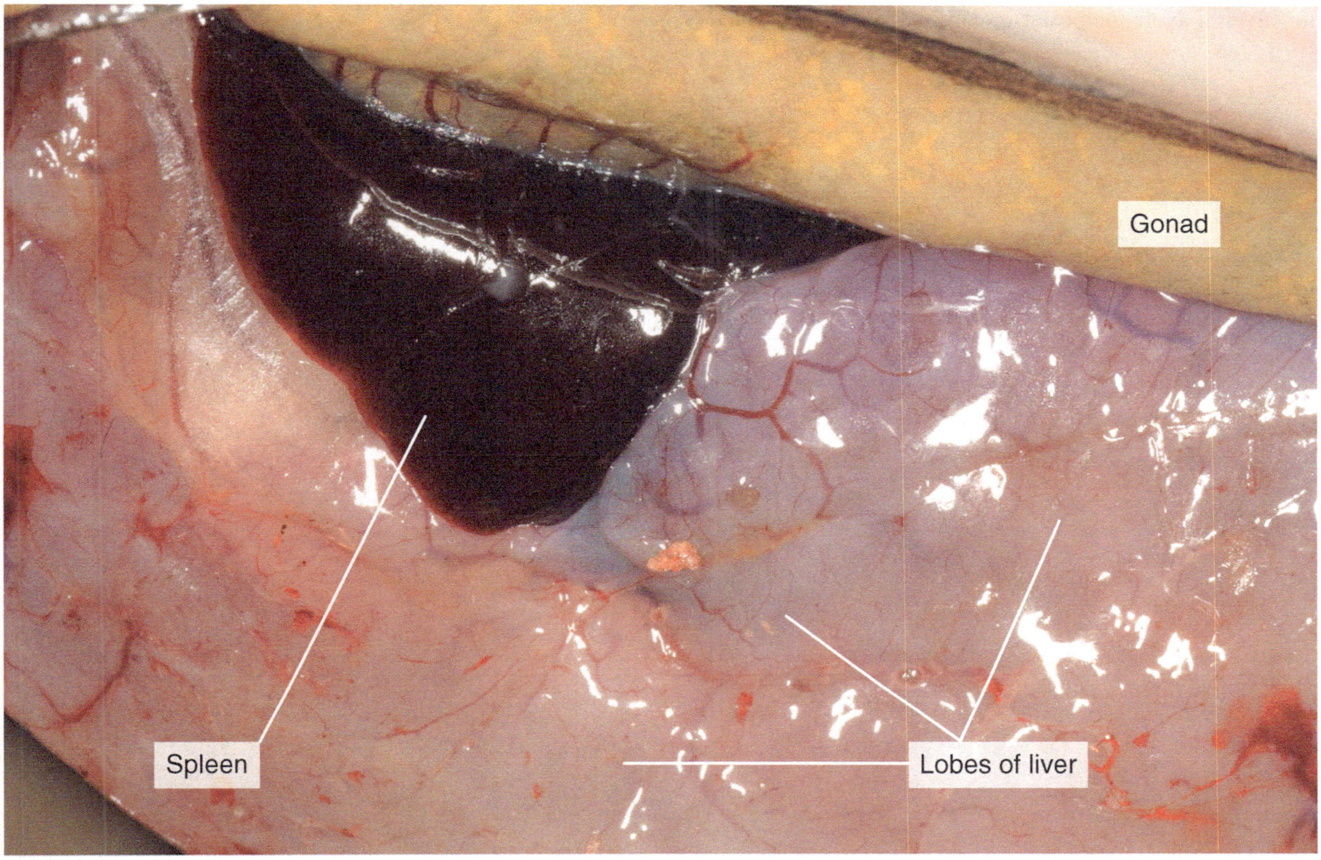

Fig. 9.28 The spleen between the gonads and the lobes of the liver

The *kidneys* (mesonephroi) are paired masses that lie against the dorsal body wall and extend the whole length of the abdomen above the swim bladder (Fig. 9.29).

They are fused anteriorly, but mostly separated by the dorsal aorta. Note the posterior cardinal veins on the surface of the kidneys. Observe the *ureter* (derivative of mesonephric or Wolffian duct) extending the short distance from the posterior end of the kidneys to the *urogenital pore* posterior to the anus (Fig. 9.26).

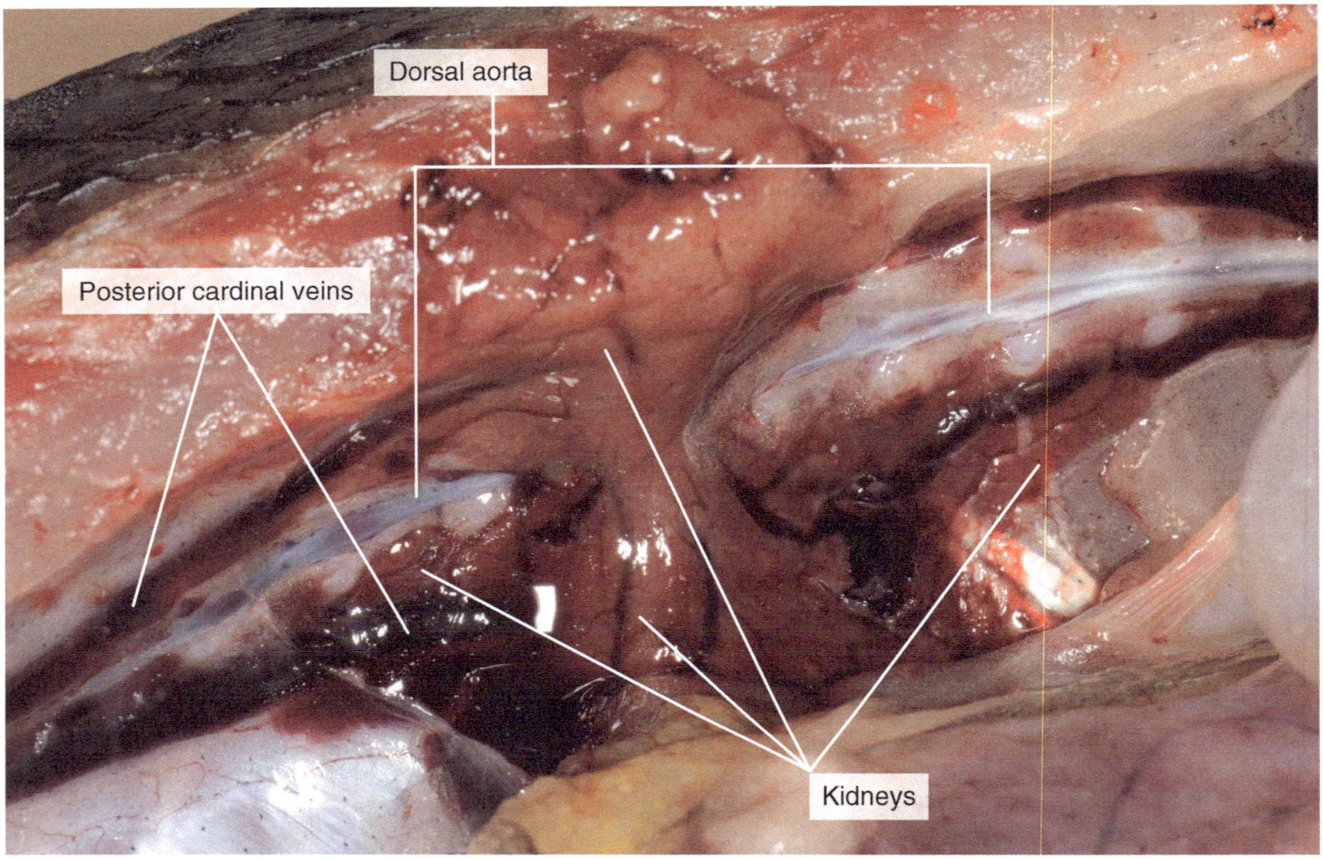

Fig. 9.29 Paired kidneys and the main blood vessels at the dorsal body wall after removal of the swim bladder

The *kidney* of fish does not show cortical and medullar parts. The excretory units of working kidney are the *nephrons*. The proximal end of developing nephrons surrounds glomeruli and forms a globular *renal corpuscle* (Fig. 9.30).

Each of them has a central *glomerulus* of capillaries and a glomerular or *Bowman's capsule*, from which the *renal tubule* originates. In the renal corpuscle, a primary filtrate is produced from the plasma, and the renal tubule modifies it to form the final urine. The Bowman's capsule has two walls: the parietal wall made of flattened epithelial cells, whereas the visceral wall is formed by cuboidal cells named *podocytes*. Between the two walls of the capsule, there is a narrow urinary space, which is continuous with the renal tubule at the *urinary pole* of capsule (Fig. 9.30, right). The glomerulus is a network of convoluted capillaries, embedded in a matrix formed by *mesangial cells* (MC). It is supplied by an *afferent arteriole* which enters the capsule opposite the urinary pole. An *efferent arteriole* leaves the corpuscle at the same point named *vascular pole* (not shown in our sections). The loose connective tissue between the excretory elements of kidney has a haematopoietic function, involved in the production of new red blood cells.

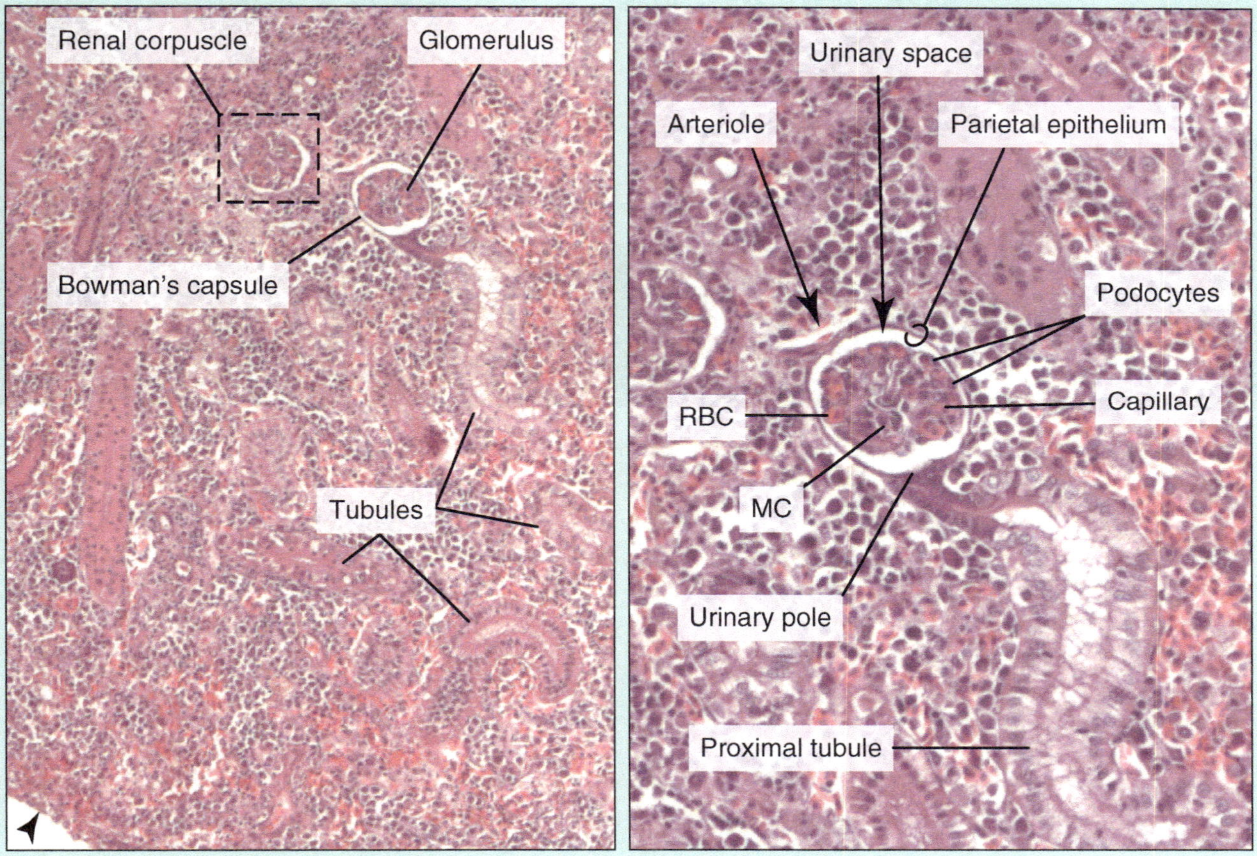

Fig. 9.30 Histological section of fish kidney (HE). *Arrowhead* surface of the kidney, *MC* mesangial cells, *RBC* red blood cells

By all these, the dissection of the internal organs is completed. Now we explore the brain, cranial nerves and the eye. Before detaching the head, try to find the largest of the *Weberian ossicles* (tripus) (Fig. 9.31).

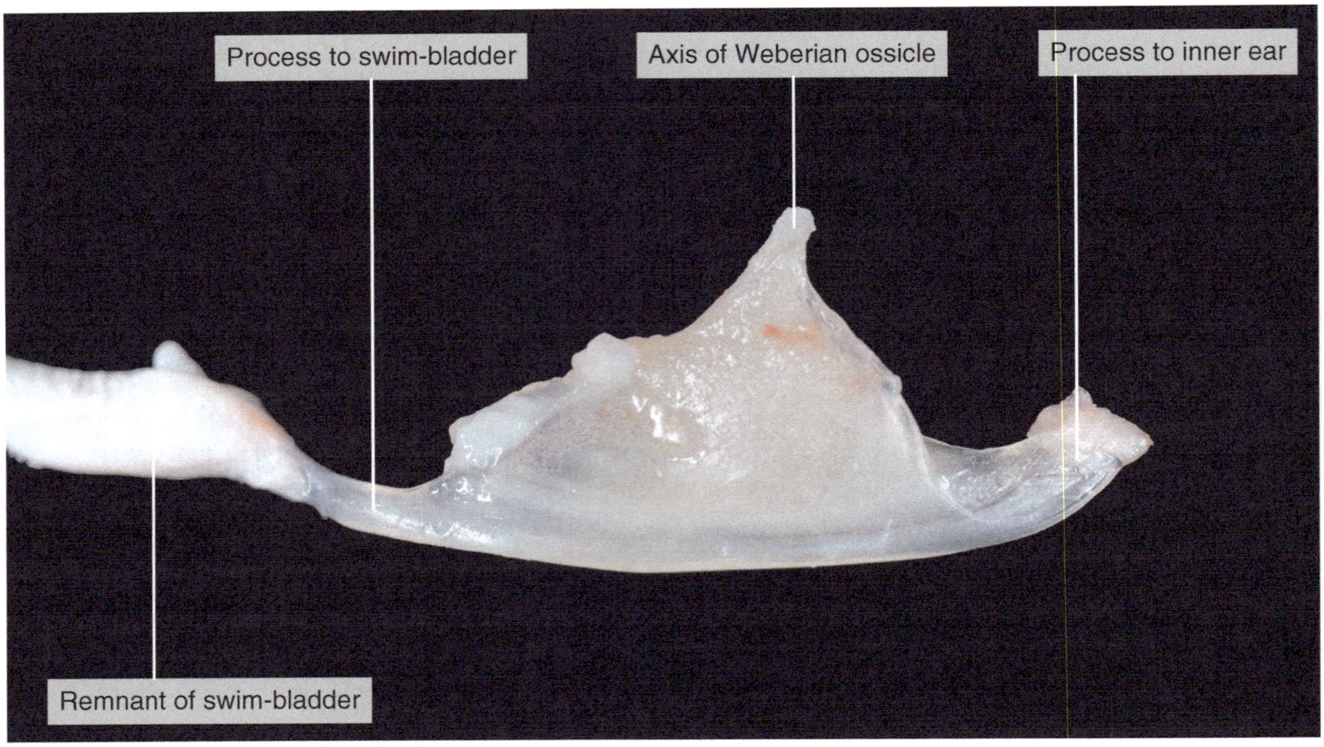

Fig. 9.31 The posterior Weberian ossicle connecting the swim bladder towards the inner ear through further ossicles

Take a strong pair of forceps and grab the edge of this bone firmly, close to the vertebral column, between the head and the swim bladder and pull it hard. The Weberian ossicles connect the swim bladder to the inner ear. The entire structure is derived from skeletal elements of the first four vertebrae. The Weberian apparatus functions by transmitting auditory signals from the swim bladder, through the Weberian ossicles, which enter the skull at the Weberian opening (Fig. 9.32) and then straight into the inner ear. The structure essentially acts as an amplifier of sound waves that would otherwise be only slightly perceivable by the inner ear structure alone. Sounds travelling through the water cause the swim bladder to vibrate, and the ossicles amplify these vibrations and transmit them to the sensitive hair cells within the liquid-filled inner ear. With the added function of the swim bladder as a resonating chamber, signals are amplified to noticeable levels.

Cut off the head of the crucian by a large pair of scissors between the skull and the first vertebra. You have to cut the large muscles holding the head horizontally as well. Identify the last gill arch, the pharyngeal jaw, the pharyngeal teeth, the Weberian openings and the foramen magnum (Fig. 9.32).

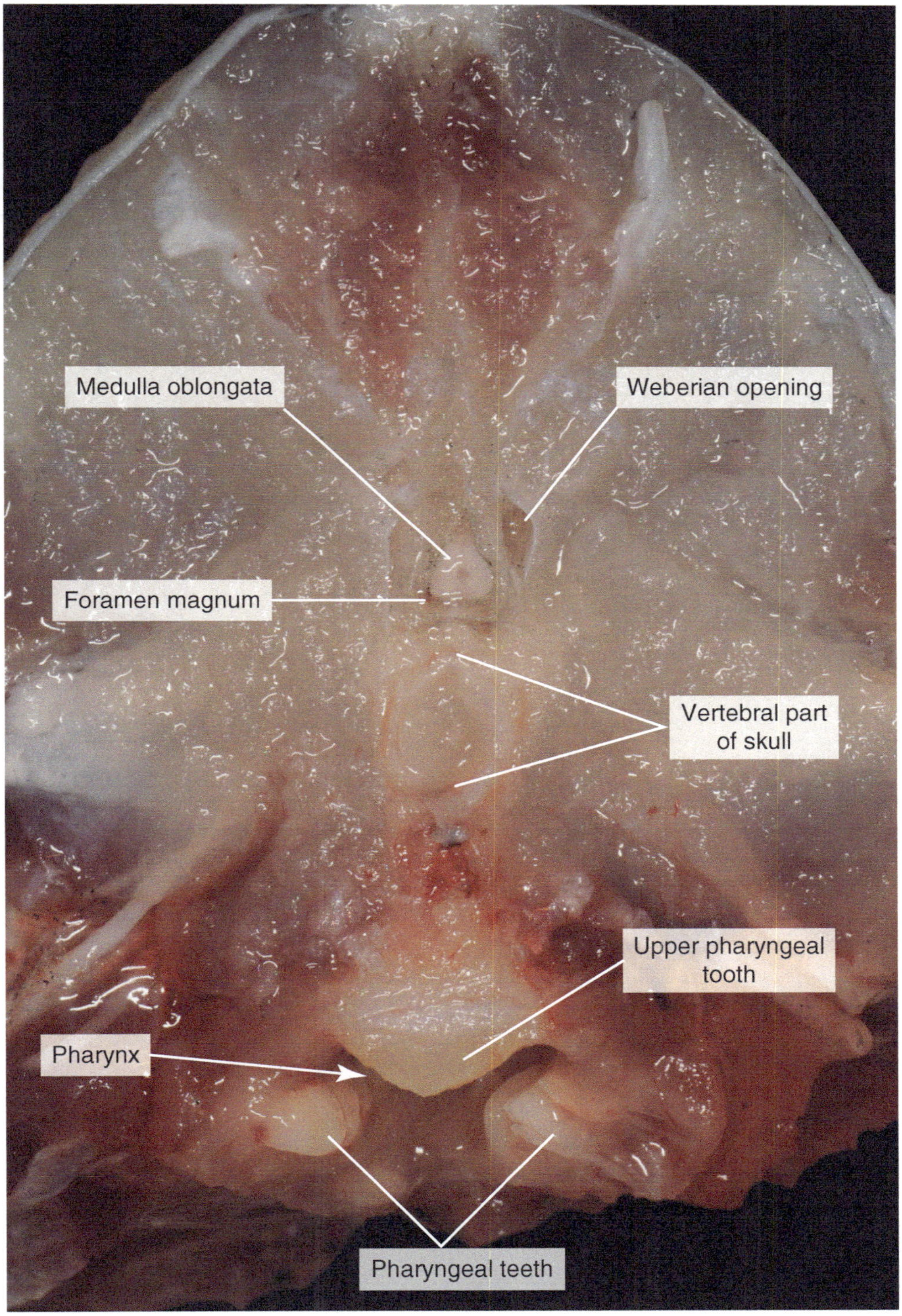

Fig. 9.32 Posterior cut surface of the detached head

The skull can be opened by a large pair of scissors but it is easier to work with a bone nibbler. Make a transverse cut connecting the nostrils, then the two parallel cuts starting also from the nostrils towards the posterior end of the skull. Lift the roof of the skull (Fig. 9.33). You may find the balance organ of the fish, the semicircular canals approaching the occipital region.

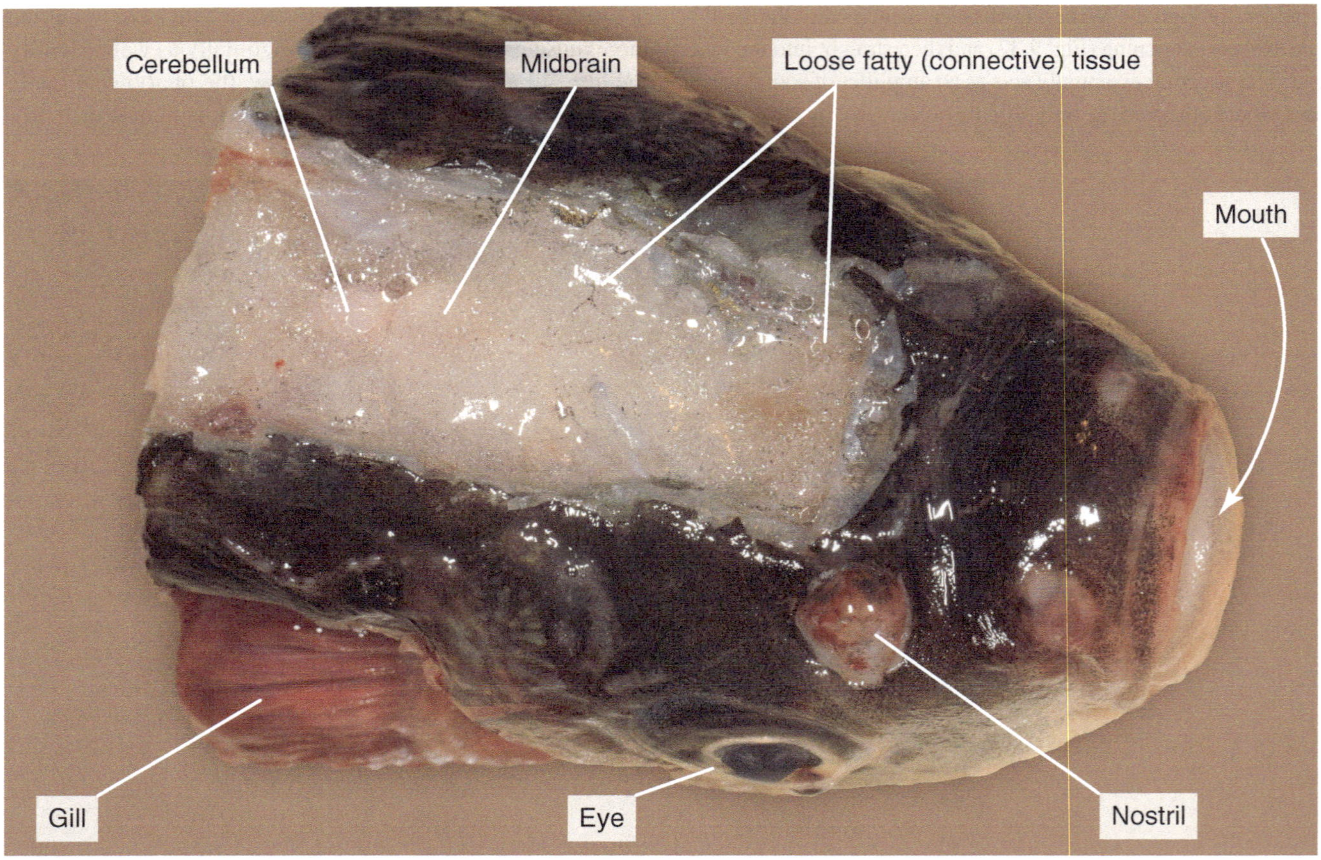

Fig. 9.33 The brain lurking in the loose fatty connective tissue after opening the skull

Remove the loose fatty connective tissue surrounding the brain very carefully. You can use small cotton pads held in forceps and a gentle tap water current. The brain is a very soft, pink organ; it is easily damaged. Fish typically have quite small brains relative to body size compared with other vertebrates. The brain consists of three regions: the fore-, mid- and hindbrain (Fig. 9.34).

The *olfactory nerves* (cranial nerve I) connect the two *olfactory bulbs* with the *forebrain* (Fig. 9.34). The forebrain is concerned mostly with olfaction. The forebrain is connected to the midbrain via the diencephalon, but this structure is partly covered by the optic lobes and not entirely visible. The diencephalon performs functions associated with hormones and homeostasis. The *pineal gland* (epiphysis) lies just on top of the diencephalon (Fig. 9.34). This structure detects light, maintains circadian rhythms and controls colour changes. The *midbrain* (mesencephalon) contains the two oval *optic lobes* that interpret nervous signals from the eyes and thus provide the sense of sight. The main function of the midbrain is to interpret messages relayed to it by the nerves, particularly those concerned with movement and the attitude of the body in the water. The *hindbrain* (metencephalon and myelencephalon) is the brain's posterior. The *cerebellum* is a single-lobed structure that is typically the biggest part of the fish brain (Fig. 9.34). It is particularly involved in the coordination of motor activities. *Vagal lobes* (large viscerosensory area of the cranial nerves IX and X) cover the most of medulla oblongata (Fig. 9.34). As well as controlling swimming and balance, the hindbrain governs respiration and osmoregulation.

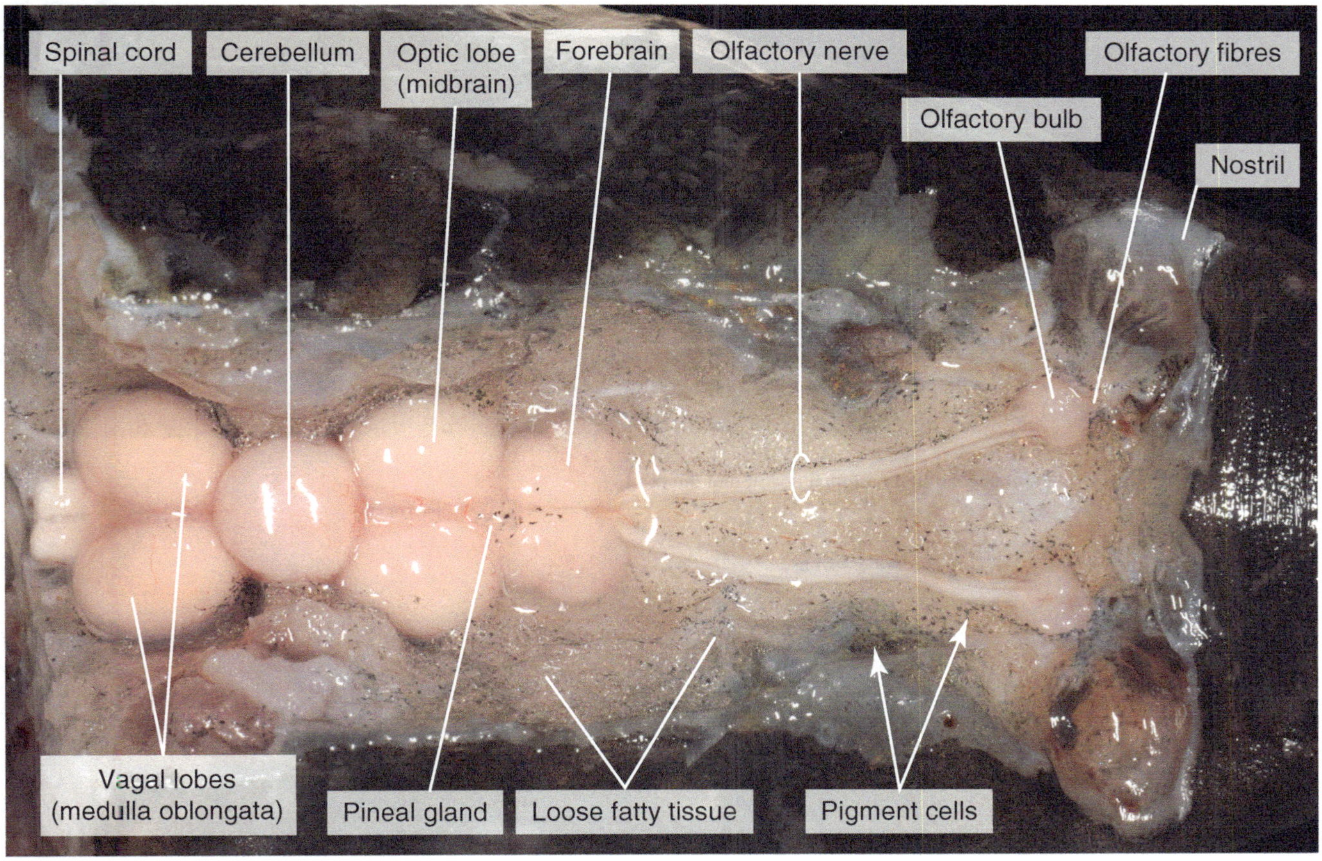

Fig. 9.34 Dorsal view of the brain of the crucian in the skull after removal of the loose fatty tissue

Next, open up the orbit: cut the connective tissue around the eye with a pair of small-angled scissors and remove the bones from the dorsal side of the orbit in the width of the eye (Fig. 9.35).

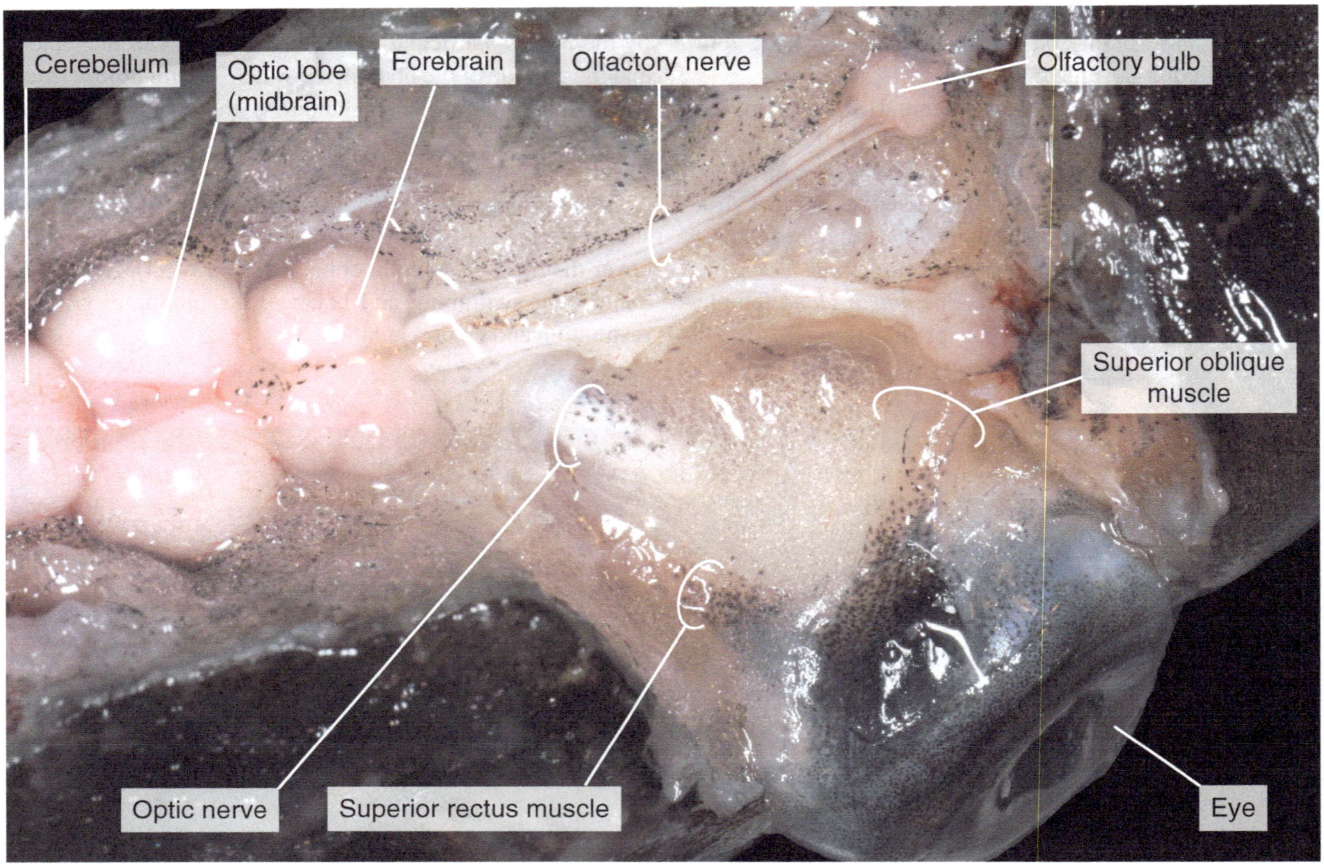

Fig. 9.35 Dorsal view of the right eyeball in its original position in the orbit

Observe the *optic nerve* (II) and some of the extrinsic eye muscles (Fig. 9.35). The system of these muscles is the same in the crucian as in all the vertebrates. It is a good opportunity to examine them here as they are readily observable. They are innervated by three cranial nerves: *oculomotor nerve* (III) innervates all but two, superior oblique muscle is innervated by the *trochlear nerve* (IV) and the lateral rectus by the *abducens nerve* (VI). These nerves are so delicate that they are not likely to be observable. Fold up the eye to examine the other eye muscles and the strong, middle branch of the *trigeminal nerve* (V) at the bottom of the orbit (Fig. 9.36).

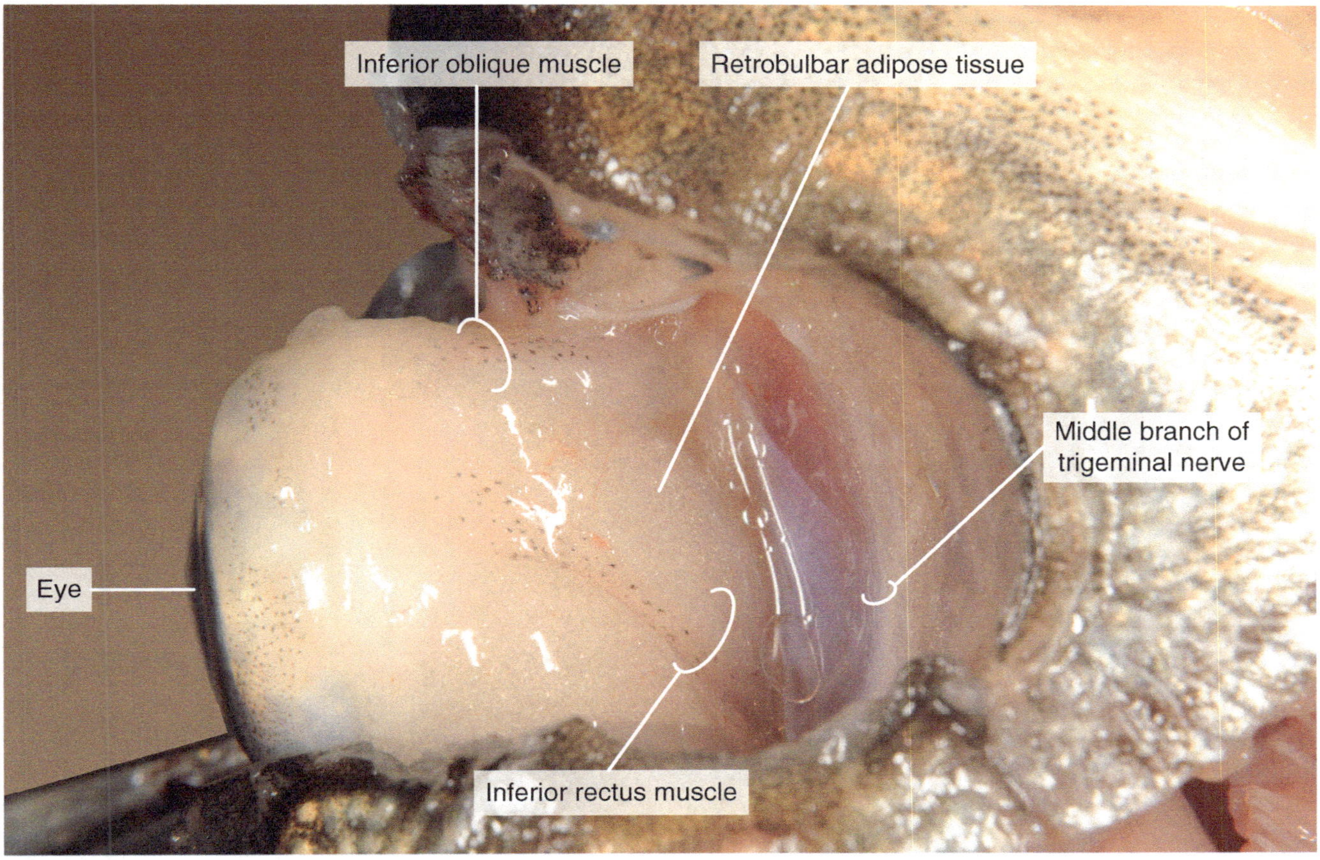

Fig. 9.36 The ventral side of the right eyeball lifted up from the orbit

Remove the bones of the skull between the orbit and the gill arches cautiously (Fig. 9.37).

Identify the common part of the trigeminal nerve (V) and behind the thinner *facial nerve* (VII) (Fig. 9.37). The *auditory nerve* (VIII) runs from the inner ear immediately to the hindbrain so it is too short to be seen. The *glossopharyngeal nerve* (IX) innervates the first, and the *vagus nerve* (X) innervates the last three the gill arches. Cut off the gill filaments from the arches and find the nerves on the dorsal surface of the arches. One strong branch of the vagus nerve turns towards to abdominal cavity and innervates the internal organs (Fig. 9.37).

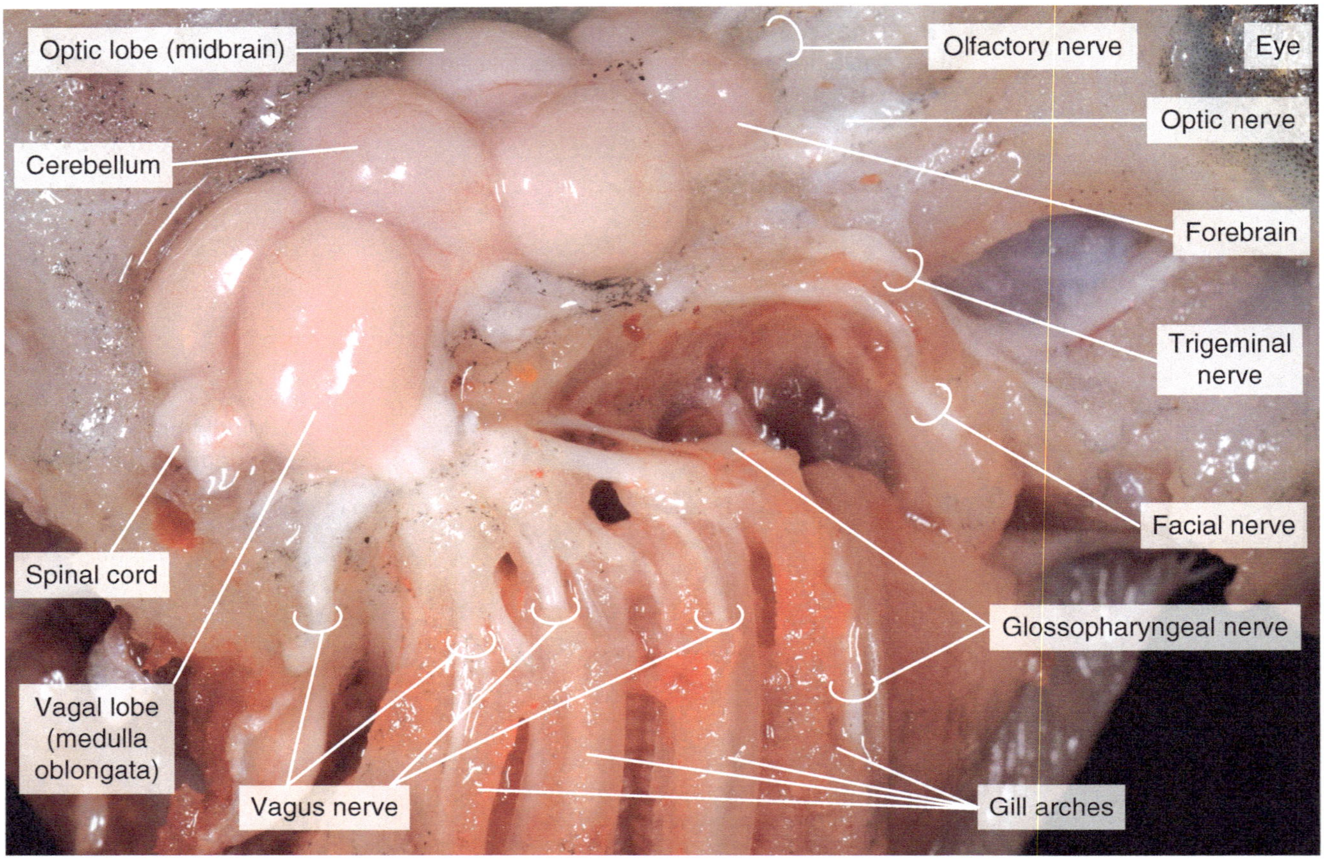

Fig. 9.37 Cranial nerves of the crucian

The crucian has well-developed eyes and have good vision. The structure of the *eye* is very similar to that of the mammals, although the accommodation system is different. Cut the eye muscles and the optic nerve and remove the eyeball from the orbit (Fig. 9.38).

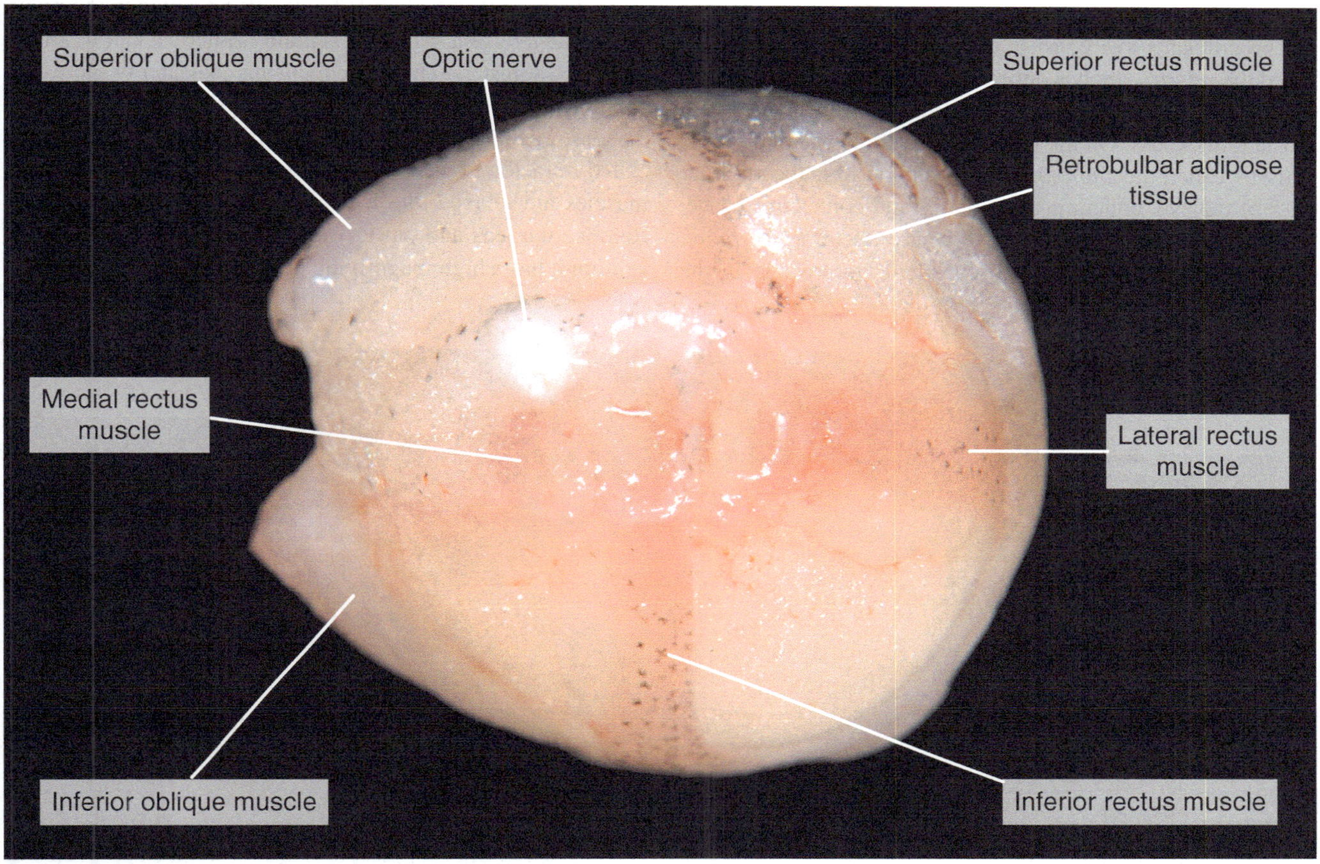

Fig. 9.38 The six extrinsic eye muscles on the rear view of the isolated right eyeball

Observe once more all the six extrinsic eye muscles on the rear view of the isolated eye. Cut the eyeball along its longitudinal axis with a sharp scalpel or razor blade. Cover the sections with water and examine them with a magnifying glass or under a dissecting microscope. Identify the structures of the eyeball and the layers of its wall (Fig. 9.39).

The crucian's eye consists of a tough, opaque outer case called the *sclera*, which is only transparent in front of the lens, where it is called the cornea (Fig. 9.39). The *cornea* has a constant thickness. On the inside, the sclera lies another layer of tissue called the *choroid* layer. The choroid is the vascular layer of the eye, containing capillaries and connective tissue. Its anterior part is the *iris*, with a centre opening, the *pupil*, through which light enters the eye. It is immovable in the crucian's eye. Inside the choroid, there is the 3rd layer; this is the light-sensitive part of the eye, called the *retina*. Retina is a thin transparent laminar structure situated at the back of the eye which does not differ fundamentally from those of other vertebrates. The space behind the lens, which makes up most of the body of the eye is filled with a fluid called the *vitreous body*. The *lens* of a fish's eye is purely spherical and its shape cannot be adjusted to facilitate focusing on nearer or more distant objects. The accommodation requires the displacement of the lens. The lens is pulled back towards the retina by muscles called retractors to accommodate to near objects, when these muscles are relaxed, the lens returns to its normal position. Remove the lens and put it over a figure on a paper demonstrating it has a high magnifying power.

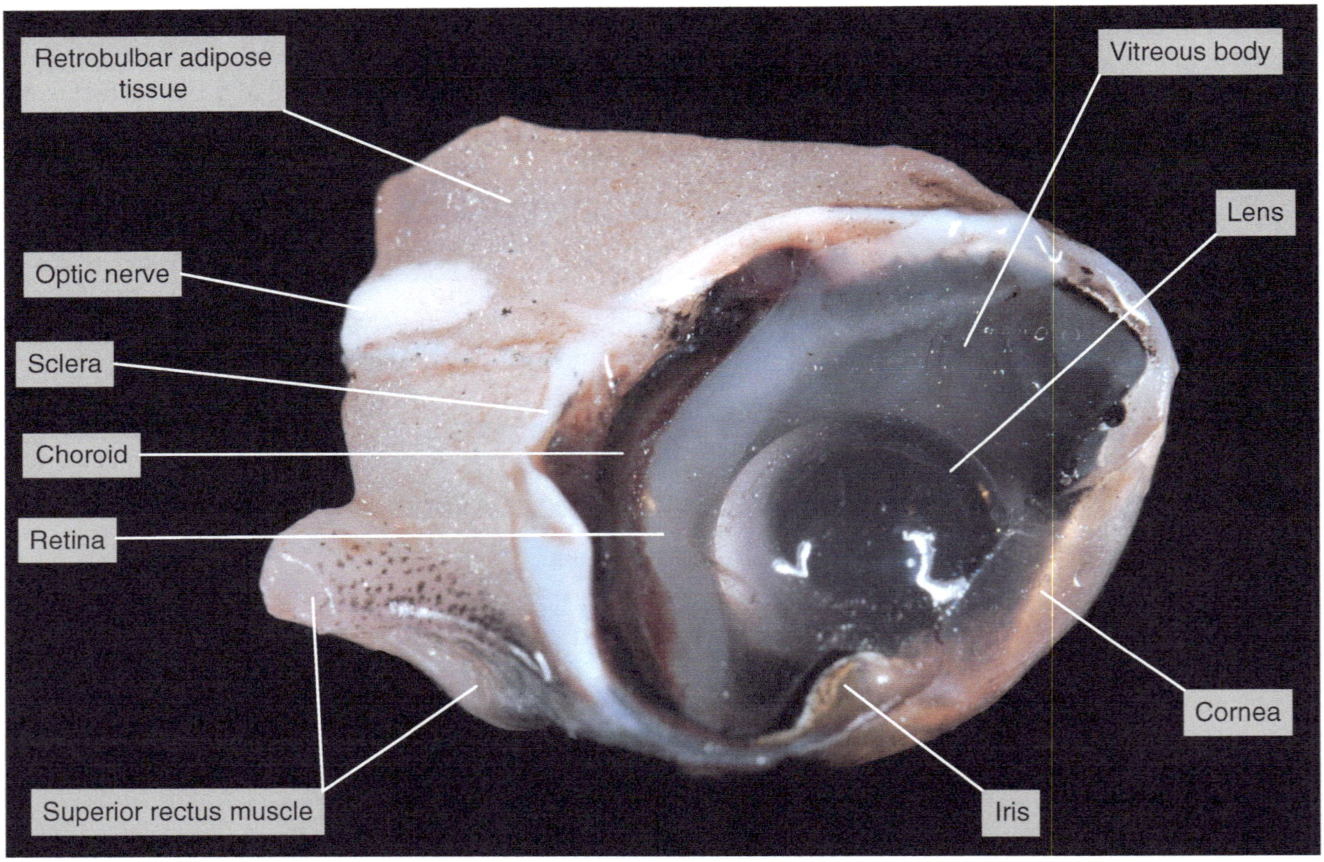

Fig. 9.39 Cross section of the crucian's eye along the vertical midline. Note the spherical lens

Dissection of a Frog (*Rana* sp.) 10

- *Availability:* Amphibians are a transition group between the aquatic and the strictly terrestrial animals. Ranid frogs (*Ranidae*) are almost worldwide in distribution. Their favourite habitats are swamps, low meadows, brooks and ponds, where they feed on insects. In their larval form, the tadpoles develop in water and herbivorous. All amphibian species are protected, so they should not be collected in the nature. We can purchase frogs for dissection from biological supply companies or from fishing firms.
- *Anaesthesia:* Put a wad of cotton wool infiltrated with diethyl ether in a large jar. Place in the frog and close the jar immediately. The dissection of the frog can be started within 20 min, when it doesn't react upon the shaking of the jar. Wash the specimen thoroughly with water.

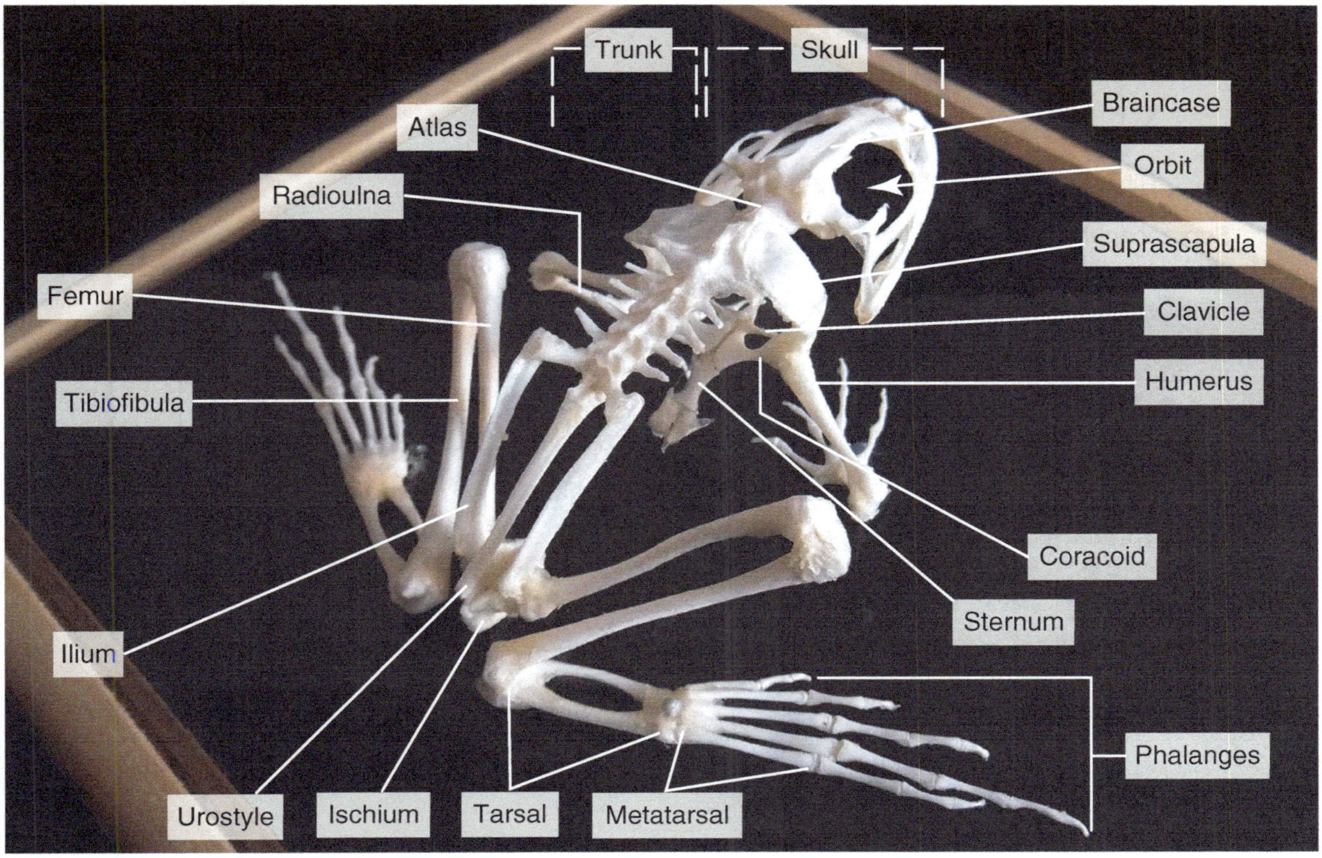

Fig. 10.1 Mounted skeleton of a frog

P. Lőw et al., *Atlas of Animal Anatomy and Histology*, DOI 10.1007/978-3-319-25172-1_10

Skeleton: The skeletal system of the frog serves as a supporting framework for the body. The *axial skeleton* includes the skull, the vertebral column and the sternum. The *appendicular skeleton* includes the pectoral and pelvic girdles and the forelimbs and hindlimbs (Fig. 10.1).

The *skull* includes the cranium or *braincase*, and the *visceral* skeleton, made up of the bones and cartilages of the jaws, the hyoid apparatus and the columellae (little bones of the ears) (Fig. 10.2).

The nasal fossae and the orbital fossae are dorsal openings where the external nares and the eyes are located. Find on the dorsal side of the skull the *nasal* bones; the single *sphenethmoid*; the long *frontoparietals*, which cover much of the brain; the *prootics*, which enclose the inner ears; and the *exoccipitals*, which surround the hind part of the brain (Fig. 10.2).

The upper jaw is formed by the *premaxillae*, the *maxillae* (these bones bear the teeth) and the *quadratojugals*. The *squamosal* supports the cartilaginous auditory capsule. The three-pronged *pterygoid* articulates with the maxillary, the prootic and quadrate cartilage (Figs. 10.2 and 10.3).

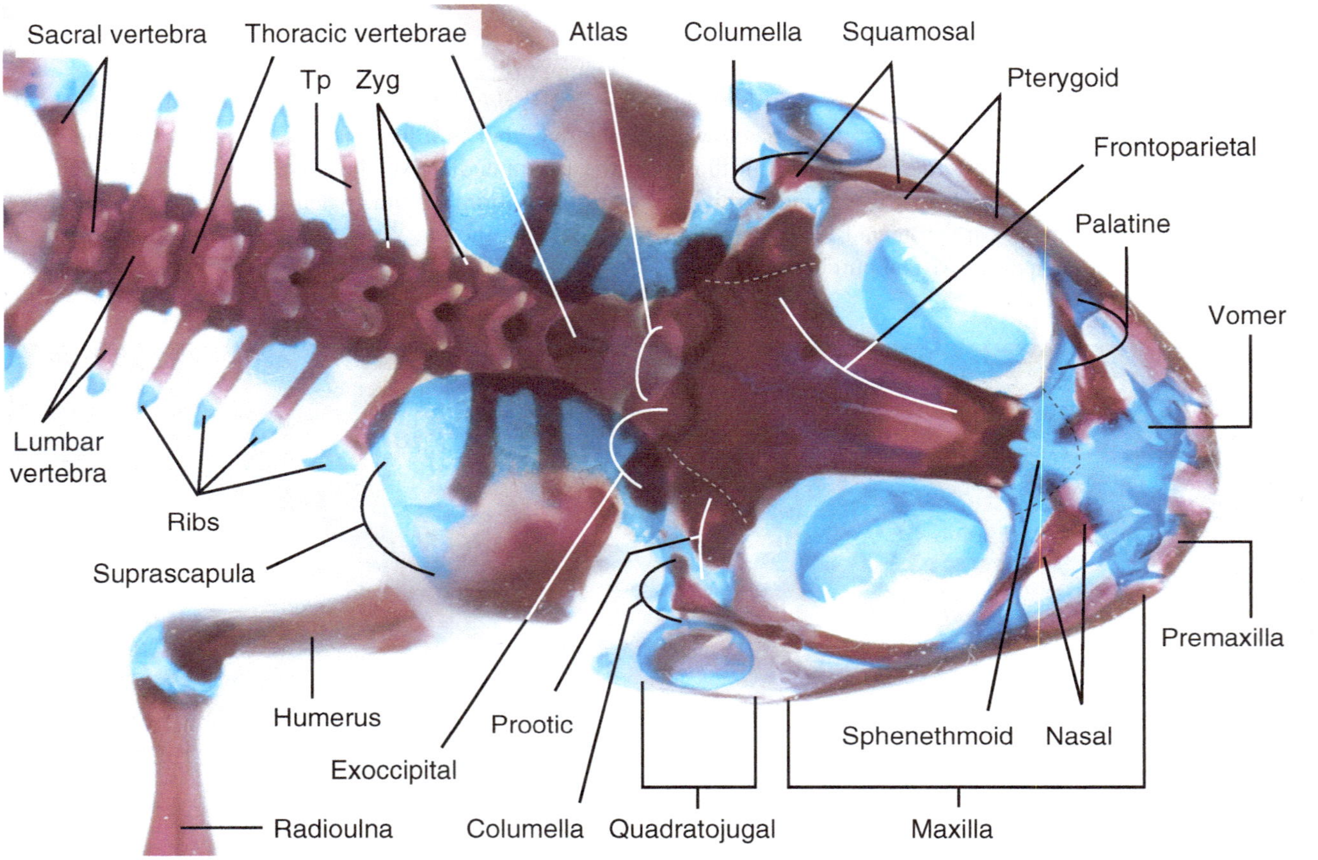

Fig. 10.2 Dorsal view of the anterior part of a frog skeleton after selective staining: cartilages are blue and bones are red (Courtesy of Zsolt Pálfia). *Tp* transverse process, *Zyg* zygapophyses

On the ventral surface of the skull, find the wing-shaped *vomers* (the vomerine teeth are projections of these bones); the slender *palatines*; and the dagger-shaped *parasphenoid*, which forms the floor of the braincase (Fig. 10.2). Because of the lack of a continuous hard, bony palate, the eyes are retractable into the mouth cavity, which has an important function in feeding and breathing.

The lower jaw (the *mandible*) consists of small *mento-meckelians*, long *dentary* bones and the *angulosplenials*. The *hyoid apparatus* lies in the floor of the mouth. It is cartilaginous and supports the tongue and larynx (Fig. 10.3).

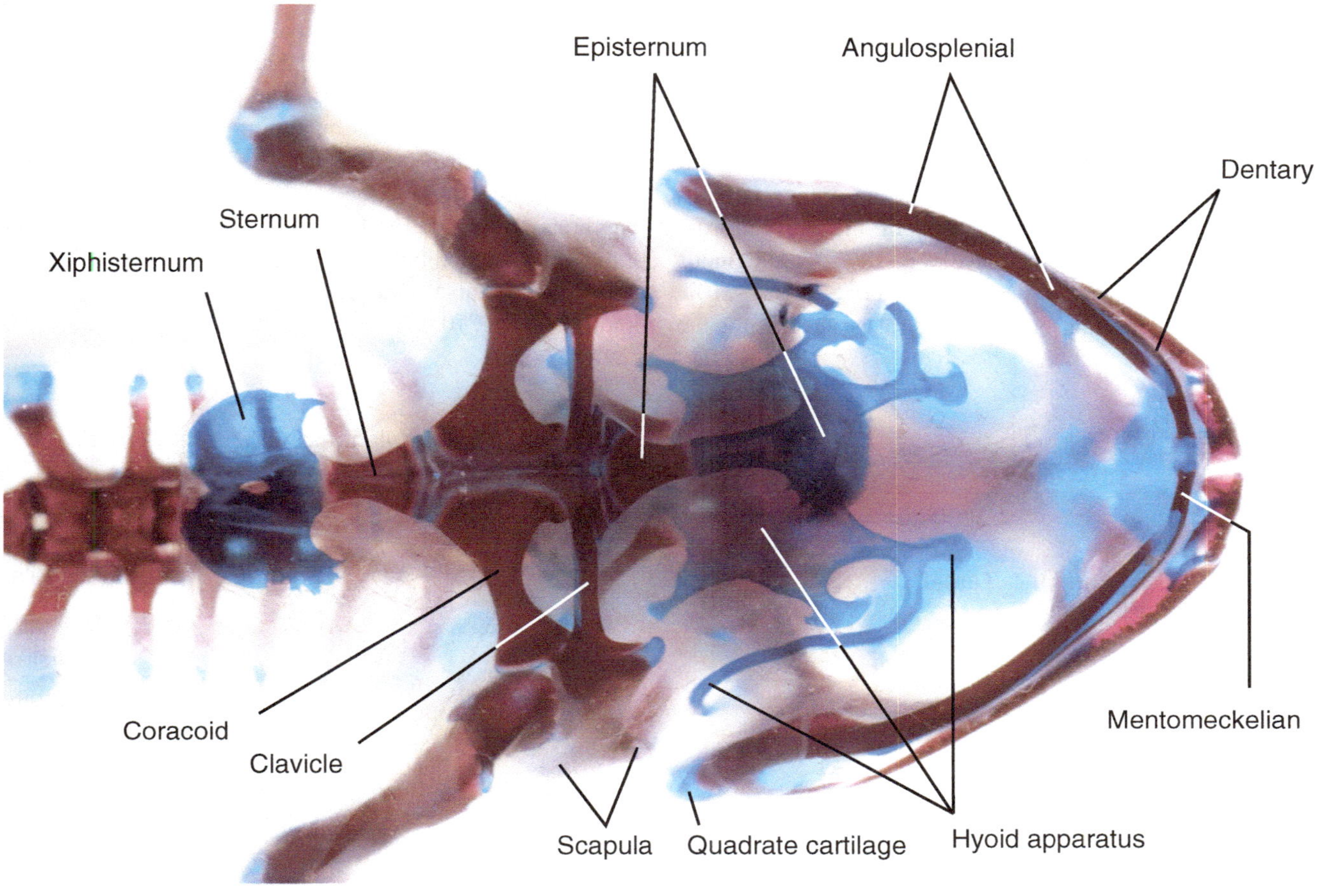

Fig. 10.3 Ventral view of the anterior part of a frog skeleton after selective staining: cartilages are blue and bones are red (Courtesy of Zsolt Pálfia)

The backbone of the frog consists of nine *vertebrae* and the *urostyle*, which probably represents the fusion of several vertebrae (Figs. 10.1, 10.2 and 10.4). The vertebral column no longer functions merely to resist change in length as it does in fish. It must also bear the weight of the body on land. Consequently, each vertebra has become more firmly attached to its neighbours through the interlocking centra (bodies of the vertebrae) and additional articulations, called zygapophyses, which develop on neural arch (Fig. 10.2). The vertebrae have a concave socket at the anterior end of the centrum; they are *procoelous* (not amphicoelous as in fish). The first vertebra, the *atlas*, articulates with the skull (Figs. 10.1 and 10.2). Vertebrae from the second to the seventh are the *thoracic vertebrae*. They are connected to the last vertebra through the eighth, the *lumbar vertebra* (Fig. 10.2). The ninth or *sacral vertebra* has transverse processes for articulation with the ilia of the pelvic girdle (Figs. 10.2 and 10.4). Frogs have very short, cartilaginous *ribs* fused to the transverse processes of vertebrae (Figs. 10.2 and 10.4). The sternum provides ventral protection for the heart and lungs and is a centre for muscular attachment. Its three parts, beginning at the anterior end, are the *episternum*, *sternum and xiphisternum* (Figs. 10.3 and 10.4).

Pectoral girdle and forelimbs: The pectoral girdle supports the forelimbs and articulates with the sternum ventrally. Each half of the girdle includes a *suprascapula* (partly cartilaginous), a *scapula* (vertical) and a *clavicle* lying anterior to the *coracoid* (Figs. 10.2, 10.3 and 10.4). The forelimb consists of the long bones: the *humerus*, the single *radioulna* (ossification of the radius and the ulna) and the bones of the wrist (carpal) and hand (metacarpal and phalanges). Find the bones of the forelimb in Figs. 10.2 and 10.4.

Pelvic girdle and hindlimbs: The pelvic girdle supports the hindlimbs. Each half is made up of the long *ilium*, the anterior *pubis* and posterior *ischium* (Figs. 10.1 and 10.4). The hindlimb is divided the same way as the forelimb; the long bones are called the *femur*, the *tibiofibula* (ossification of the tibia and the fibula) and the bones of the ankle (tarsal) and foot (metatarsal and phalanges). See the bones of the hindlimb in Fig. 10.4.

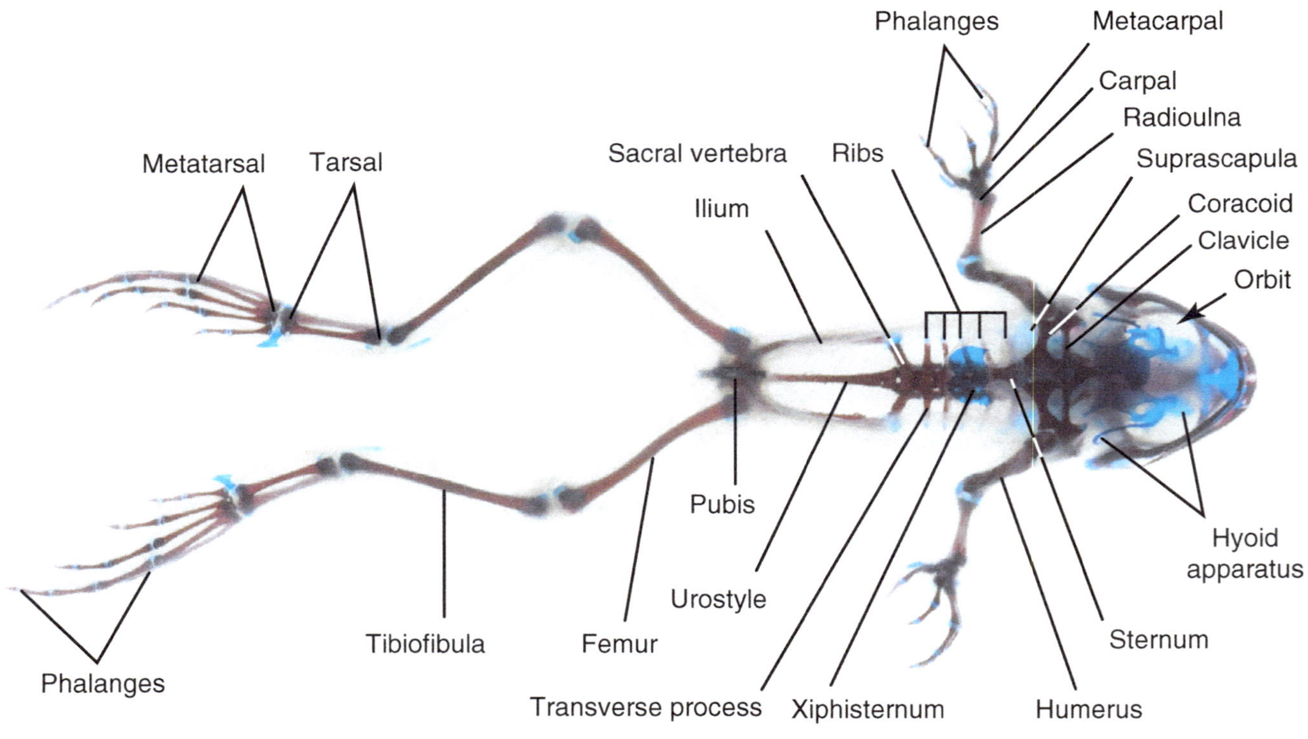

Fig. 10.4 Ventral view of a frog skeleton after selective staining: cartilages are *blue* and bones are *red* (Courtesy of Zsolt Pálfia)

Rinse the frog with water then first observe the external features of the animal before dissection. Observe the *head*, the *trunk* and the *appendages* (Fig. 10.5).

Fig. 10.5 Dorsal view of a male frog

Note the difference in the pigmentation between the dorsal and ventral side (Fig. 10.6).

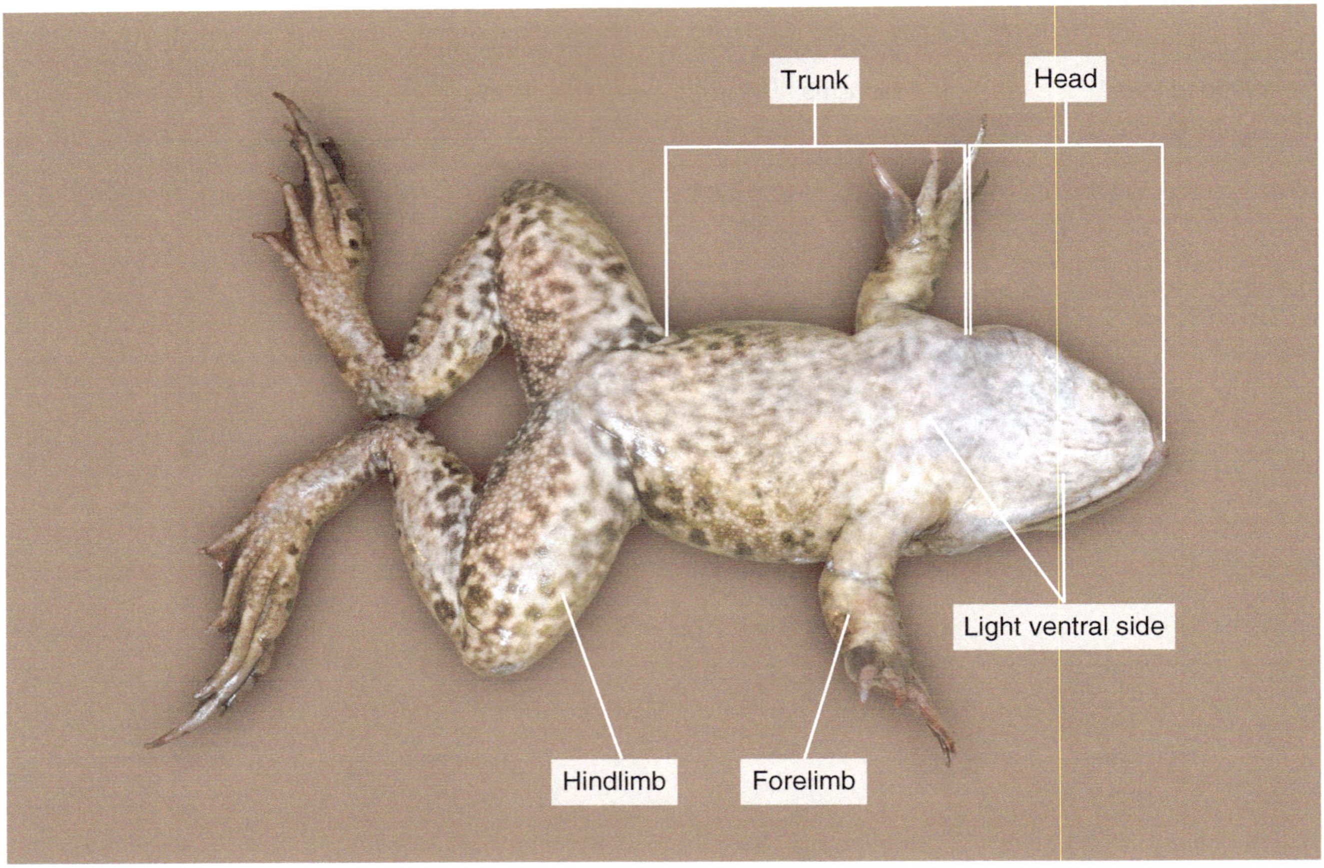

Fig. 10.6 Ventral view of a male frog

On the *forelimbs*, identify the arm, forearm, wrist, hand and digits. Identify the sex of the frog. The easiest way to quickly tell the difference between male and female frogs is to look at their forelimbs (Figs. 10.7 and 10.8).

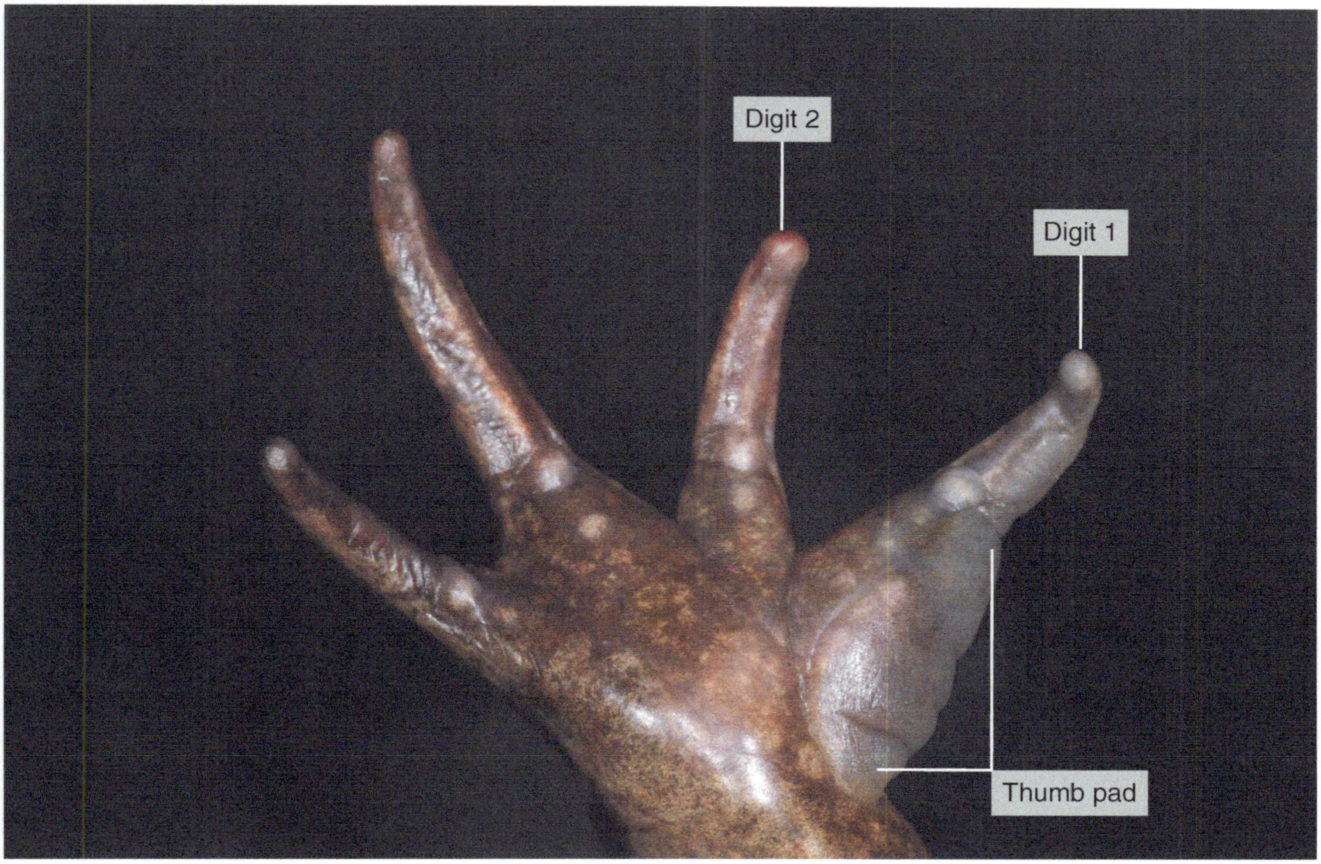

Fig. 10.7 The forelimb of a male frog with the thumb pad

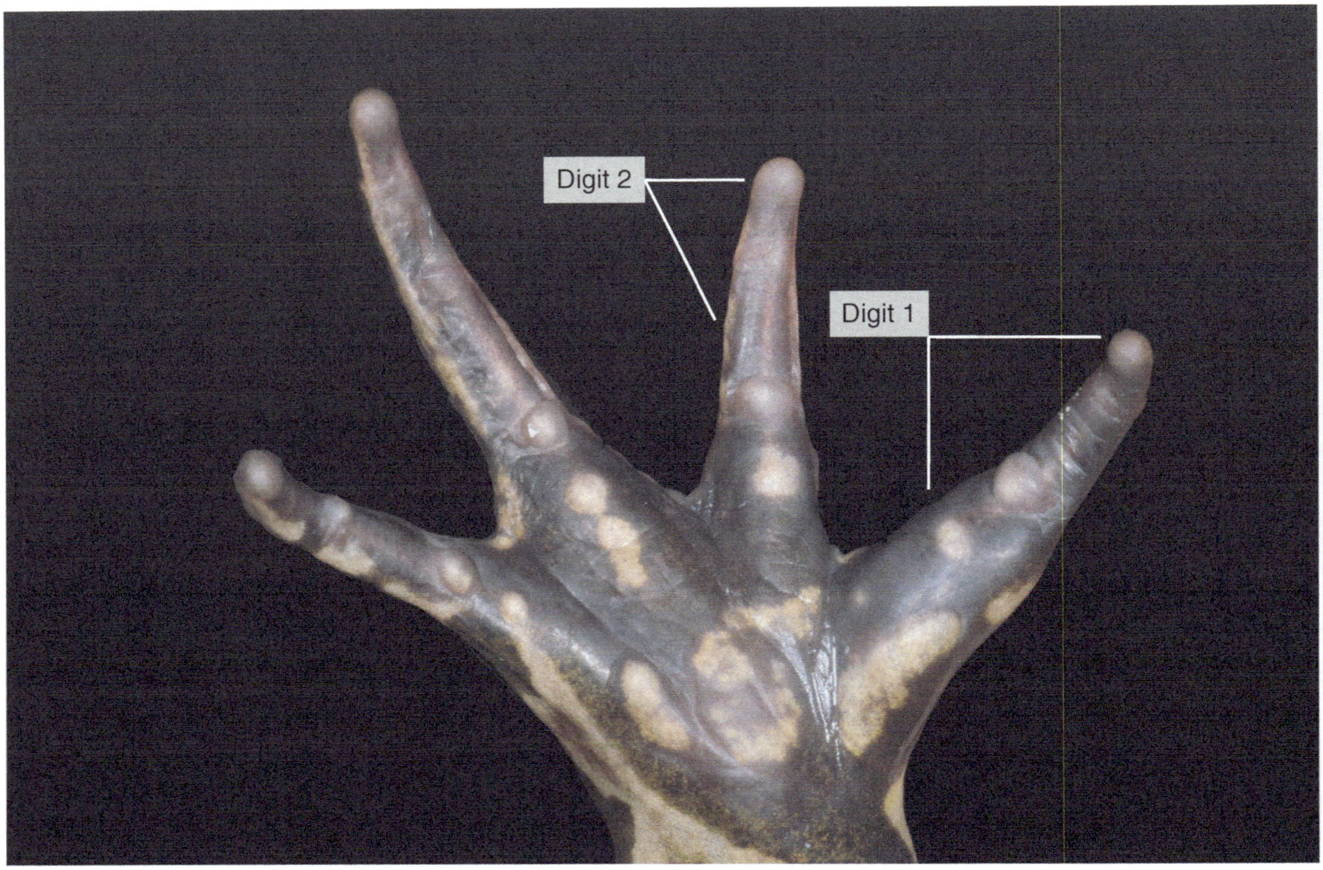

Fig. 10.8 The forelimb of a female frog

The forelimb of male frogs should have a fatter thumb pad, and the thumb should look bulbous and fatter than the thin fingers of the female frog. During the breeding season, the first toe (thumb) of the male is enlarged into a *nuptial pad* for clasping the female, which help to release the eggs during mating. On the *hindlimbs*, find the thigh, shank, ankle, foot and digits. Observe the *webbed toes* (Fig. 10.5). Find the *cloacal opening* at the posterior end of the body (Fig. 10.9).

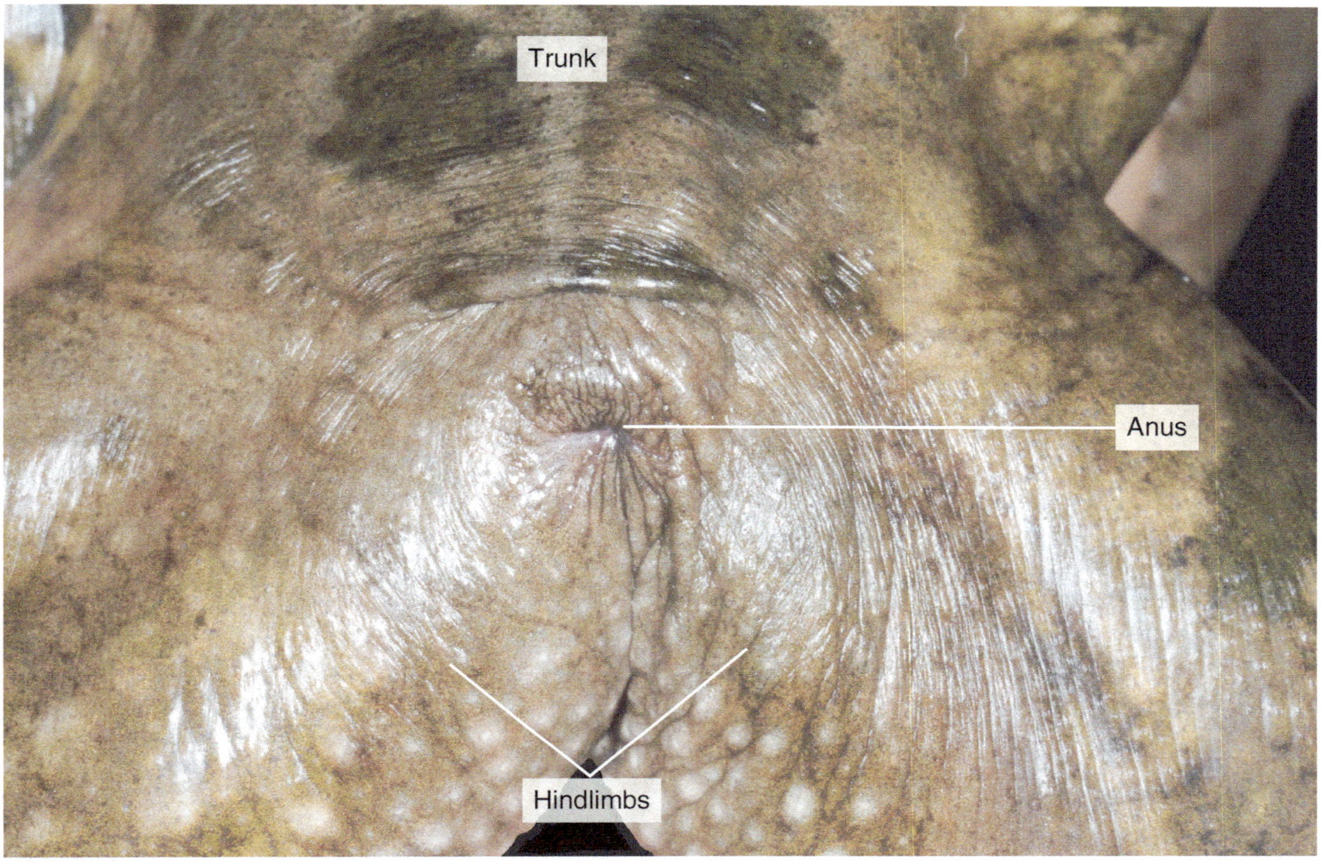

Fig. 10.9 The opening of cloaca is shifted a little towards the dorsal side

Examine the head. The *external nares* are the technical term for the frog's nostrils, which are used for breathing and should be far forward, above the mouth opening (Fig. 10.10).

The *eyes* are protected by a thin, transparent *nictitating membrane*, so the frog can see under water (Fig. 10.10). The *tympanic membrane* (eardrum) is a circular region of tightly drawn skin located just behind the eyes. The frog has no external ear, only middle and inner ears. Open the *mouth* widely to examine the inside (Figs. 10.11 and 10.12).

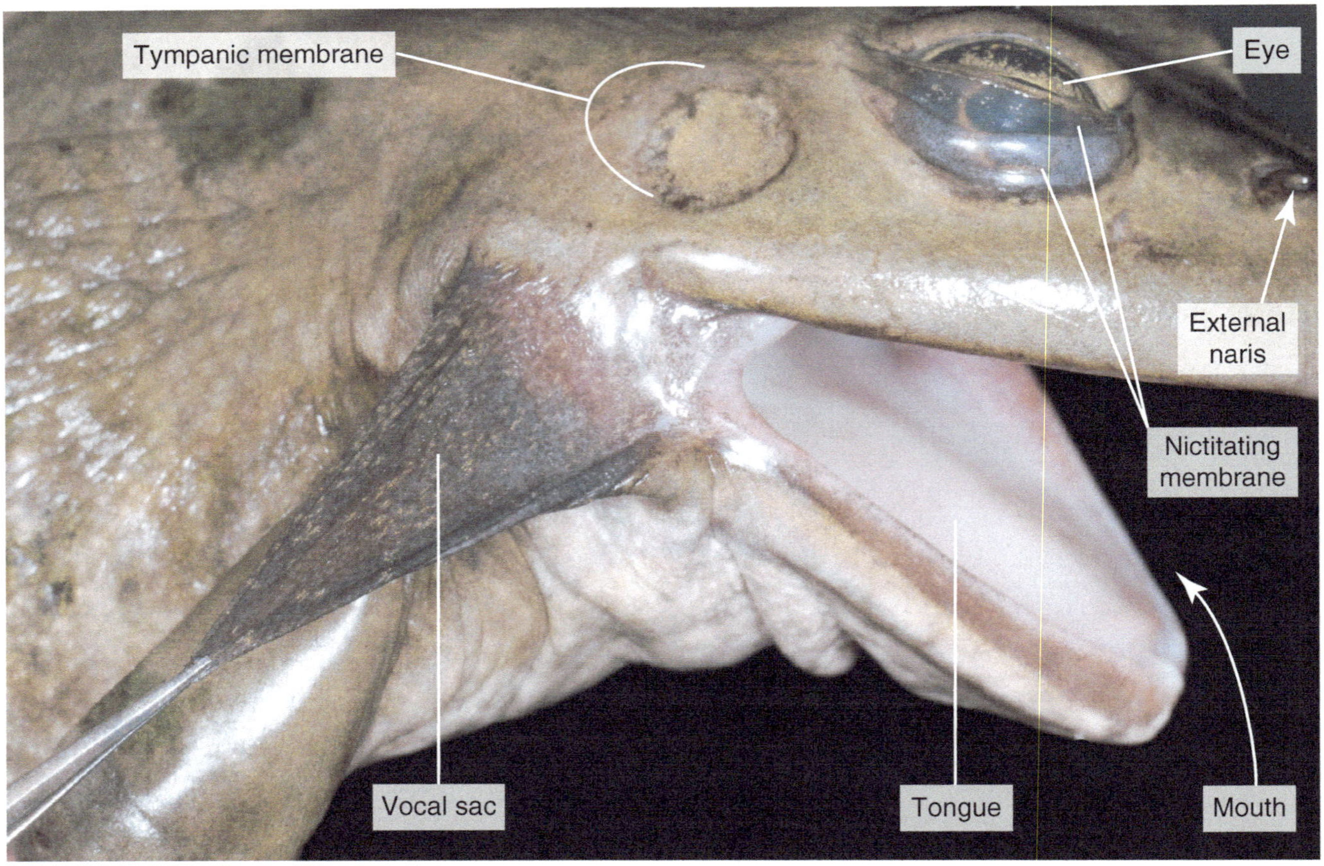

Fig. 10.10 The head of a male frog with pulled vocal sac

Feel the *maxillary teeth* along the upper jaw and the *vomerine teeth* in the roof of the mouth, though both are used to secure prey and keep it secure in the mouth. Find the *internal nares* (singular naris) in the roof of the mouth behind the vomerine teeth, and note how they connect with the external nares (Fig. 10.11). Note how the ridge on the lower jaw fits into a groove in the upper jaw to make the mouth closure airtight. This is important in the frog's respiratory movements. The *Eustachian tubes* are to the left and right of the back of the throat and are used to equalise pressure between the middle ear and the environment (Fig. 10.11). Through the wide Eustachian tube, the columella can be spotted also. In male frogs, openings on the floor of the mouth slightly anterior to the Eustachian tubes lead to *vocal sacs*, which, when inflated, serve as resonators to intensify the mating call (Fig. 10.10). Examine the *tongue*, which is quite large and stretchy, and because of its sticky dorsal surface, it is used for capture of the prey (Fig. 10.11). It is attached in the front of the mouth (Fig. 10.12).

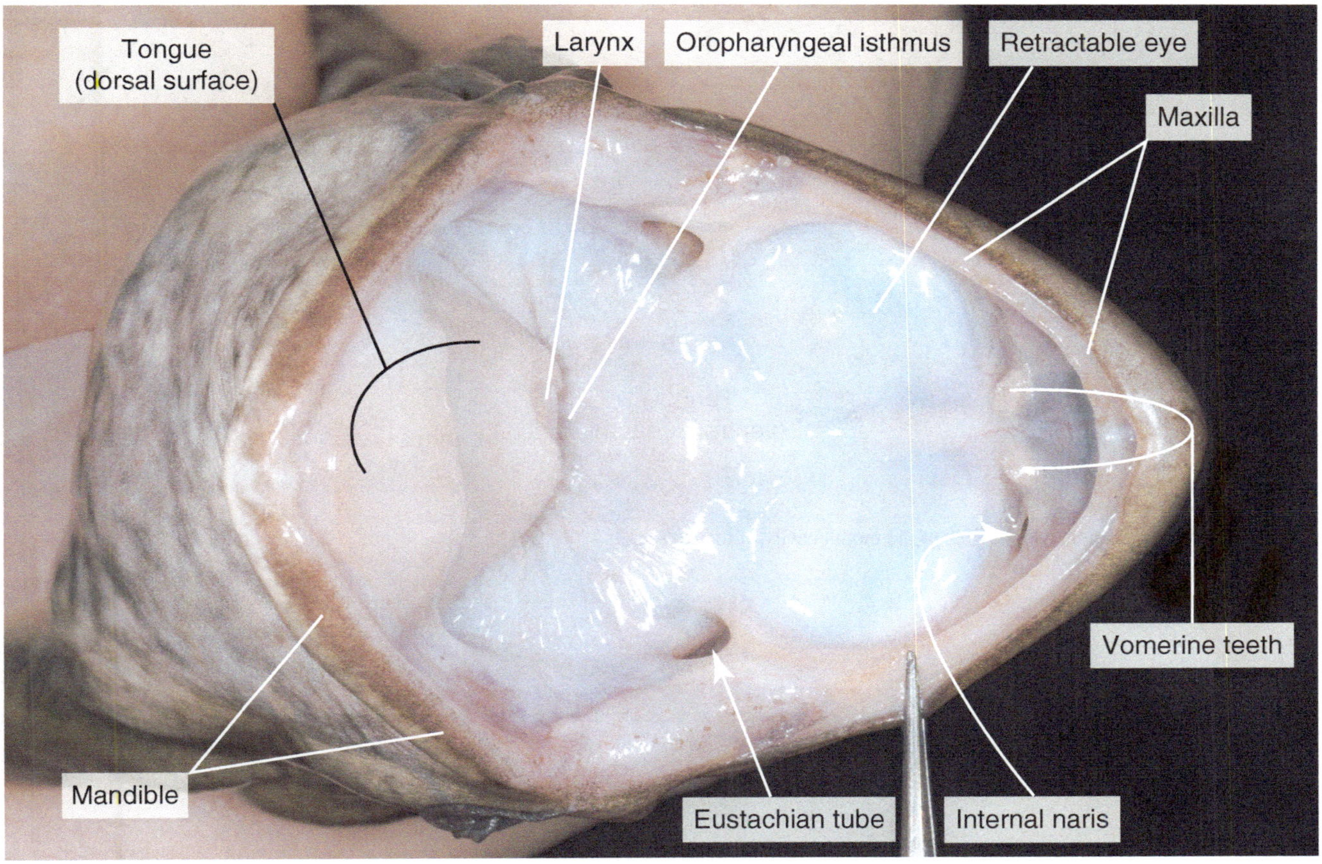

Fig. 10.11 Structures on the dorsal side of the mouth cavity of the frog. Because of the lack of a continuous hard, bony palate, the eyes are retractable into the mouth cavity

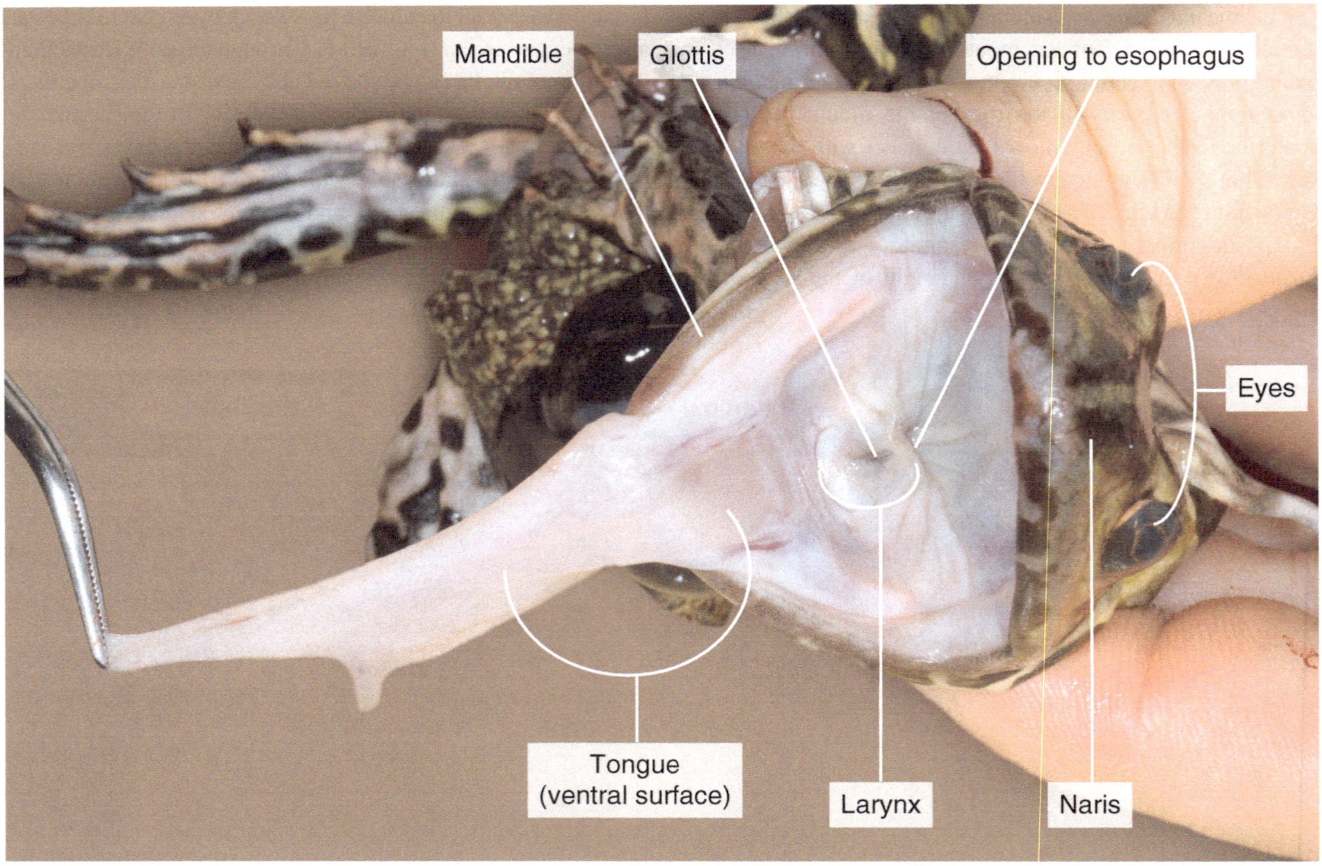

Fig. 10.12 Structures on the ventral side of the mouth cavity of the frog

The frog *tongue* has two significantly different surfaces (Fig. 10.13). The ventral surface is folded and covered by ciliated columnar epithelium with mucous goblet cells (GCs). The dorsal surface forms papillae, and it is covered with non-ciliated columnar epithelium without goblet cells. The *papillae* are fingerlike filiform and wider fungiform. There is a *gustatory disc* (GD) on the top of the latter, which is formed by tall, regular, light columnar cells. Several tubular *lingual glands* are visible beneath the dorsal surface: they secrete protein-rich, sticky material for catching the prey. Lingual glands open onto the surface between papillae. The propria contains lymphoid cells, nerve branches and blood vessels (BVs), and it is attached to the striated muscle network (MU) of the tongue (Fig. 10.13).

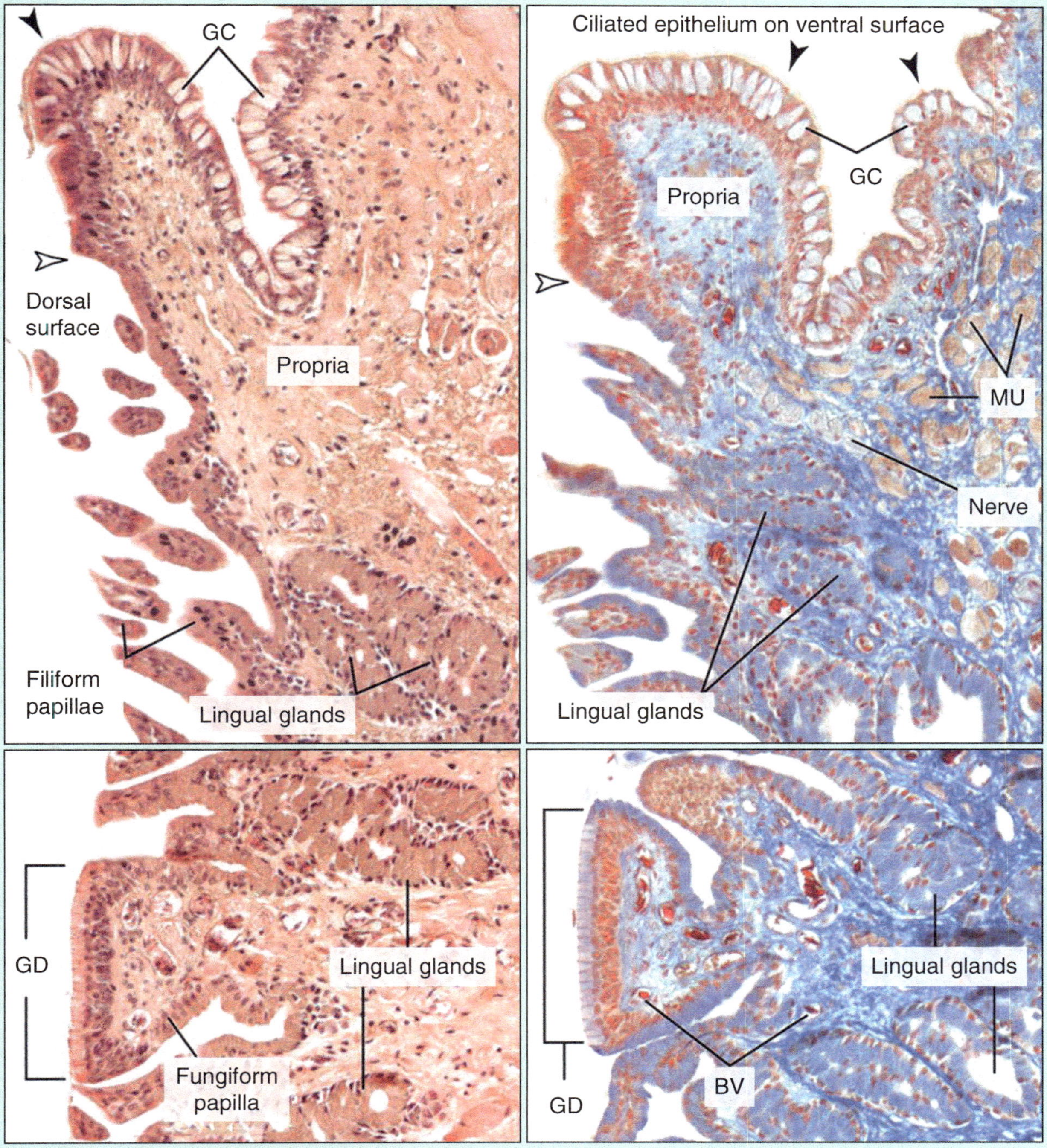

Fig. 10.13 Histology of the surfaces of the frog's tongue (*left*: HE, *right*: Azan). *Black arrowheads* ciliated columnar epithelium, *white arrowheads* dividing line between dorsal and ventral surface, *BV* blood vessels, *GC* goblet cells, *GD* gustatory discs, *MU* striated muscles

You should be able to see the opening of the oesophagus and a slight elevation in the floor of the mouth, containing the *glottis*, a slitlike opening to the *larynx*. Probe through the glottis into the pharynx (Fig. 10.14).

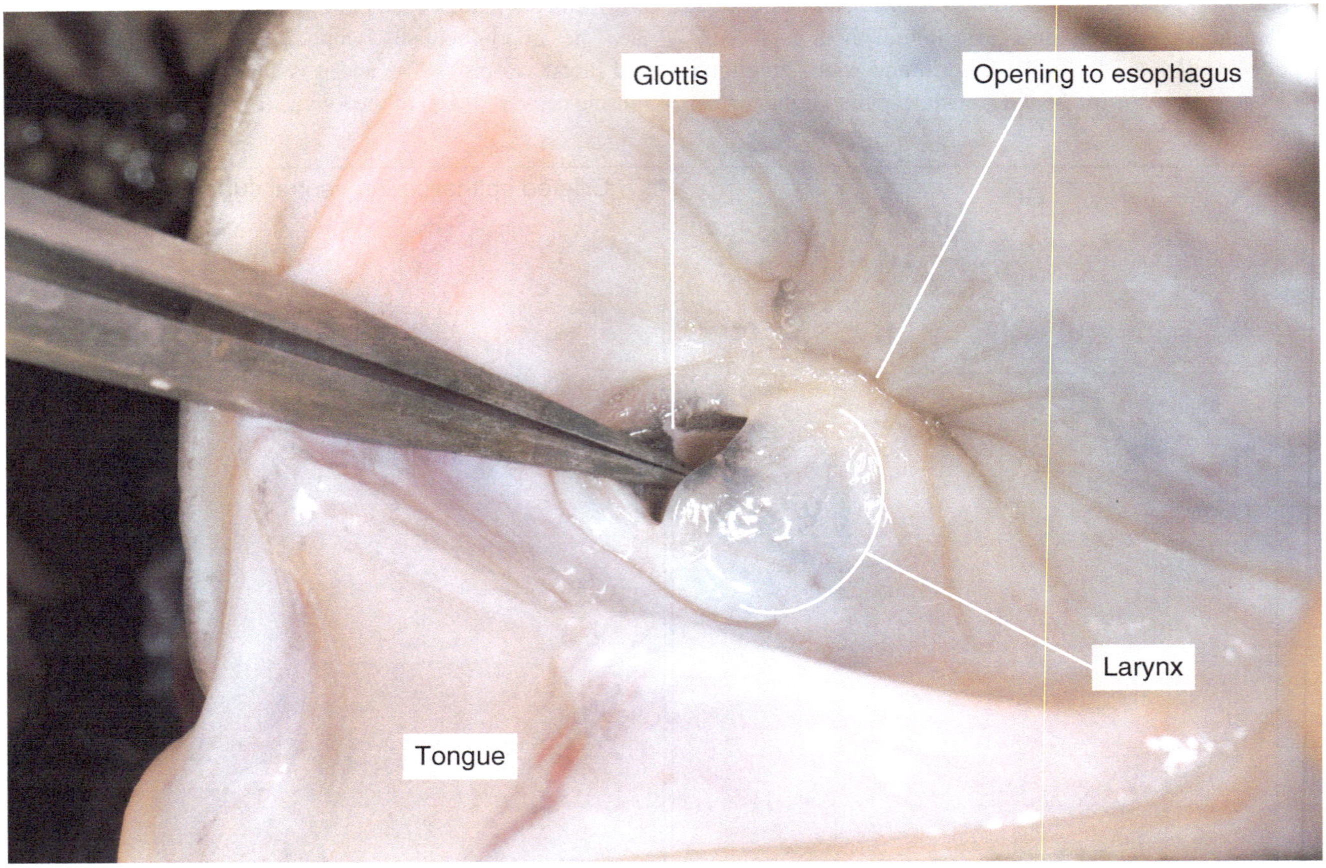

Fig. 10.14 Posterior part of the floor of the mouth with the glottis and the opening of the oesophagus

To start the dissection, lift the skin of the frog away from the body wall; cut around the lower trunk region in a girdle (Fig. 10.15).

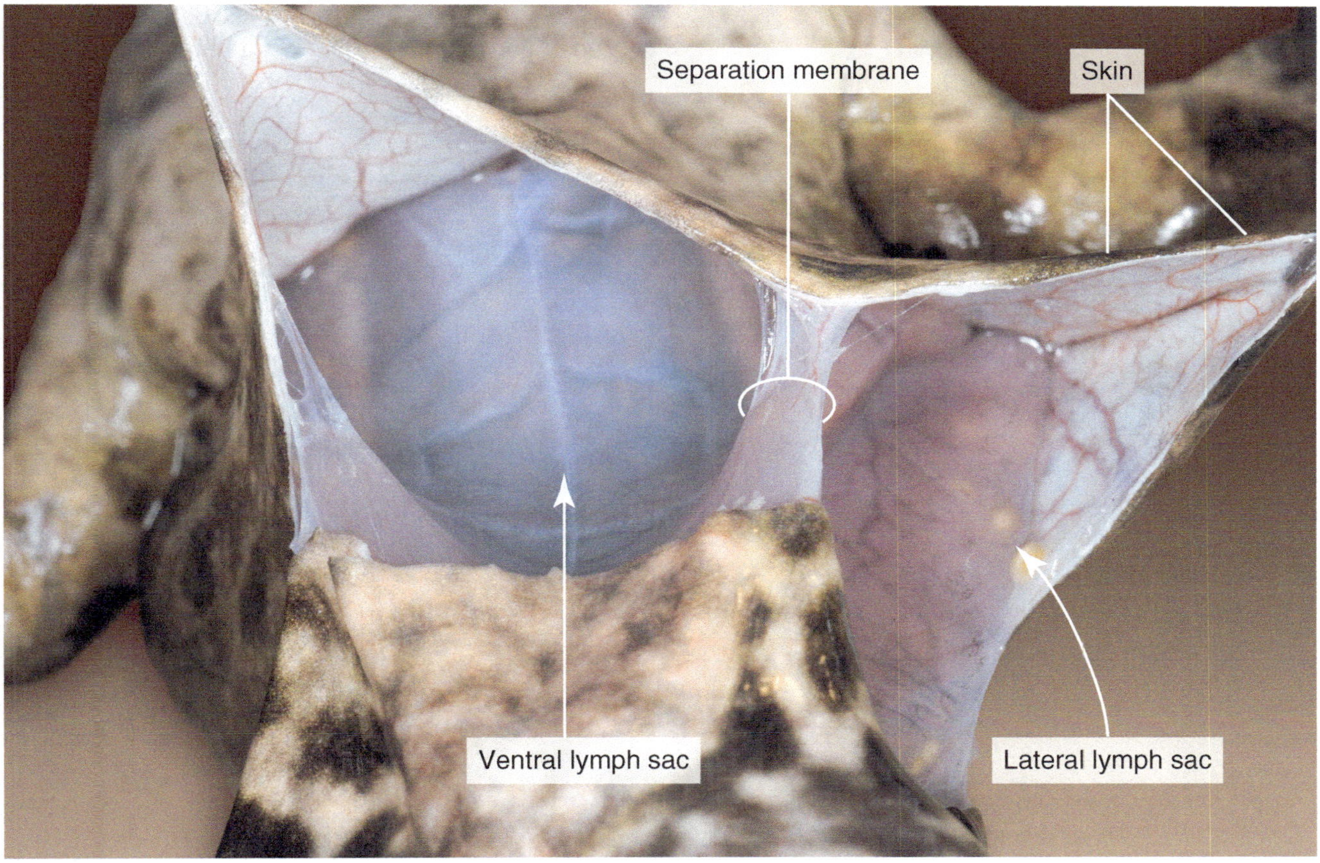

Fig. 10.15 The special lymphatic spaces, the lymph sacs underneath the frog's skin

The four (ventral, two lateral and dorsal) *lymph sacs* will be opened this way which are situated under the frog's skin. They prevent the skin from drying as liquid reservoirs. You have to cut carefully the separation membranes between the lymph sacs. Turn the specimen on the ventral side, pinch the dorsal skin and observe the skin nerves reaching it (Fig. 10.16).

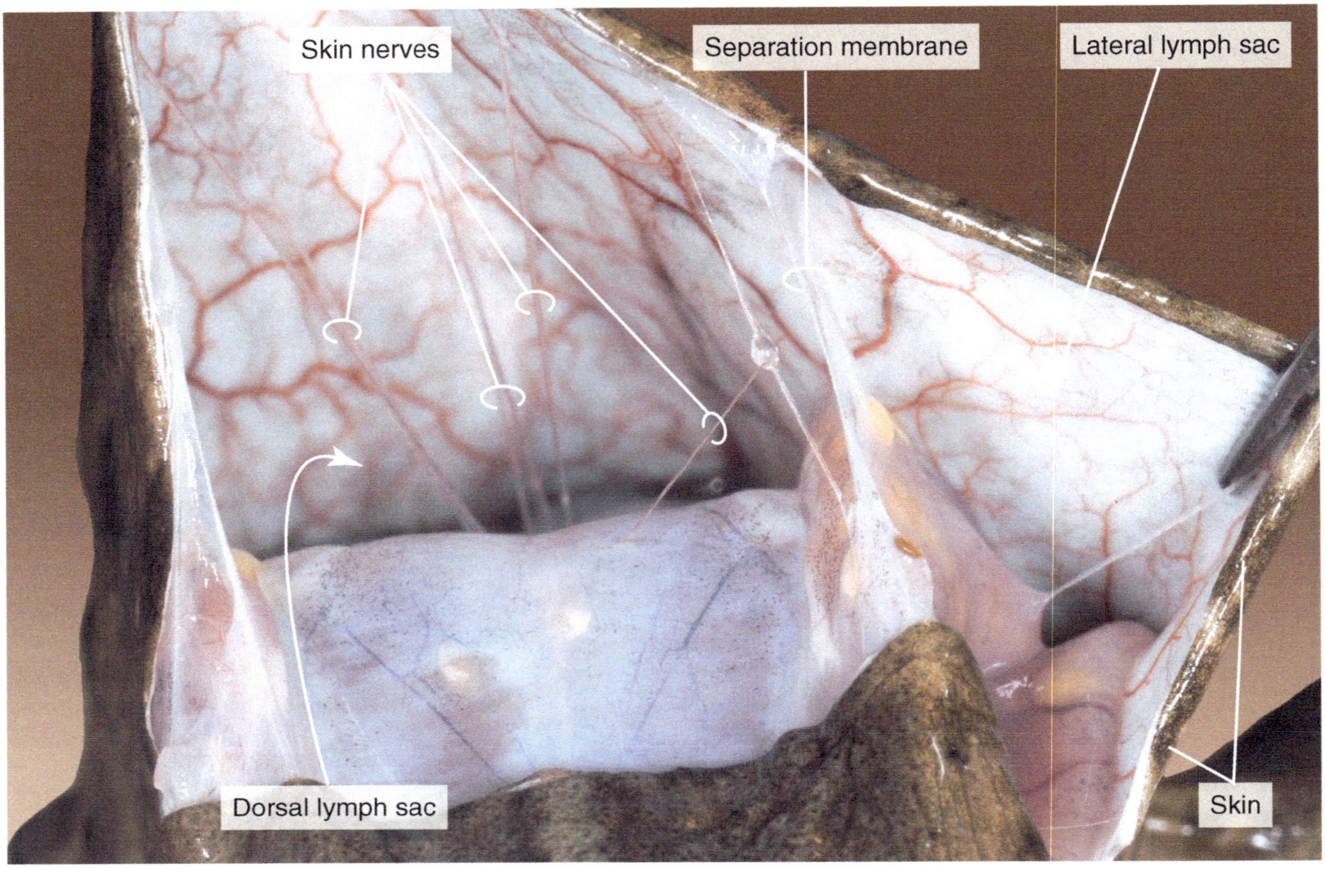

Fig. 10.16 The skin nerves span the lymph sac supplying the skin

The *epidermis* of the frog skin is a slightly keratinized, stratified squamous epithelium (Fig. 10.17).

Its *germinal layer* (GL) is conspicuous because it is composed of proliferating, columnar shaped stem cells. The differentiating daughter cells become flattened by the time they reach the surface. The corneal layer is composed with only one cell layer. *Melanophores* (MEs) frequently immigrate into the epithelium. Some capillaries may occur in the epidermis because of the respiratory function of the skin. The *dermis* contains two types of subepithelial glands, blood vessels and melanophores. The upper, smaller glands are *mucous glands* (MGs) with cuboidal-columnar secretory epithelial cells: they secrete hygroscopic mucous layer onto the surface to keep it moist. They are surrounded by melanophores. The bigger glands are *poison glands* (PGs) lined with flattened epithelial cells and filled with granular secretion. They occupy lower layer of dermis. These glands are not evenly distributed on the body surface of the frog; they are concentrated into two rows in the dorsal skin (Fig. 10.5). They have a protective function as their irritant secretion makes the frog distasteful for most predators.

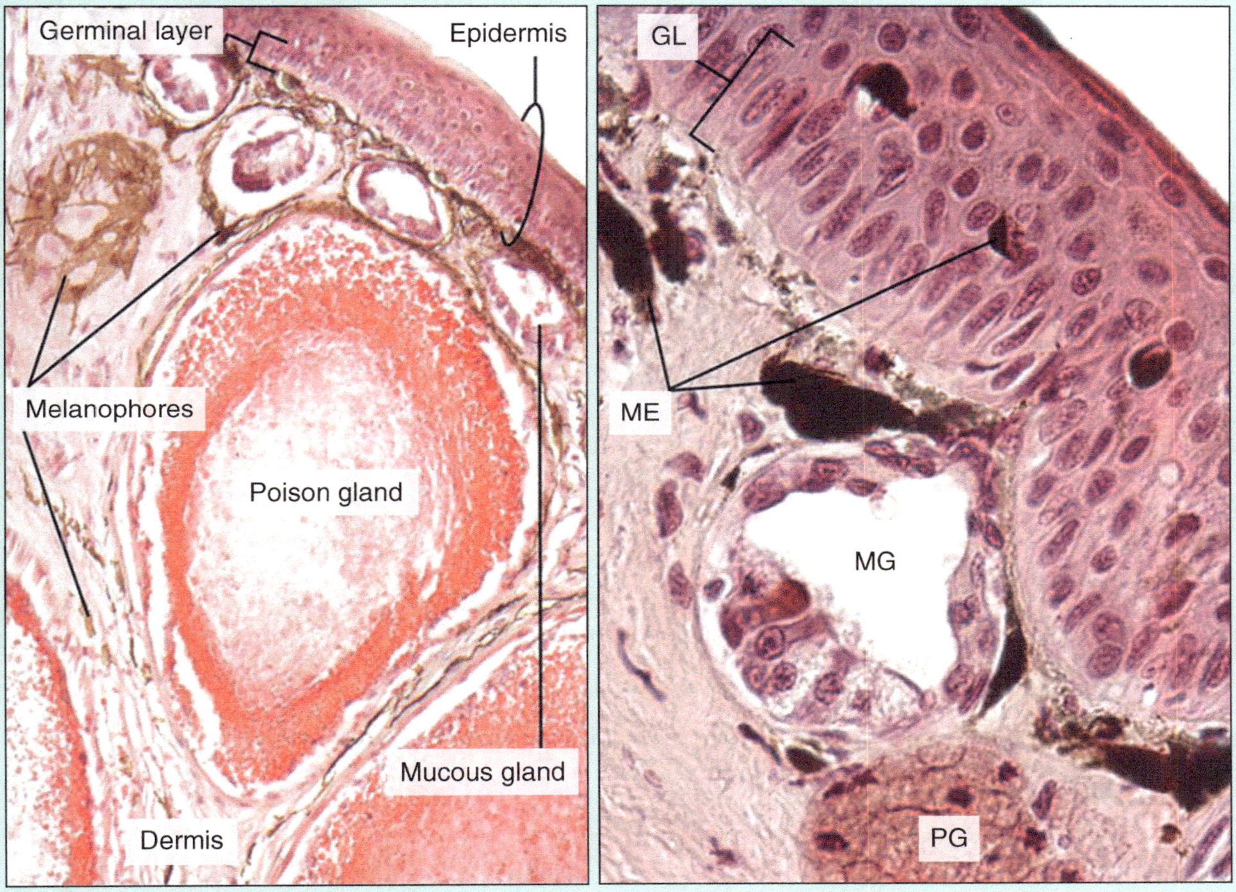

Fig. 10.17 Histological sections of the frog's skin (HE). *GL* germinal layer, *ME* melanophores, *MG* mucous gland, *PG* poison gland

Now place your specimen in the dissection pan. The frog should be lying on its dorsal side with its belly facing up. In the majority of the following figures demonstrating the phases of the dissection, the *anterior* (head) region of the frog is towards the *upper* or *right* margins. In those exceptional cases where the animal is in a different position and the anterior (head) region faces to the left, this is indicated in the legend of the figure. Make a median ventral incision from the girdle up to the mandibular edge and deflect the skin laterally, to show the *linea alba* running along the midventral surface of the abdomen between the *rectus abdominis* muscles (Fig. 10.18).

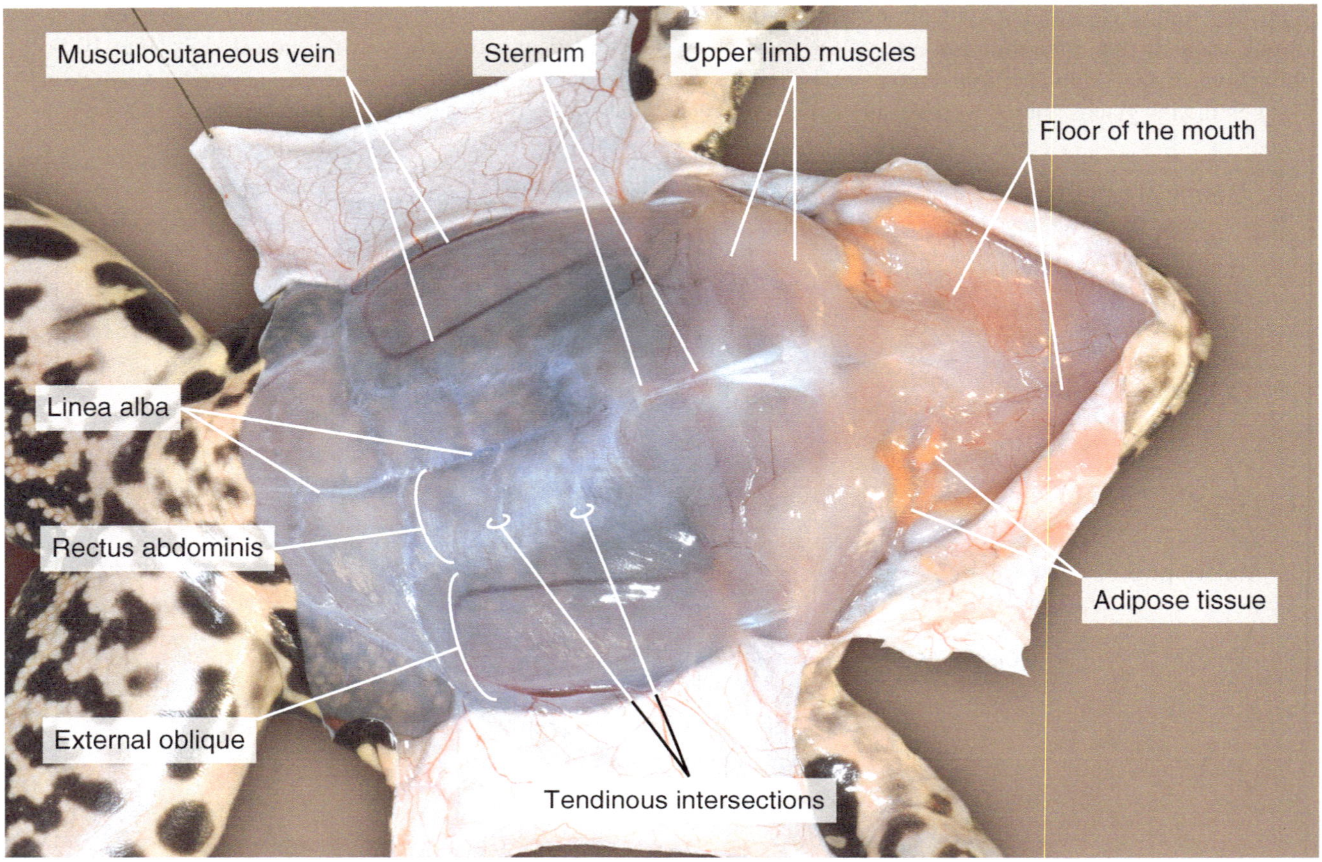

Fig. 10.18 Under the cut and deflected skin appear the abdominal musculature and the double bow of the musculocutaneous veins

Take care not to damage the musculocutaneous veins which join the body wall from the skin (Fig. 10.19).

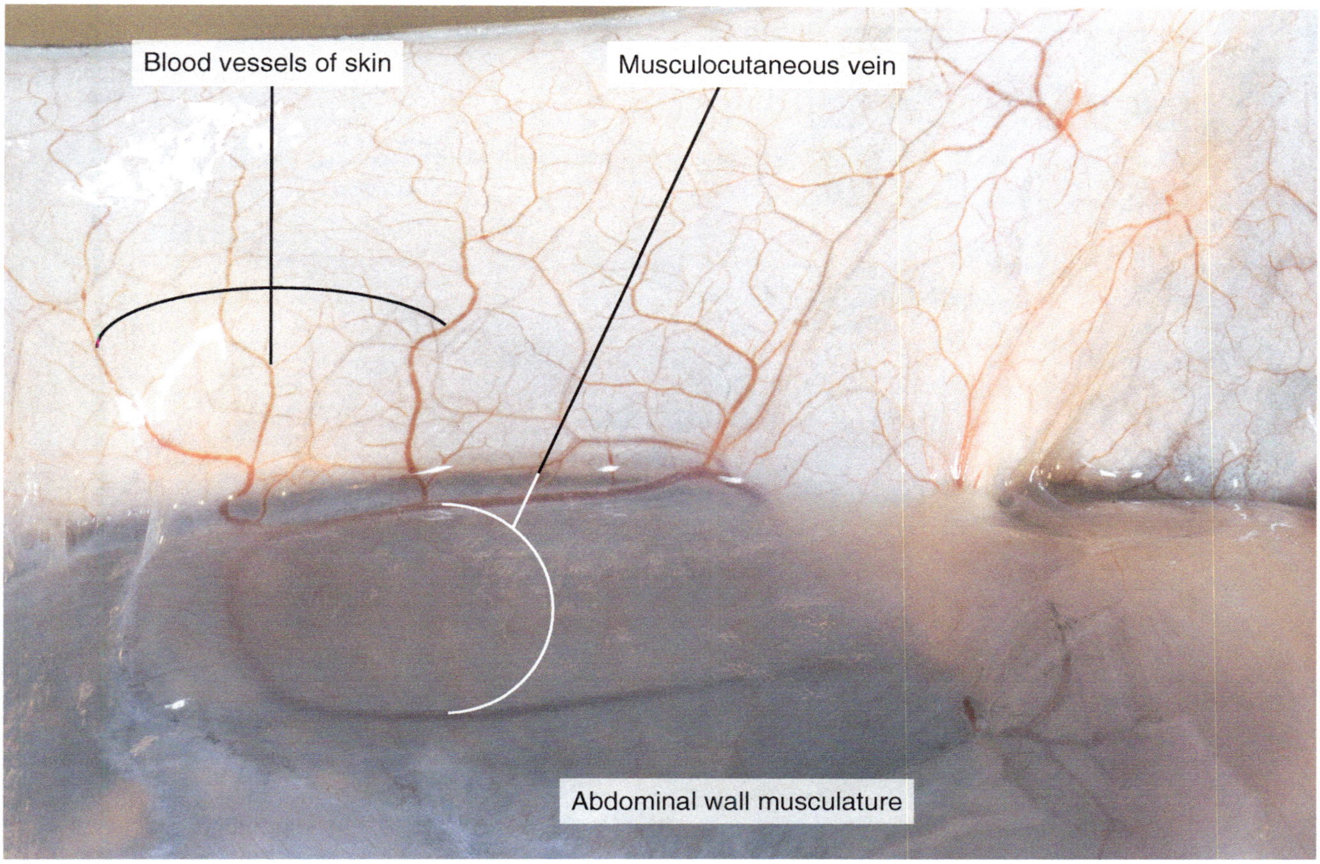

Fig. 10.19 The musculocutaneous vein collects the oxygenated blood from the skin then turns under the body wall

Use forceps to lift the abdominal muscles away from the body cavity. Cut the ventral abdominal muscle parasagittally, on the right side of the ventral abdominal vein with a small pair of scissors. Take care not to damage the underlying structures (Fig. 10.20).

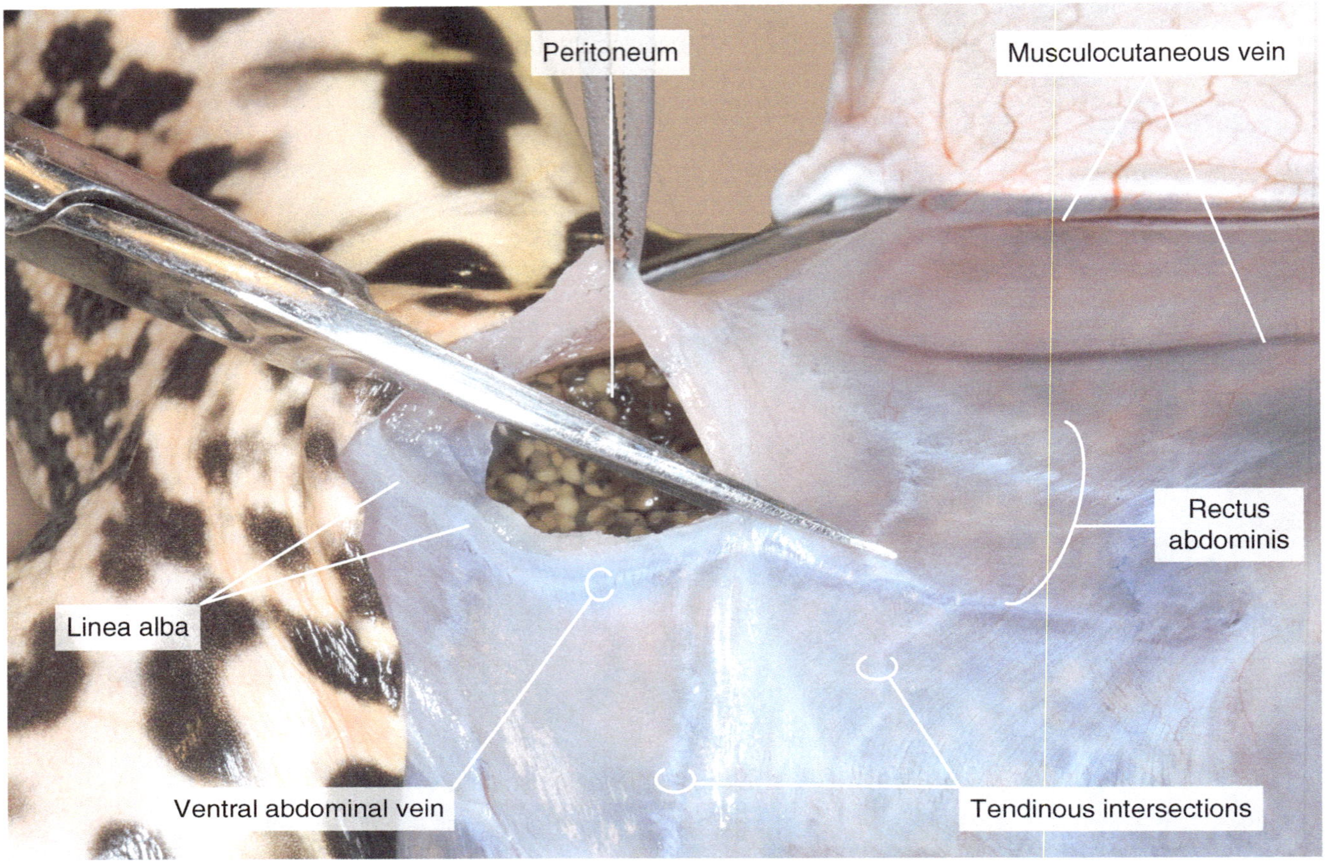

Fig. 10.20 Incisure made on the ventral abdominal muscle on the right side of the ventral abdominal vein

Take particular care where the ventral abdominal vein enters the liver anteriorly and where it divides into the pelvic veins posteriorly (Fig. 10.21).

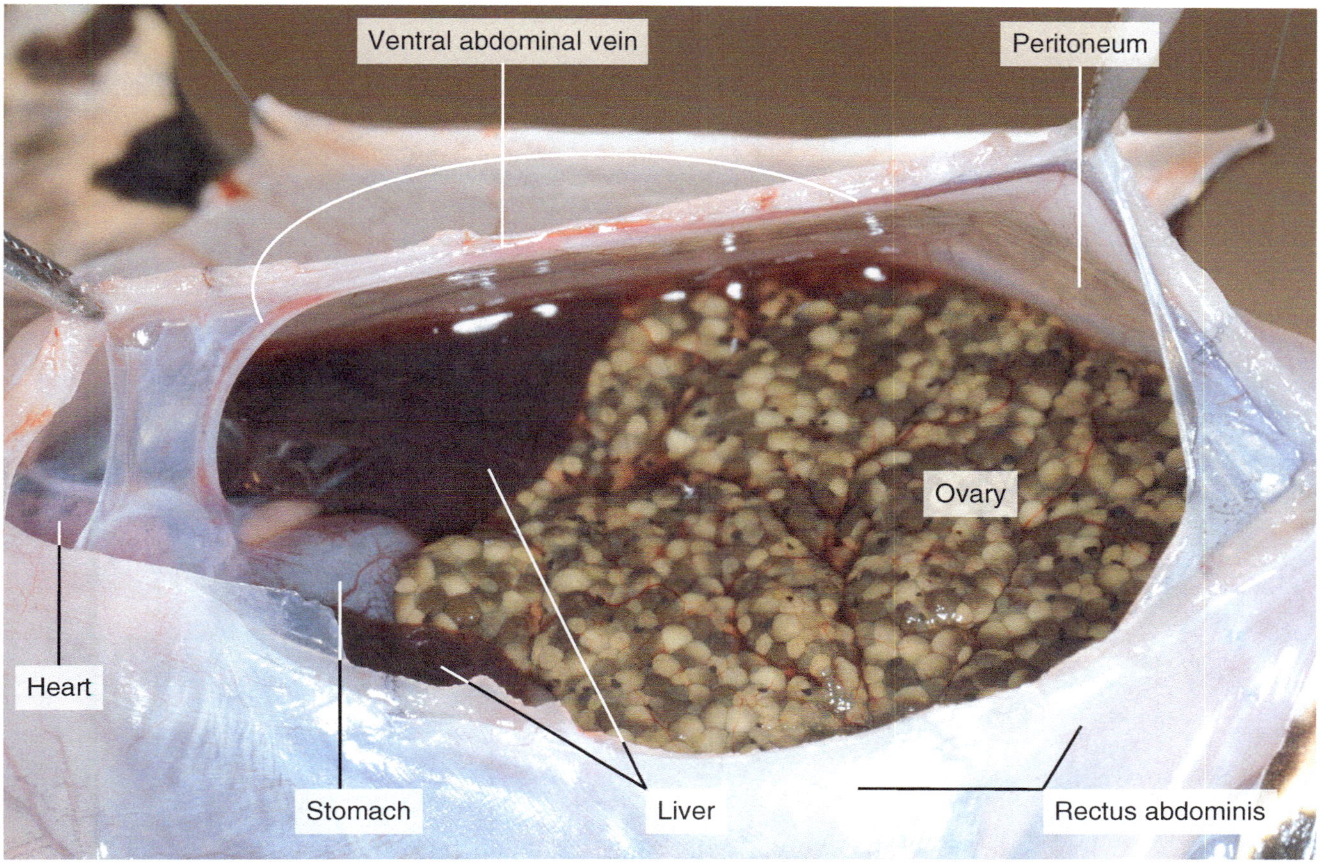

Fig. 10.21 The ventral abdominal vein enters the liver anteriorly (Anterior end is to the left.)

Lift the pectoral girdle by the xiphisternum clear by gently cutting the connective tissue. Keep the scissors horizontal and the pectoral girdle lifted in order to avoid damage to the underlying heart and blood vessels. Cut through the midsection of the pectoral girdle. Lift up the right half of the pectoral girdle and gently free it from the underlying connective tissue, taking care not to damage the underlying veins. Then cut the pectoral girdle close to its connection with the humerus (Fig. 10.22).

To fold the body wall on the side, cut the ventral abdominal vein at the liver.

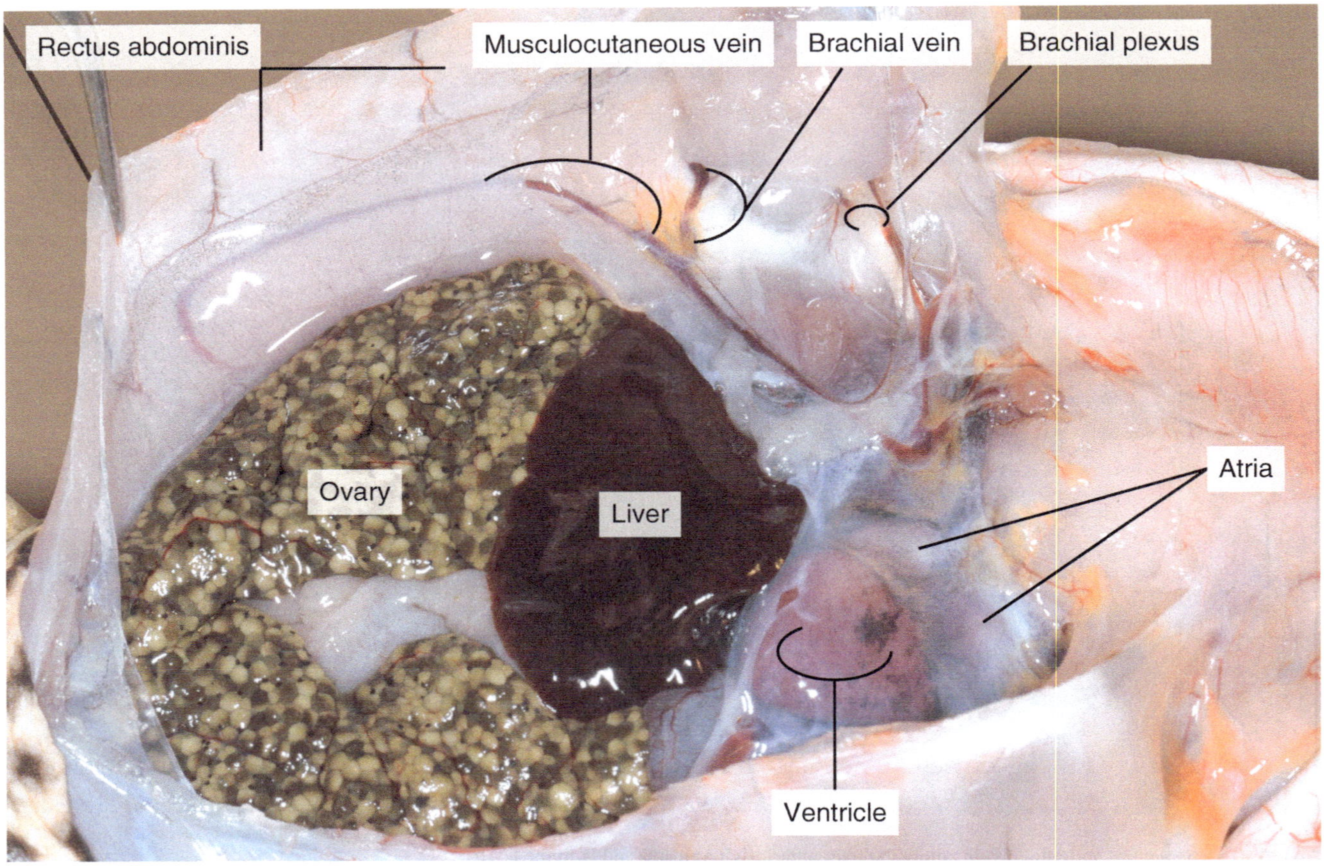

Fig. 10.22 Disclosure of the connection of the musculocutaneous vein with the brachial vein (right part of the pectoral girdle is removed)

First display the anterior venous system. Clear the tissue from the anterior veins to display this system more clearly. Dissect carefully and do not cut and injure the blood vessels (Fig. 10.23).

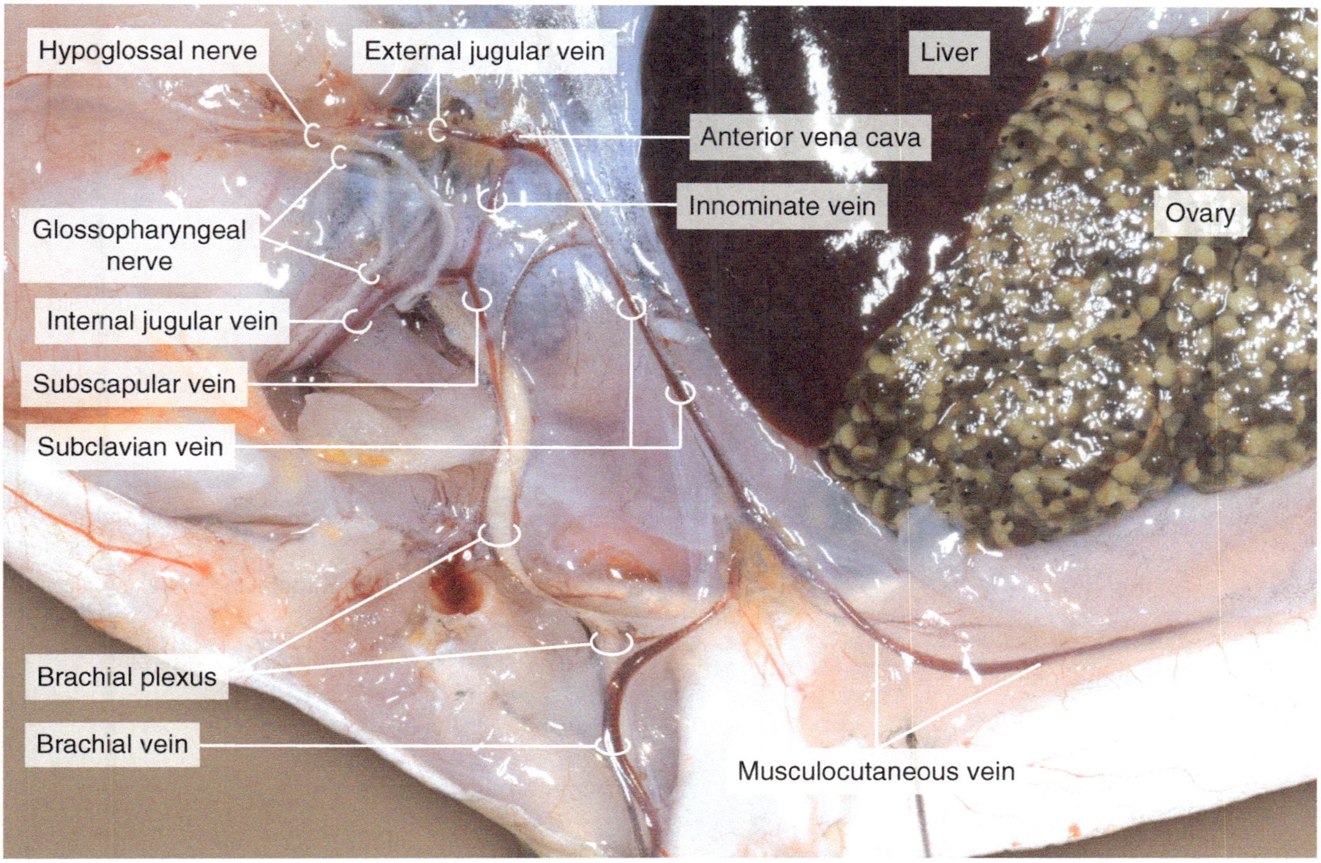

Fig. 10.23 The connections within the anterior venous system (Anterior end is to the left.)

Find the *subclavian vein* which receives blood from the arm's *brachial vein* and the *musculocutaneous vein* coming from the ventral body wall. The *innominate vein* is made up from the *subscapular vein* from the shoulder running next to the brachial plexus and the *internal jugular vein* from the brain (Fig. 10.24).

The *anterior vena cava* is formed by the union of the *external jugular vein* from the tongue and floor of the mouth and the innominate vein. Find the *thyroid gland* and the *glossopharyngeal* and *hypoglossal nerves* next to the external jugular vein (Fig. 10.24).

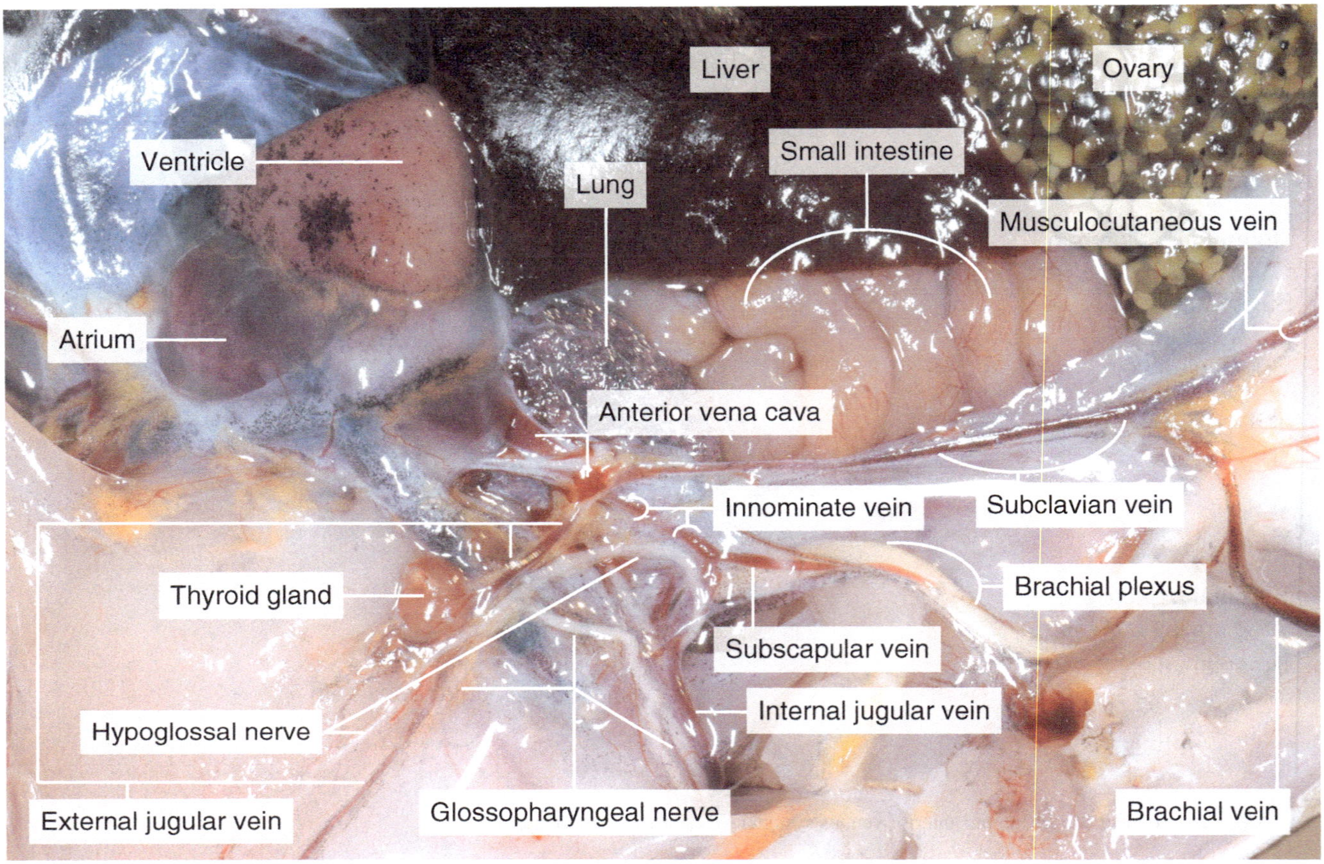

Fig. 10.24 The heart with the anterior venous system (Anterior end is to the left.)

Carefully remove the *pericardium*, the sac that contains the heart. Lift the heart with forceps and observe the thin-walled *sinus venosus*, formed by the convergence of three large veins, the two *anterior vena cavas* and one *posterior vena cava* (Fig. 10.25).

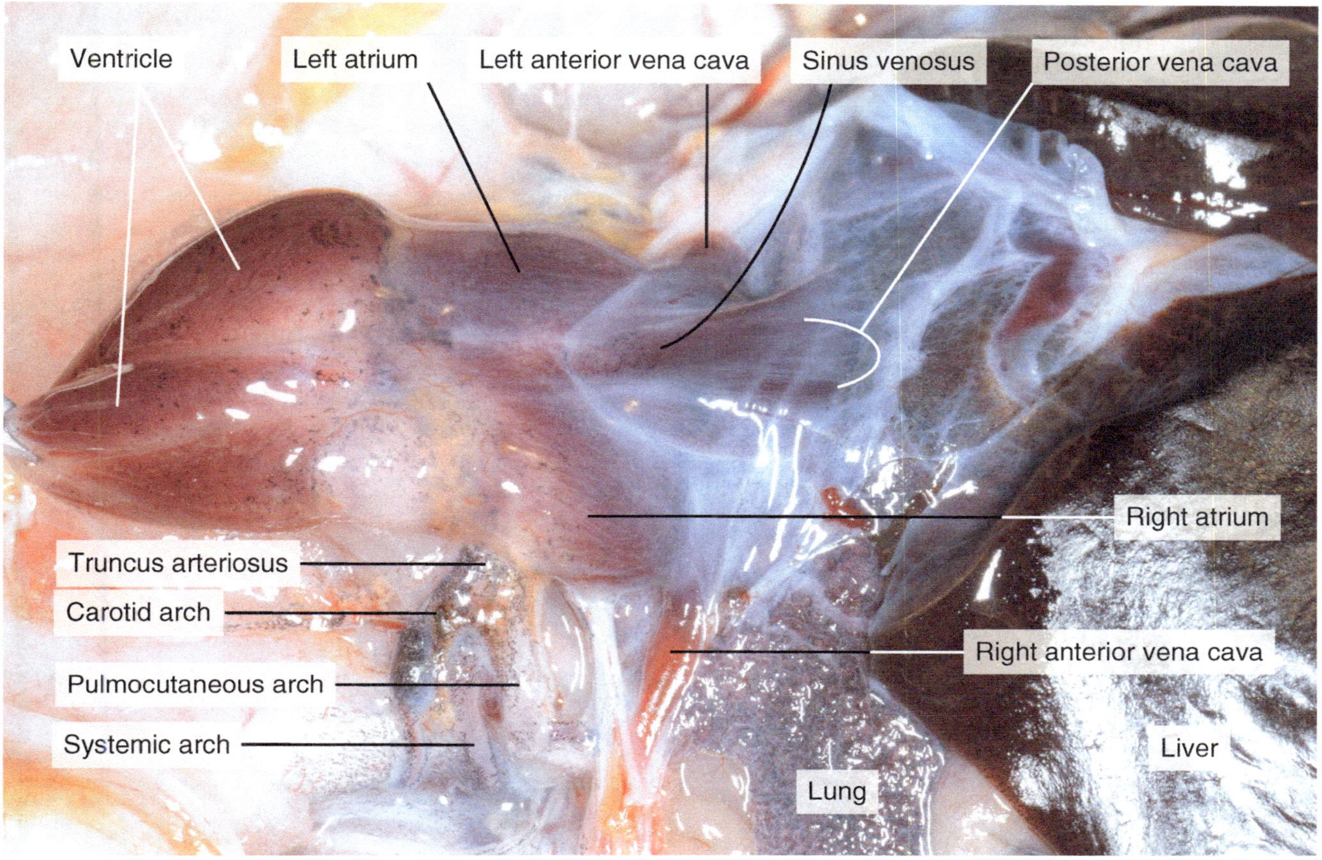

Fig. 10.25 Dorsal side of the heart with sinus venosus formed by the convergence of three large veins, the two atria and the ventricle. The ventricle of the heart is lifted up towards the head to uncover the three large veins (Anterior end is to the left.)

Fold the *heart* back again and identify the thick-walled conical *ventricle*, the thin-walled *left* and *right atria* and the *conus arteriosus*, arising from the ventricle and dividing to form the *truncus arteriosus* on each side. Locate the three main arterial arches: *carotid*, *systemic* and *pulmocutaneous arches* which arise from the truncus arteriosus (Figs. 10.25 and 10.26).

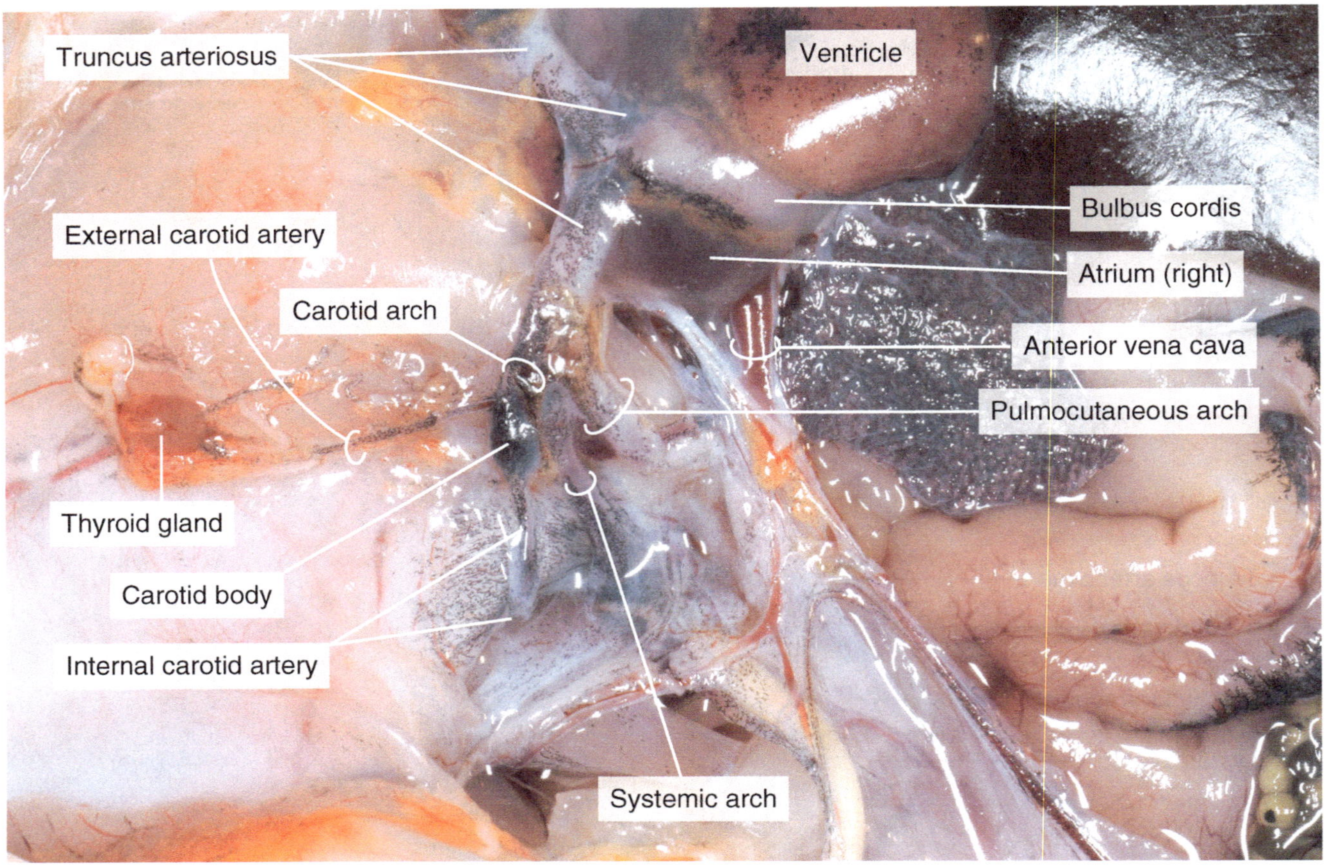

Fig. 10.26 The heart, the right truncus arteriosus and the aortic arches (Anterior end is to the left.)

The *carotid body* (glomus caroticum) is located near bifurcation of the carotid artery. It detects changes in the composition of blood flowing through it, mainly the partial pressure of oxygen, but also of carbon dioxide. The walls of the arteries contain pigment cells, unlike those of the veins, which absorb the infrared rays and help to warm up the blood.

Optionally remove the heart and make a frontal section of it with a sharp scalpel or razor blade (Fig. 10.27).

Find the opening from the sinus venosus into the right atrium. The amphibian ventricle is a spongy structure subdivided by crypts that may prevent the free mixing of the two blood streams (Fig. 10.27). In the three-chambered amphibian heart, there is a separate pulmonary system, but some mixing of venous and arterial blood can occur. The carotid and systemic arches receive oxygenated blood primarily from the left atrium through the left side of the ventricle. The vessels to the lung and skin are fed from the right atrium through the right side of the ventricle. To facilitate this, the *bulbus cordis* (conus arteriosus), which receives blood from the ventricle, is divided by the twisted *spiral valve* (Fig. 10.27). This valve partially separates the *systemic and carotid cavity* and the *pulmocutaneous cavity* in the entire length of the conus. The structure and the functioning of the heart are such that the oxygenated blood is directed mainly into the systemic and carotid cavity while the venous blood into the pulmocutaneous cavity.

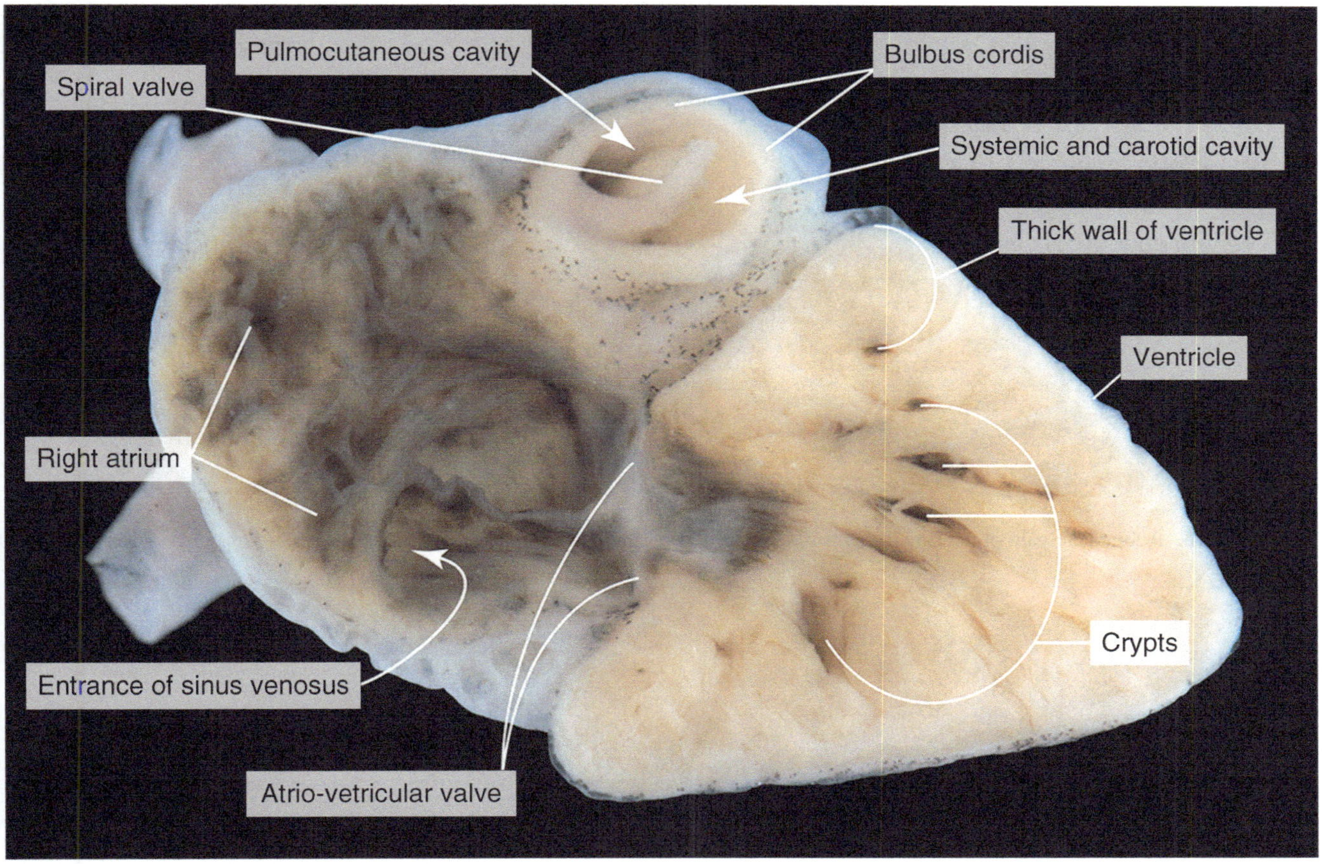

Fig. 10.27 Parasagittal section of the frog's heart (formalin-fixed specimen)

Observe the abdominal organs in their original position (Fig. 10.28).

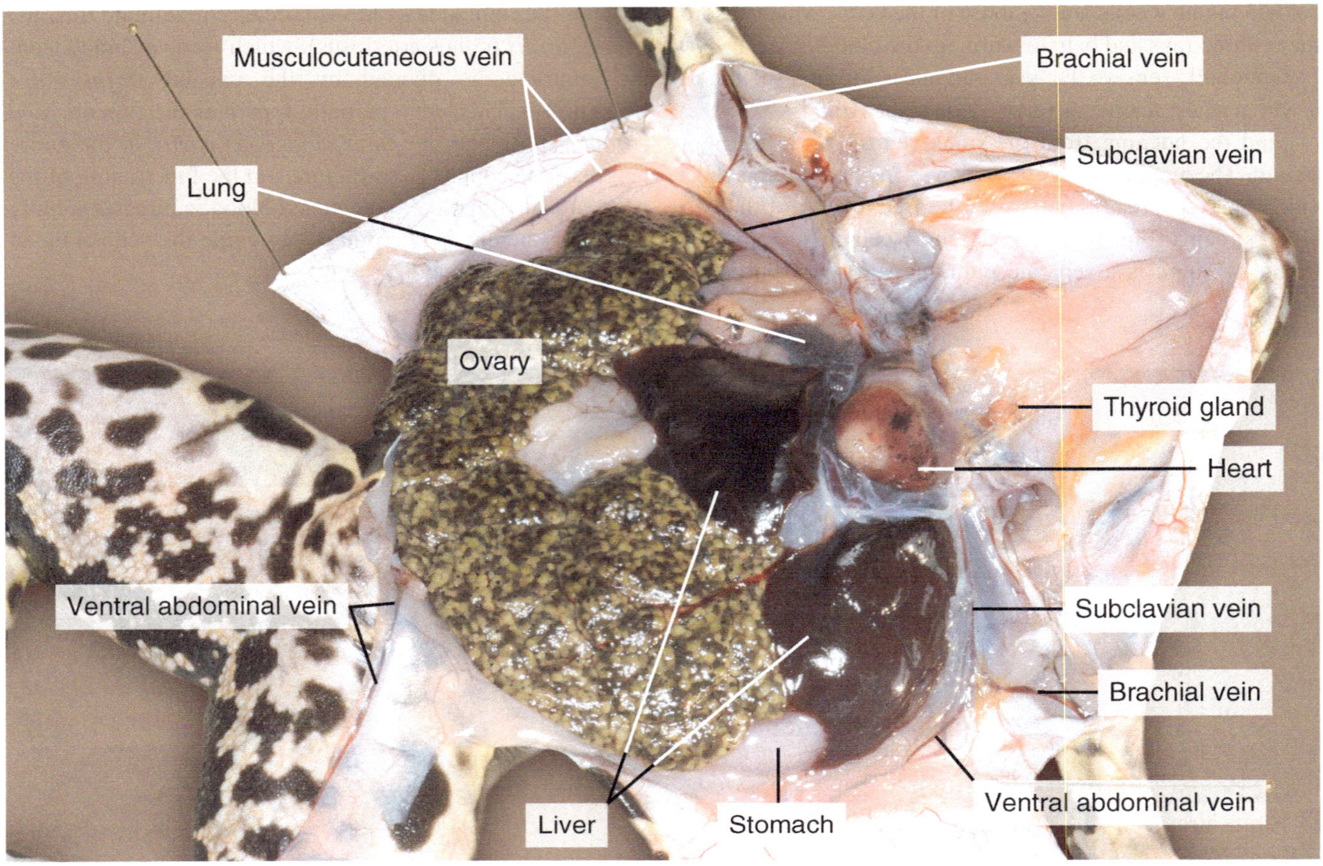

Fig. 10.28 The abdominal organs of the frog in their original position

The *lungs* are very small if they are deflated. However, they are very conspicuous if they are inflated with air (Fig. 10.29).

The lungs are hollow sacs with thin walls, through which the borders of the *alveoli* (singular alveolus) are visible (Fig. 10 29). If they are shrunk, try to blow them up with a straw inserted into the glottis.

The frog has no developed ribs, intercostal muscles and diaphragm. Thus it draws air into its mouth cavity through nares by closing the glottis and depressing the floor of its mouth. Then, by closing the nares and raising the floor of its mouth cavity, it forces the air from the mouth through the glottis into the lung. Air is expelled from the lungs by the contraction of the muscles of the body wall and the elastic recoil of the stretched lung. Because the respiratory surface of their lungs is too small, the pulmonary respiration is not sufficiently effective and amphibians breathe through their skin (cutaneous respiration) as well. Some respiration also occurs through the mucous lining of the mouth.

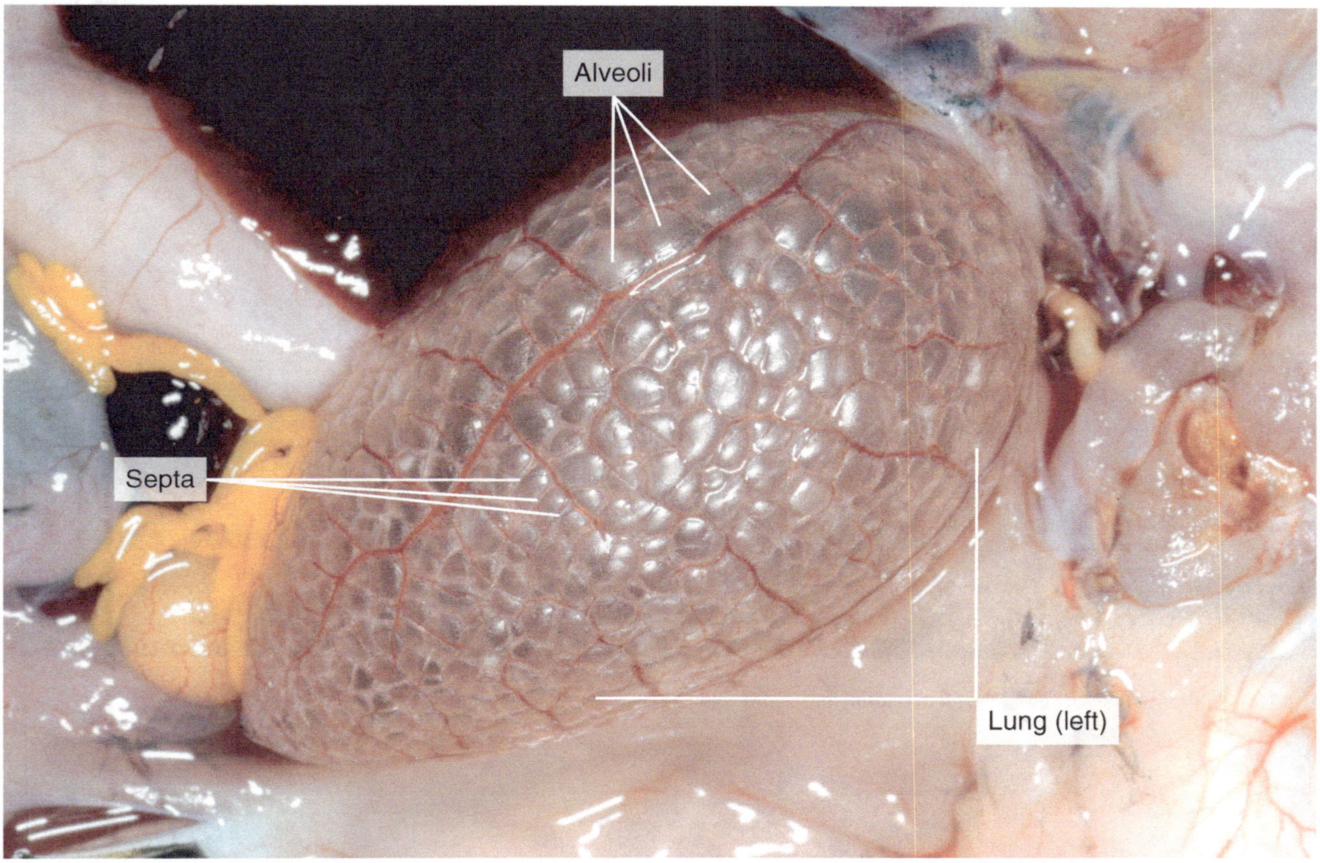

Fig. 10.29 The left lung of the frog inflated with air

Optionally remove the lungs together with the larynx. With a sharp scalpel or razor blade, cut the glottis mid-sagittaly. Cover with water and examine it under a dissecting microscope. Find the *vocal cords*, the *laryngotracheal sac* and the very short bronchus connecting each lung to the *larynx* (Fig. 10.30).

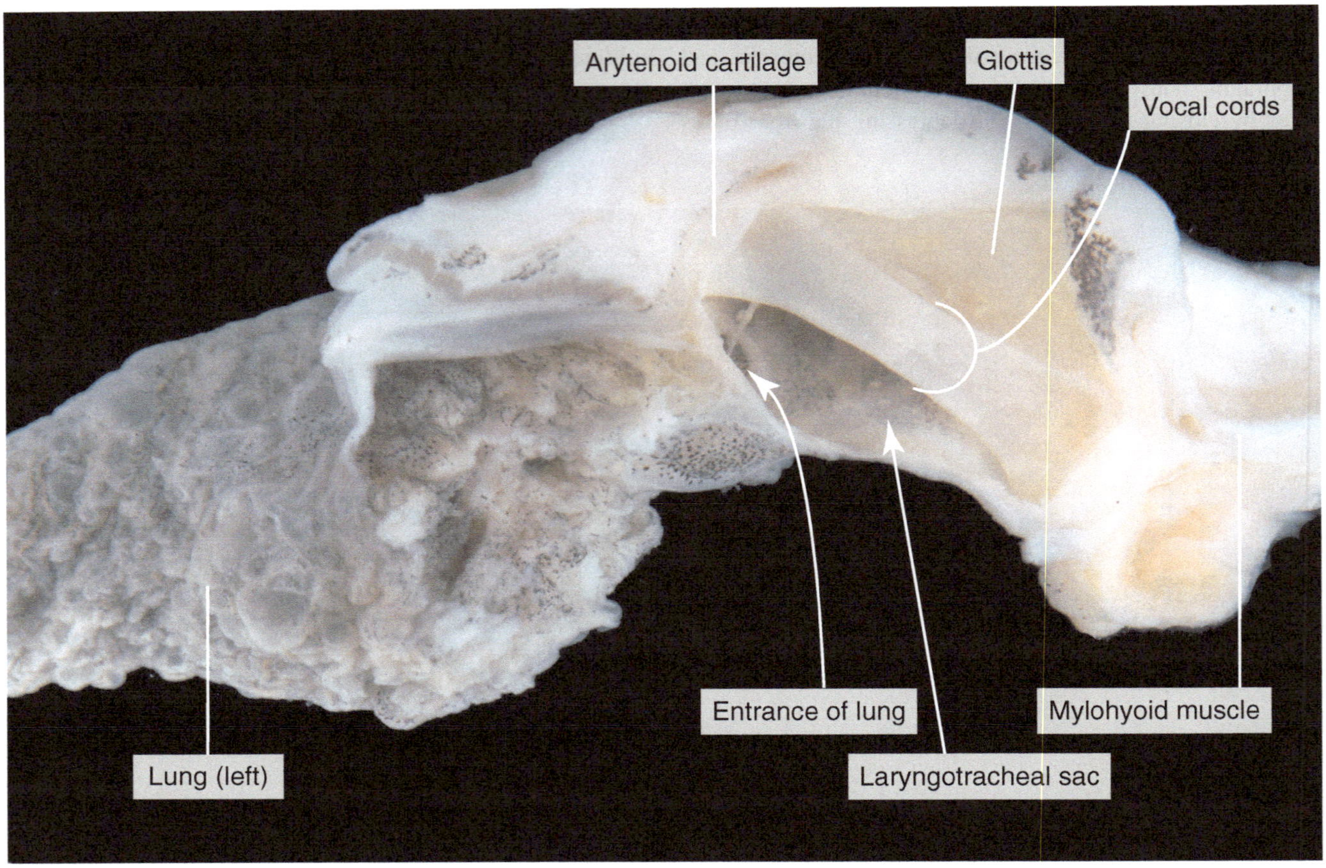

Fig. 10.30 Mid-sagittal section of the glottis (formalin-fixed specimen)

Slit open a lung and observe its internal structure. Note the little pockets or *alveoli* between the *septa* in the lining (Fig. 10.31).

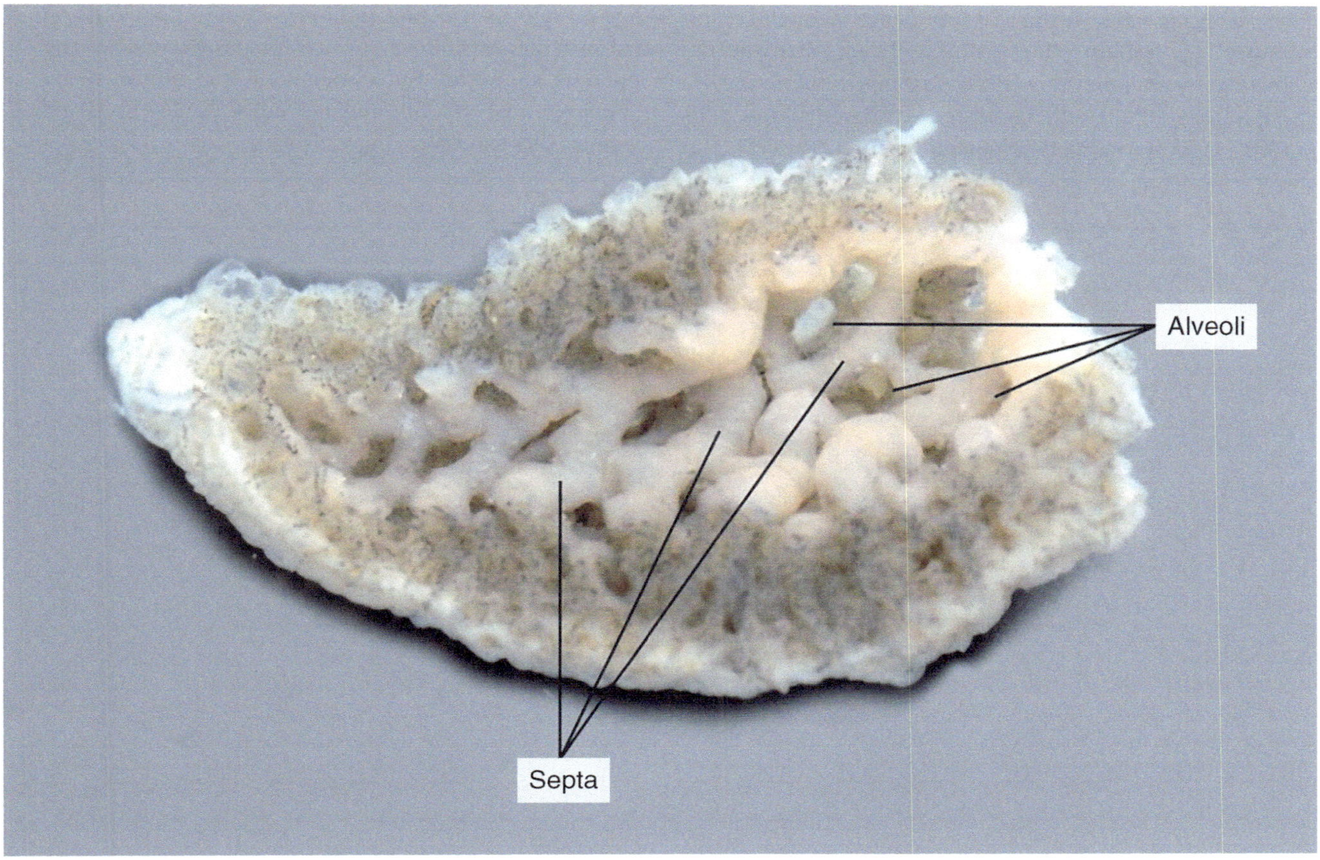

Fig. 10.31 Longitudinal section of the lung (formalin-fixed specimen)

The inner space of the frog's *lung* is subdivided into smaller cavities by connective tissue septa of the wall. The cavities enclosed by septa are called as *alveoli* (Fig. 10.32).

Alveolar walls (surface of septa) are lined with respiratory epithelium (Fig. 10.32, black arrowheads) formed by *pneumocytes* (PCs). Just below them, several capillaries filled with nucleated, oval red blood cells (RBCs) form a network. Nuclei of pneumocytes (PCNu) nestle in grooves between capillaries. Pneumocytes have thin cytoplasmic parts which cover the neighbouring capillaries to form a respiratory surface (Fig. 10.32, white arrowheads). Pneumocytes have two functions: they secrete a surfactant layer to reduce surface tension and provide respiratory surface for gas exchange. The septa are supported by connective tissue infiltrated by melanophores, smooth muscle cells (SMCs), lymph vessels (LVs) and blood vessels (BVs).

Fig. 10.32 Histological sections of the frog's lung (*top*: Azan, *bottom*: semithin section with toluidine blue staining). Top micrographs: Septa and alveoli after normal fixation. Bottom micrograph: Appearance of a septum after perfusion. *Black arrowheads* respiratory epithelium. *White arrowheads* respiratory surface. *BV* blood vessels, *LV* lymph vessel, *PCNu* nuclei of pneumocytes, *RBC* red blood cells, *SMC* smooth muscle cells

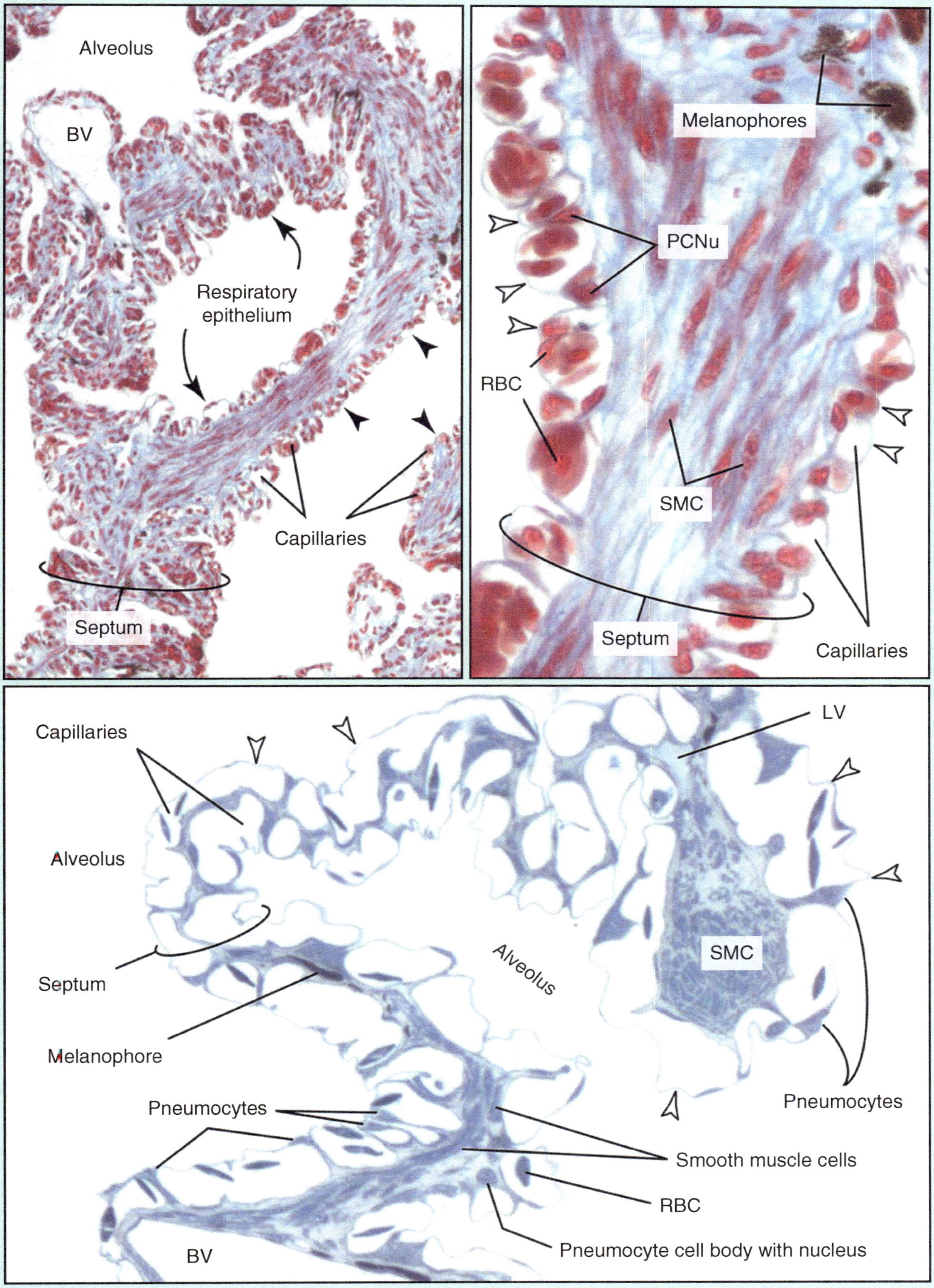
Alveolus
BV
Respiratory epithelium
Capillaries
Septum
Melanophores
PCNu
RBC
SMC
Septum
Capillaries
Capillaries
Alveolus
Septum
Melanophore
Pneumocytes
BV
LV
Alveolus
SMC
Pneumocytes
Smooth muscle cells
RBC
Pneumocyte cell body with nucleus

Lift the heart, liver and lungs to see where the wide *oesophagus* empties into the *stomach* (Fig. 10.33).

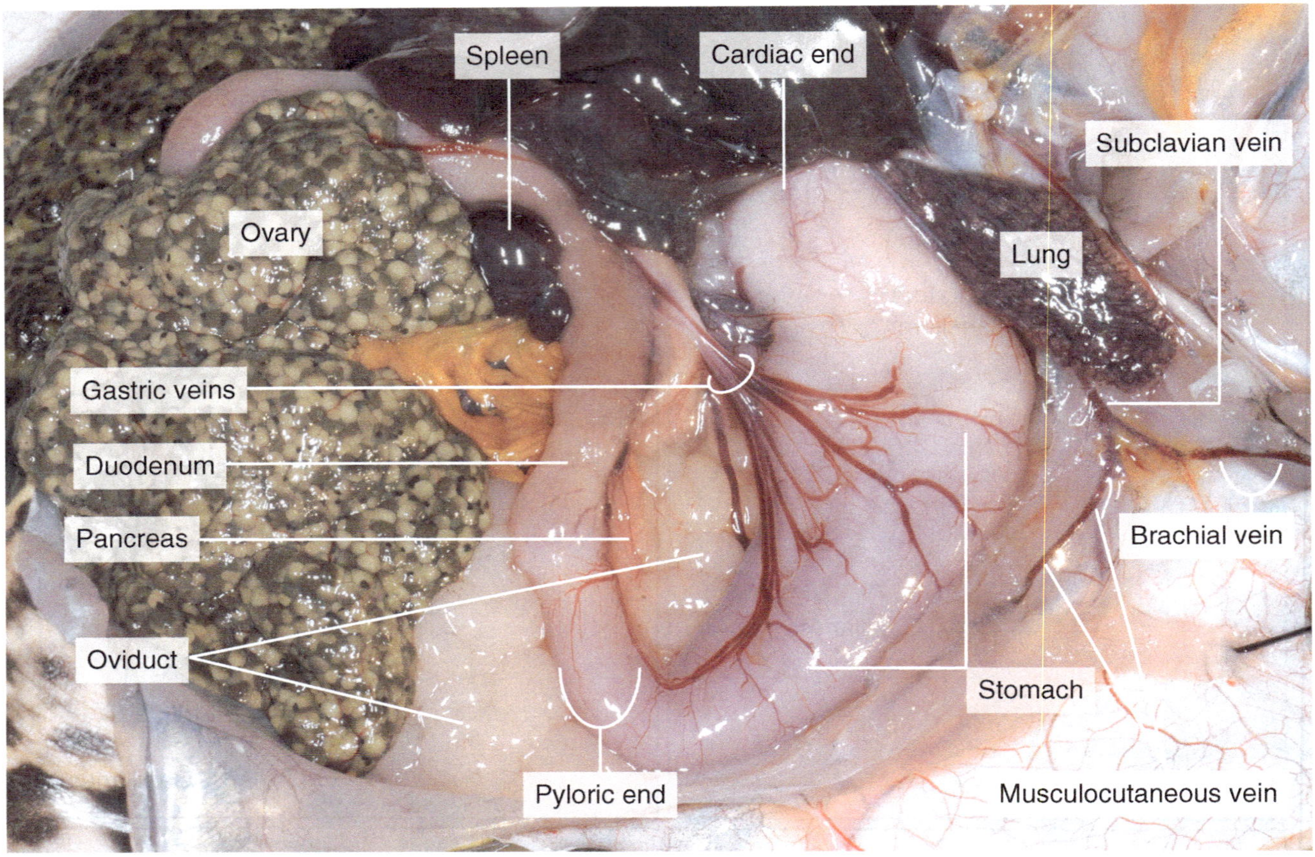

Fig. 10.33 The stomach of a female frog and the neighbouring organs

A *pyloric valve* controls the movements of food from the stomach into the *duodenum*, first part of the small intestine (mid-gut) (Fig. 10.33). Note the blood vessels in the *mesentery*, which hold the stomach and small intestine in place. Displace the alimentary canal to the animal's right side to show the mesentery with the *gastric vein*, which gives one branch of the *hepatic portal vein* (Fig. 10.34).

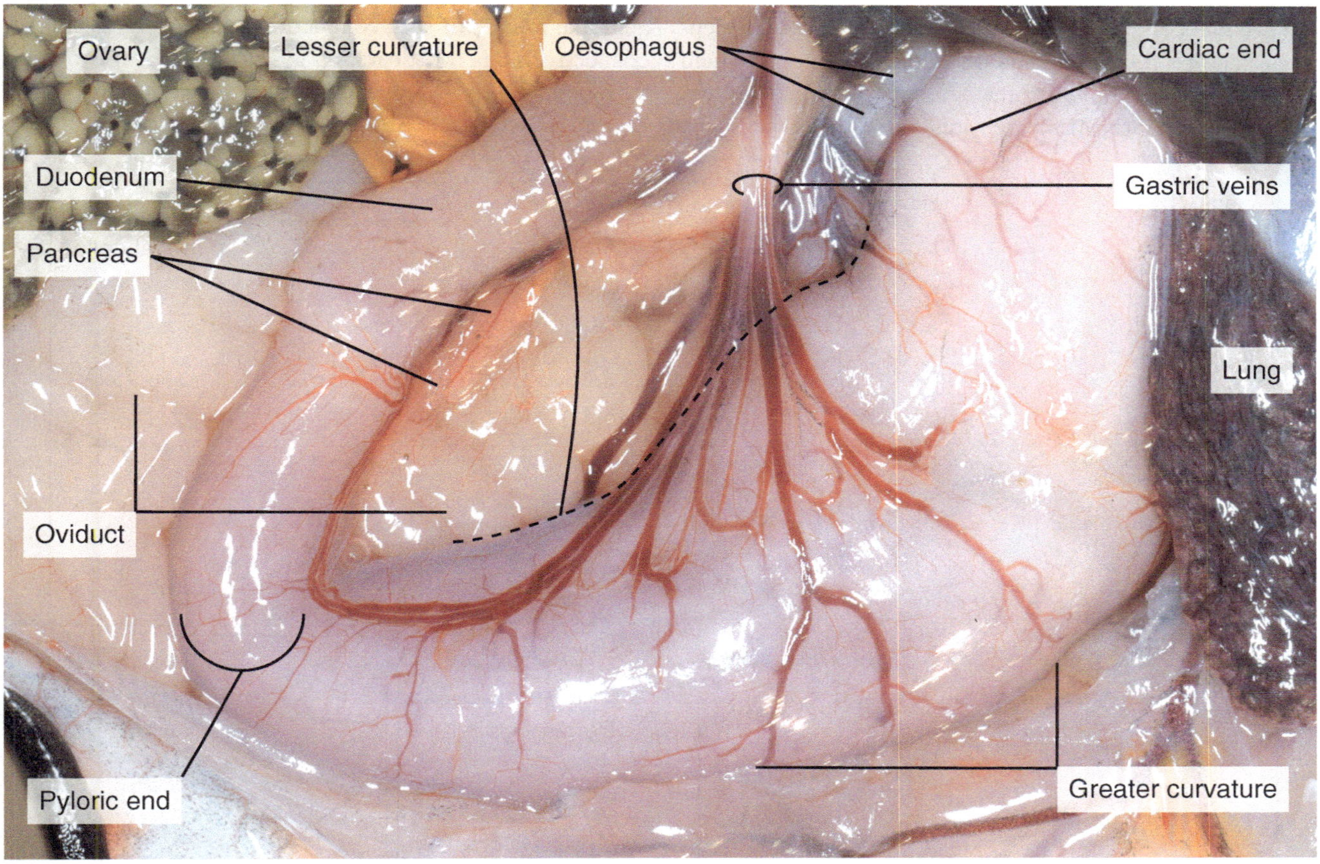

Fig. 10.34 The stomach of the frog and the mesentery with the gastric vein. *Dashed line* indicates the lesser curvature of the stomach

The histological section of frog's *stomach* shows folded mucosa with columnar epithelium and gastric glands (Fig. 10.35).

On the surface, *gastric pits* are formed, where the tubular *gastric glands* open (Fig. 10.35, arrows). Glands have two cell types: *parietal cells* secrete enzymes (pepsin) and produce hydrochloric acid to guarantee an acid condition favourable for the action of pepsin. The *neck mucous cells* (NMCs) are situated at the upper neck region of glands. They secrete protective, PAS-positive material, as the *surface mucous cells* (SMCs) of the mucosal epithelium do (Fig. 10.35, bottom right). PAS positivity of mucous cells of neck region indicates their protective function against self-digestion and irritation of acid like in case of surface mucous cells. Propria forms connective septa for separation gastric glands – it is rich of blood vessels. Mucosa terminates with its deepest *smooth muscle layer* (SML) joining to submucosa.

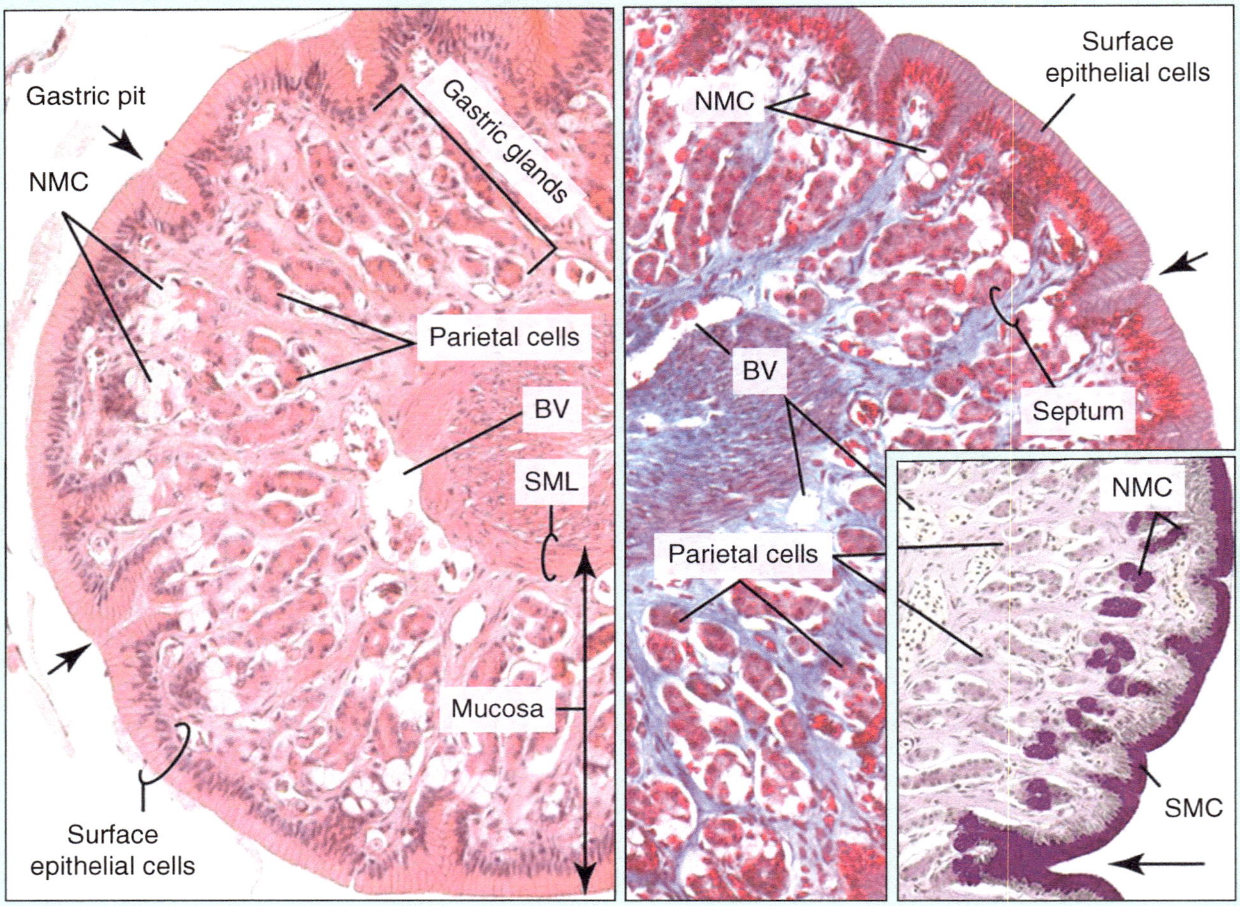

Fig. 10.35 Histological sections of the mucosa of frog stomach (*left*: HE, *right*: Azan, *inset*: PAS). *Arrows* gastric pits, *BV* blood vessels, *NMC* neck mucous cells, *SMC* surface mucous cells, *SML* smooth muscle layer

The *liver* is the largest gland in the body and secretes bile, which is carried by a small duct to the *gall bladder* for storage. Find the gall bladder between the right and median lobes of the liver (Fig. 10.36).

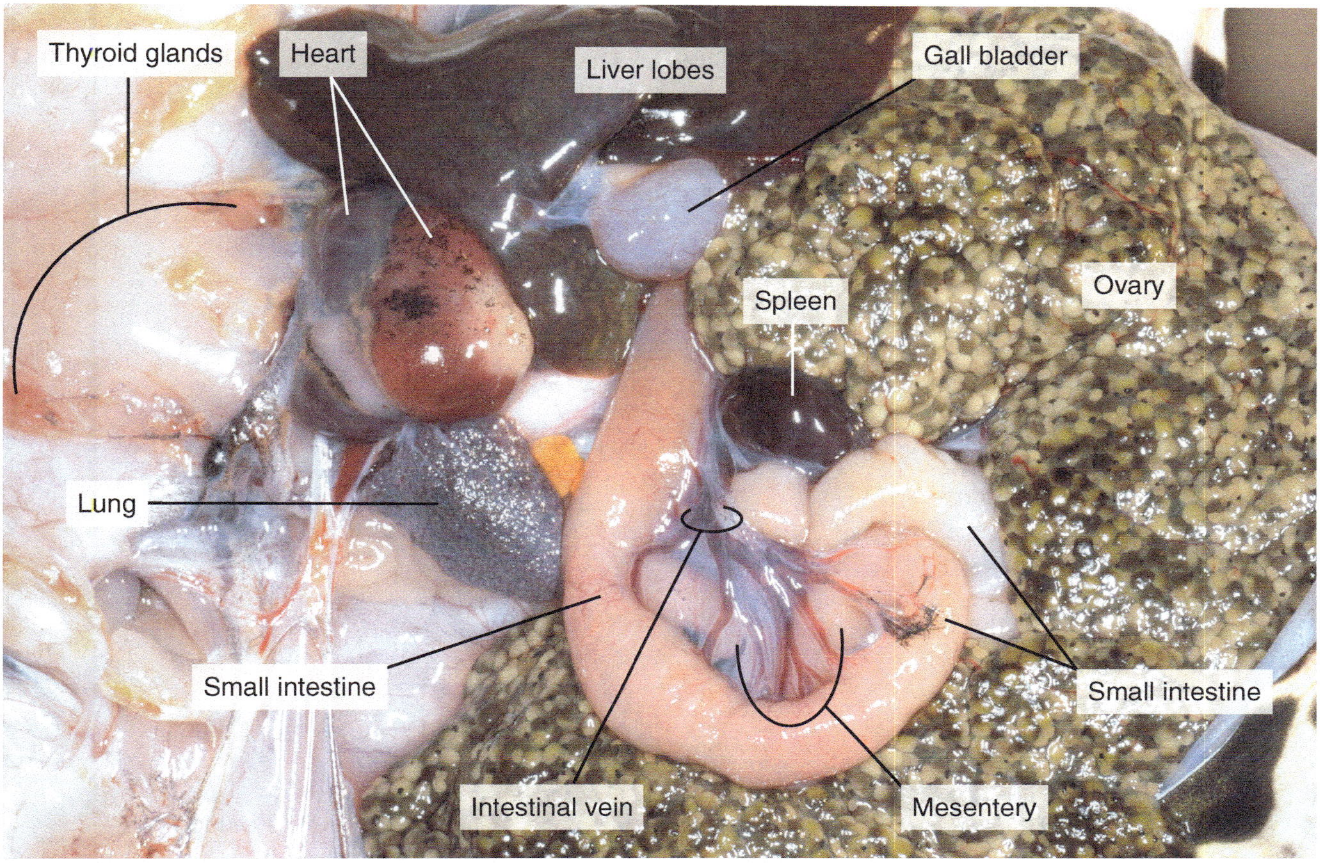

Fig. 10.36 Lobes of the liver and the gall bladder (Anterior end is to the left.)

Bile is carried away in a tube called the common *bile duct*, into which pancreatic juice, a digestive juice from the pancreas, also flows. The *pancreas* is thin and inconspicuous, lying in the mesentery between the stomach and the duodenum (Figs. 10.34 and 10.37). The contents of the common bile duct flow into the small intestine, where most of the digestion and absorption of nutrients into the bloodstream takes place. The *large intestine* (colon) may also absorb water, although its main task is to collect indigestible materials and form into faeces. The *rectum* then passes it into the *cloaca*, the common exit chamber of the digestive, excretory and reproductive systems (Fig. 10.37).

Observe the *spleen* close to the small intestine (Figs. 10.33 and 10.37).

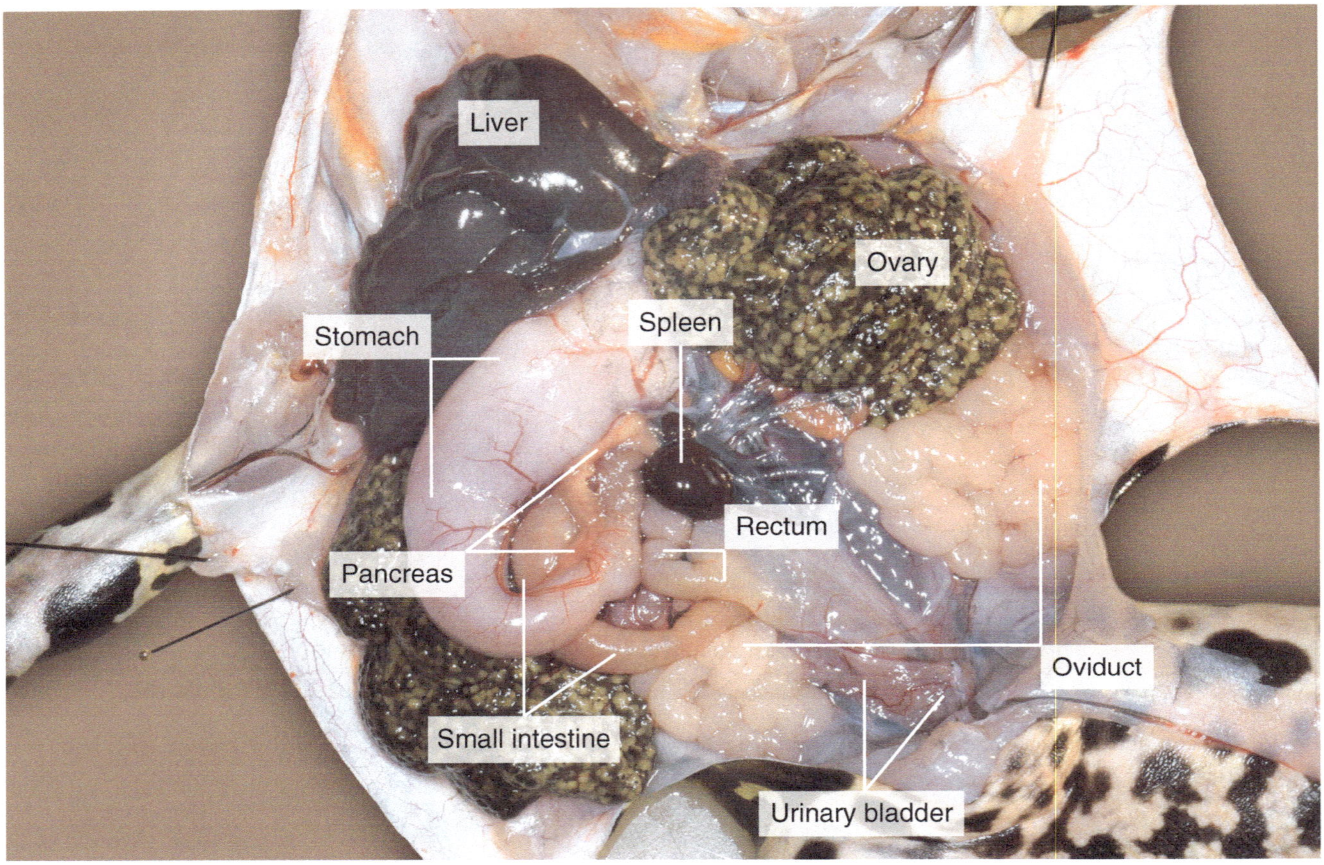

Fig. 10.37 Internal organs of a female frog with parts of the alimentary canal

Now study the *urogenital system*. Functionally, the urogenital system is comprised of two systems, the *excretory* (or urinary) *system* and the *reproductive* (genital) *system*. However, because some features function in both systems, they are usually considered together. In a female specimen, identify the *ovaries* and the adjoining fat bodies (Fig. 10.38).

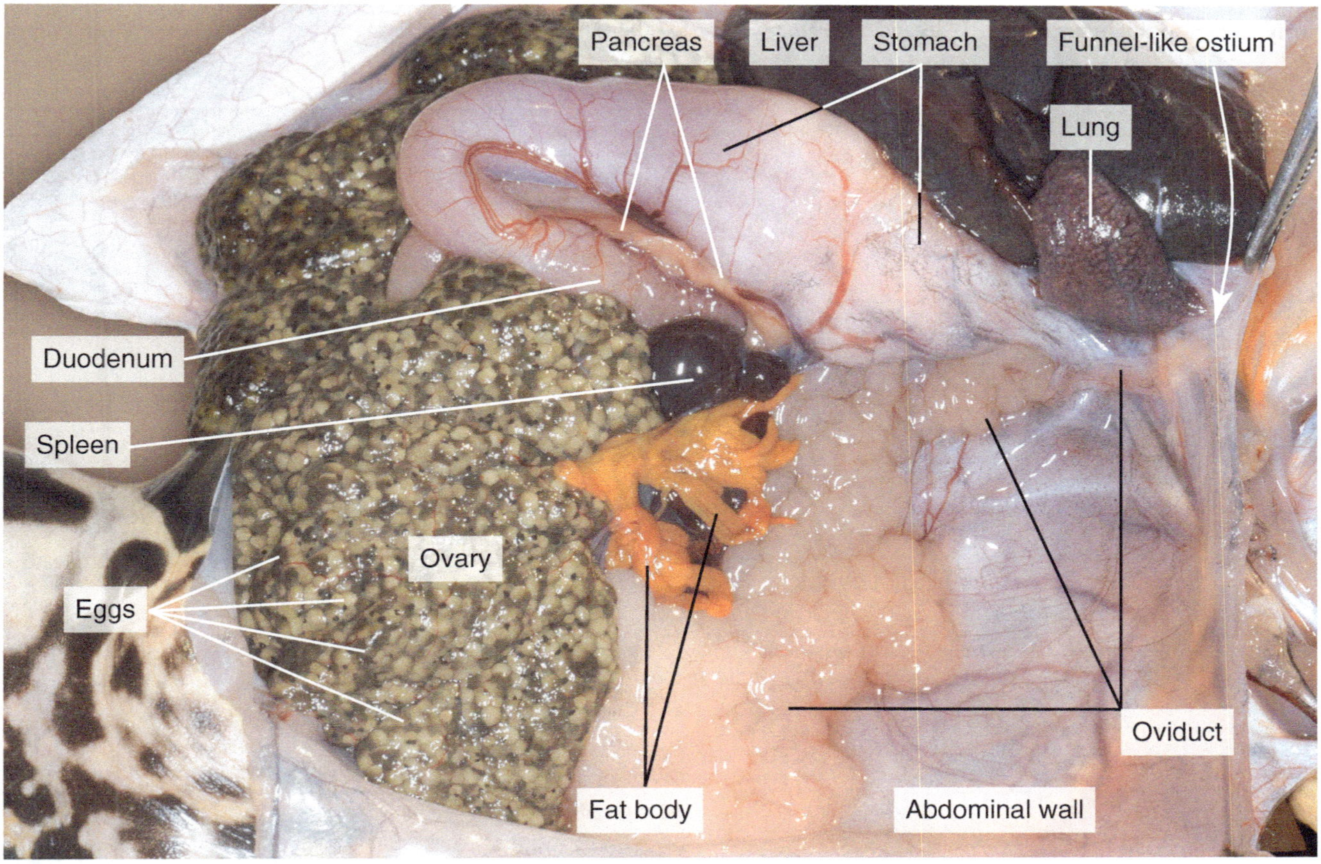

Fig. 10.38 The reproductive organs of a female frog

They are attached by mesenteries to the dorsal wall of the coelom. In winter and in early spring, the ovaries are distended with eggs. If the animal was killed in the summer or early fall, the ovaries will be small and pale. Convoluted *oviducts* begin with funnel-shaped *ostia* (singular ostium) anteriorly, dorsal to the lungs (Figs. 10.38 and 10.39).

Eggs are released from the ovary into the coelom, carried in coelomic fluid to the ostia, and then down the oviducts by ciliary action to the outside through the cloaca. At amplexus (the copulatory embrace), the male clasps the female and fertilises the eggs externally as they are laid in the water.

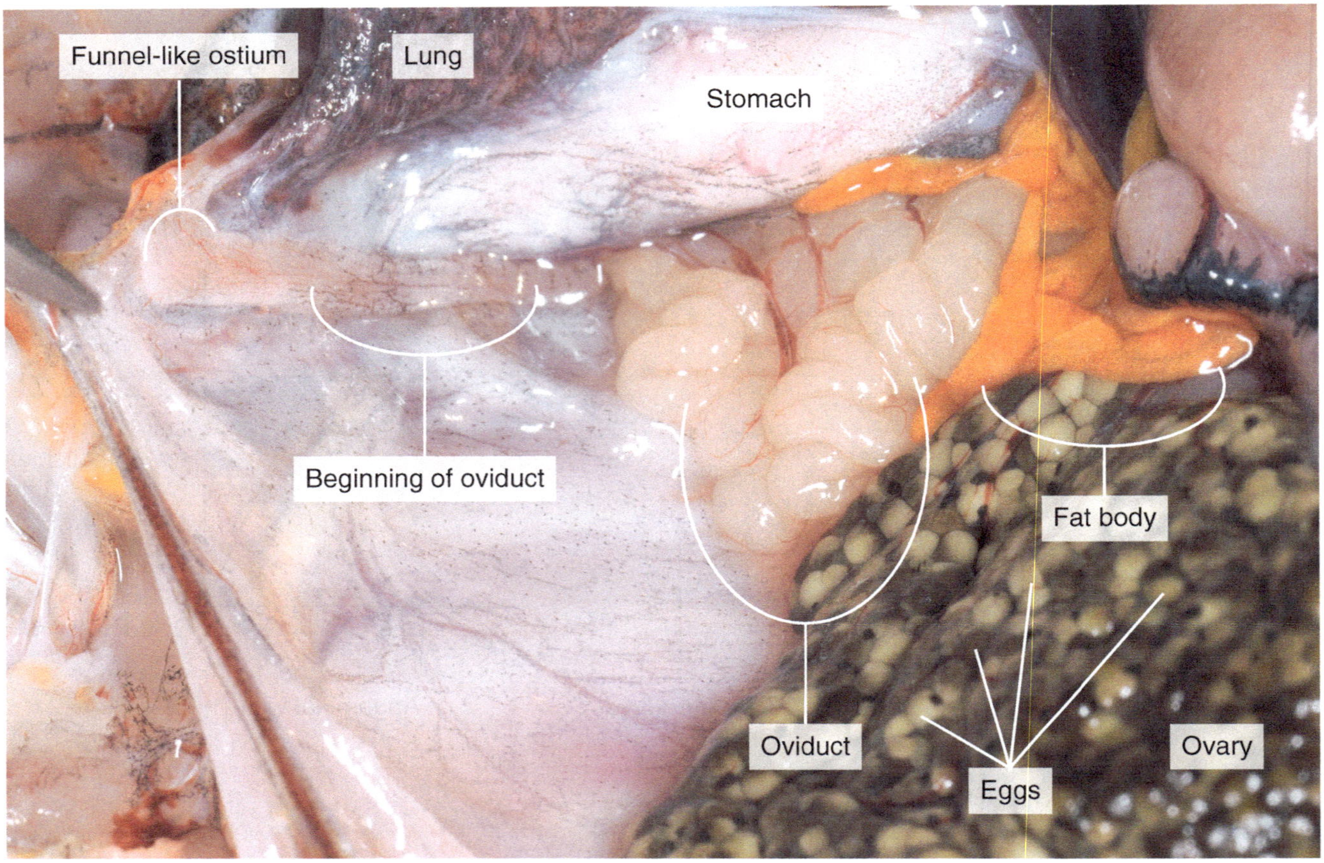

Fig. 10.39 Ostium, the beginning of the convoluted oviduct (Anterior end is to the left.)

In male animals, find a pair of large, bright orange *testis* and attached orange fat bodies (Fig. 10.40).

They lie close to the ventral side of the kidney. Sperms pass from the testis into some of the kidney tubules and then are carried by the ureter to the cloaca and to the outside. Thus, male ureters serve also as genital ducts (vas deferens).

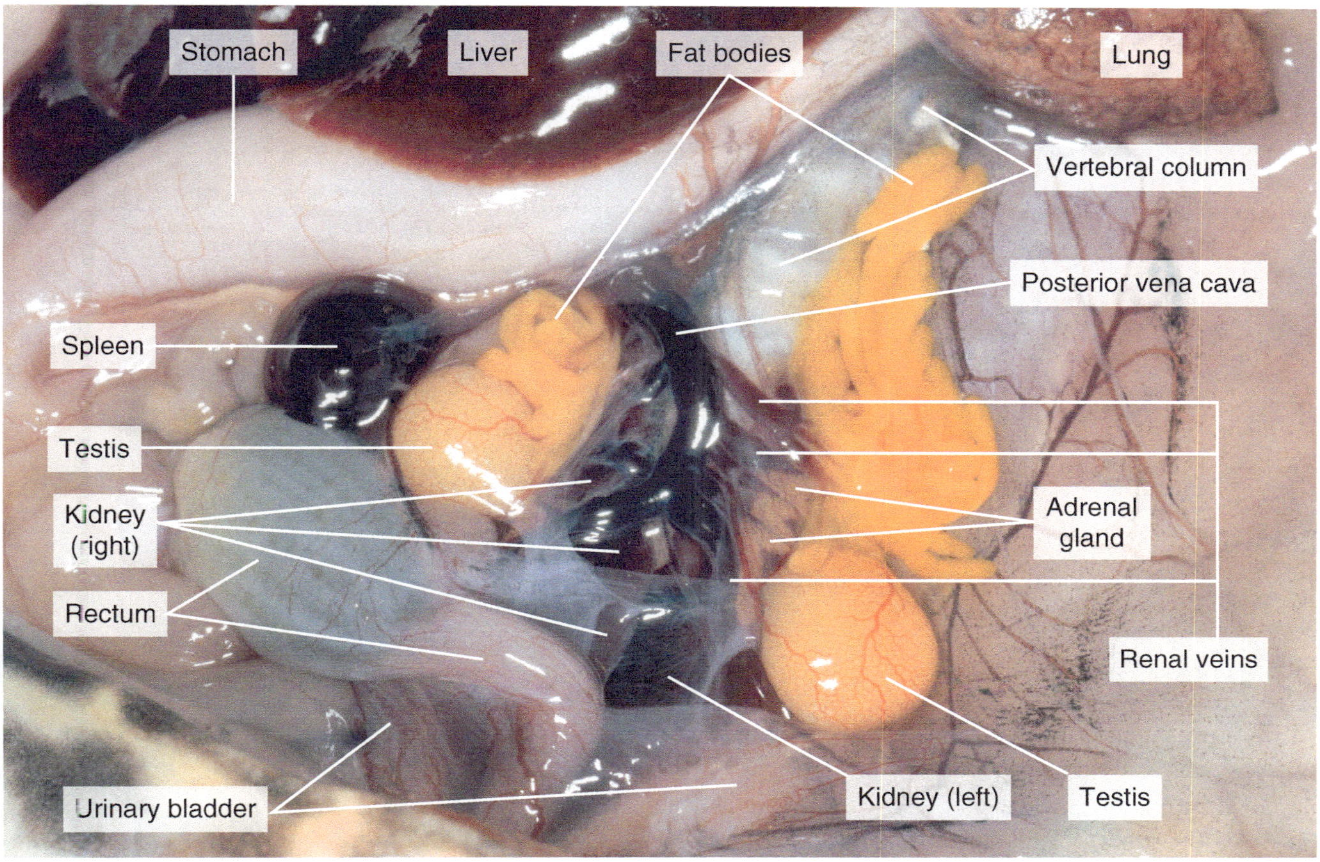

Fig. 10.40 The testes of a male frog with the adjoining fat bodies

Fold up the examined viscera; carefully dissect away the dorsal abdominal *peritoneum* to display the kidneys and ventral branches of spinal nerves. Do not disturb the dorsal aorta and systemic arches (Fig. 10.41).

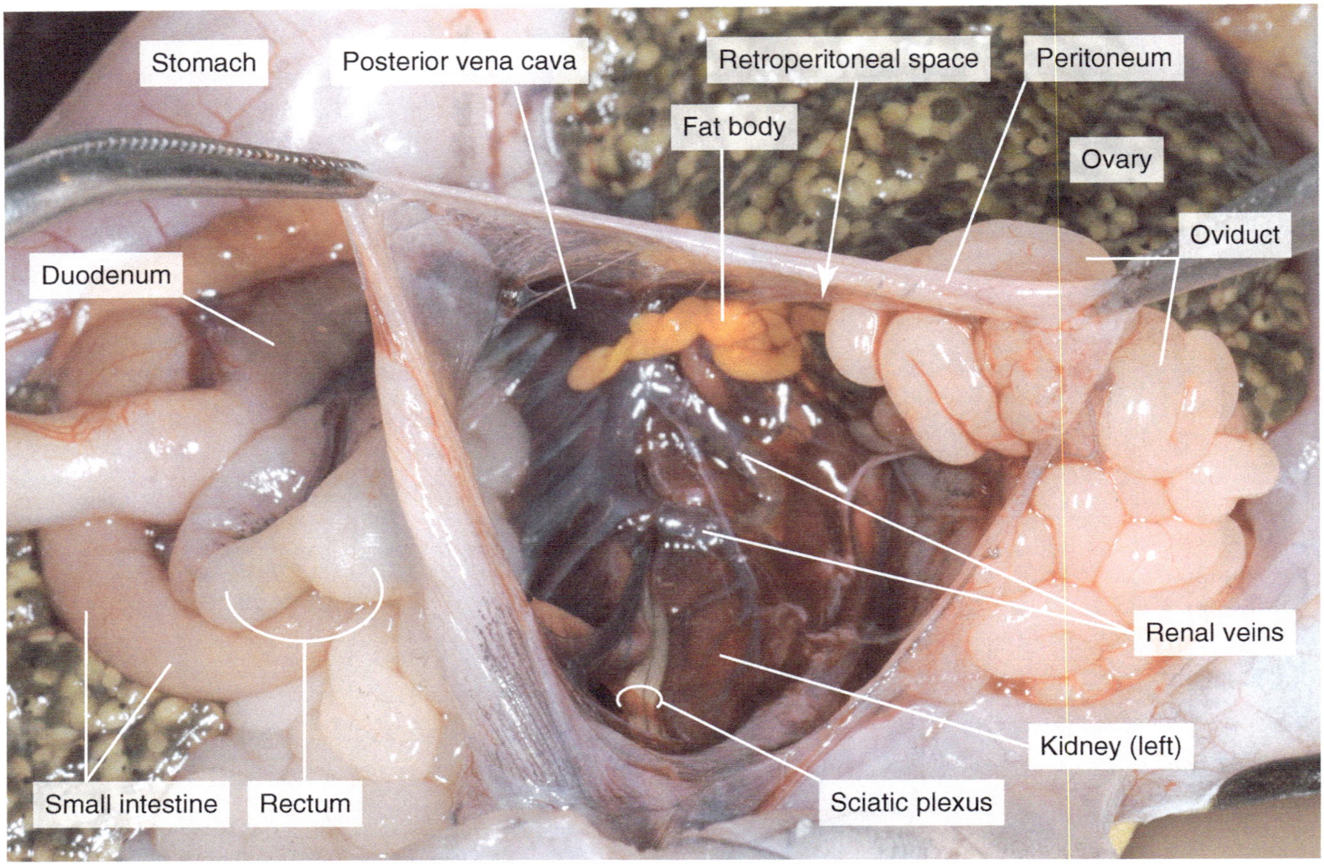

Fig. 10.41 The opened retroperitoneal space behind the peritoneum

The *kidneys* (mesonephroi) excrete the urine from the blood. They rid the body of metabolic wastes, and they maintain proper water balance in the body and a general constancy of content in the blood. The kidneys lie close to the dorsal body wall, separated from the coelom by a thin peritoneum (retroperitoneal position) (Fig. 10.41). The *ureters* connect the kidneys with the *cloaca*. The *urinary bladder*, when collapsed, appears as a soft mass of thin tissue just ventral to the large intestine (Figs. 10.40 and 10.44). It is bilobed and empties into the cloaca. The urinary bladder can store urine from which water can be taken by the absorptive walls. The *adrenal glands*, a bright yellow stripe on the ventral surface of each kidney, are endocrine glands. Vivid orange *fat bodies* attached to the kidneys, but lying in the coelom, are for energy storage (Fig. 10.42).

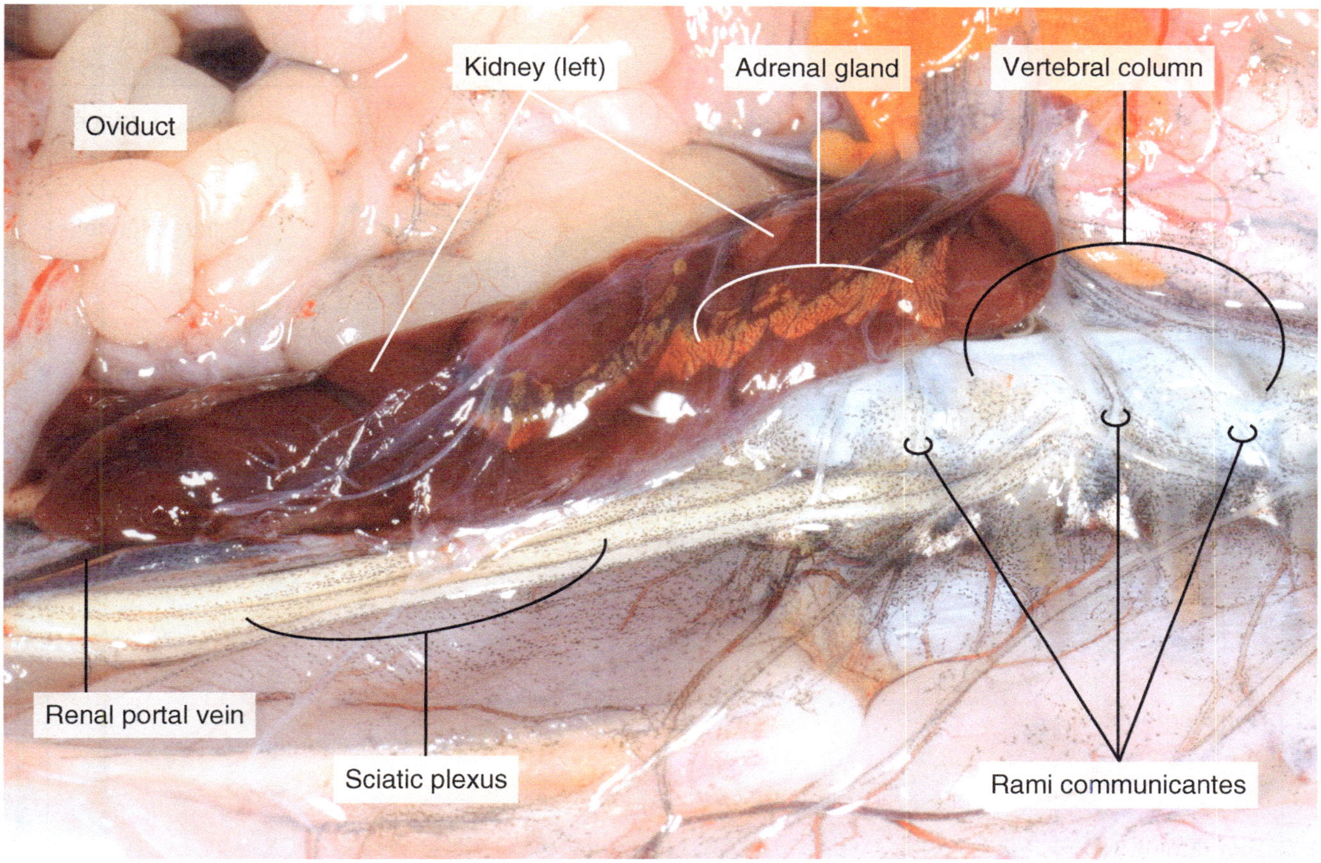

Fig. 10.42 The kidney with the adrenal gland on its ventral surface

Histological composition of frog's *kidney* shows the exclusive excretory function of the organ. Its nephrons contain *renal corpuscles*, in which *Bowman's capsules*, *glomeruli* (capillary networks) and mesangial cells are seen (Fig. 10.43).

Histology of Bowman's capsule traces the general vertebrate pattern: its parietal wall is formed by flattened cells, whereas visceral wall is composed of cuboidal *podocytes* (Fig. 10.43). Vascular pole is the point, where glomerular afferent and efferent arterioles enter and leave the glomerulus, respectively. The urinary space, a narrow cavity between the two walls of Bowman's capsule, opens into the nephric tubule at urinary pole. The *convoluted tubule* has several histologically different segments and collateral branches opening onto the ventral surface of the kidney which is immersed in a lymph sac. This opening is a *ciliated funnel*, which propels lymph from the lymph sac situated next to the vertebrate column into the nephric duct (Fig. 10.43, right).

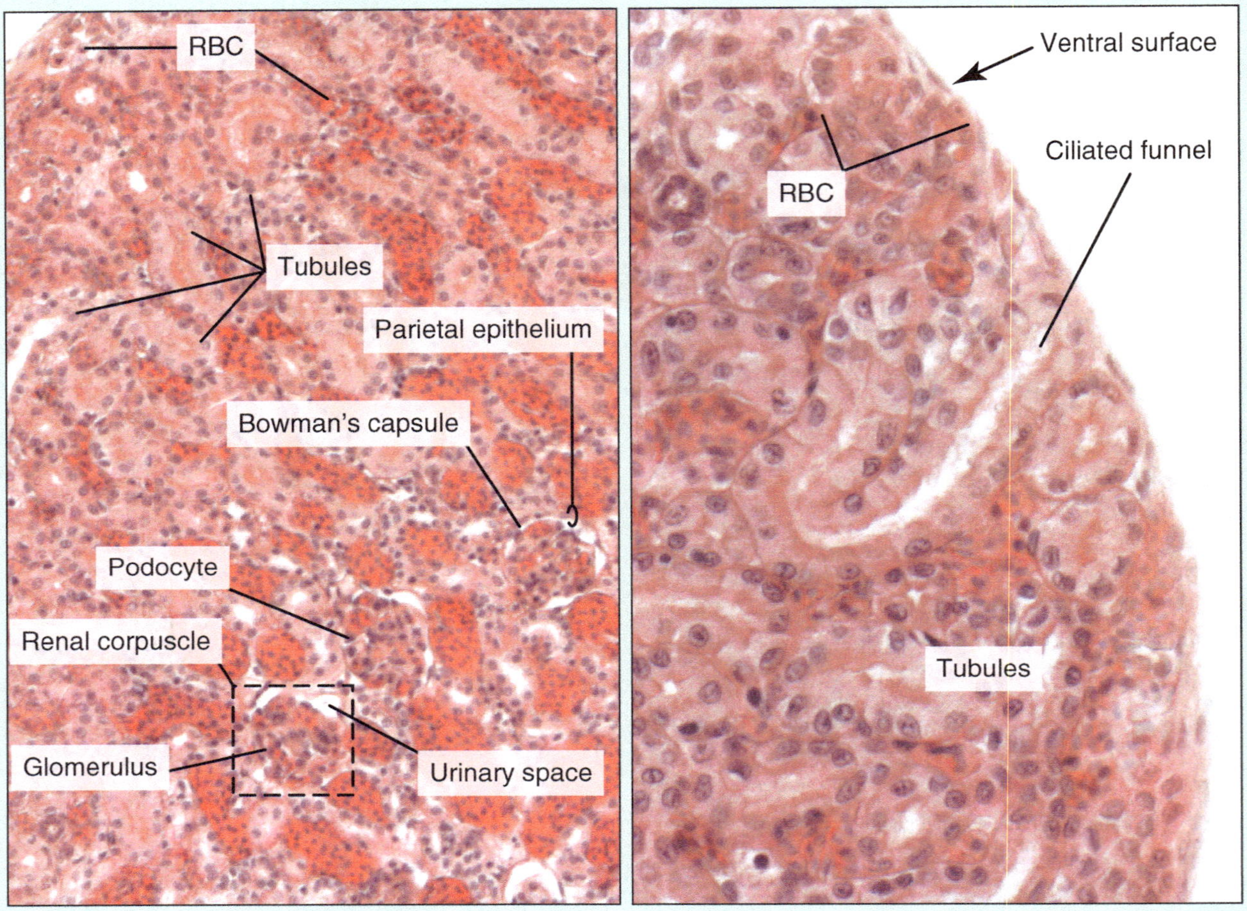

Fig. 10.43 Histological sections of a frog's kidney (HE). *Left panel*: Three renal corpuscles (one is framed in a *dashed square*) with Bowman's capsules and capillaries and several convoluting nephron tubules. *Right panel*: Collateral branch of a frog's nephron. *RBC* red blood cells

Note that in frog, like in fish, the kidneys have a *portal circulation*. Much of the blood from the hindlimbs passes through this renal portal system. The *renal portal vein*, found along the outer margin of each kidney, is formed by the union of the *sciatic* and *femoral veins* (Fig. 10.44).

Blood from the kidneys is picked up by the *renal veins* and carried to the *posterior vena cava* (Fig. 10.44). The renal portal system is absent in mammals.

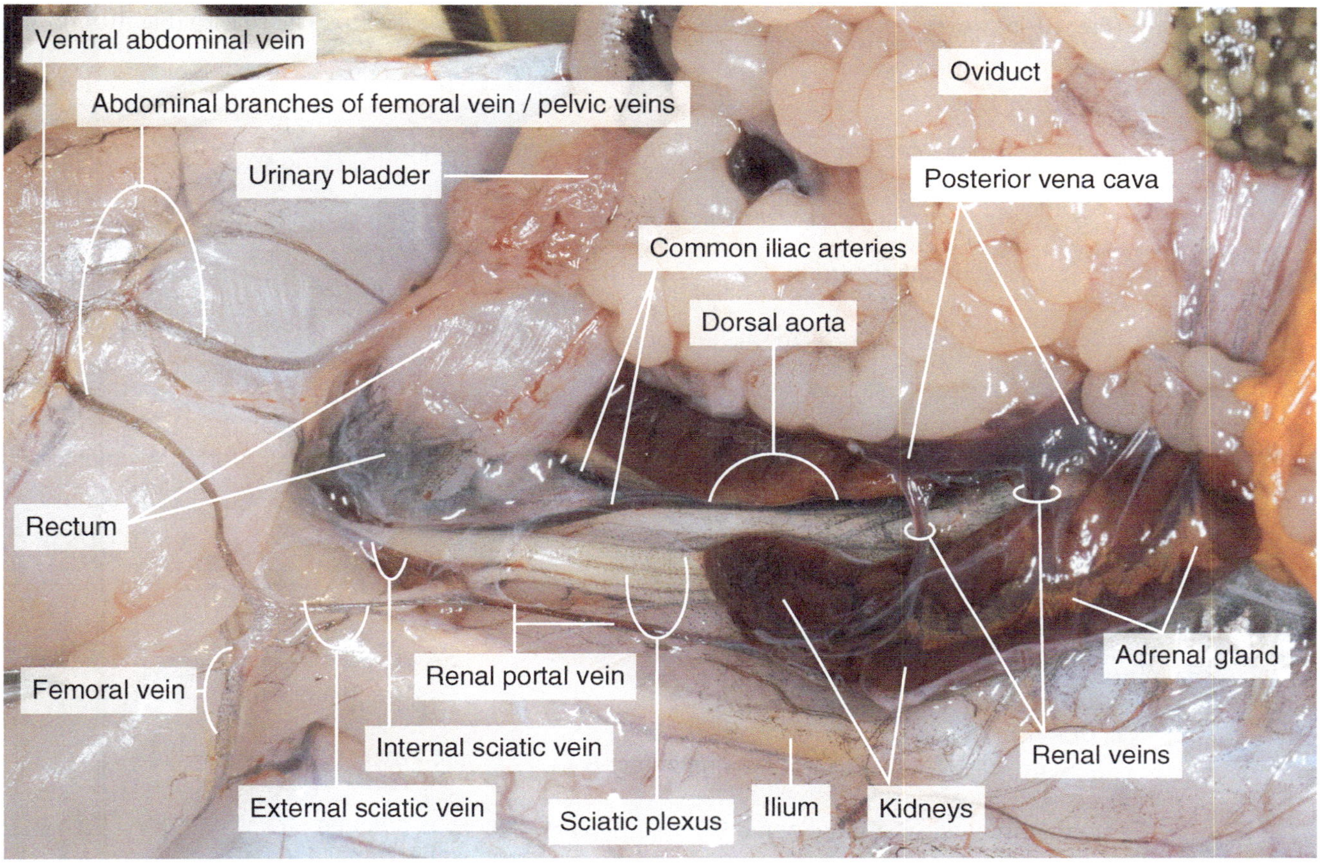

Fig. 10.44 The veins of the renal portal system

The *nervous system* is composed of the central and the peripheral nervous system. The *peripheral nervous system* is made up of the paired cranial and spinal nerves and the autonomic nervous system. Ten pairs of *spinal nerves* emerge through small openings between the vertebrae and appear as white threads in the dorsal body wall. There are bright white *calcareous patches* around the *spinal ganglia* where the spinal nerves emerge (Fig. 10.45).

The first three nerves on each side form the *brachial plexus* of the neck, shoulder and arm (Figs. 10.22, 10.23 and 10.45). A plexus is a network of nerves that interchange fibres. The last three (spinal nerve 8, 9 and 10) form the *sciatic plexus* to the leg (Figs. 10.44 and 10.45). The sciatic plexus on each side gives off to the hindlimb a femoral nerve and a large *sciatic nerve*, the largest nerve in the body. Cut through the pelvic girdle, spread the legs apart and trace the nerves into the leg (Fig. 10.45).

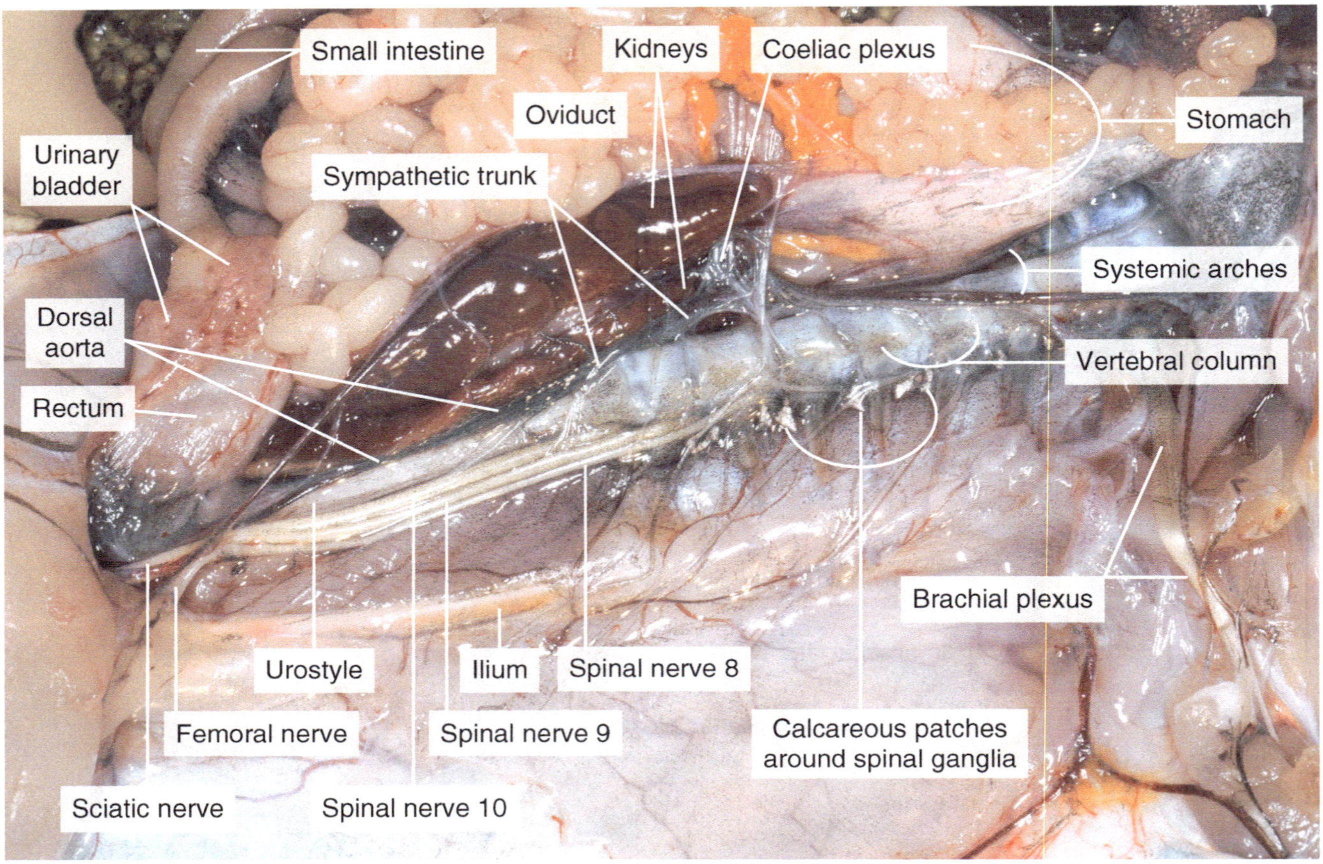

Fig. 10.45 The peripheral nervous system: spinal ganglia and nerves, and the autonomic nervous system: nerves and ganglia (Viscera and the left kidney are folded to the right side.)

The autonomic nervous system is divided into sympathetic (thoracolumbar) and parasympathetic (craniosacral) system. These two systems differ in their actions and in places where the efferents leave the central nervous system. The thoracolumbar system consists of a pair of sympathetic nerve cords together with their ganglia and nerve fibres (coeliac plexus), which innervate viscera (Figs. 10.45 and 10.46). Trace the *sympathetic nerve trunks*, like small, whitish cords on either side of the systemic arches and dorsal aorta. Note the autonomic ganglia and the *rami communicantes*, which connect the ganglia with the spinal nerves (Fig. 10.46).

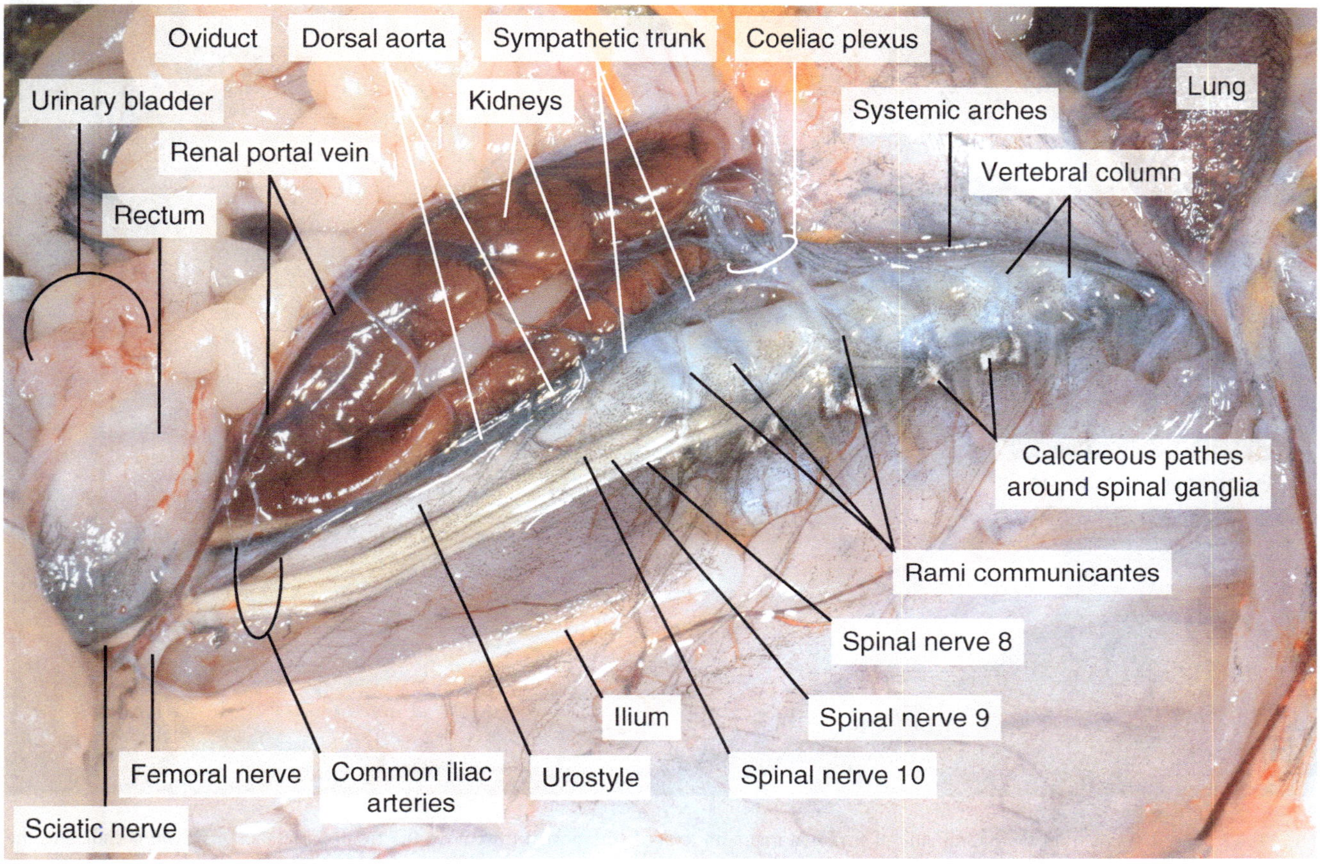

Fig. 10.46 The vertebral column with the dorsal aorta and the sympathetic trunk (Viscera and the left kidney are folded to the right side.)

The central nervous system consists of the spinal cord and the brain. For the examination of the brain, cut off the head of the frog by a large pair of scissors between the skull and the first vertebra. Strip away the skin from the midline of the head down to the eardrum–eye–nostrils line. Make two parallel cuts through the skull by a small pair of scissors beginning from the lateral side of foramen magnum towards the nostrils (Fig. 10.47).

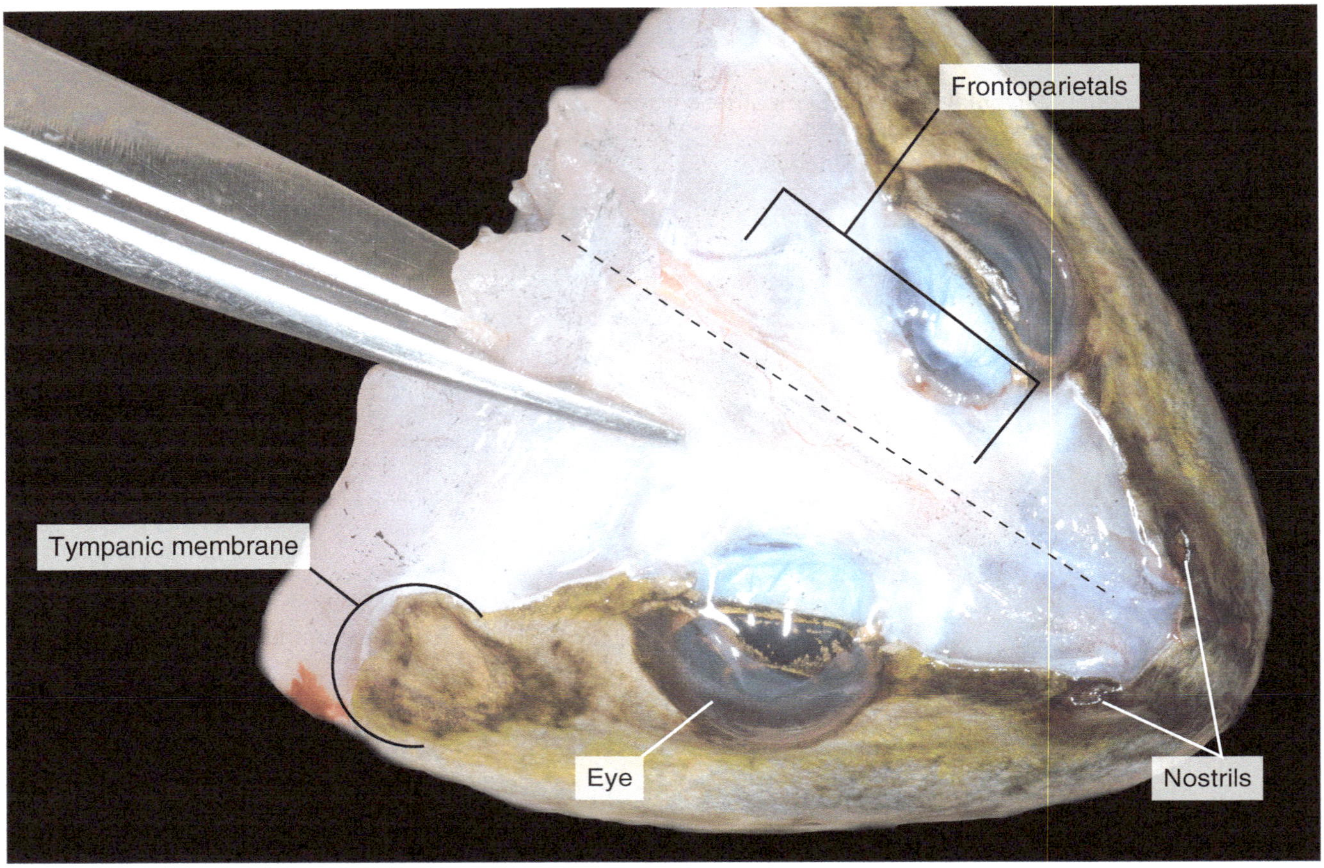

Fig. 10.47 The first cut from the lateral side of foramen magnum towards the nostril to remove the frontoparietals. *Dashed line* indicates the mid-sagittal plane

Be careful not to injure the delicate brain tissue beneath.
Lift the roof of the skull (Fig. 10.48).

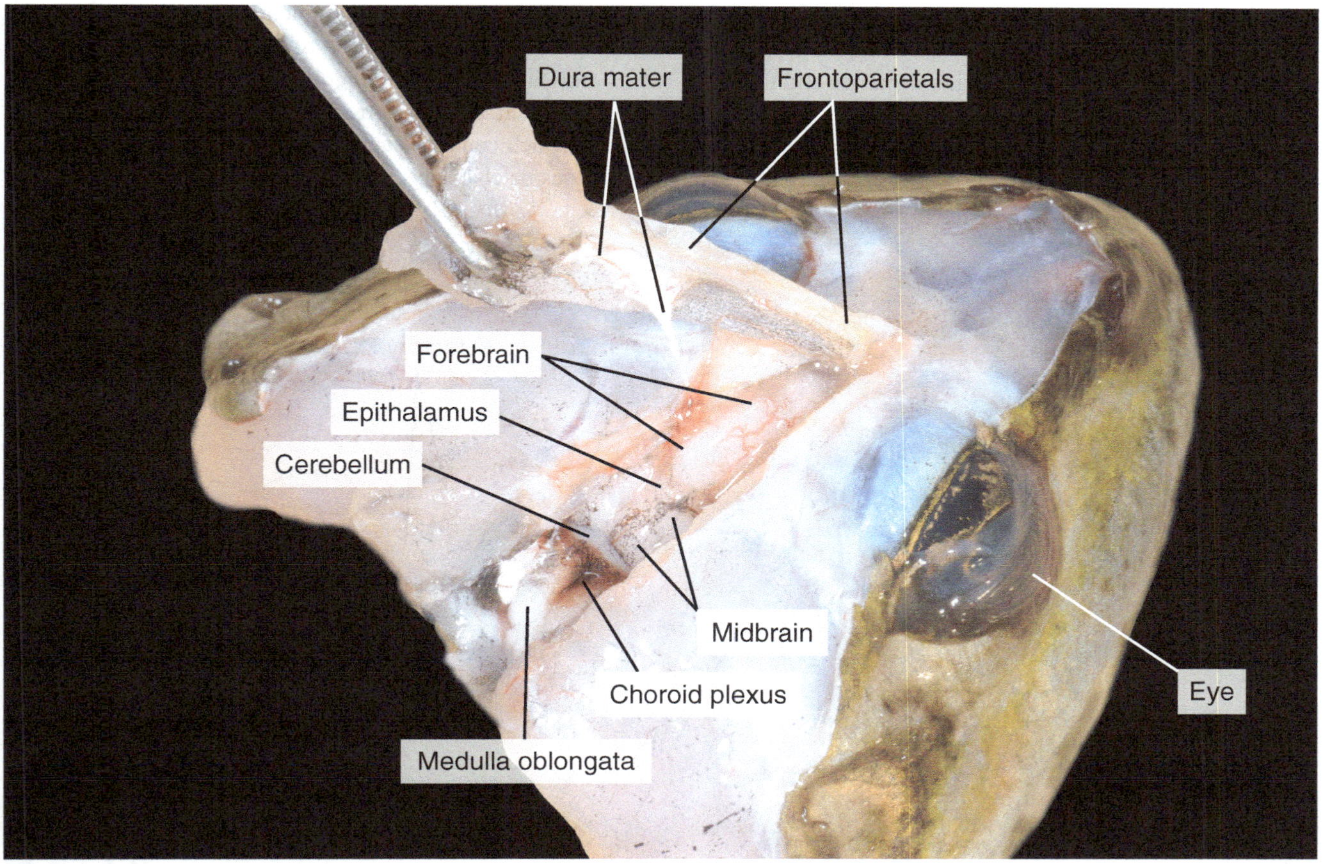

Fig. 10.48 Opening the braincase of the frog's skull with the removal of the frontoparietals

The central nervous system is enclosed in two membranes called *meninges*. The tough *dura mater* usually clings to the cranial wall and neural canal (Fig. 10.48). The thinner *pia mater* adheres to the brain and spinal cord. Remove dura mater from the brain and identify the following parts: the *forebrain* consists of the cerebral hemispheres and the diencephalon (Fig. 10.49).

The cerebral hemispheres constrict anteriorly to form the *olfactory lobes*, from which the *olfactory nerves* (cranial nerve I) extend to the nares. The diencephalon is a depressed region behind the hemispheres. The *pineal gland* (epiphysis), ascending dorsally is a rudimentary "third eye". The *midbrain* bears on the dorsal side two prominent *optic lobes* (Fig. 10.49). The hindbrain consists of the narrow, underdeveloped *cerebellum* and the *medulla oblongata*. The anterior roof of the medulla is thin, membranous and infolded to form the *choroid plexus*, which contains capillaries and produces the liquor. A calcareous patch encapsulates the *inner ear* (Fig. 10.49).

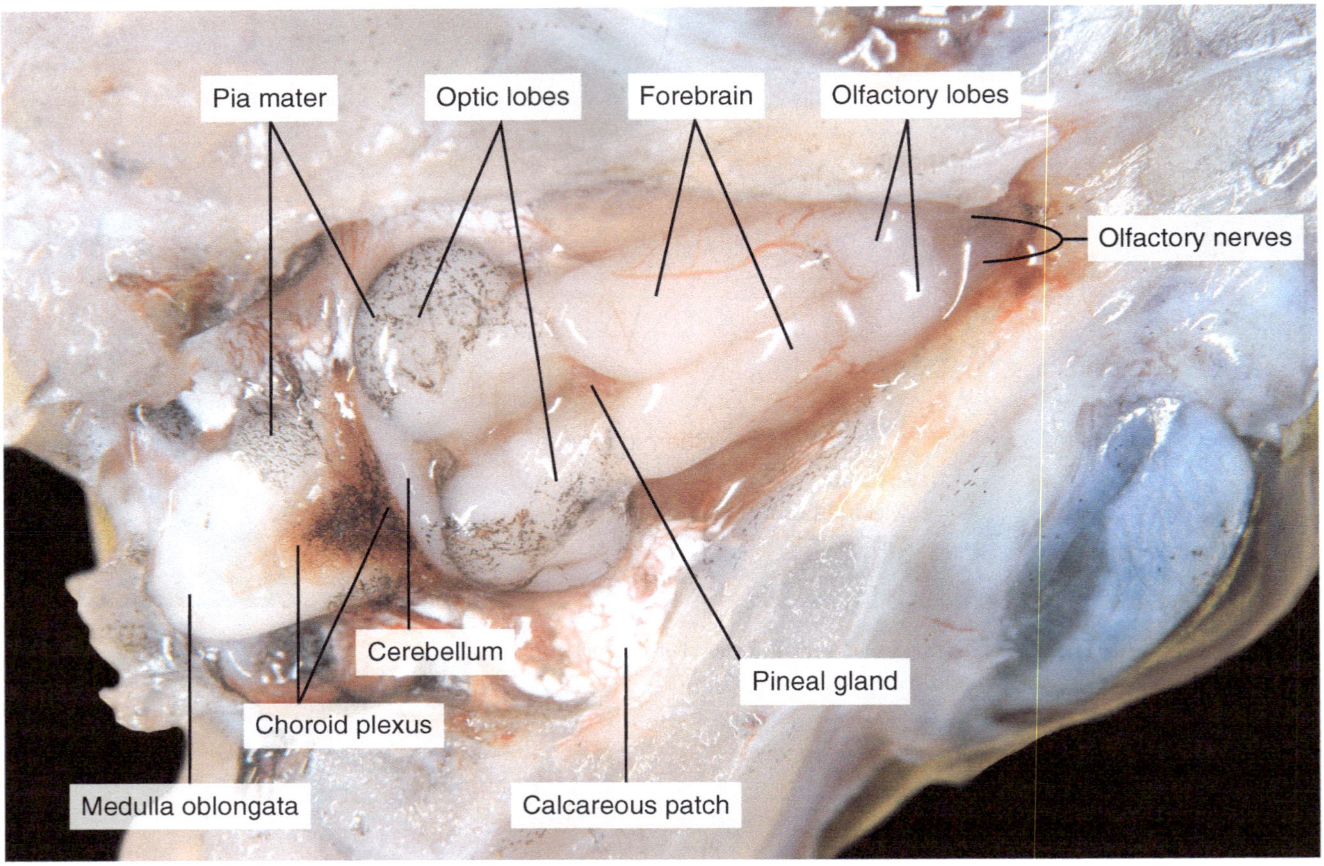

Fig. 10.49 Dorsal view of the brain and inner ear surrounded by a calcareous patch

Finally, take a closer look at the frog's middle *ear*. The frog lacks an external ear. The *tympanic membrane* (eardrum) is the external boundary of the middle ear. Cut around the periphery of the tympanic membrane and lift it up to locate the tiny ear bone, the *columella*, which transmits vibrations from the tympanic membrane to the inner ear, where auditory nerve picks them up (Fig. 10.50, compare with Fig. 10.2).

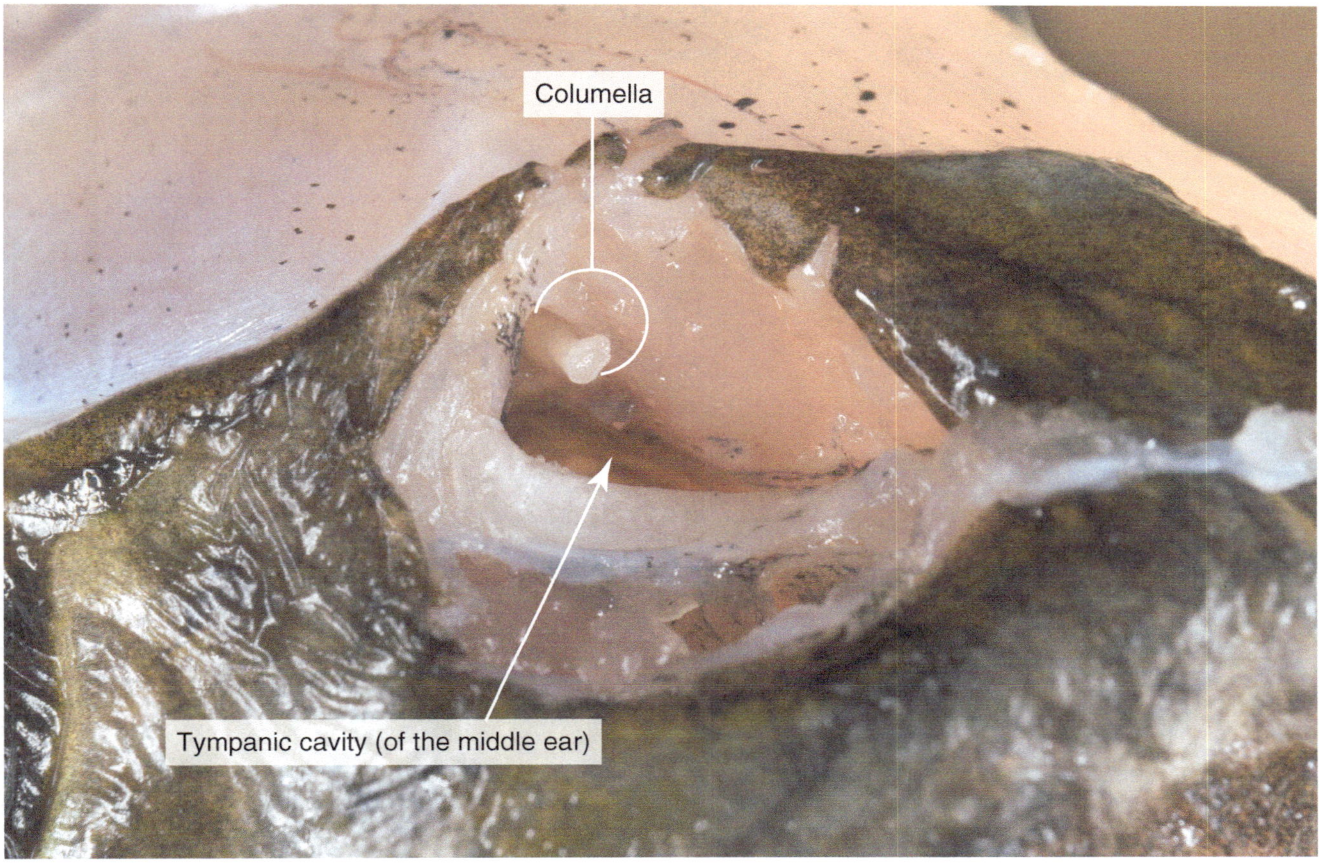

Fig. 10.50 The columella, the tiny ear bone, after removal of the tympanic membrane

Dissection of a Chicken (*Gallus domesticus*)

11

- *Availability*: Birds differ from other vertebrates by their adaptations to flight. These adaptations have limited further structural variation so that although over 8,600 species of birds are known, structural differences between the groups are small. The domestic fowl belongs to the genus *Gallus* which includes four species. It is very probable that more than one ancestor has contributed to existing breeds. The domestic fowl therefore may be of polyphyletic origin so the name *Gallus domesticus* may be used. We can purchase 4–5 months old chicken, the most suitable for dissection, from small firms breeding domestic fowl.
- *Anaesthesia*: Inhalation anaesthesia such as diethyl ether or chloroform can be used. Place the chicken in an appropriate plastic crate, drop a cotton wad infiltrated with diethyl ether next to it and pull over a large garbage bag. Close the bag immediately. When, after about 20 min, it doesn't move upon rocking the crate, and doesn't show any reflexes, we can begin to wash it. Dissection can be started after drying.

Skeleton: The skeletal system of birds is similar in all species, a characteristic feature is the presence of *pneumatic bones* (containing air spaces) which are often in direct connection with the air sacs and therefore with the respiratory tract.

Skull: The skull is divided into ethmoidal (olfactory), orbital (optic), otic (auditory) and occipital regions and the cranial vault (calvaria). The bones of the skull fuse soon after hatching so their borders are not visible in the adult. The ethmoidal region is formed by the bones nasal, mesethmoidal and vomer. The two orbits are very large cavities, a characteristic of all birds. Each orbit is bounded dorsally by frontal, alisphenoid, orbitosphenoid and anterodorsally by lacrimal and posteriorly by the zygomatic process. Orbit is incomplete on the ventral side. The two orbits are separated by the interorbital septum. Each otic region is formed largely by the prootic bone and the squamosa. The calvaria is formed by parietals and frontals; the ventral part is by basisphenoid and parasphenoid bones. In the occipital region, supraoccipital, exoccipitals and basioccipital bones are present. Beneath the foramen magnum there is a single occipital condyle (monocondylic skull). Upper and lower jaws form the beak. Each half of the upper jaw is formed by intermaxilla, maxilla, quadratojugal and jugal bones. The inner arcade of the upper jaw forms the roof of buccopharyngeal cavity which consists of processes of intermaxilla and maxilla, palatine, pterygoid and quadrate. Each half of the lower jaw is formed by articular, angular, supraangular and dental. Both jaws lack teeth. The hyoid apparatus of the chicken which supports the tongue and larynx consists of remnants of the 2nd and 3rd gill arches (Fig. 11.1).

P. Lőw et al., *Atlas of Animal Anatomy and Histology*, DOI 10.1007/978-3-319-25172-1_11

Axial skeleton: There is great fusion of the vertebral elements of birds due to their modifications for flight and bipedal locomotion. The forces acting on the wings require a firm thoracic region, while the impact of landing as well as the problems of balance results in a fusion of lumbar and sacral vertebrae (Figs. 11.1 and 11.2). The only freely articulating vertebrae are in the cervical and coccygeal regions. The centra of birds are saddle shaped (*heterocoelous*), rather than flat as in mammals.

Cervical vertebrae: The *atlas* is a small ring-like bone forming a ball and socket joint with the occipital condyle of the skull which permits great movement of the head. The *axis* or 2nd cervical, has an odontoid process (dens axis), as well as dorsal and ventral spines. It is much larger than the atlas. The remaining cervical vertebrae are progressively larger and bear conspicuous posteriorly directed spines called *cervical ribs* (pleurapophyses) (Fig. 11.1). These are characteristic of most cervical vertebrae. Note the transverse foramen piercing the base of each cervical rib. Often the last one or two cervicals are fused to the thoracics.

Thoracic vertebrae: The seven elements making up this region are usually fused with one another for the greater part of the region. The *ribs* are movably attached to the vertebrae (Figs. 11.1 and 11.2).

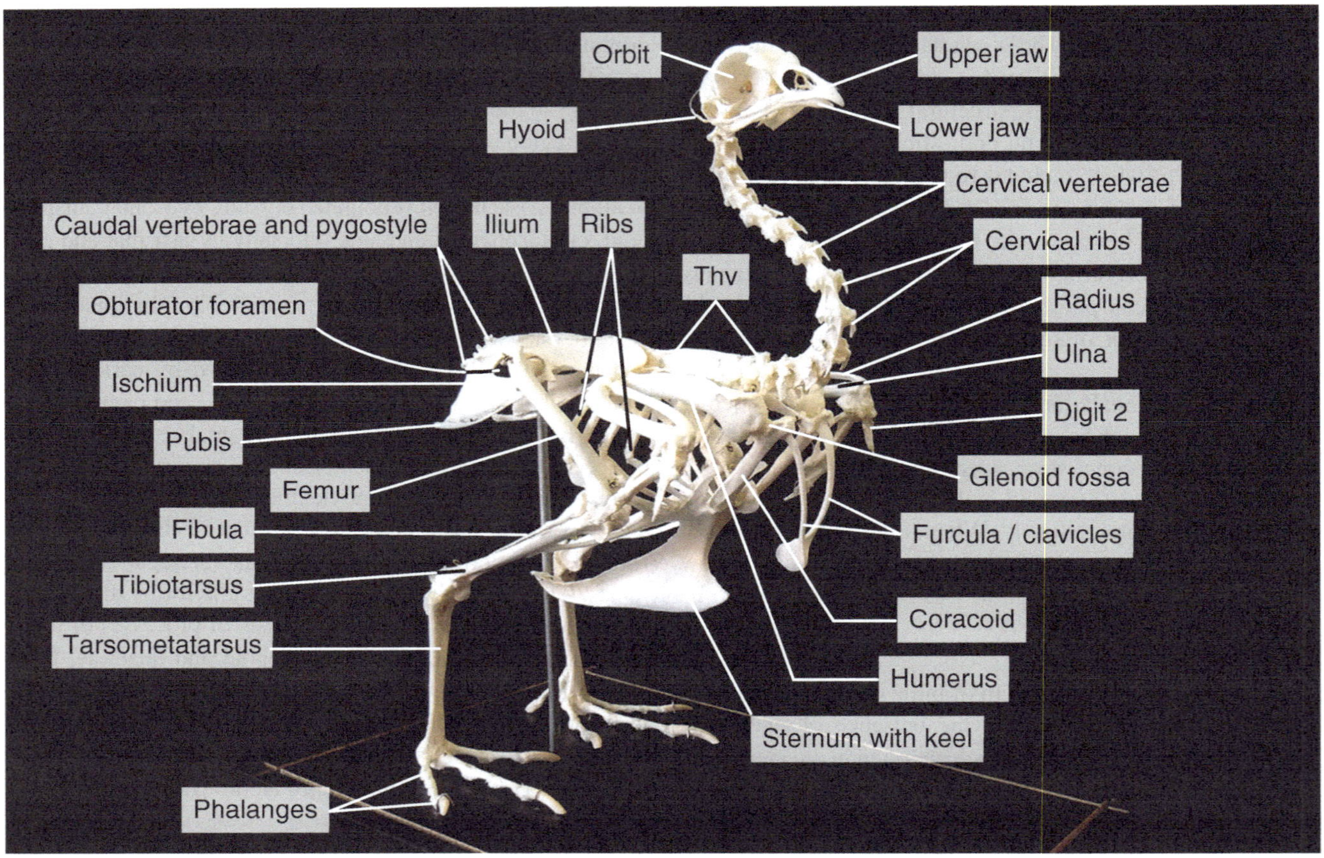

Fig. 11.1 Mounted skeleton of the chicken (*Gallus domesticus*) *Thv* thoracic vertebrae

Lumbar and sacral vertebrae: The 14 lumbar and sacral vertebrae are completely fused with one another as well as with the last few thoracics and first coccygeal. This complex is known as the *synsacrum* and is firmly joined to the ilia on each side (Fig. 11.2).

Coccygeal vertebrae: The caudal vertebrae are free except for the first which is fused with the sacrals. The terminal coccygeal vertebra represents a fusion of several elements and is known as the *pygostyle* (Fig. 11.2). This bone supports the preen gland and affords attachment for the tail feathers (Fig. 11.7).

Thorax: Of the seven pairs of ribs, the first, second and sometimes the seventh do not reach the sternum. Those ribs which articulate with the sternum consist of two elements, a vertebral and a sternal part. An *intercostal joint* between the vertebral and sternal portion permits the sternum to be lowered and raised in respiration (Fig. 11.2). Note the posteriorly directed *uncinate processes* of the vertebral ribs which strengthen the thoracic cage and provide attachment for muscles which support the shoulder blade (Fig. 11.2).

Sternum, pectoral girdle and forelimb: The *sternum* of the chicken is elongate and well ossified with middorsal pneumatic openings. The large midventral keel or *carina* serves for the attachment of the flight muscles (Figs. 11.1 and 11.2). The medial, anterior projection is known as the rostrum or manubrium, while the lateral anterior projections are the *costal processes*. Between the rostrum and the costal process of each side, the sternum articulates with the *coracoid* (Figs. 11.1 and 11.2). The long, posterior lateral projections of the sternum are the *xiphisternal processes*, the anterior pair of which overlap and support the last three ribs.

The *pectoral girdle* consists of the blade-like *scapula*, the *coracoid*, a stout bone reaching the sternum and the "wish bone" or *furcula* (Figs. 11.1 and 11.2). The furcula really consists of the *two clavicles* united ventrally by the *interclavicle*. The scapula, coracoid and clavicle form the *shoulder joint* for the humerus (glenoid fossa) (Fig. 11.1). There is a passage over that for the supracoracoid tendon of the deep pectoral muscle.

The forelimb or *wing* is adapted for flight by a reduction and fusion of the distal elements. The stout *humerus* has an ovoid *head* and prominent greater and lesser *tuberosities*. The greater tuberosity bears on its posterior side the *pneumatic foramen* for the passage of air into the humerus. In the forearm, the *ulna* is the larger bone which has at its proximal

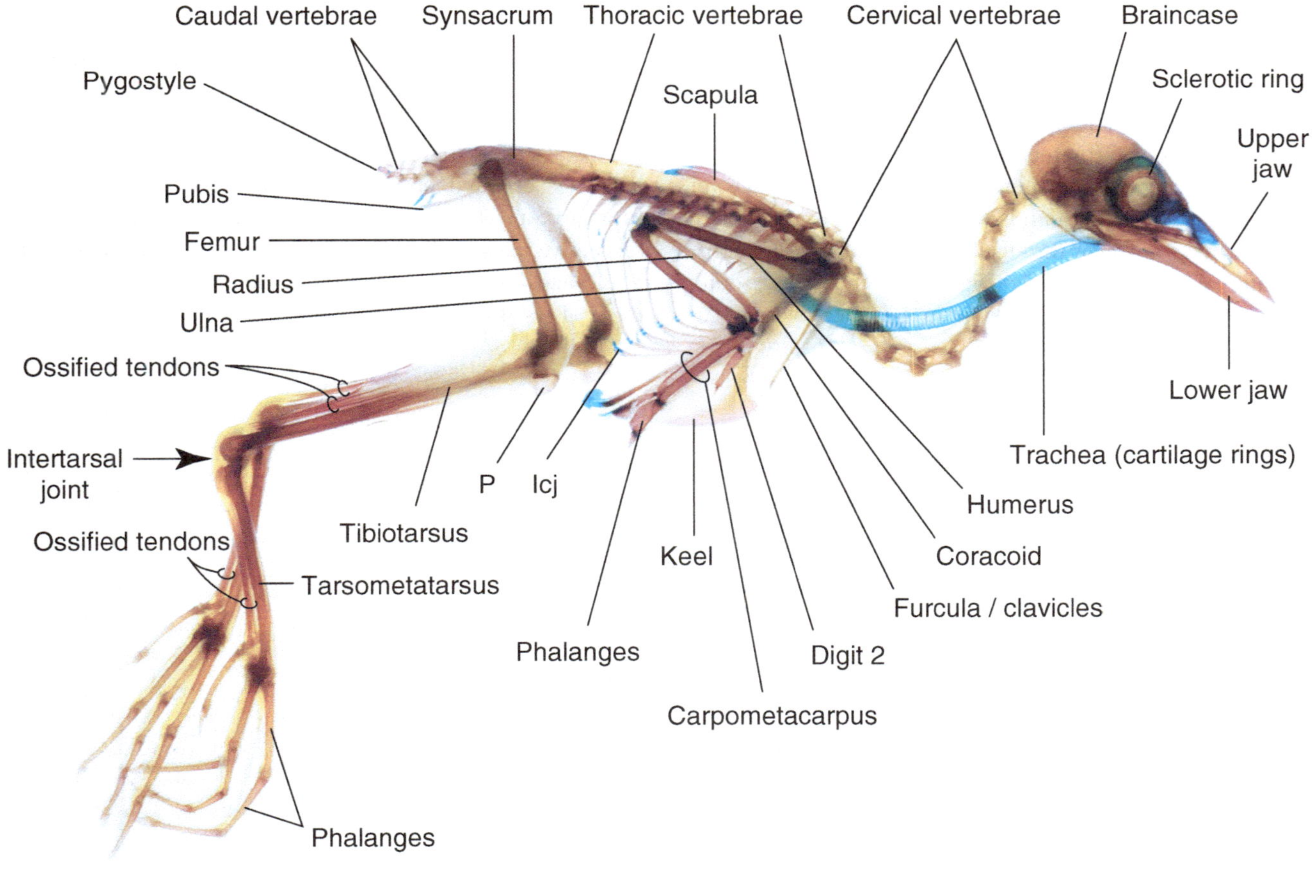

Fig. 11.2 The skeleton of the little crake (*Porzana parva*) after selective staining: cartilages are blue and bones are red (courtesy of Viktor Kis). *Icj* intercostal joint, *P* patella

end the *olecranon process*. The slender *radius* is almost as long as the ulna (Figs. 11.1 and 11.2).

Developmentally, the carpal elements (wrist) fuse with the metacarpals to form the *carpometacarpus*. This region is greatly reduced and consists of only two separate bones, a radial carpal (radiale) and an ulnar carpal (ulnare) in the adult. There are two phalanges on digit 2, two or three phalanges on digit 3 and one phalanx on digit 4 (Fig. 11.2).

The pelvic girdle and hind limb: Since the sternum of birds functions to support the viscera, there is little development of the pubis (Figs. 11.1 and 11.2). In the chicken as well as in other birds, the two halves of the "girdle" are widely separated ventrally which better permits the passage of relatively large eggs through the pelvic outlet. The loss of strength resulting from this lack of fusion of the ventral portions of the girdle is compensated for by the strong fusion of the ilia along their dorsal borders to the vertebral column.

The bones of the pelvic girdle correspond to those of mammals in a general way being composed of three parts: the *ilium*, *ischium* and *pubis* meeting at the *acetabulum* where the head of the *femur* articulates (Figs. 11.1 and 11.2). The anterolateral face of the ilium is concave to receive the gluteal muscles. The ventral surface of the ilium is concave and divided into three regions: an anterior region which contains the anterior lobe of the kidney, a medial in which lies the sciatic plexus and middle lobe of the kidney and a posterior which contains the posterior lobe of the kidney.

The *ischium* is smaller than the ilium but fused with it except at the sciatic foramen for passage of the sciatic nerve. The *pubis* is reduced to a thin splint-like bone along the ventral border of the ischium. It is separated from the ischium by the *obturator foramen* through which passes the tendon of the internal obturator muscle (Fig. 11.1). The ilium, ischium and pubis are separate and distinct in the chick.

Since the weight of the bird is balanced on the hindlimbs, the centre of gravity has been moved posteriorly and ventrally. The stresses of landing and walking are transferred to the elongated ilia which in turn are fused with the synsacrum.

The *femur* is somewhat bent and has a small head which fits deeply into the open acetabulum. The distal end of the femur forms the trochlea anteriorly and the two condyles posteriorly. Between the medial and lateral ridges of the trochlea fits the triangular *patella* which provides attachment for the quadriceps and patellar ligaments (Fig. 11.2). The *tibiotarsus* is the largest bone of the chicken, composed of the tibia fused distally with the proximal row of tarsals. At its proximal end is the expanded articular surface of the condyles of the femur. The splint-like bone articulating with the lateral condyle of the femur is the *fibula* whose distal third has atrophied. The distal end of the tibiotarsus bears the medial and lateral malleoli for articulation with the tarsometatarsus. The *tarsometatarsus* represents the fusion of the 2nd, 3rd and 4th metatarsals and the distal tarsal bones. Separate tarsal bones do not exist in adult birds since the proximal tarsals have fused with the tibia and the distal tarsals have fused with the metatarsus resulting in an *intertarsal joint* (Figs. 11.1 and 11.2).

There are usually *four digits*, three directed anteriorly and one posteriorly, so that the weight is borne in tripod fashion. The 5th metatarsal and digit are absent. A remnant of the 1st metatarsal is seen as a small projection on the medial side of the distal end of the tarsometatarsus. Its digit with two phalanges is directed posteriorly. Metatarsals 2, 3 and 4 are fused except at their distal ends. Each articulates with its respective clawed digit. The number of phalanges on digits 1, 2, 3 and 4 runs in regular progression: 2, 3, 4 and 5, respectively (Figs. 11.1 and 11.2).

Before dissection, examine the external features of the chicken. Observe the *head*, the *trunk*, the *tail* and the *appendages* (Fig. 11.3).

The *skin* or integument of the chicken consists as in mammals of two parts: epidermis and dermis. The *epidermis* is generally thin and the *dermis* is much thicker. (See later histology of the skin.) The epidermis gives rise to many structures which we commonly speak of as integumental derivatives. In the chicken, such structures would be the feathers, glands, scales, beak, claws, spurs, wattles, combs and ear lobes.

Feathers are the characteristic epidermal structures of birds, derived from reptilian scales. They function as an insulation to maintain the high body temperature (40.6–42.8 °C or 105–109 °F) and are necessary for flight. Feathers are not attached to birds in a random manner over the entire body of the bird. Instead they are usually found in linear tracts, called *pterylae*. The spaces on the bird's body without feather tracts are referred to as *apteria* (Fig. 11.3). The densest area for feathers is often on the bird's head and neck. On a plucked chicken, these tracts can be seen to consist of dorsal, humeral, femoral, ventral, head, alar (on wing), crural and caudal tracts.

Three types of feathers are distinguished: the *contour feathers* (pennae), the *down feathers* (plumules) and the *filoplume*.

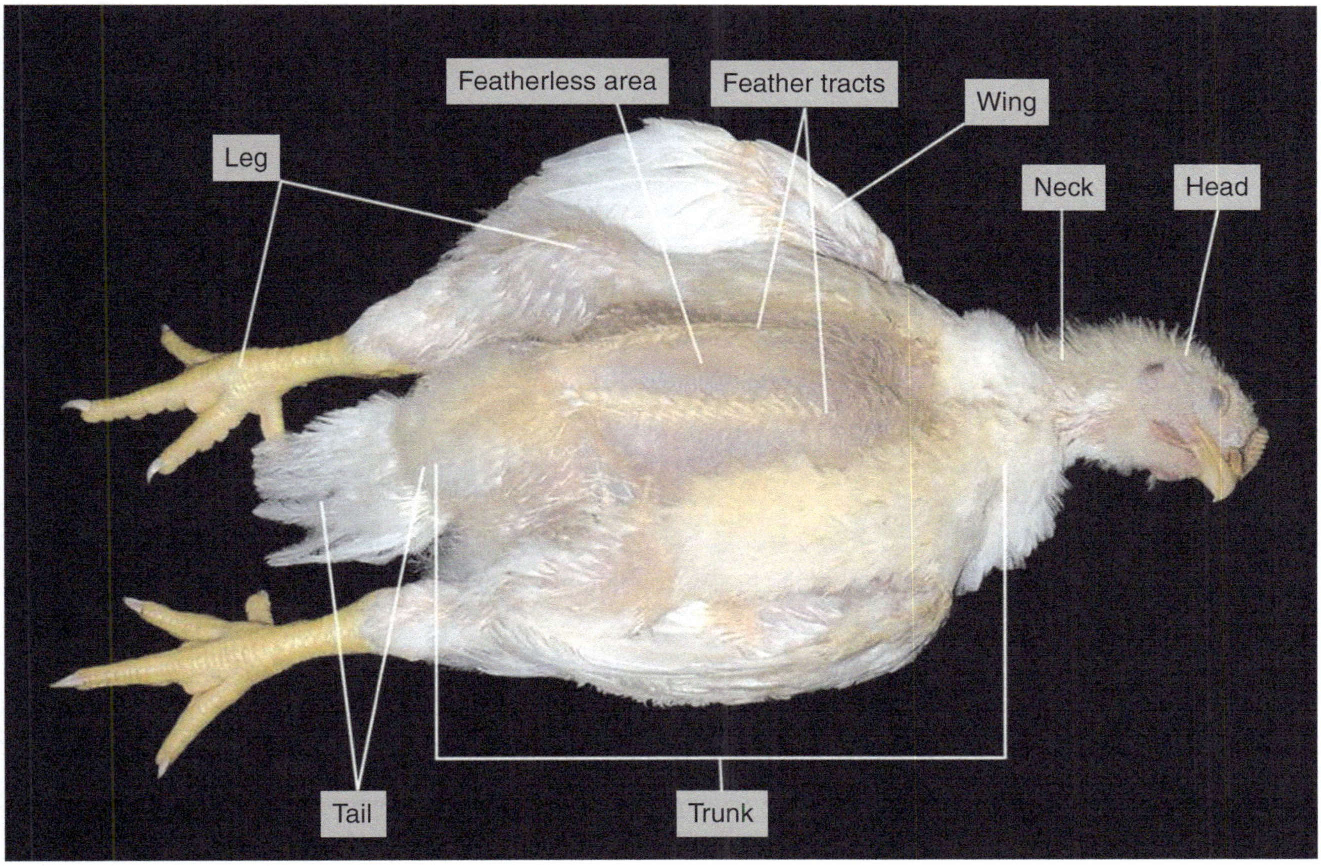

Fig. 11.3 Ventral view of juvenile chicken

The *contour feathers* are the large feathers which sheath the body as well as the wings and tail. The feathers of the wings are the *remiges*, while those of the tail are the *retrices*. The remiges consist of *primaries* (on fused third and fourth metacarpals and digits), *secondaries* (along ulna) and *tertiaries* (along humerus) (Fig. 11.4).

A tuft of feathers arising from the 2nd digit forms the *alula* or bastard wing (ala spuria). The *patagium* is an elastic fold of skin on the bird's wing extending from the shoulder to the wrist making up the leading edge of the inner wing.

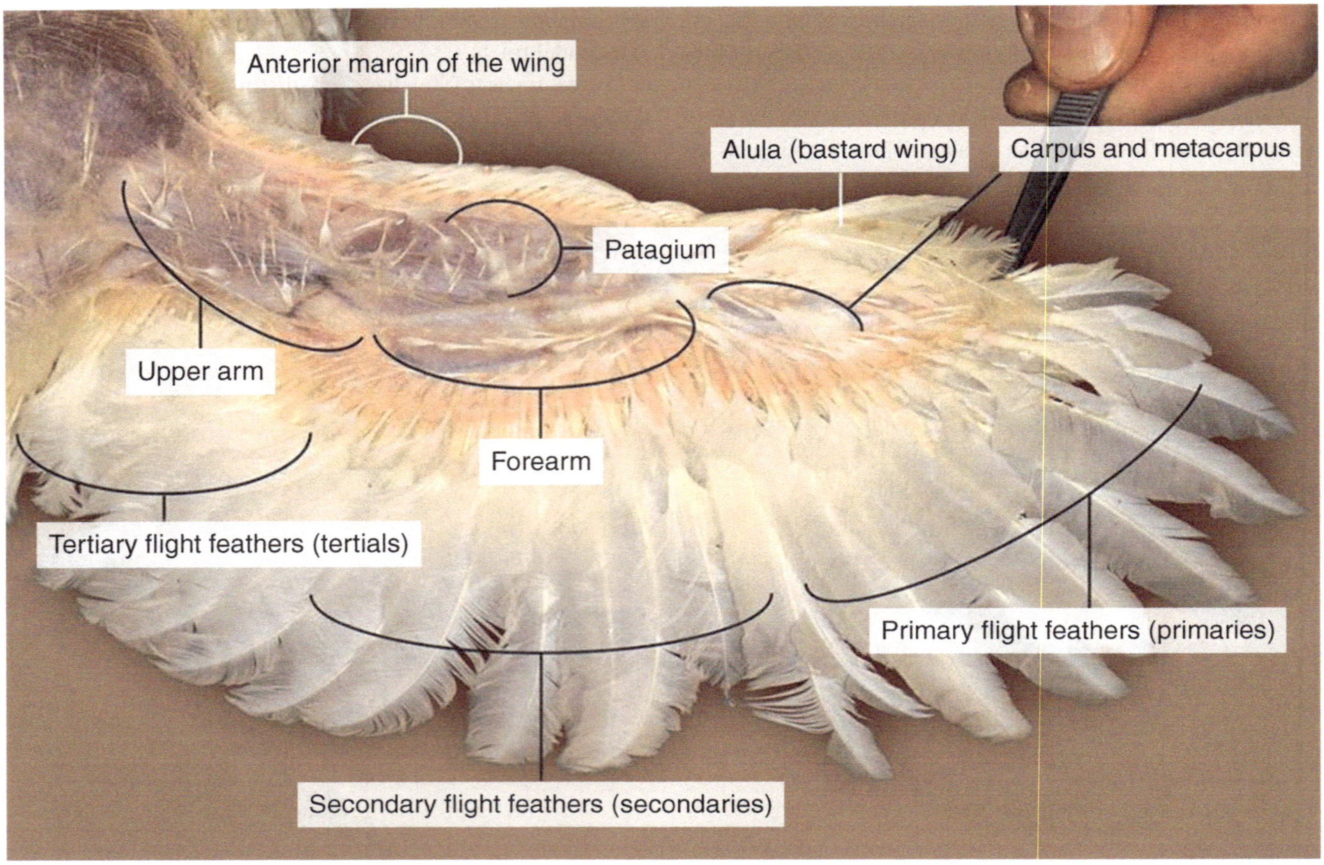

Fig. 11.4 The forelimb of the birds is modified to a wing

Pluck a large wing feather and note the basal portion or *calamus* (quill), a hollow cylinder filled with a pithy material which is the remains of mesodermal tissue (Fig. 11.5).

At the base of the calamus is a small opening, the inferior *umbilicus* through which a mesodermal papilla is extended during development. At the distal end of the calamus lies the expanded portion of the feather, the *vane* (vexillum). The axis of the feather is the *shaft* (rachis), and in some feathers a small secondary axis at the vane base is called the *aftershaft* (hyporachis). From each side of the shaft extend the closely arranged *barbs*. At first, these elements appear to form a continuous sheet, but closer inspection (magnified) shows *barbules* which arise on each side of the barbs and by a system of hooks and notches interlock with the barbules of adjacent barbs (Fig. 11.5, insert). A disruption of this interlocking arrangement such as "splitting" the vane with your fingernail can be repaired by pulling the vane between your fingers in a direction parallel to the barbs. This function is performed by the bird when "preening".

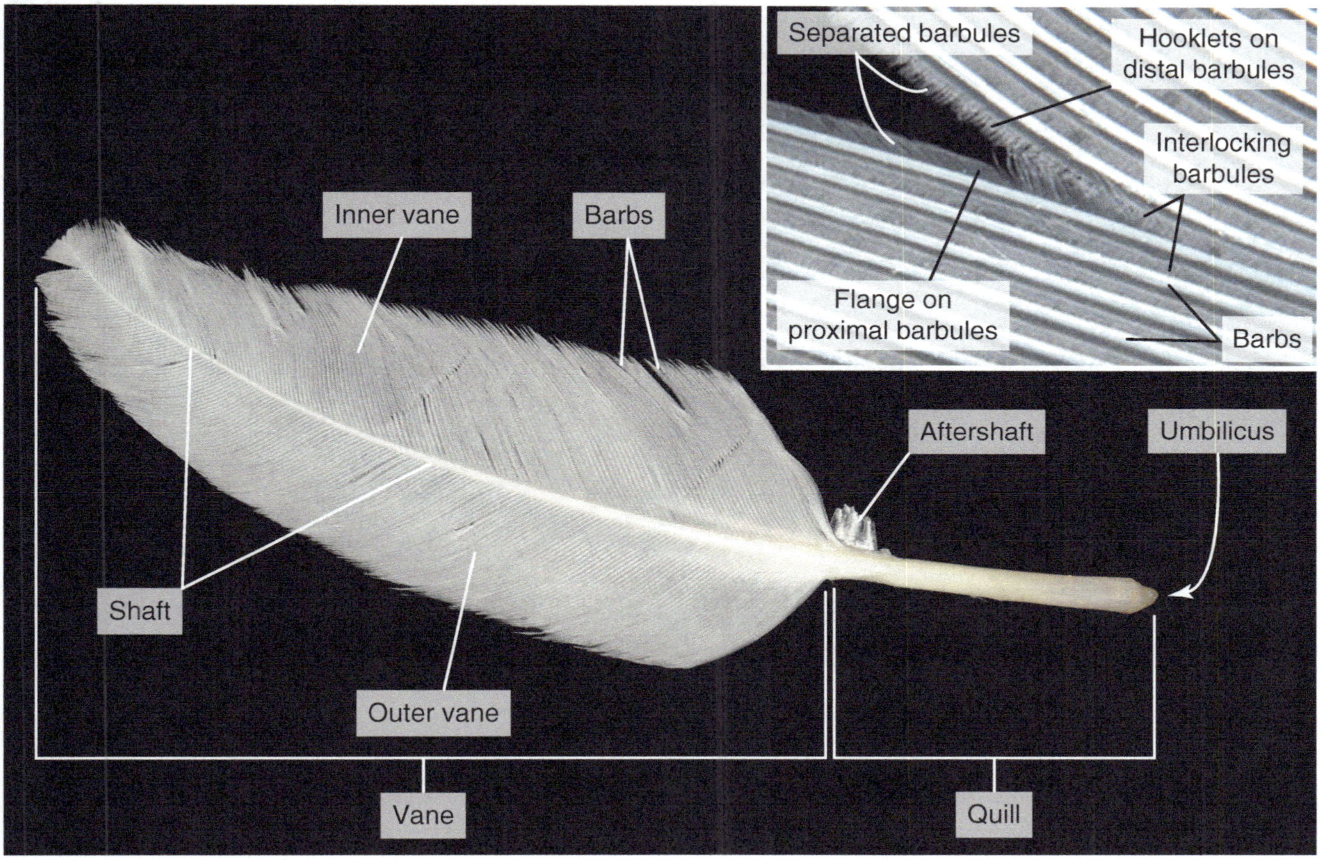

Fig. 11.5 The contour feathers are large feathers which sheath the body as well as the wings and tail. Insert: In the vane of contour feather barbules which arise on each side of the barbs interlock by a system of hooks and notches

In *down feathers*, the rachis is either missing completely or substantially reduced in length (Fig. 11.6).

The barbules lack hooks, which combined with the lack of rachis results in a very soft and fluffy feather. Without the hooks, the barbs and barbules create a puffy tangle of insulating air pockets. These form the body covering of a chick and underlie the contour feathers of the adult.

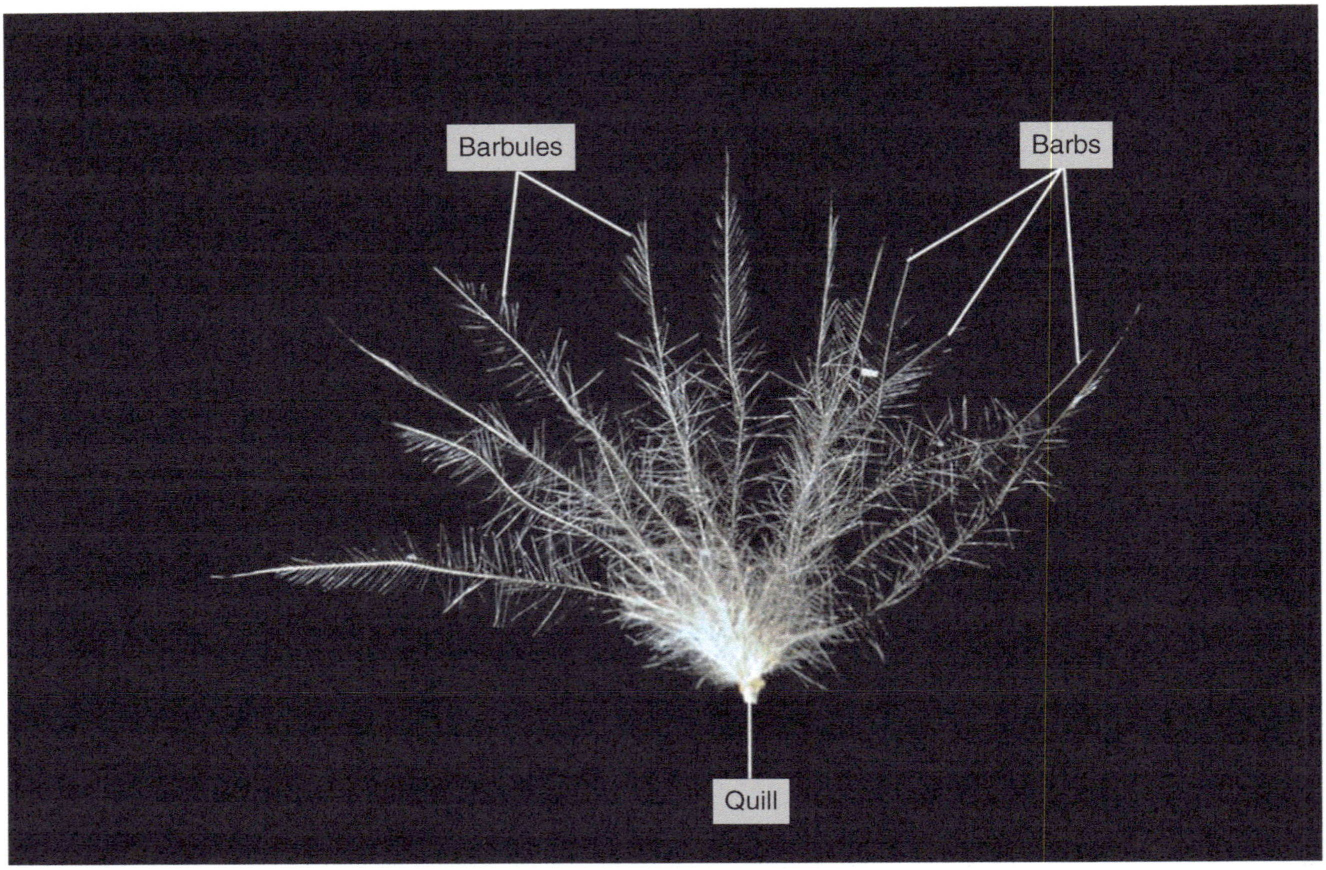

Fig. 11.6 A down feather looks fluffy because of the lack of rachis and hooks on barbules

The simplest feathers, the *filoplumes*, are small hair-like structures with a slender shaft terminating in a tiny tuft of barbs (Fig. 11.7). These relatively stiff feathers lack specific feather muscles but have sensory receptors next to the base of the feathers. Filoplumes lie under the contour feathers.

Feather replacement continues throughout life, and in many birds, it is a seasonal phenomenon called moulting. Once a year, the chicken moults completely, usually in late summer or fall. The hen almost always ceases laying at this time.

Skin glands are rare in birds and the only conspicuous gland is the *preen gland* (uropygial gland or oil gland) (Fig. 11.7).

This is a compound alveolar structure divided into two lobes, each the size of a pea, by a median septum and located above the pygostyle. Its function is to produce a waterproofing material which the bird spreads on its feathers by means of the beak (called preening). Normally, the numerous tubules of the gland pass into two ducts and open on a single *uropygial papilla* (Fig. 11.7). Free the gland from the overlying feathers and examine it.

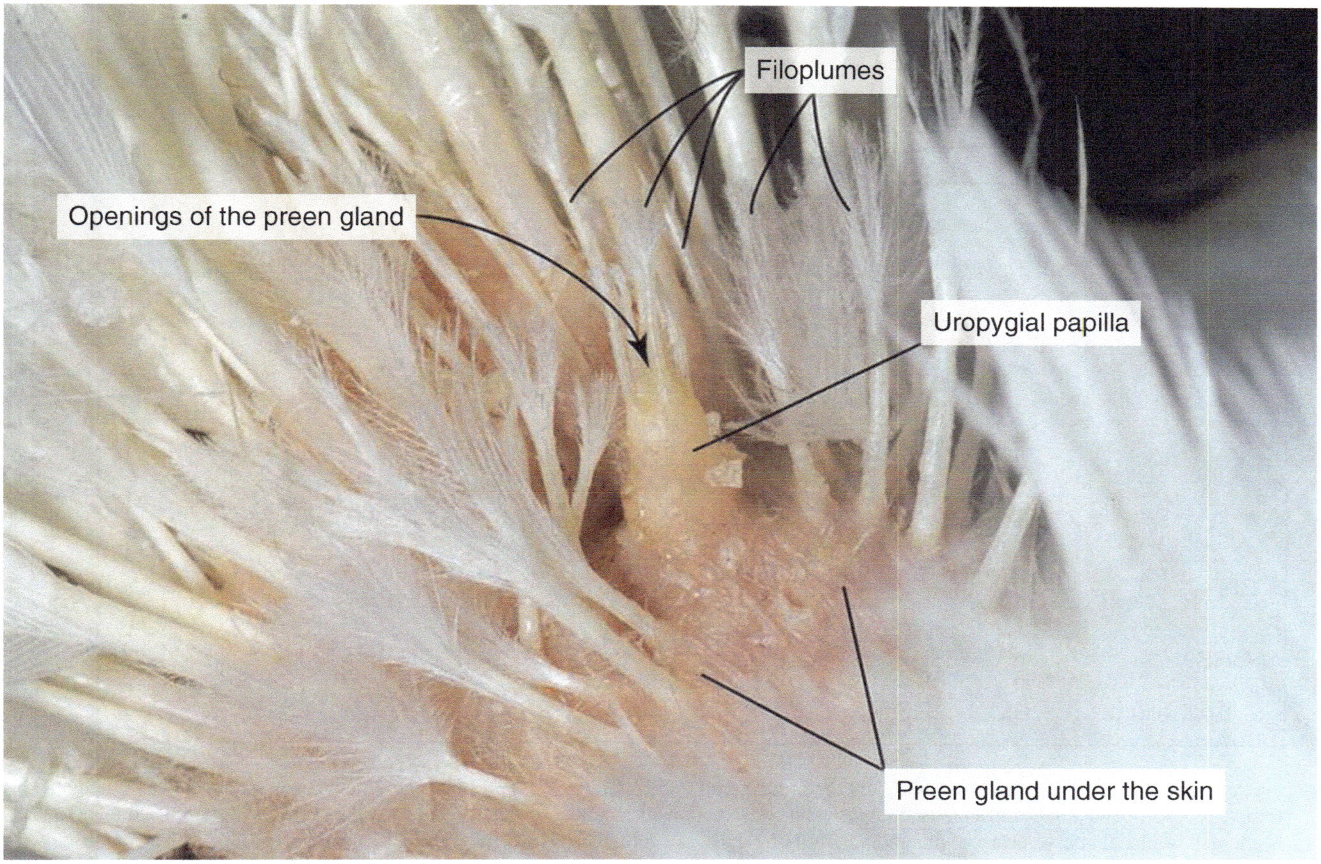

Fig. 11.7 The preen gland is the only gland in a bird's skin

The *skin* of birds has a thin keratinized squamous epithelium with feather follicles, which are arranged in groups. Follicle is a fold of the epidermis, which is caused by a growing dermal papilla (DP). The detailed structure of feathers and feather follicles are rather complex – we mention here some structural elements only with a simple background. There is a dermal papilla (pulp) and epidermal collar on the bottom of the follicle (Fig. 11.8, left).

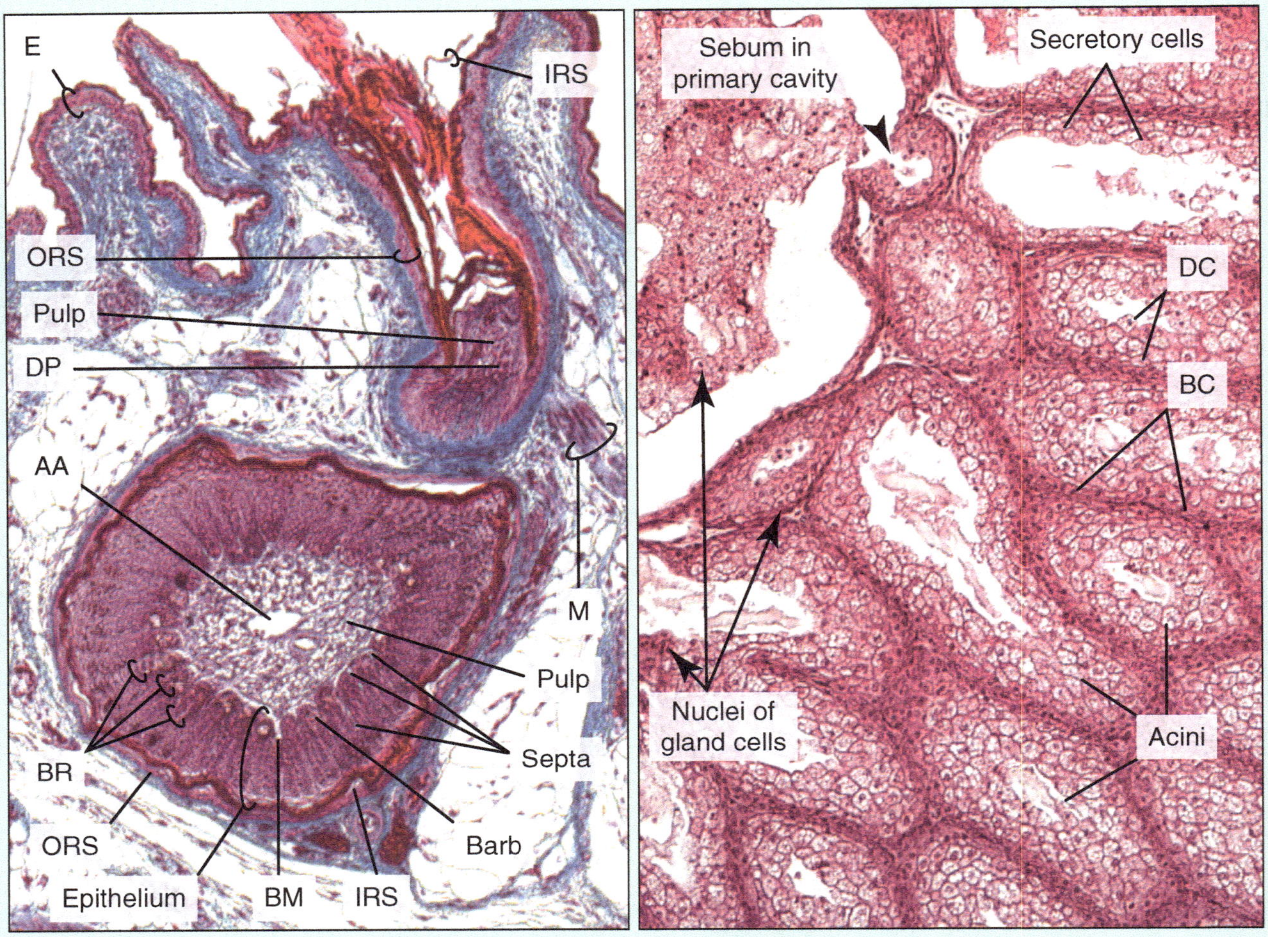

Fig. 11.8 Histological sections of the skin and the preen gland (*left*: Azan, *right*: HE). Left micrograph: two feather follicles; the upper follicle was sectioned almost through its longitudinal plane – thereby the connection between its dermal papilla and the surrounding dermis is not seen. *AA* axial artery, *BM* basal membrane, *BR* barb ridges, *DP* dermal papilla, *E* epidermis, *IRS* inner root sheath, *M* feather muscles, *ORS* outer root sheath. Right micrograph: *arrowhead* tubular gland opening, *BC* basal cells, *DC* differentiating cells

Dermal papilla continues towards the surface and forms a cylinder-shaped loose connective tissue matrix, called pulp (Fig. 11.8, left). There is an axial artery (AA) in the centre of the pulp that provides nourishment to growing feather. The pulp matrix forms radial ridges (septa), which divide the thick epithelium into stripes called barb ridges (BR). The epithelium has a basement membrane lying on the pulp, a basal layer, an intermediate and an outer layer. The intermediate layer produces the morphological elements of the vane (rachis, barbs and barbules). The outer epidermal layer forms the inner feather sheath (inner root sheath, IRS) by keratinisation. This sheath becomes ruptured by the growing feather. The outer root sheath (ORS) is continuous with the thin epidermis. Feather muscles (M) are similar to the arrector pili muscles in mammals; they attach to the feather follicle and are capable of elevating or lowering entire groups of feathers (Fig. 11.8, left).

The *preen gland* is a composition of two lobes containing their own primary cavity (Fig. 11.8, right). Each cavity is surrounded by tubular glandular acini opening into it. Each lobe has connective tissue capsule which gives septa for separating the acini. The glandular epithelium is composed of multilayered epithelium, in which the basal cells (BC) proliferate and produce differentiating cells (DC) migrating towards the lumen of the acini. These cells synthesise preen oil which is stored in cytoplasmic vesicles. The routine histological procedures extract the secretion, so the cytoplasm of these cells becomes foamy. At the end of their differentiation, these cells liberate the secretion by *holocrine* mechanism: the entire cell disintegrates to form the secretion, which accumulates in the primary cavity of the gland. Left and right primary cavities have a common efferent (primary) duct leading to the surface (not shown).

Note the *horny scales* on the chicken's leg (podotheca) which are homologous with the scales on the feet and tails of rodents (Fig. 11.9). They represent reptilian scale vestiges.

The *claw* forms a protective covering for the tips of the digits. The base of the claw has a germinative layer protected by a fold of skin, and from this matrix, the keratinized claw grows outwards over the dermis to be continually worn away at the tip (Fig. 11.9).

Fig. 11.9 The leg of a chicken

The toes are flexed by superficial and deep flexor tendons which constitute the "perching mechanism". When the leg is bent as the bird roosts, flexion of the tarsometatarsus stretches the flexor tendons as they pass over the intertarsal joint, automatically bending the toes around the perch (Figs. 11.2 and 11.10).

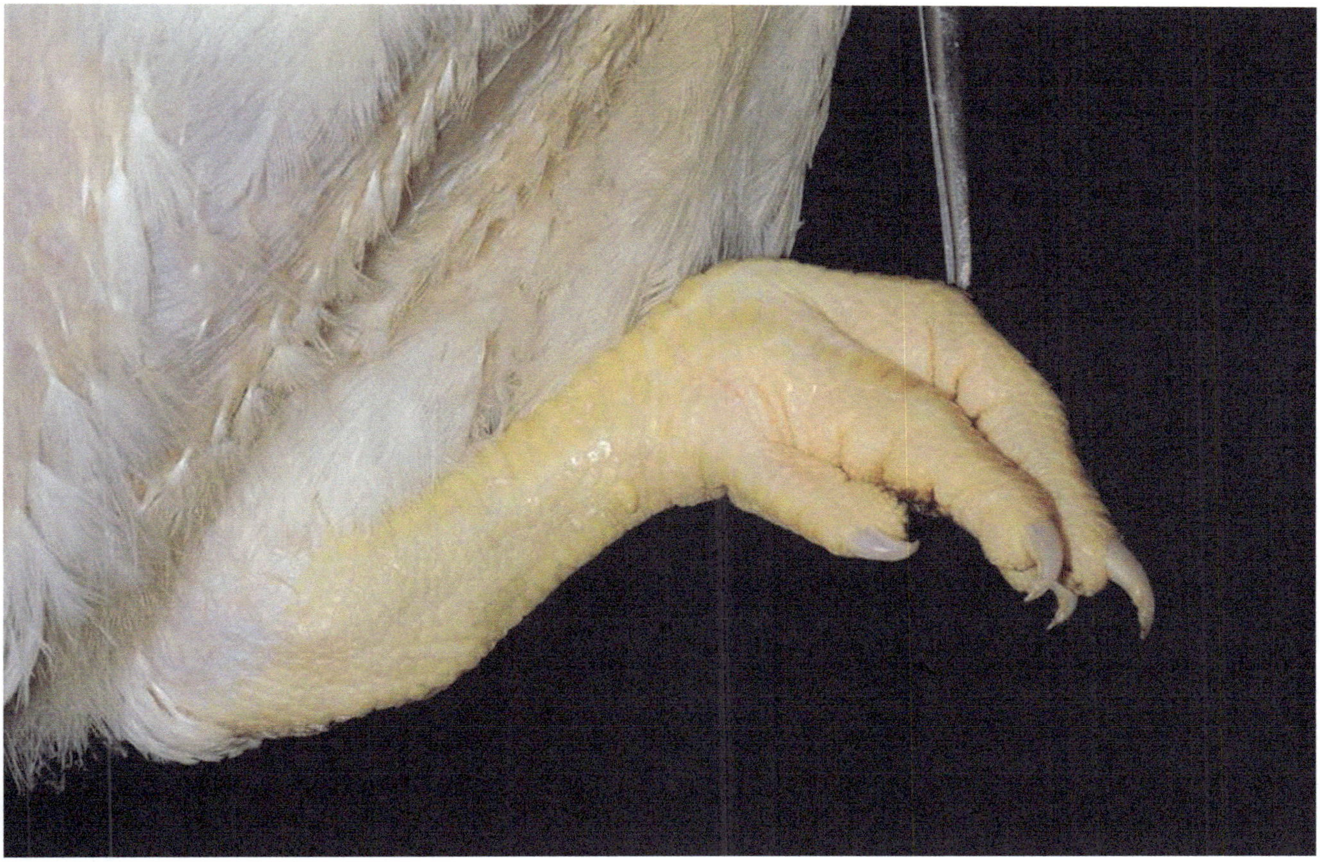

Fig. 11.10 The perching mechanism of the leg of a chicken

Examine the outer structures on the head: the comb and wattles, the eyes with eyelids and nictitating membranes, the opening of the outer ear and the beak with nostrils and mouth (Fig. 11.11).

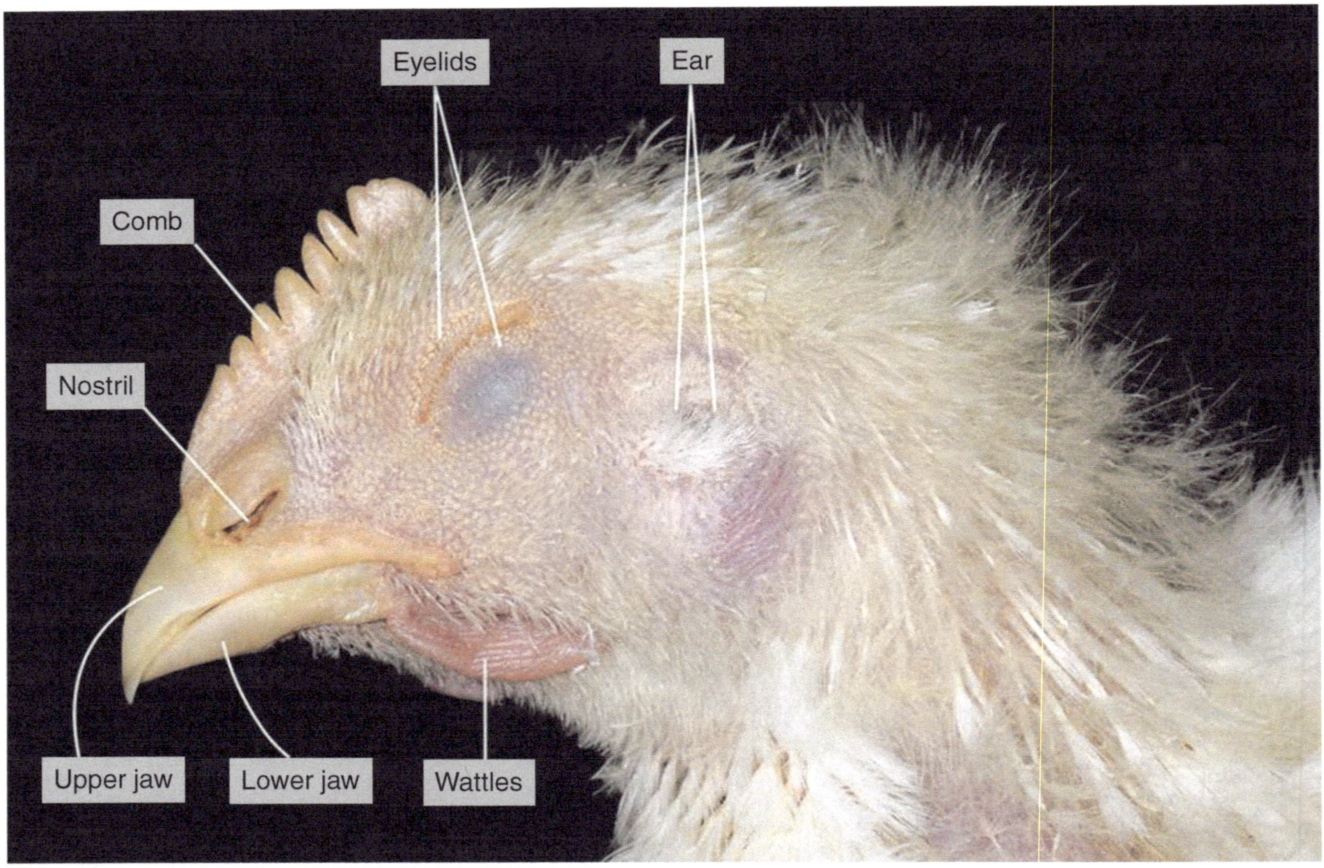

Fig. 11.11 The head of a chicken

The bill or *beak* is cornified epidermis (rhamphotheca) which is constantly being proliferated. The *eyelids* of a bird are not used in blinking (Fig. 11.12). Instead the corneal surface of the eye is kept clean and lubricated by the *nictitating membrane*, a third concealed eyelid that sweeps horizontally across the eye like a windscreen wiper (Fig. 11.12).

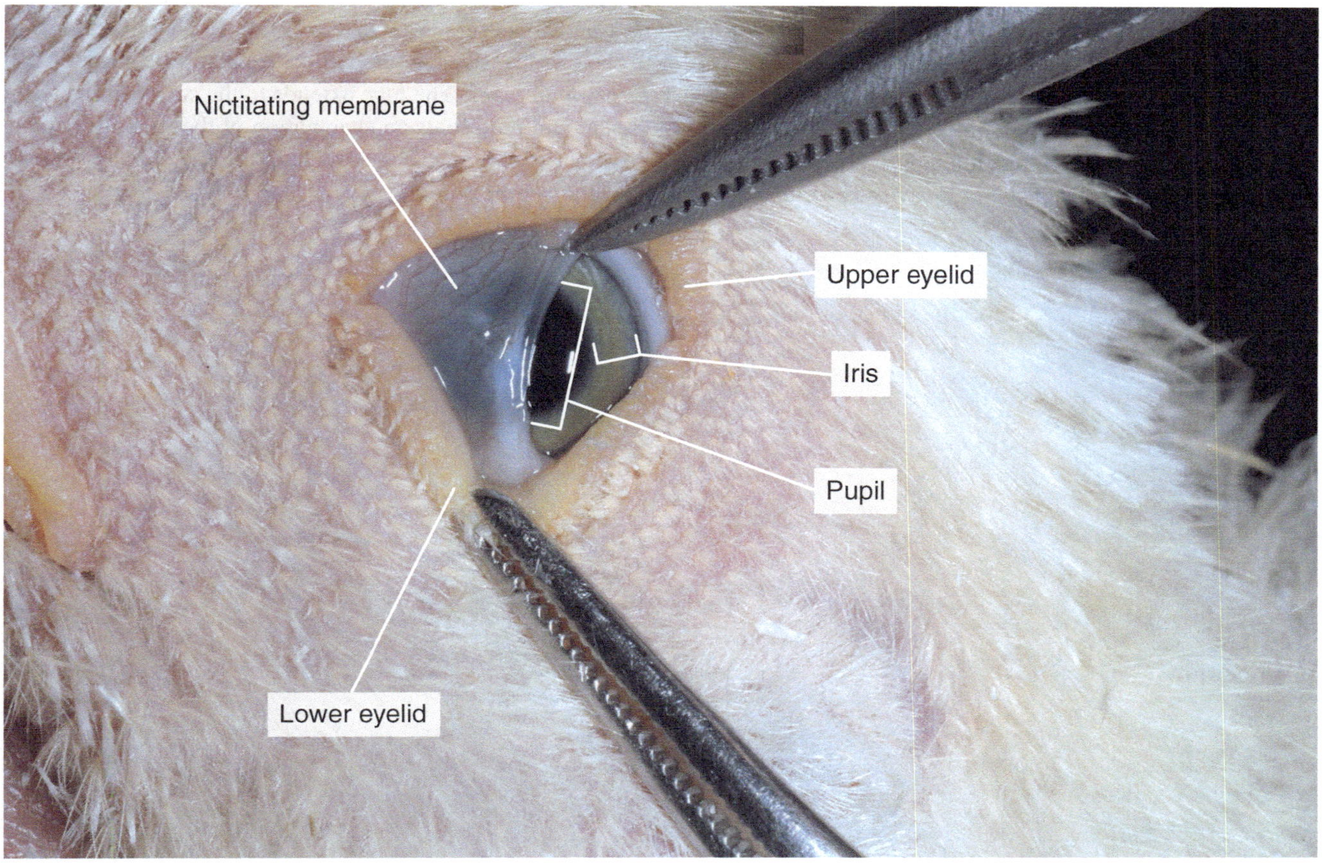

Fig. 11.12 The nictitating membrane moves perpendicular to the eyelids

The nictitating membrane also covers the eye and acts as a contact lens in many aquatic birds when they are under water. Eye is also cleaned by tear secretions from the lacrimal gland and protected by an oily substance from the Harderian gland which coats the cornea and prevents dryness. Open the *mouth* widely and examine the roof of the oral cavity first (Fig. 11.13).

An anterior *oral cavity* bounded by the horny beak, a posterior *pharynx* and the opening of the *choana* (internal naris) can be seen. Teeth are lacking although rudiments are found in the embryo. The roof of the oral cavity is formed by the *palatal folds* which have free fimbriated margins and correspond to the hard palate of mammals (Fig. 11.13). These folds are separated from one another by a median slit resembling the anomalous condition of mammals known as "cleft palate". Insert a probe through the *nostrils* and verify its connection with the choana. On the roof of the pharynx posterior to the choana, identify the slitlike common opening of the *Eustachian tubes*. The tube extends to the cavity of the middle ear and represents an evagination of the first gill pouch. Posterior to the Eustachian opening are the slightly fimbriated ends of the *palate*.

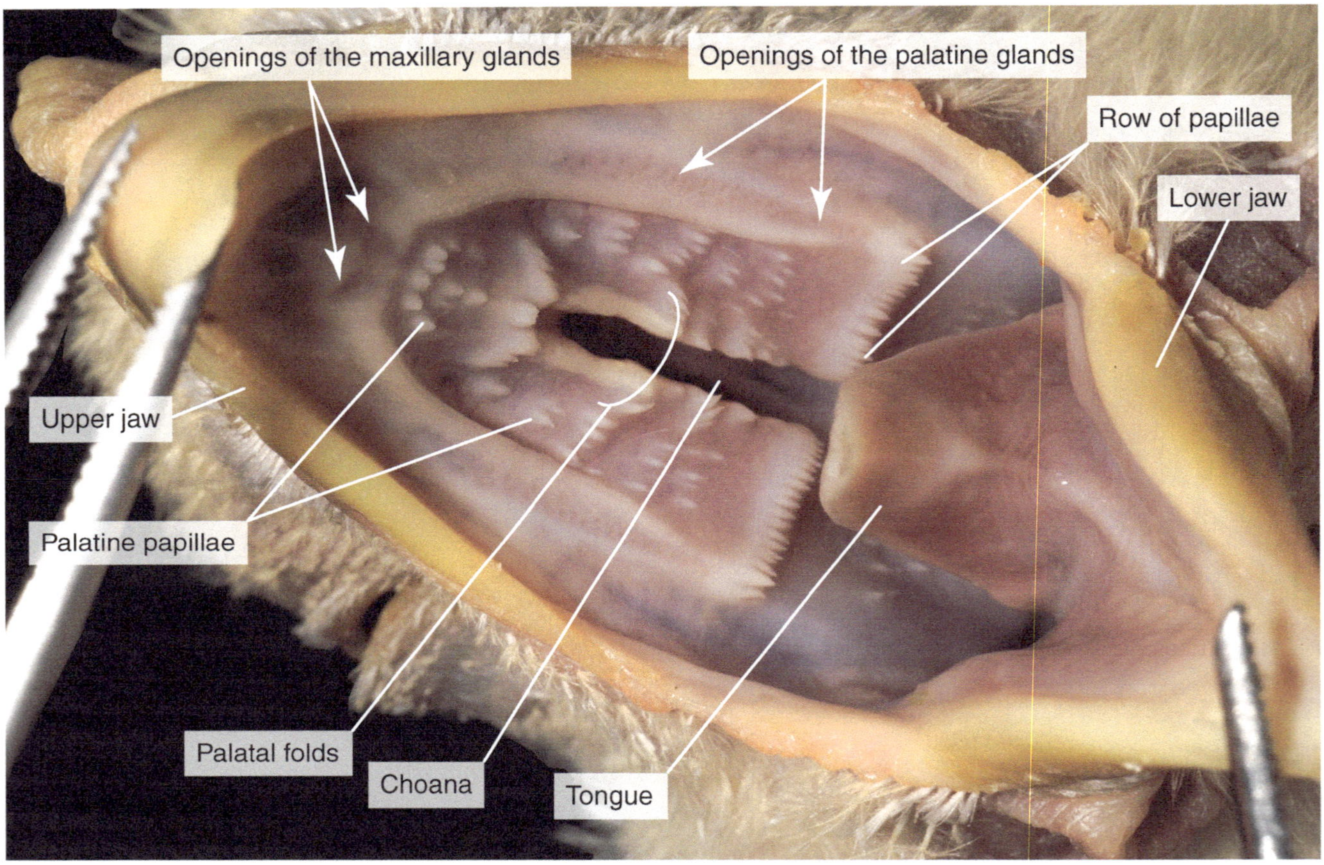

Fig. 11.13 The roof of the oral cavity

On the floor of the oral cavity, note the laryngeal prominence with the *glottis* or opening to the trachea (Fig. 11.14).

The larynx of the chicken is morphologically similar to that of mammals but no vocal cords are present. Sound is produced in the *syrinx* to be dissected later at the bifurcation of the trachea (Fig. 11.31). The floor of the oral cavity bears the pointed *tongue* which has a fimbriated posterior border (lingual papillae) (Fig. 11.14). Within the muscle of the tongue is the entoglossal process of the hyoid apparatus. You can open the mouth wider and examine the structures in one go by cutting through the angles of the jaws and down the oesophagus. In the majority of the following figures demonstrating the phases of the dissection, the *anterior* (head) region of the chicken is towards the *upper* or *right* margins. In those exceptional cases where the animal is in a different position and the anterior (head) region faces to the left, this is indicated in the legend of the figure.

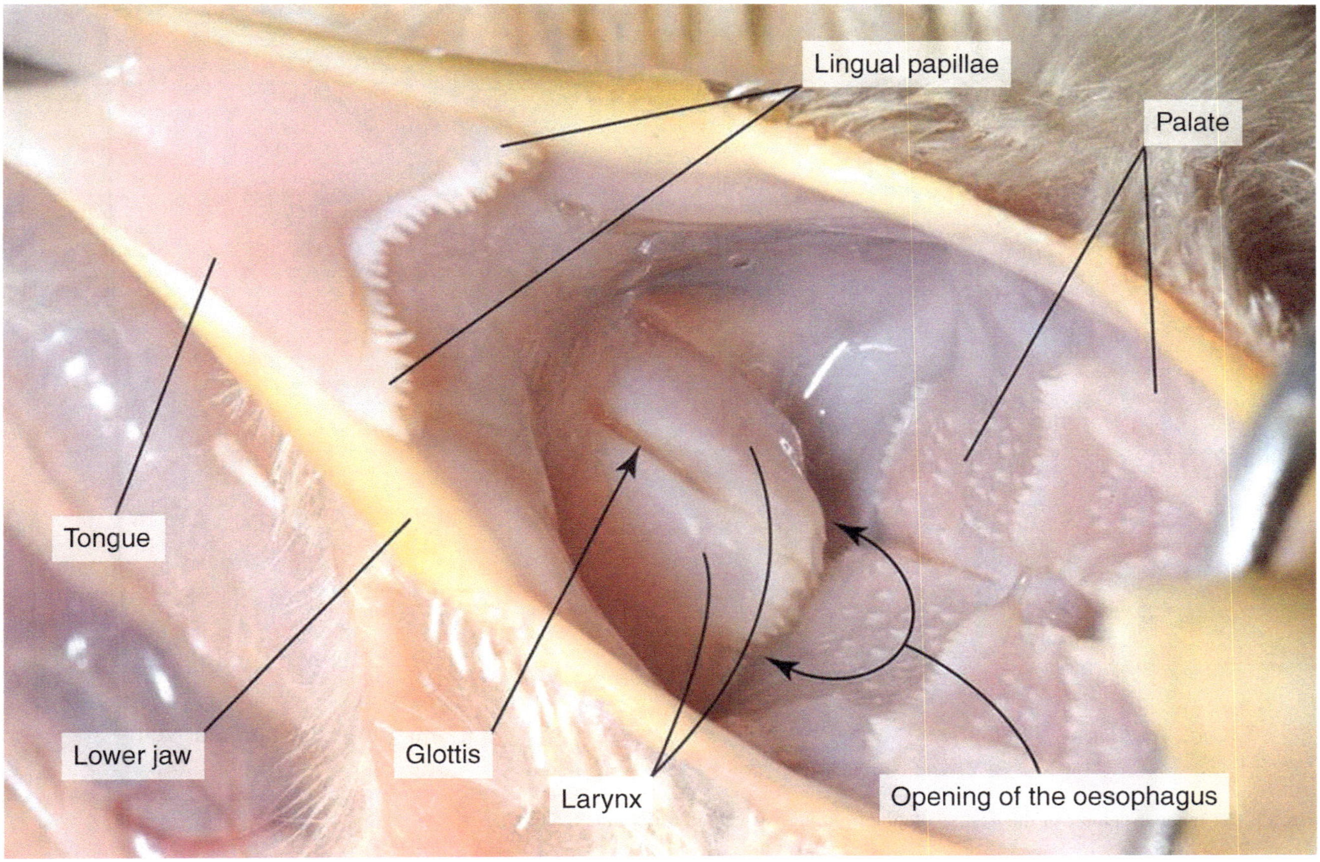

Fig. 11.14 The floor of the oral cavity

Make a median, ventral, longitudinal incision of the skin of the neck being careful not to cut the thin-walled crop and reflect the skin (Fig. 11.15).

The *trachea* is stiffened by broad cartilaginous rings (Fig. 11.15). The *oesophagus* which opens above and behind the glottis lies dorsal and to one side of the trachea. Trace the oesophagus posteriorly and note the great expansion into a crop anterior to the sternum. The *crop* is mainly a storage place and is best developed in the grain-eating birds (Figs. 11.15 and 11.18). Free the crop from the pre-sternal region. The *jugular veins* of birds show a great asymmetry: the right jugular vein is wider than the left (Fig. 11.15).

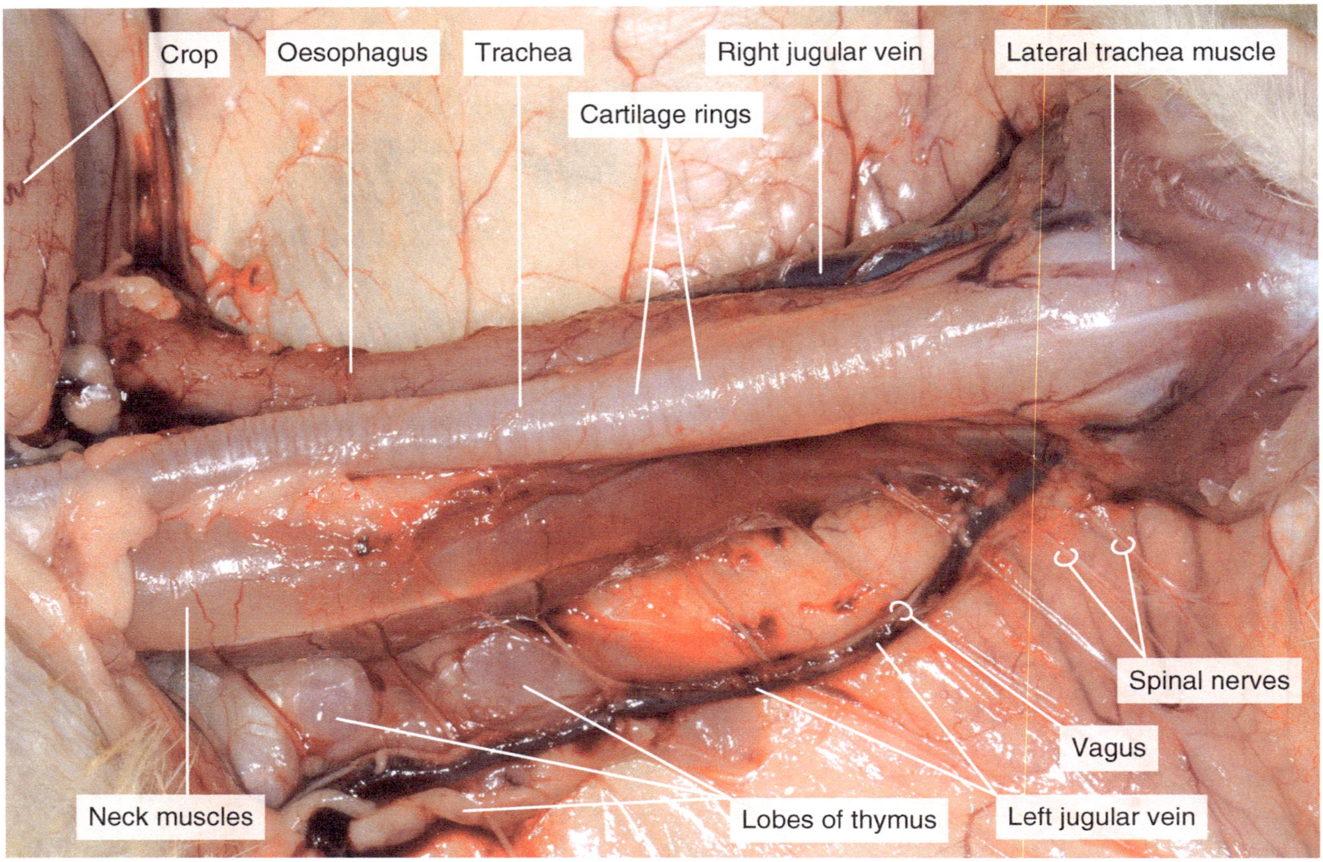

Fig. 11.15 Inner structures of the neck. Note the right jugular vein is wider than the left

Observe the segmental *spinal nerves* innervating the neck muscles, the lobes of the *thymus* and close to the thorax the *thyroid gland* (Figs. 11.15 and 11.16).

The *common carotid arteries* penetrate the ventral cervical muscles and should be traced to their emergence near the skull (Fig. 11.16). Each divides into an external carotid artery supplying oesophagus, palate and head and an internal carotid artery to the brain.

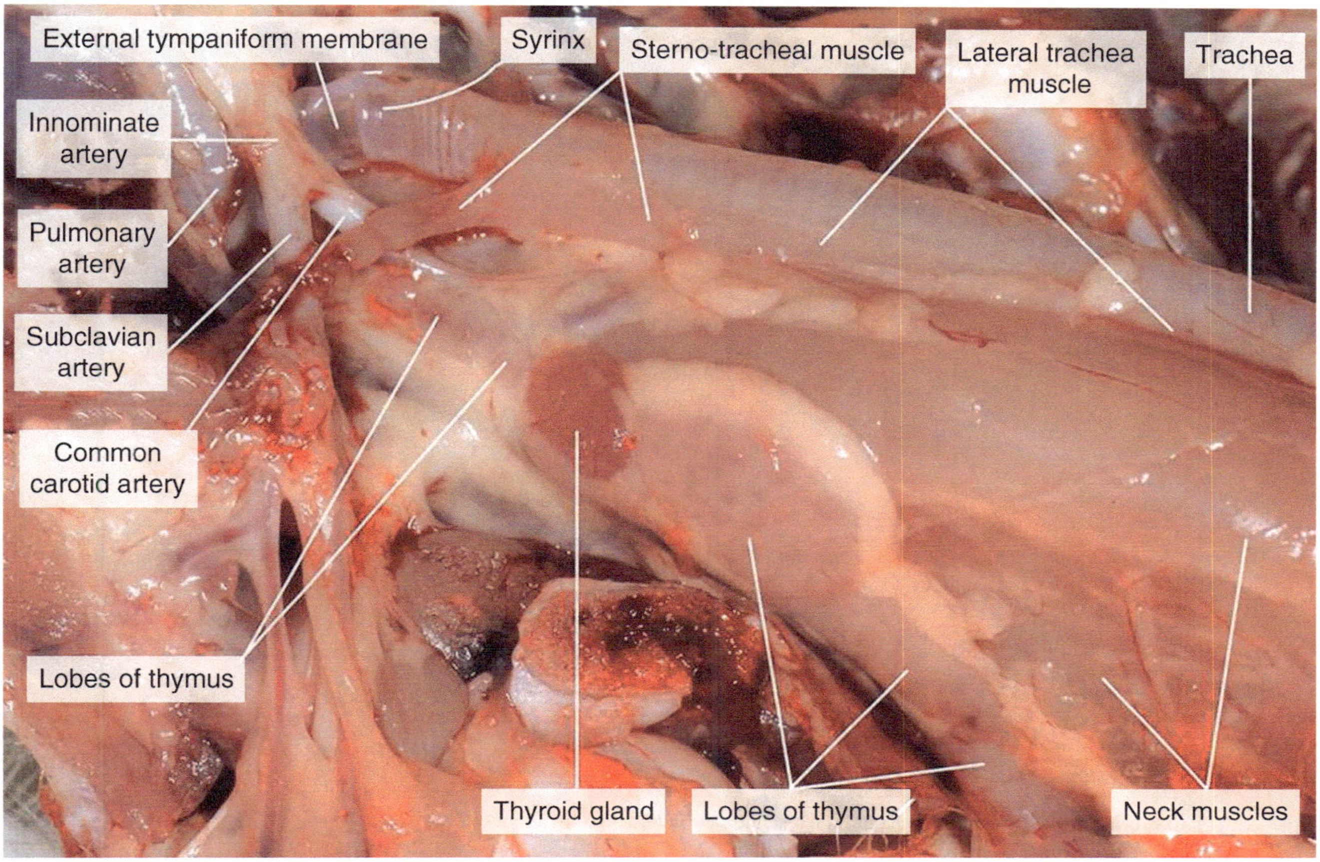

Fig. 11.16 Lobes of the thymus, the thyroid gland and tracheal muscles in the neck region of the chicken

Examine the syrinx muscles which originate on the sternum or the lateral surface of the trachea and insert on the trachea or the hyoid bone: *sterno-tracheal*, *lateral trachea* and *tracheo-hyoid muscles*. They take part in the sound production (Figs. 11.16 and 11.17).

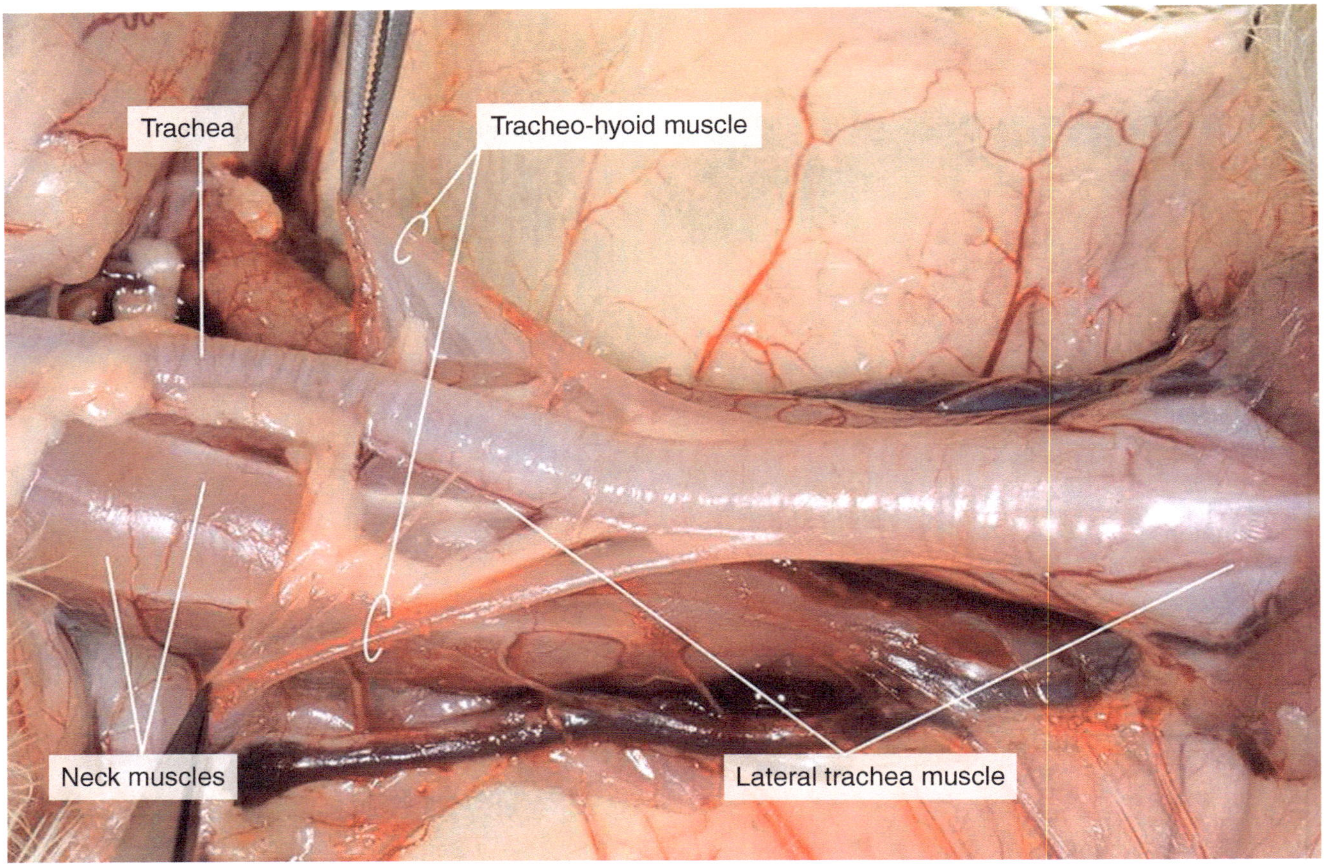

Fig. 11.17 Two overlapping syrinx muscles: lateral trachea muscles and tracheo-hyoid muscles

Remove the skin of the breast by a midventral incision from the gular region to the cloaca being careful not to incise the thin-walled crop. The large muscle of the breast which is now exposed is the *superficial pectoral muscle* (musculus pectoralis) (Fig. 11.18).

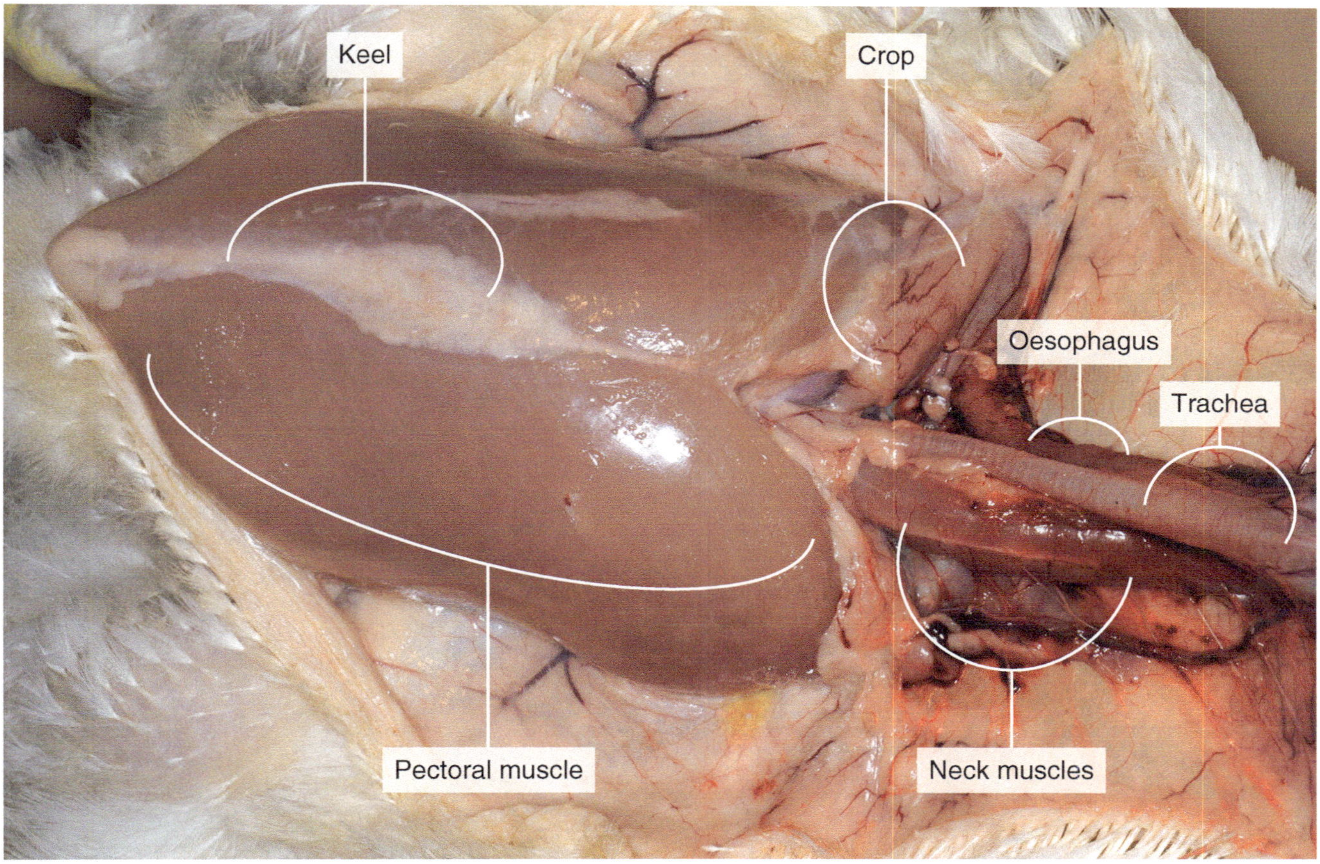

Fig. 11.18 The large superficial pectoral muscle. The edge of the keel is covered by adipose tissue

Its origin is on the sternal carina, posterior xiphisternal process, the clavicle, the sternoclavicular ligament and the sternal ribs. Its insertion which will be seen later is on the proximal portion and lateral surface of the humerus. It adducts and depresses the wing for the powerful "downbeat".

Transect the superficial pectoral along its origin on the sternal carina and reflect it laterally. This exposes the *deep pectoral muscle* (musculus supracoracoideus) which originates on the lateral surface of the sternum, the clavicle and the medial side of the coracoid (Fig. 11.19).

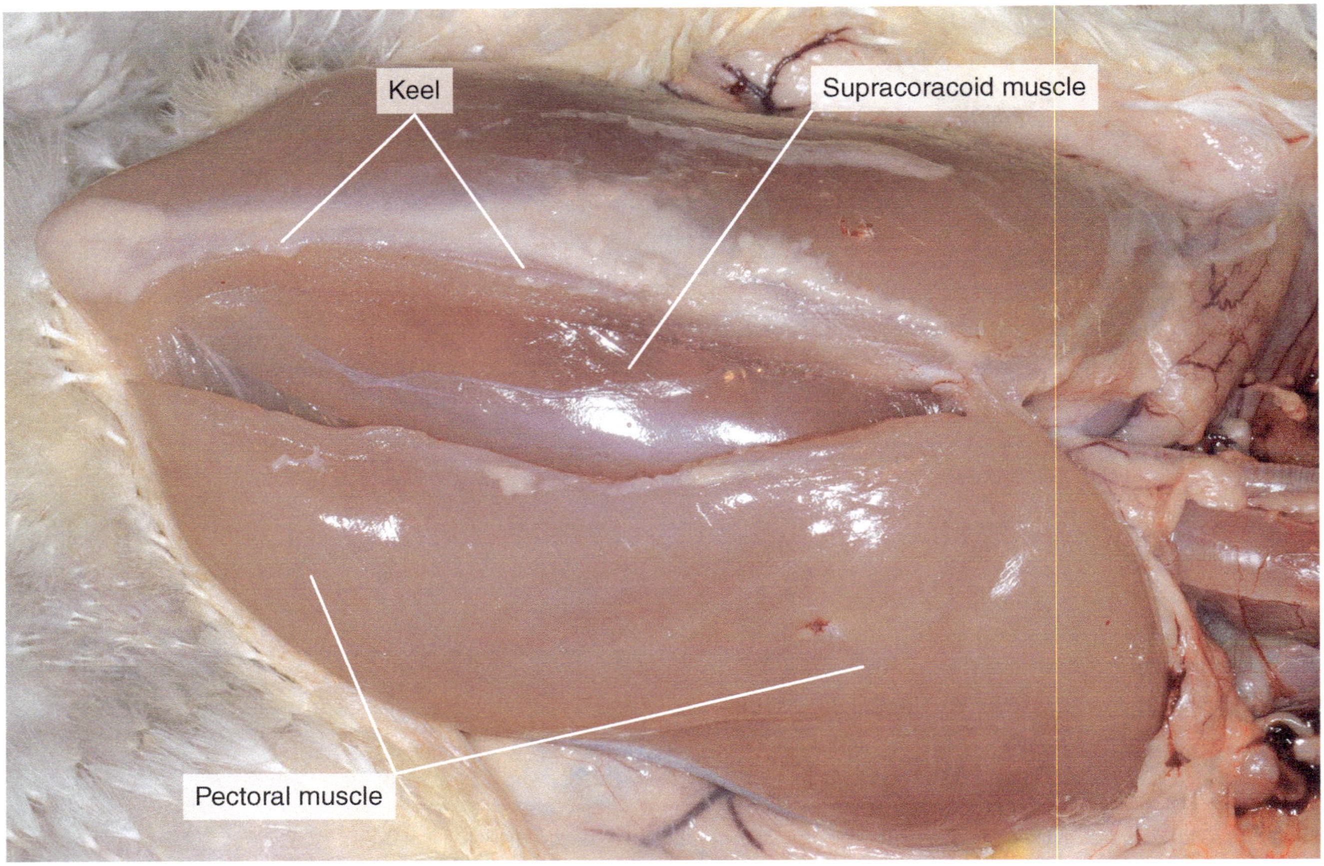

Fig. 11.19 The two major pectoral muscles responsible for wing movement: superficial pectoral muscle and deep pectoral muscle

It inserts by means of a tendon which passes through the shoulder joint to attach mediodistal to the humeral crest. The action of the deep pectoral muscle is abduction and elevation of the wing (Fig. 11.20).

Free the entire origin of the superficial pectoral muscle and transect it in the axilla. Free the deep pectoral muscle at its origin. Grab each muscle with one hand and demonstrate the function of flight muscles by pulling them alternately.

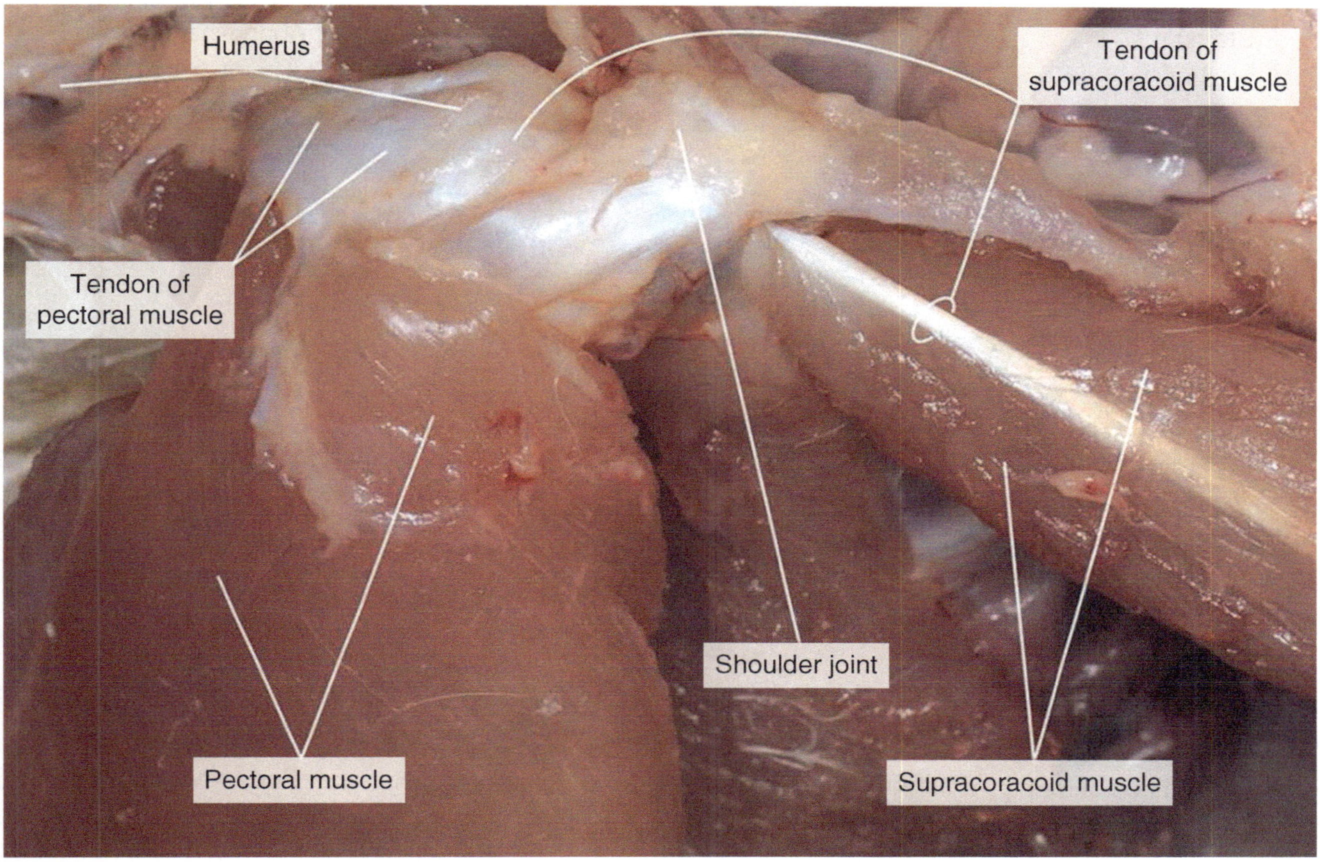

Fig. 11.20 Pectoral muscle inserts on the proximal lateral surface of the humerus, while supracoracoid muscle inserts by means of a tendon which passes through the shoulder joint to attach mediodistal to the humeral crest (In the picture, the muscles are not in their natural position. Anterior end is to the left)

Cut through the abdominal wall along the posterior margin of the sternum, insert scissors and carefully reflect the sternum by cutting through the sternal ribs. The viscera will have to be handled with care since the blood vessels are to be studied later. As you reflect the sternum, note that the ventral ligament of the stomach is continuous with the *falciform ligament* of the liver (Fig. 11.21).

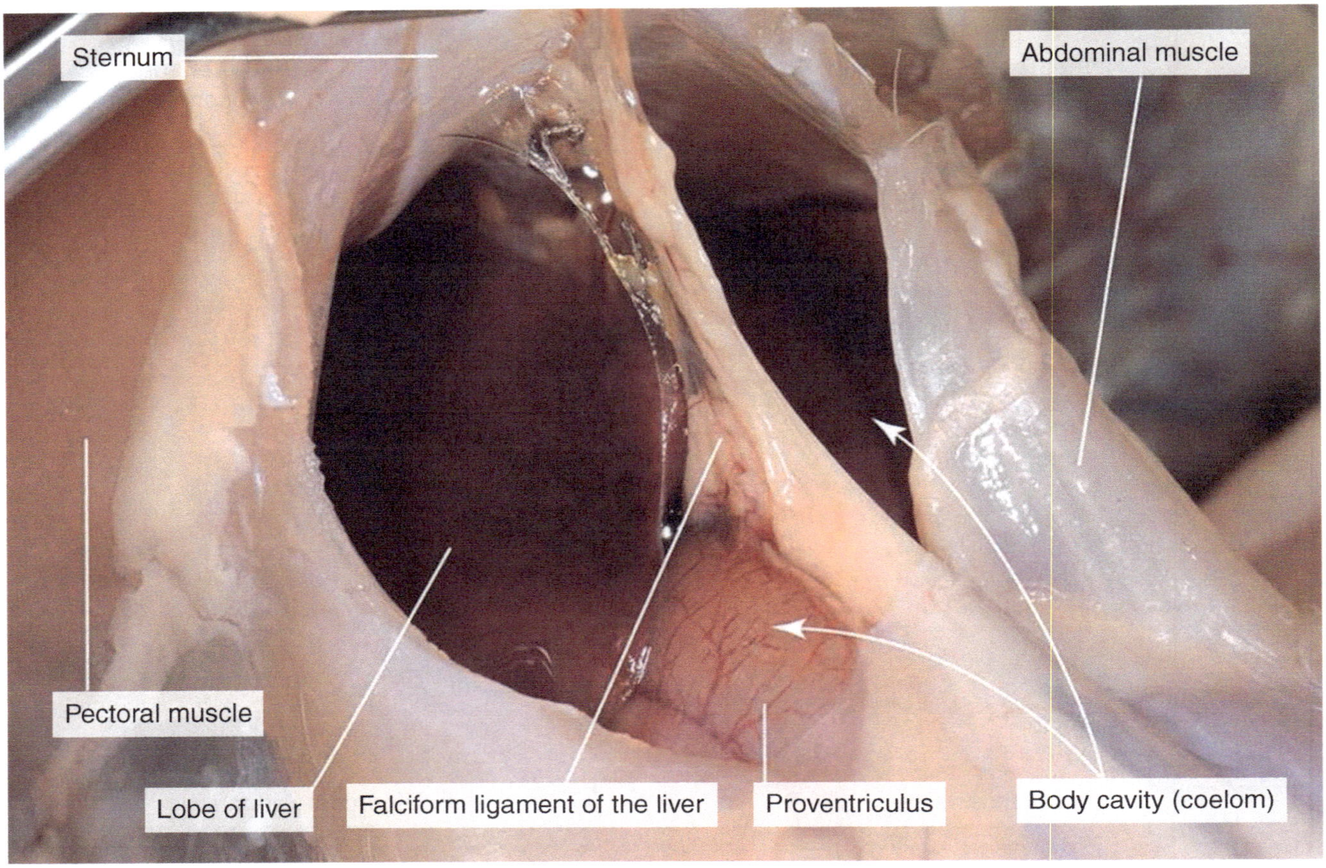

Fig. 11.21 The coelomic cavity revealed by opening the abdominal wall (anterior end is to the left)

Raise the left lobe of the liver and note the *gastro-hepatic ligament* passing between the stomach and liver. These ventral ligaments partially divide the peritoneal cavity (coelom) into a large right portion and a smaller left portion. Lift the sternum and attached parts vertically and observe the air sacs between the organs (Figs. 11.22 and 11.23).

Shift the lobes of the liver towards the middle to expose the *oblique septum* and the anterior thoracic air sac behind it (Fig. 11.23).

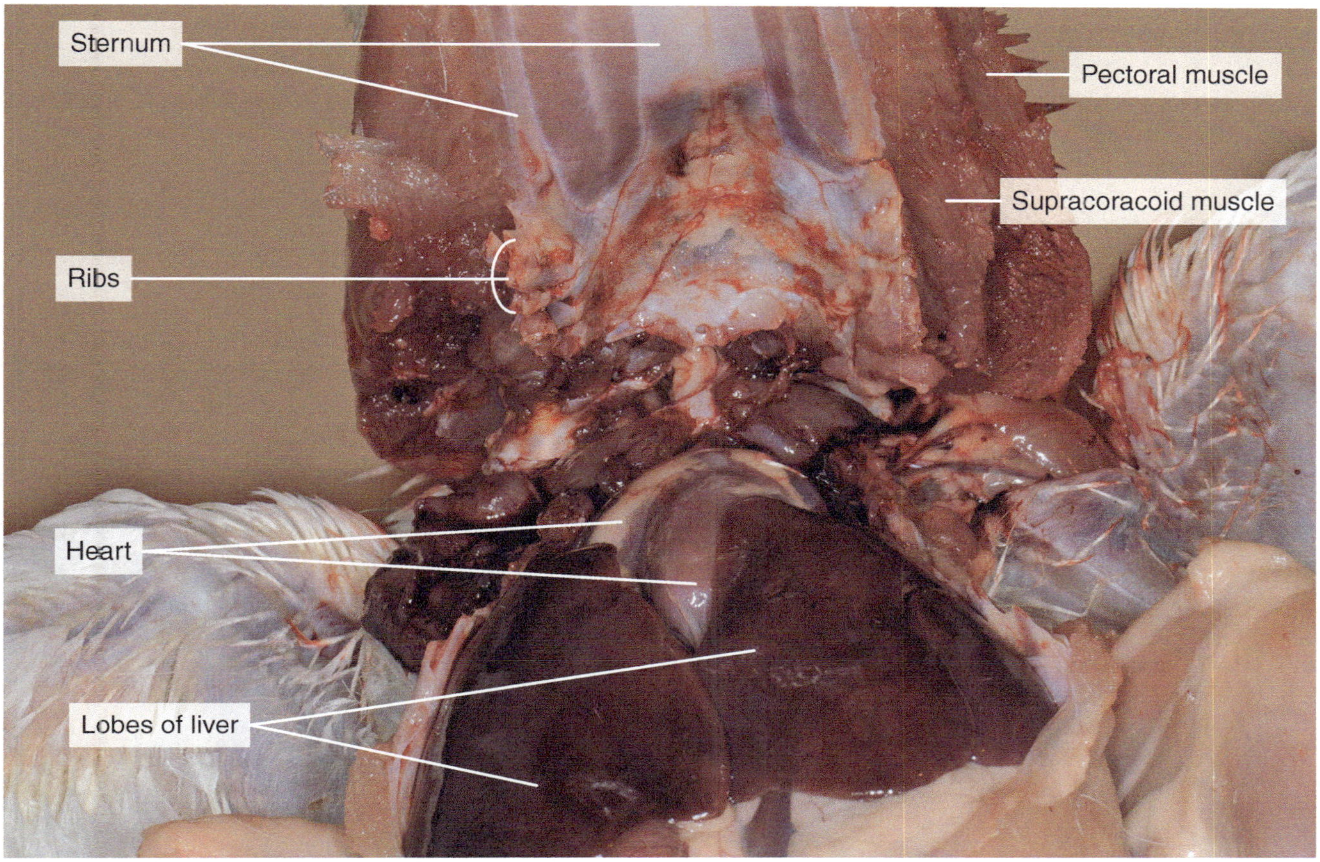

Fig. 11.22 The lifted sternum with anterior part still attached to the body

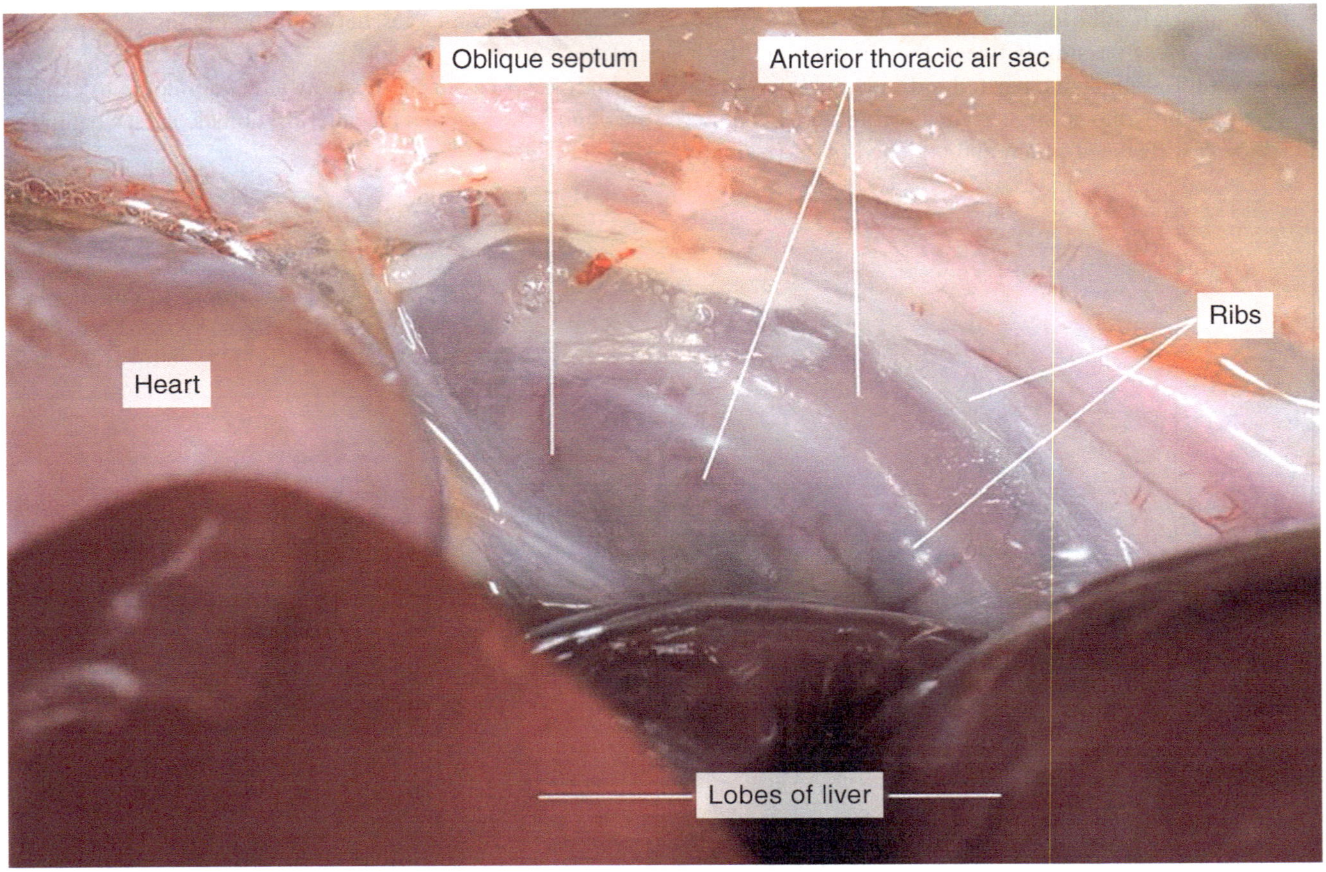

Fig. 11.23 The anterior thoracic air sac behind the oblique septum, ribs are visible through them (anterior end is to the left)

The oblique septum is a membrane formed by the attachment of the parietal peritoneum and the wall of the anterior thoracic air sac. The pericardial sac rests anterior and between the lobes of the liver and is attached by the *coronary ligament* (Fig. 11.24).

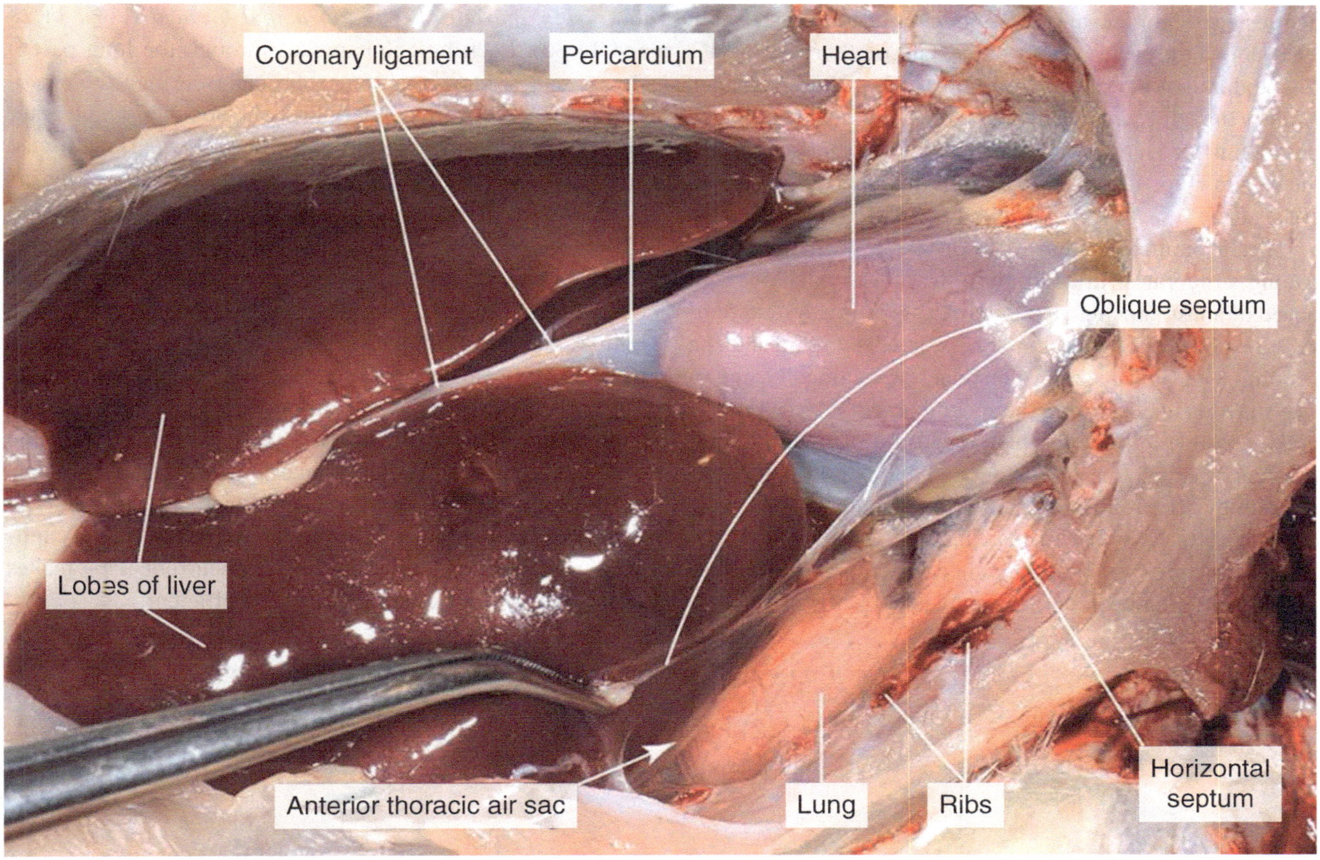

Fig. 11.24 The position of oblique and horizontal septa on the two sides of the thoracic anterior air sac

From the points where the pericardial sac meets the lateral body wall, note the *oblique septum* and the *horizontal septum*. They are homologous to the diaphragm of mammals, separating the pleural from the peritoneal cavities (Fig. 11.25).

Enclosed between the oblique and the horizontal septa are the posterior thoracic air sacs, and anterior to these, lying to each side of the pericardium are the anterior thoracic air sacs. The abdominal air sacs are on both side of the viscera and slightly dorsal. Lying against the dorsal thoracic wall are the paired lungs covered ventrally by the horizontal septum and the pleura (Fig. 11.25). Cut into the lung and note its solid consistency.

The respiratory system of birds is the most remarkable among vertebrates. In addition to the *lungs* which are composed of anastomosing air capillaries, there is a system of *air sacs* located beneath the skin or between the viscera and connecting with air spaces in the bones. Functionally, the air sacs may aid in decreasing the specific gravity of the bird and make possible a more complete exposure of lung tissue for respiratory exchange. The extent and number of air sacs varies with the different species of birds. The chicken is usually considered to have five pairs of air sacs connected with the lung. These air sacs are thin walled. The lungs and air sacs play an important role in heat regulation. Birds are covered with isolating feathers and they lack sweat glands, so they get rid of the excess heat by evaporation on the extended surfaces of lungs and air sacs.

A short description of the respiratory current is as follows: Air enters the *glottis*, passes down the *trachea*, through the *syrinx* and into the two *principal bronchi* (Figs. 11.28 and 11.30). Each bronchus passes through the lung parenchyma and is known as a mesobronchus. Along the *mesobronchus*, there are secondary bronchi, some of which enter the air sacs while others enter the *parabronchi*. The ultimate divisions of the bronchial circuits are the anastomosing *air capillaries* which exit along the parabronchi. Intermingled with the network of air capillaries in the lung parenchyma is a network of blood capillaries for the gaseous exchange. Upon expiration which is an active phase of avian respiration (unlike mammals), the air in the air sacs passes by way of recurrent bronchi into the parabronchial tubes of the lung. These *recurrent bronchi* are outgrowths of the air sacs and are present on all of the air sacs with the exception of the cervicals. Avian gas exchange takes place within air capillaries (not in alveoli, as in mammals) which are extensions of the parabronchial lumen. The air capillaries and blood capillaries are arranged so that flow is crosscurrent. This makes the avian gaseous exchange extremely efficient.

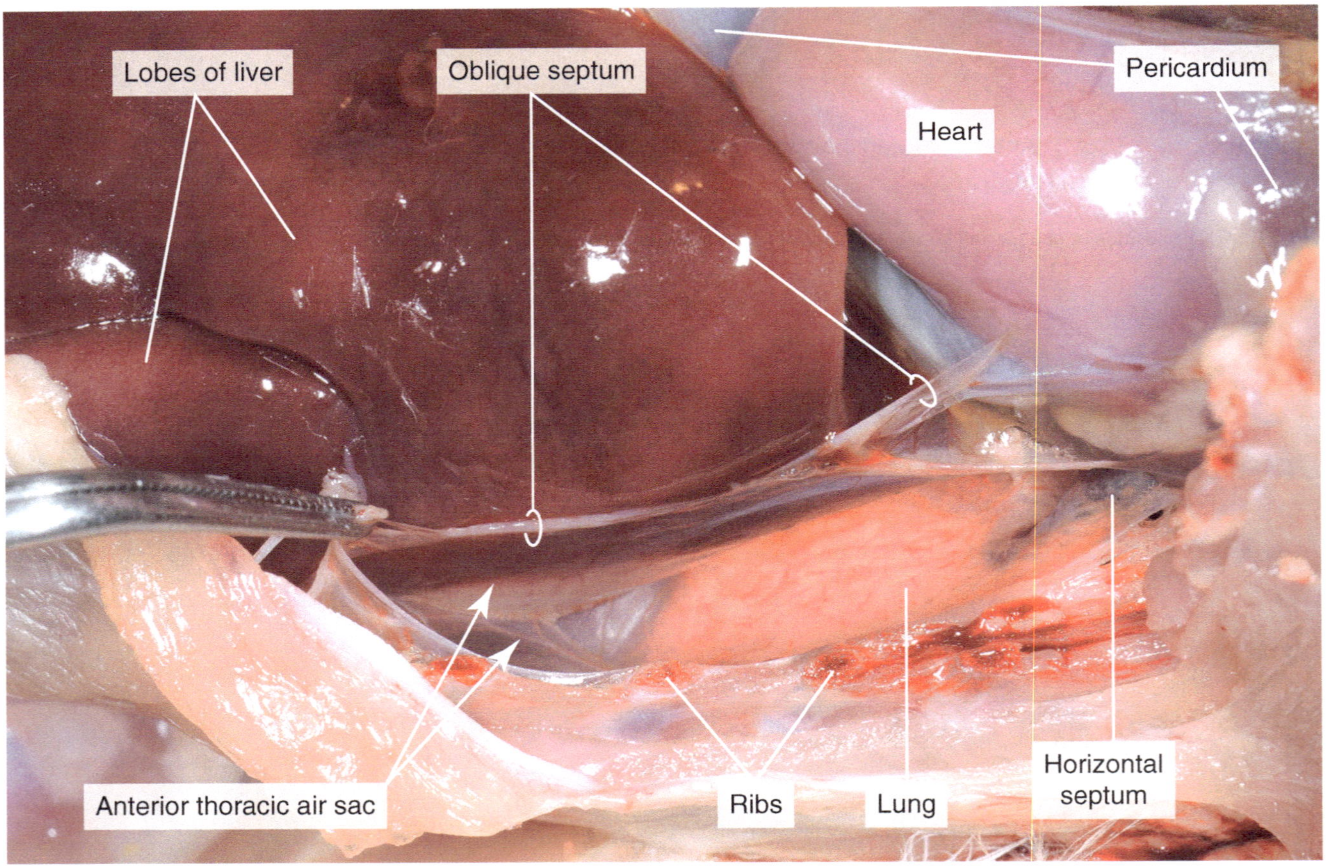

Fig. 11.25 The position of oblique and horizontal septa on the two sides of the thoracic anterior air sac

Bird *lung* has a unique structure featured by its special air-conducting system formed by parallel oriented bronchi called *parabronchi* (PB) and their radial branches named *air capillaries* (Fig. 11.26).

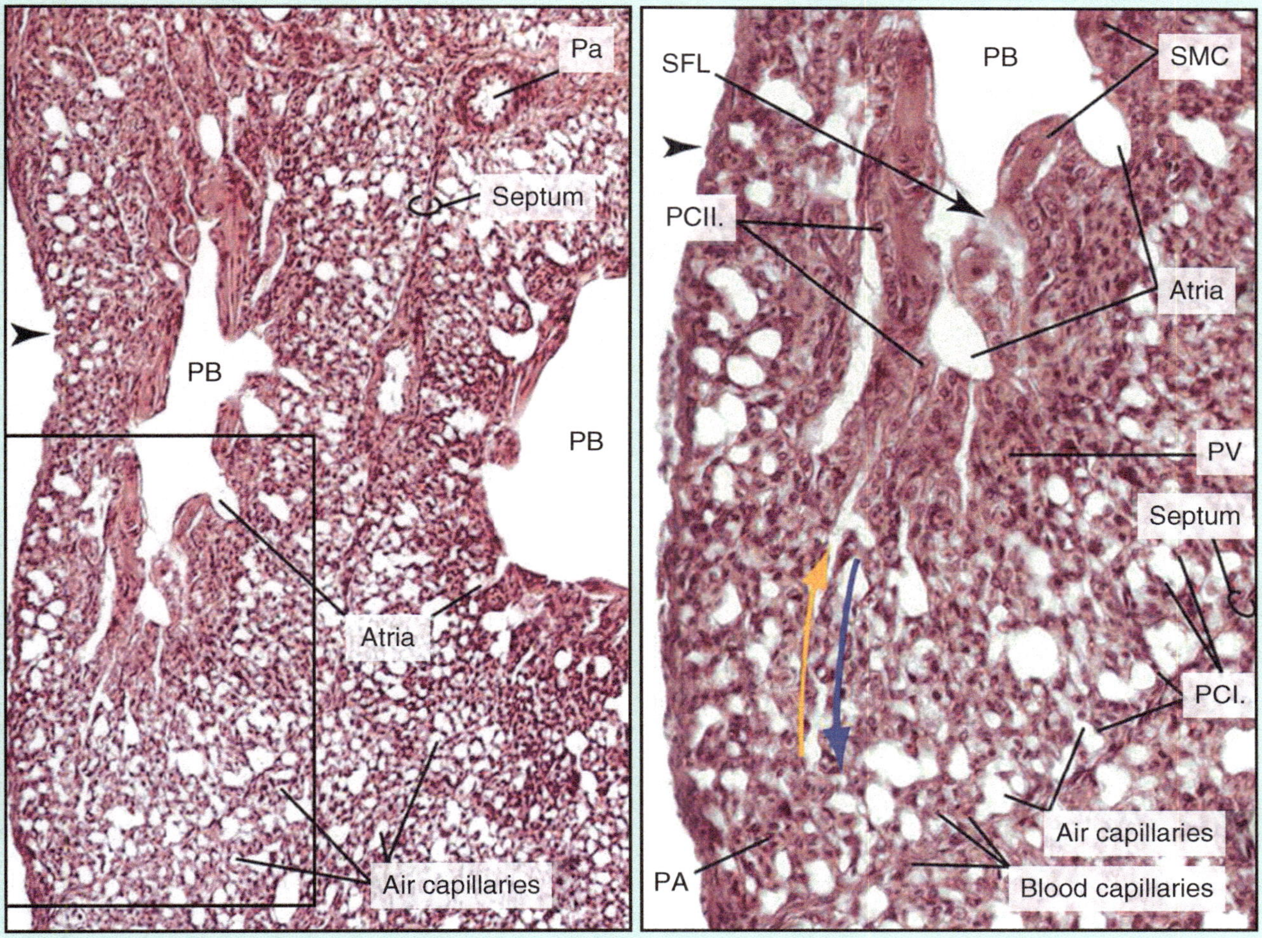

Fig. 11.26 Histological section of the lung of a bird (HE). Left side: A parabronchial unit; *arrowhead* missing visceral pleura, *Pa* pulmonary artery, *PB* parabronchus. Right side: Higher magnification of the framed area from the left micrograph. *Arrowhead* missing visceral pleura, *blue arrow* air flow, *orange arrow* blood flow, *PA* pulmonary artery, *PCI* pneumocytes type I, *PCII* pneumocytes type II, *PV* pulmonary vein, *SFL* surfactant layer, *SMC* smooth muscle cells

Parabronchi are surrounded by a layer of *smooth muscle cells* (SMC) for controlling airflow through the system (Fig. 11.26, *right*). Air capillaries open from *atria*, which are sac-like protrusions of parabronchi (Fig. 11.26). Epithelial lining of the proximal (beginning) segment of air capillaries is formed by *type II pneumocytes* (PCII.). These cells produce and secrete the surfactant layer (SFL) which reduces the surface tension and keep air capillaries open against capillary action. The respiratory surface is composed of flattened squamous epithelial cells called *type I pneumocytes* (PCI.). These cells form the lining of the air capillaries after the beginning segment. Air capillaries orient radial around their parental parabronchus, and they end at the connective tissue septum which surrounds the parabronchial unit. This architecture is shown by the lung of birds flying occasionally. (Air capillaries of adjacent parabronchial unit are connected in birds flying regularly and taking long distances.) Branches of the pulmonary arteries (Pa) are found in the septa between parabronchial units and give blood capillaries joining to the air capillaries. The blood flows in blood capillaries (Fig. 11.26, *right*, orange arrow) and the airflow (Fig. 11.26, *right*, blue arrow) in the air capillaries are opposite. This phenomenon ensures the sufficient effectivity of the gaseous exchange for flying. Smaller branches of the pulmonary veins (PV) are formed by collecting blood capillaries in the vicinity of the parabronchi.

Remove the clavicles and coracoids to expose the base of the heart (Fig. 11.27).

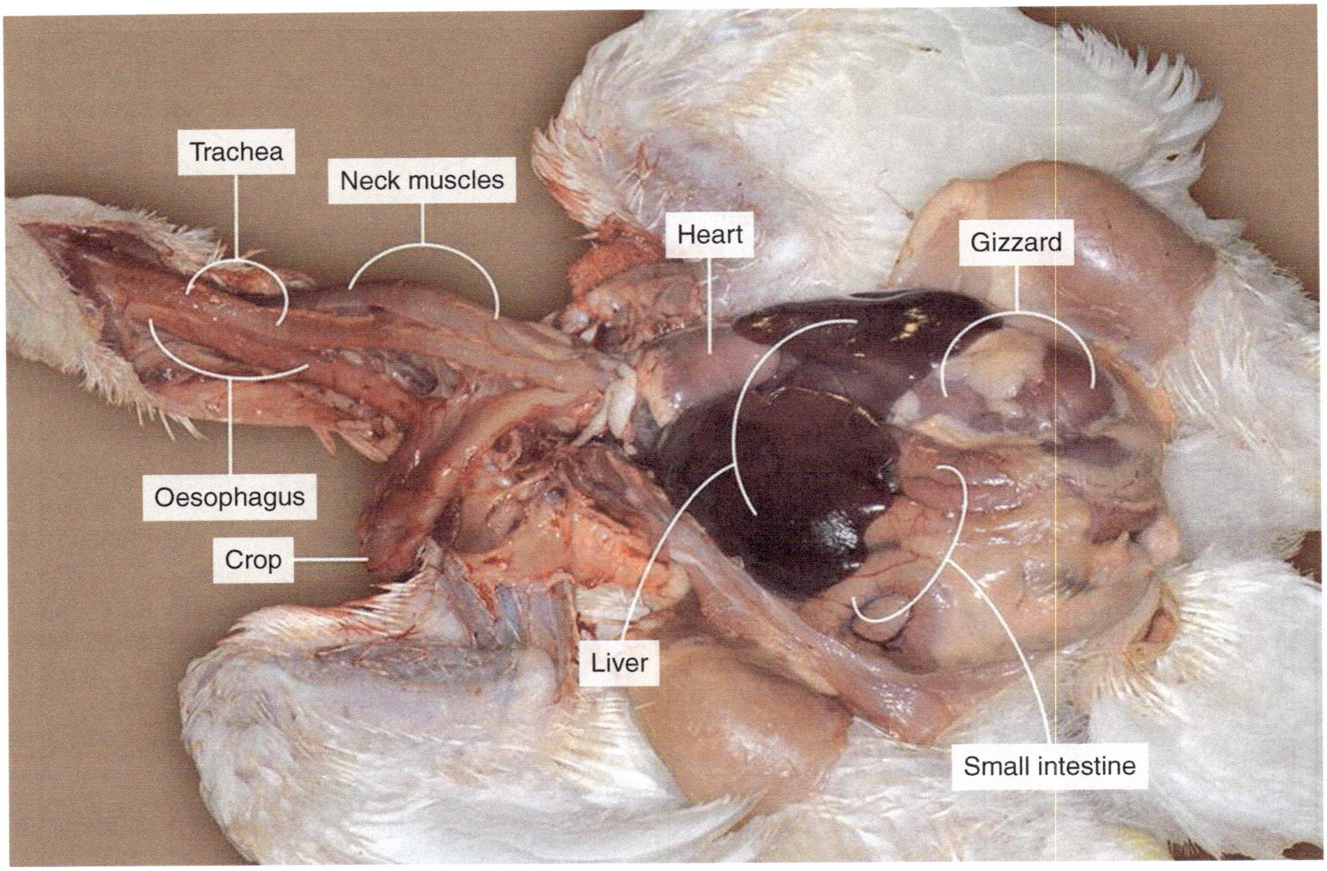

Fig. 11.27 Internal organs of the chicken after removal of the sternum, ribs, pectoral muscles and the abdominal wall (anterior end is to the left)

The large vessels seen leaving the heart on the cranioventral surface are from left to right: *pulmonary artery*, which soon forks into a large left and a smaller right branch; left and right *innominate arteries*; and *aorta*. Trace one of the innominate (brachiocephalic) arteries; it gives rise to an anterior *common carotid artery* and a lateral *subclavian artery* (Fig. 11.28).

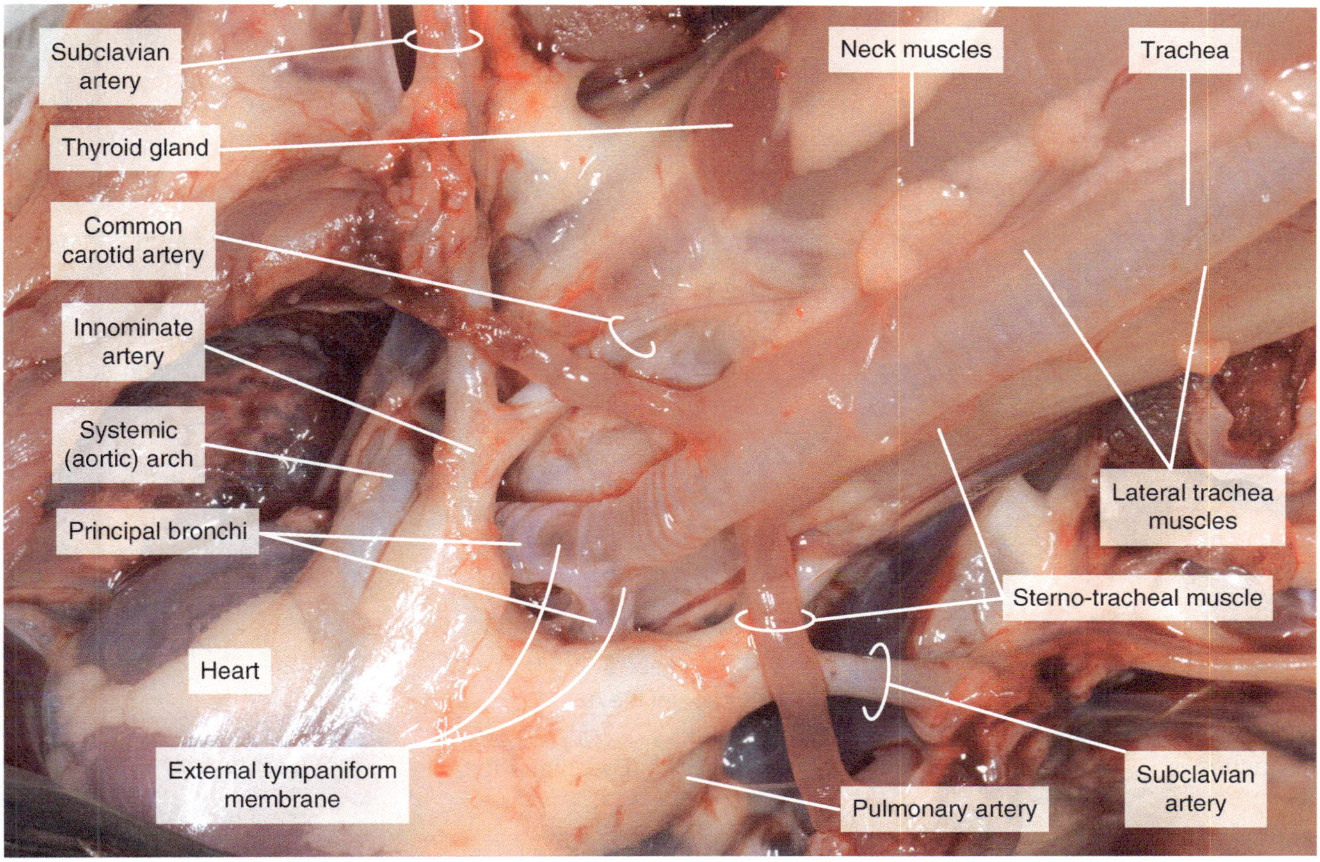

Fig. 11.28 The large vessels leaving the heart from left to right: pulmonary artery, left and right innominate arteries and aorta

The *subclavian artery* gives rise to the internal thoracic artery which supplies the inner surface of the rib cage, the *pectoral arteries* to the superficial and deep pectoral muscles and the *axillary artery* which continues into the wing as the brachial artery (Figs. 11.29, 11.30 and 11.32).

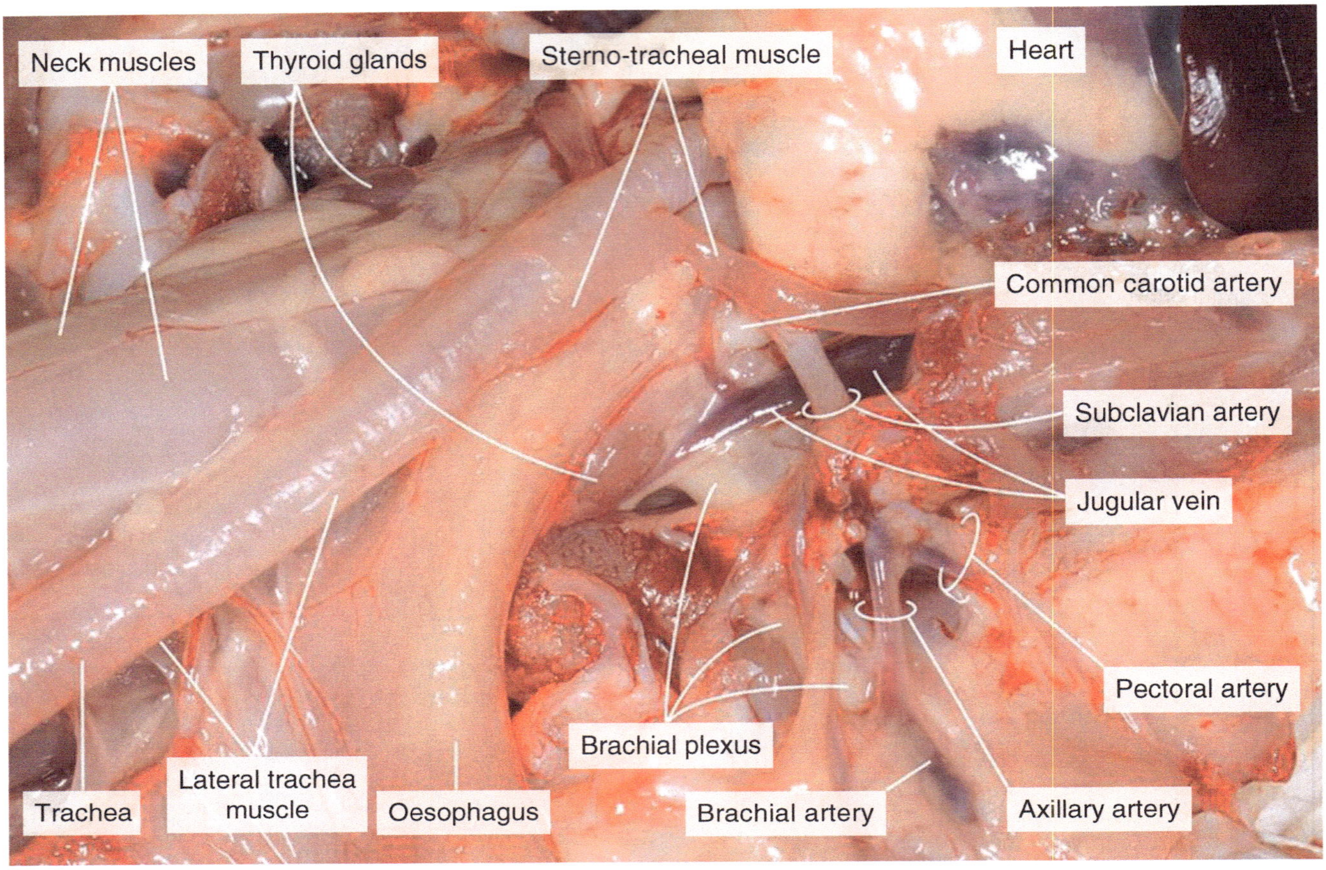

Fig. 11.29 The branches of the subclavian artery and the brachial plexus (anterior end is to the left)

The axillary artery supplies a small branch to the shoulder muscles and a twig to the trachea. The *brachial plexus* is formed by the ventral branches of the last three cervical nerves and the first thoracic. On the ventral surface of the wing, the ulnar and median nerves can be found. The *syrinx* is the sound-producing organ of birds sometimes referred to as the posterior larynx (Fig. 11.30).

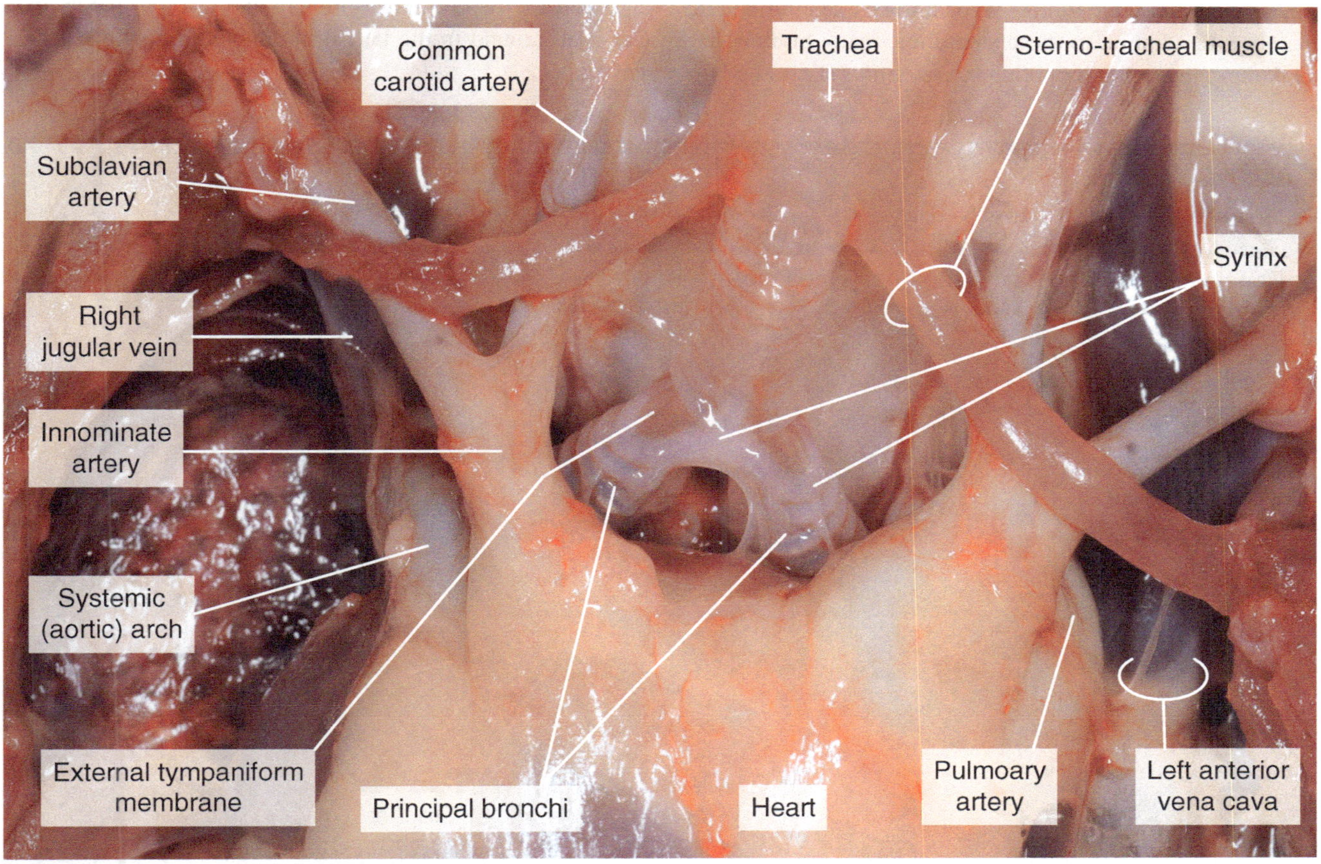

Fig. 11.30 Original position of the syrinx behind the large vessels leaving the heart

It is a modification of the trachea at its bifurcation to form bronchi, dorsal to the base of the heart. In ducks and some other birds, the syrinx is greatly enlarged but in the chicken it is relatively simple. The characteristic structures are two *external* and two *internal tympanic membranes*, so placed that they are caused to vibrate when air is passed through the chamber. In addition to these structures, many birds have a *semilunar membrane* supported by a *pessulus* (Fig. 11.31).

With the aid of the figure, try to imagine some of these structures without removing the syringe and cutting any vessels. The external tympanic membranes and sterno-tracheal muscles are visible without further dissection (Figs. 11.28 and 11.30). In the end of the dissection, you can cut the trachea and the two principal bronchi and take out the syrinx. Cut it all around in the lateral midline with a pair of small scissors.

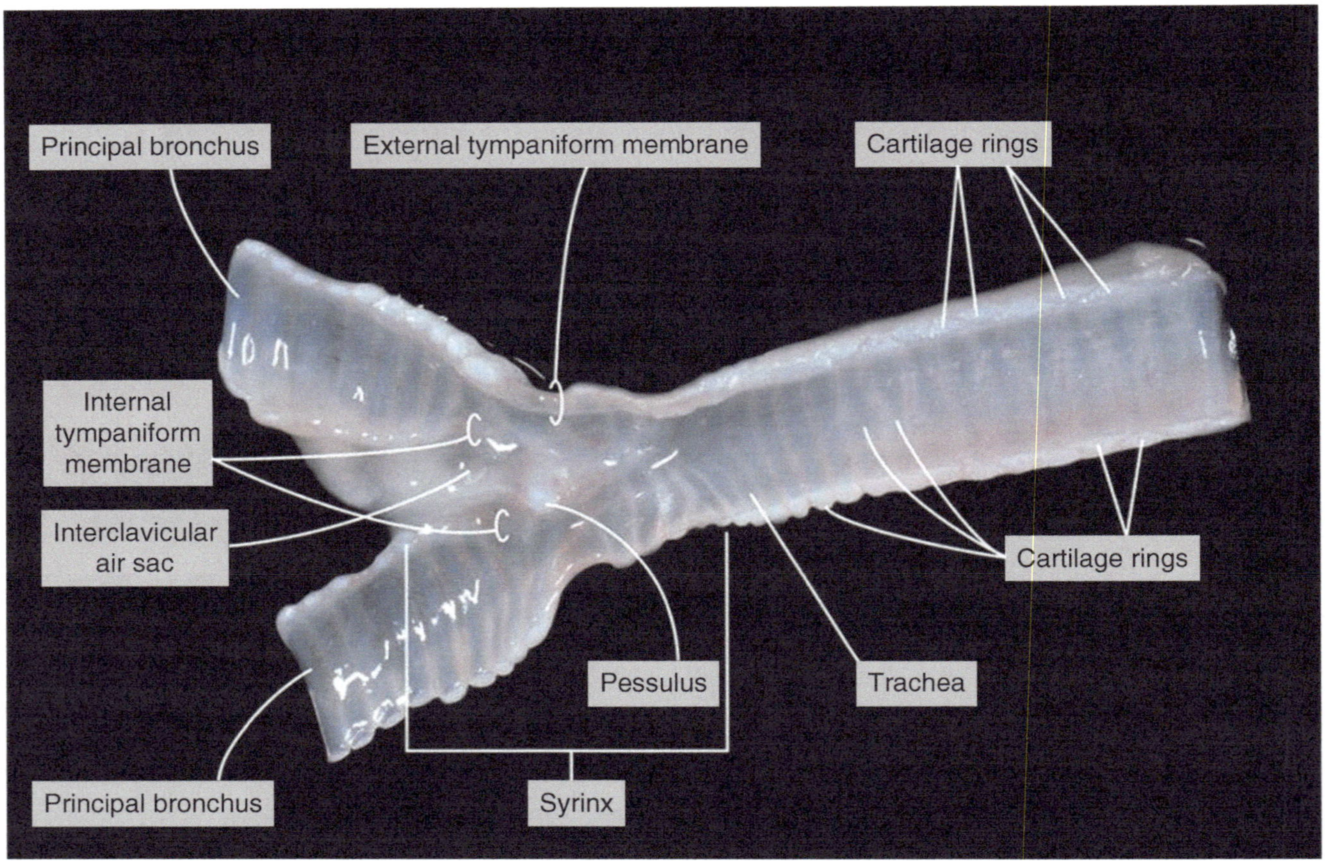

Fig. 11.31 Horizontal, longitudinal section of the syrinx

Remove the pericardial sac and carefully free the aorta and innominate arteries from the adhering adipose tissue (Fig. 11.32).

The paired *pulmonary veins* should be seen entering the left atrium anterior to the entrance of the *posterior vena cava*. To see these, reflect the apex of the heart anteriorly.

Now cut the walls of the air sacs and in the abdominal cavity remove the remnants of the abdominal wall and expose the internal organs (Fig. 11.33).

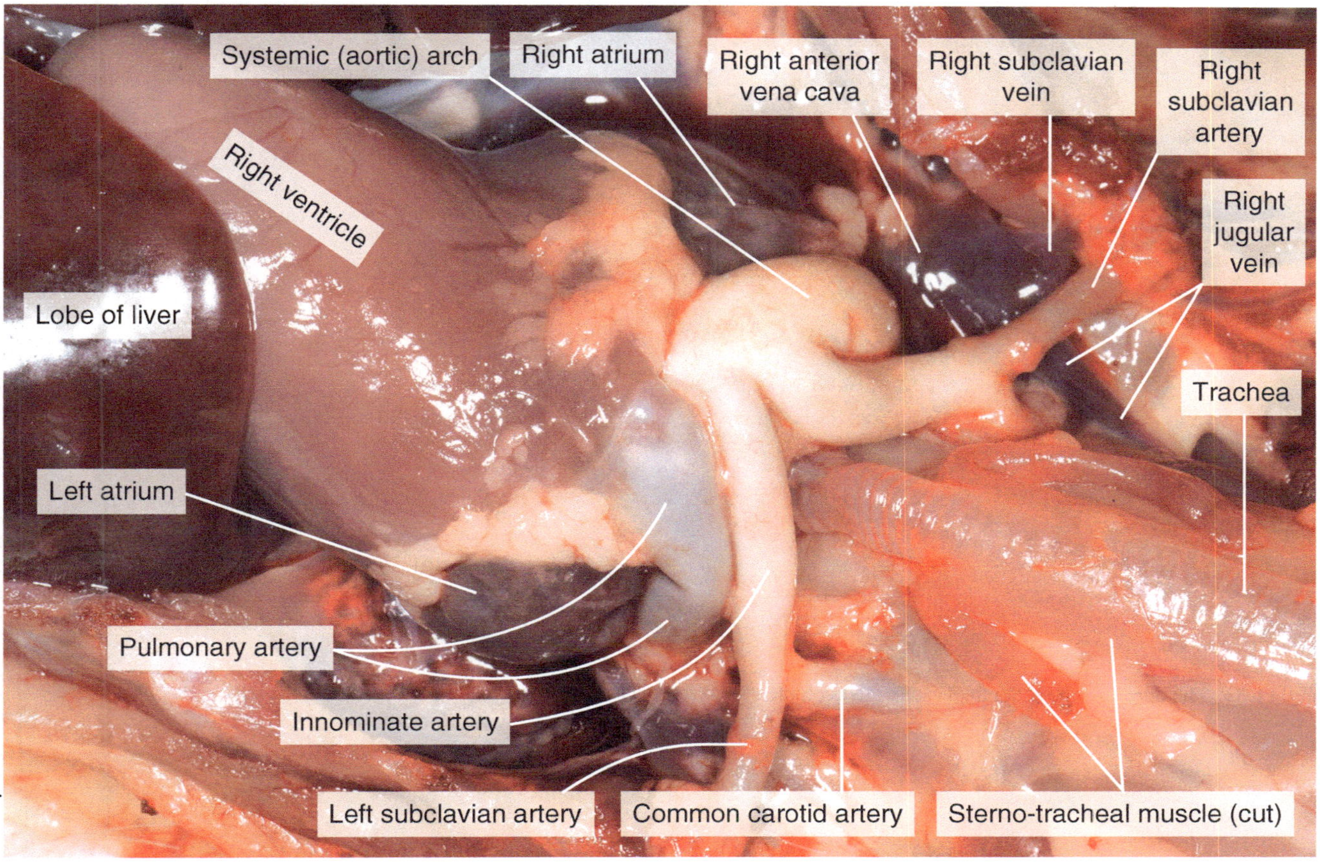

Fig. 11.32 The large vessels leaving the heart from left to right: pulmonary artery, left and right innominate arteries, and aorta after the removal of adhering adipose tissue

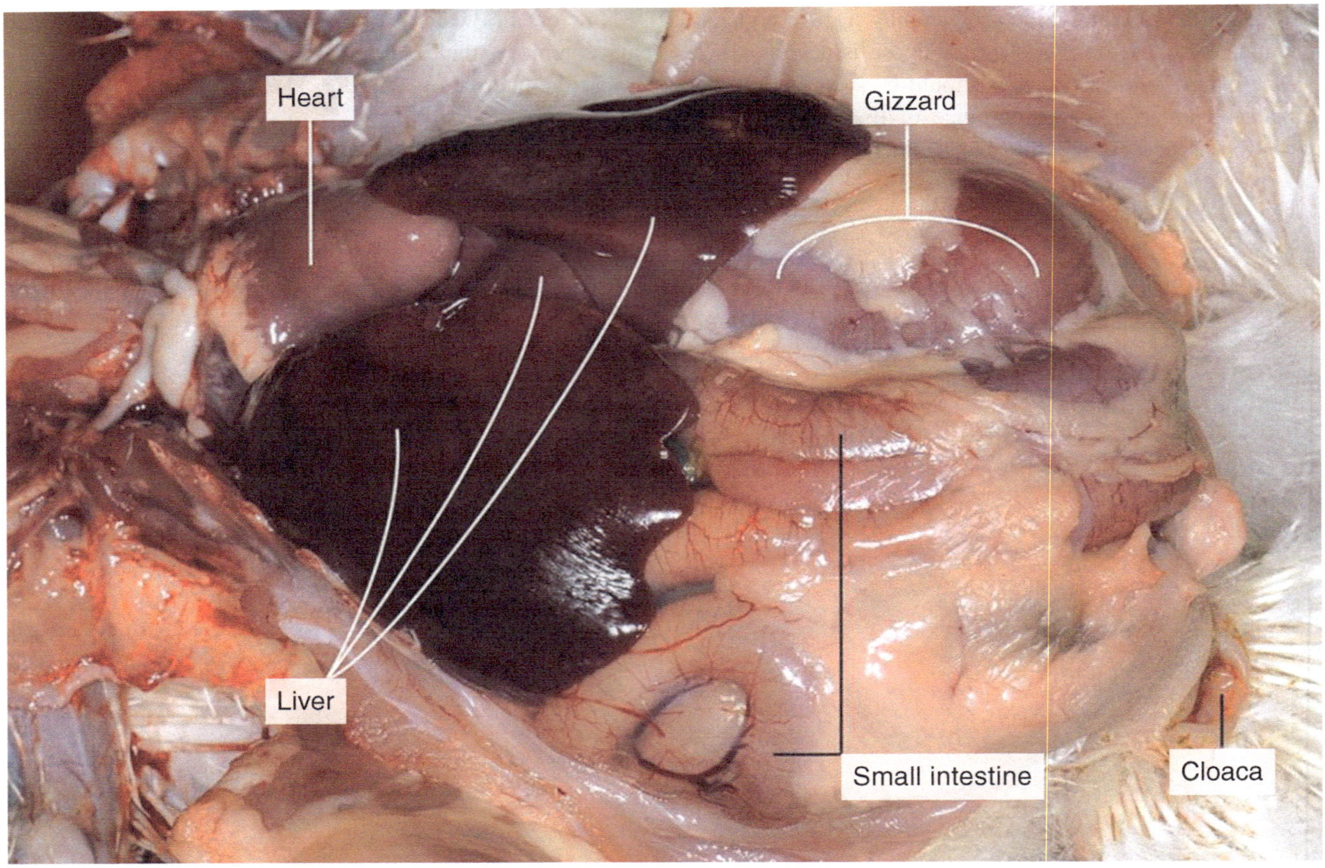

Fig. 11.33 The organs of the abdominal cavity after complete removal of the abdominal wall (anterior end is to the left)

Note the *gizzard* (ventriculus) which functions to grind the food as would teeth. Anterior to the gizzard and dorsal to the left lobe of the liver is the *proventriculus* which is the glandular portion of the stomach (Fig. 11.34).

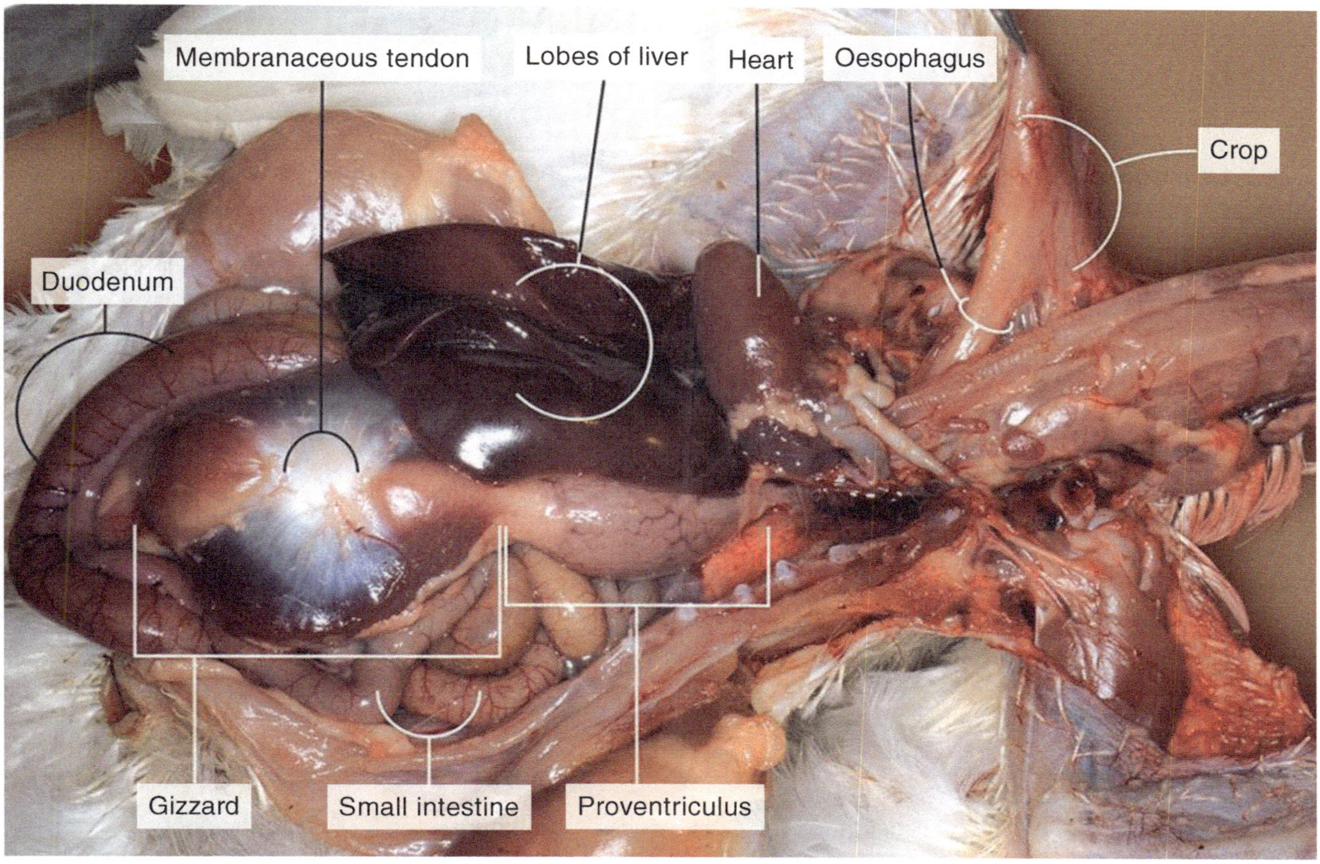

Fig. 11.34 Organs of the anterior part of the alimentary canal

The *liver* consists of right and left lobes of nearly equal size. On the right side dorsal to the liver, note the large *gall bladder* which is lacking in many other species of birds (Fig. 11.35).

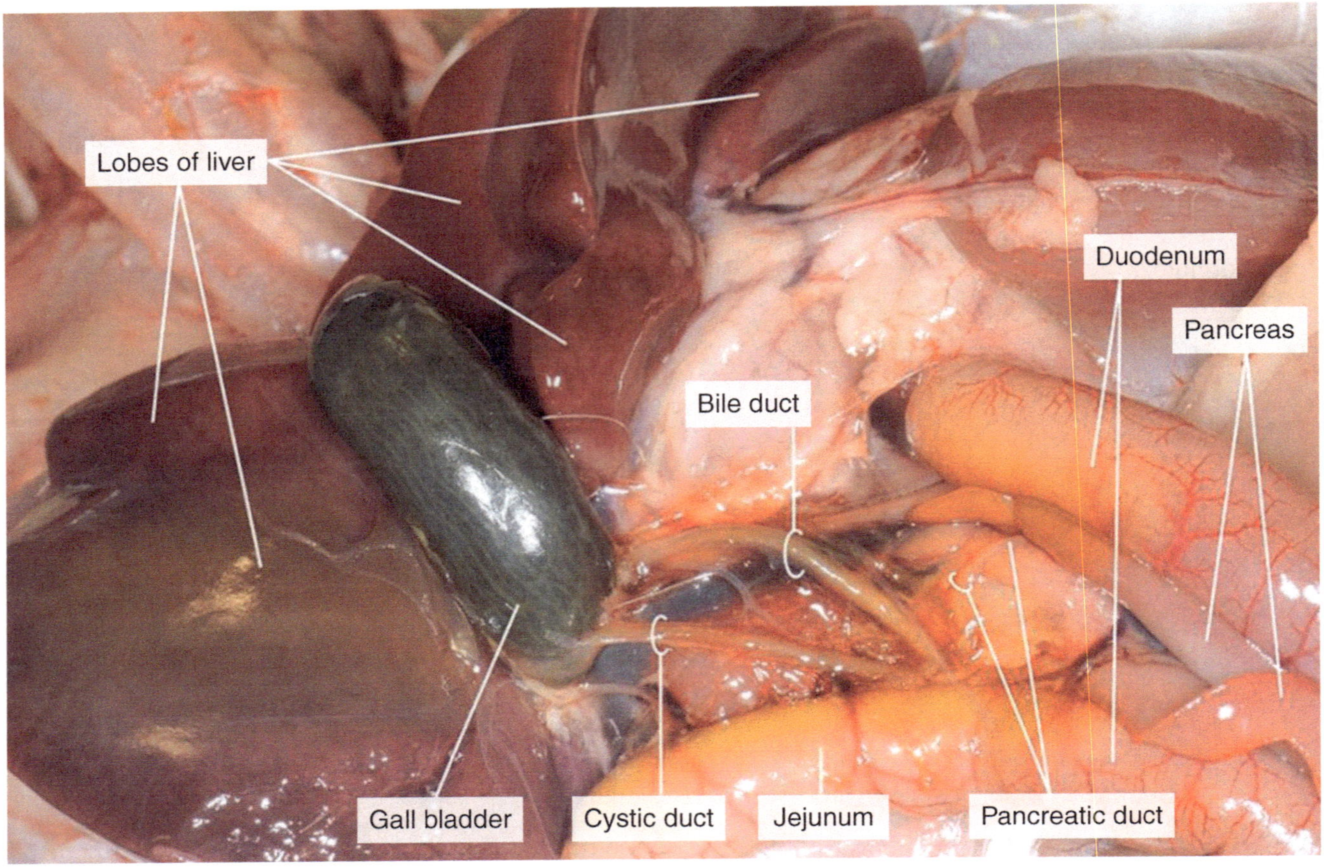

Fig. 11.35 Lobes of the liver and the large gall bladder (anterior end is to the left)

Near the junction of the proventriculus and gizzard, the small intestine arises. The *duodenum* makes a long "U"-shaped loop posteriorly and its beginning is attached to the liver by the hepatoduodenal ligament (Fig. 11.36).

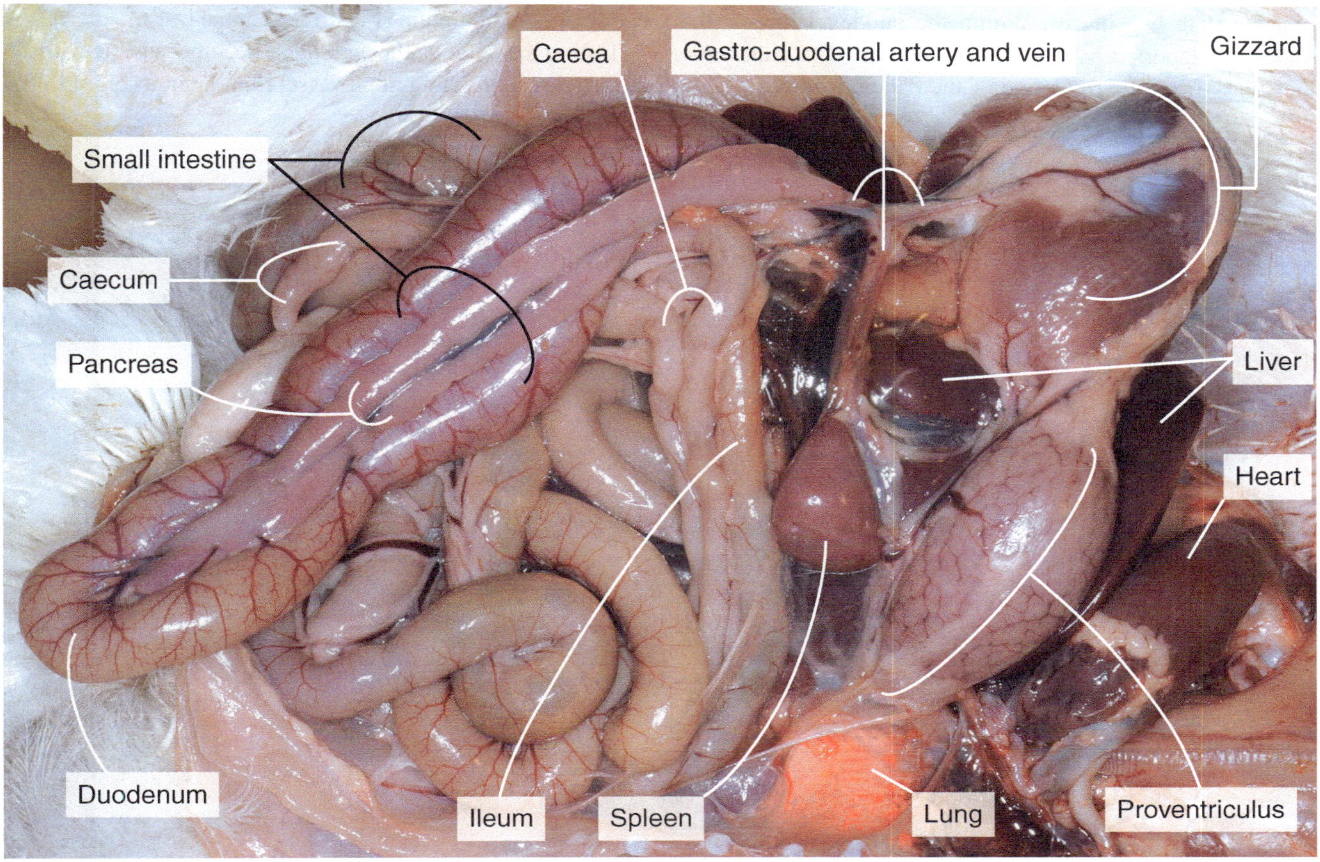

Fig. 11.36 The loop of the duodenum with the pancreas

The small intestine consists of three sections: the *duodenum*, the *ileum* and the *jejunum*. Find the large chestnut-like *spleen* anterior to the gizzard and to the right of the proventriculus (Fig. 11.36). Note how the duodenal loop is bent upon itself in two places. Within the duodenal loop is the *pancreas* supported by the mesoduodenum. The pancreatic ducts pass from the right side of the pancreas into the right limb of the loop. At the base of the right limb, find the large *cystic duct* from the gall bladder, a large *bile duct* from the liver and three good-sized *pancreatic ducts*. All of these ducts enter the duodenum at the same point at the border of duodenum and jejunum (Figs. 11.35 and 11.37).

Lifting the *jejunum* the *mesentery* becomes visible with all the arteries, veins and lymph vessel running in it (Fig. 11.38).

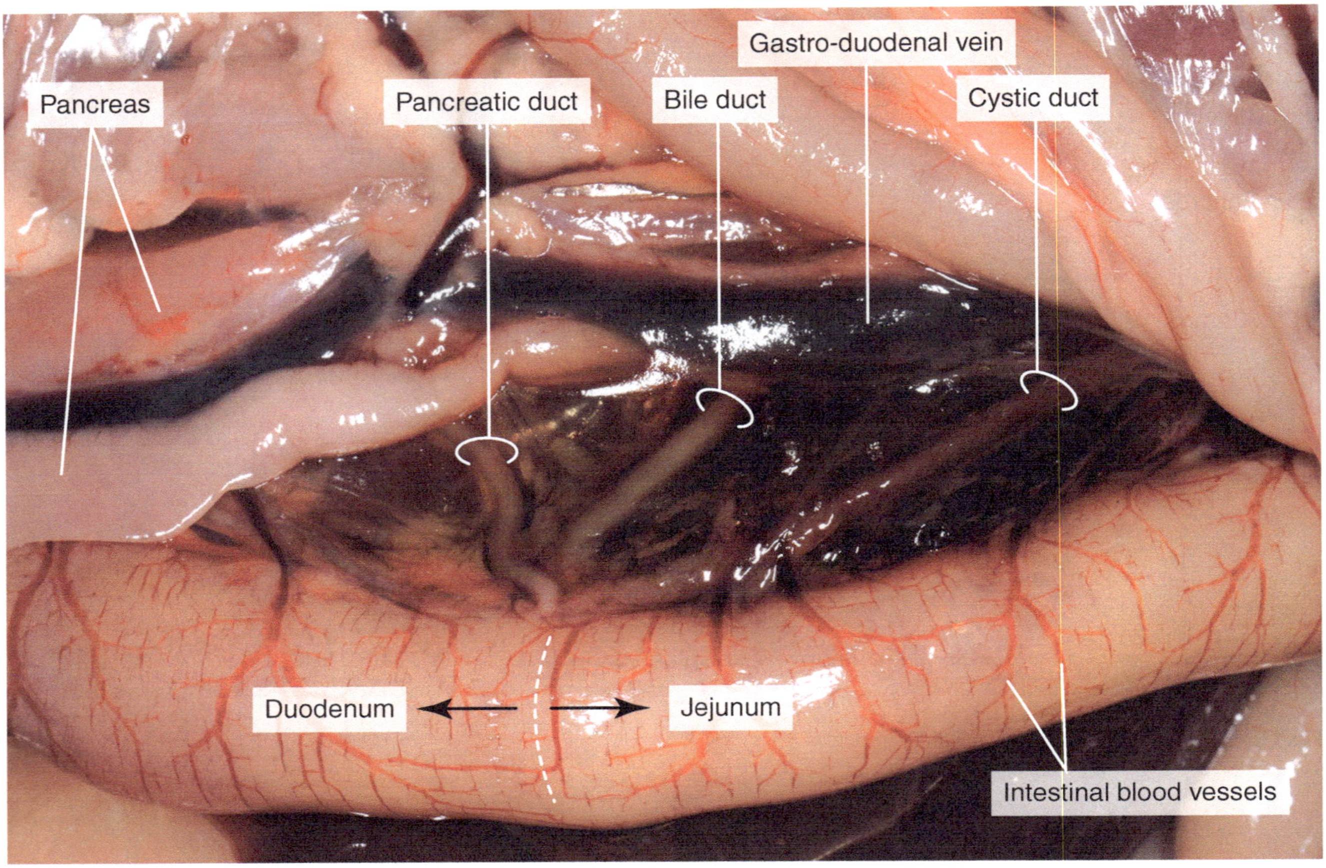

Fig. 11.37 Bile and pancreatic juice enter the gut through several ducts at the border of duodenum and jejunum. Note the blood vessels supplying the gut

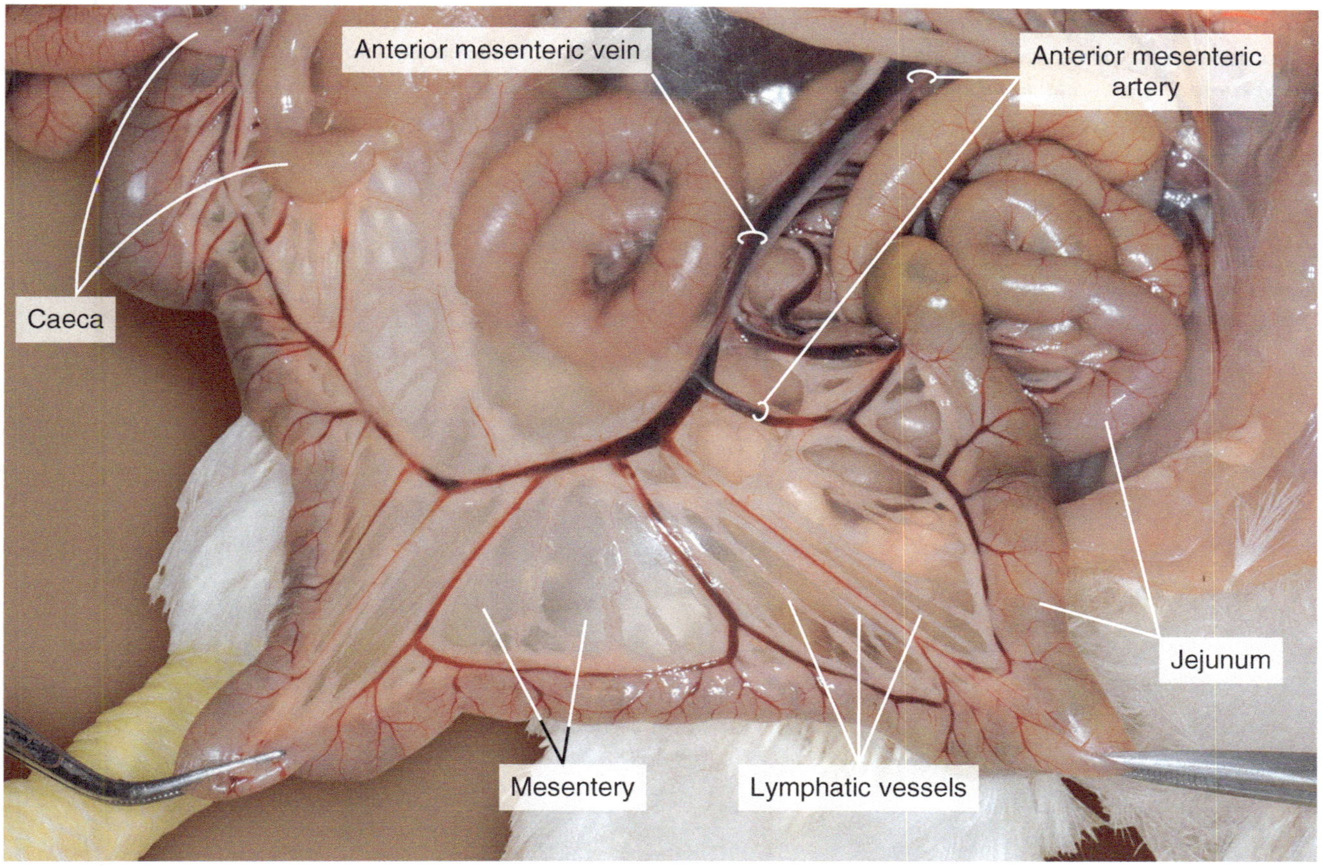

Fig. 11.38 The jejunum with the mesentery after reflecting the abdominal organs to the left. Note adipose tissue surrounding the blood and lymphatic vessels

Note the *anterior mesenteric vein*, which gives the major part of the *hepatic portal vein* (Figs. 11.38 and 11.48). The *coccygeo-mesenteric vein* joins here as well (Fig. 11.42). Find the *vitelline caecum* in the middle of the jejunum. This is a vestige of the embryonic vitelline sac (Fig. 11.39).

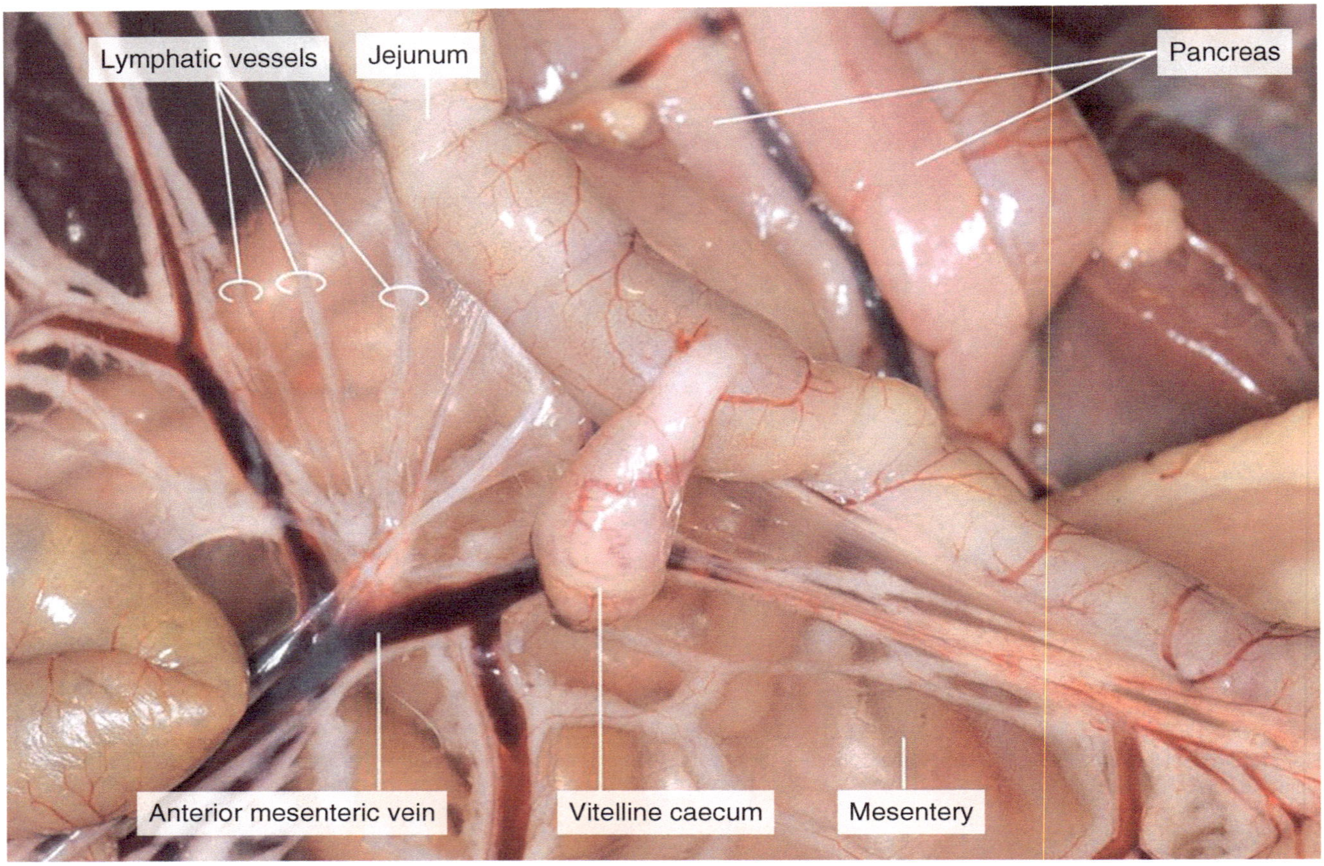

Fig. 11.39 The vitelline caecum in the middle of the jejunum

Trace the small intestine posteriorly and note the greatly coiled condition with the resultant fusions of the mesentery. The small intestine becomes *large intestine* without any noticeable change in diameter. The division between the two is marked by the long paired *colic caeca* which are bound to the small intestine and project anteriorly (Fig. 11.40).

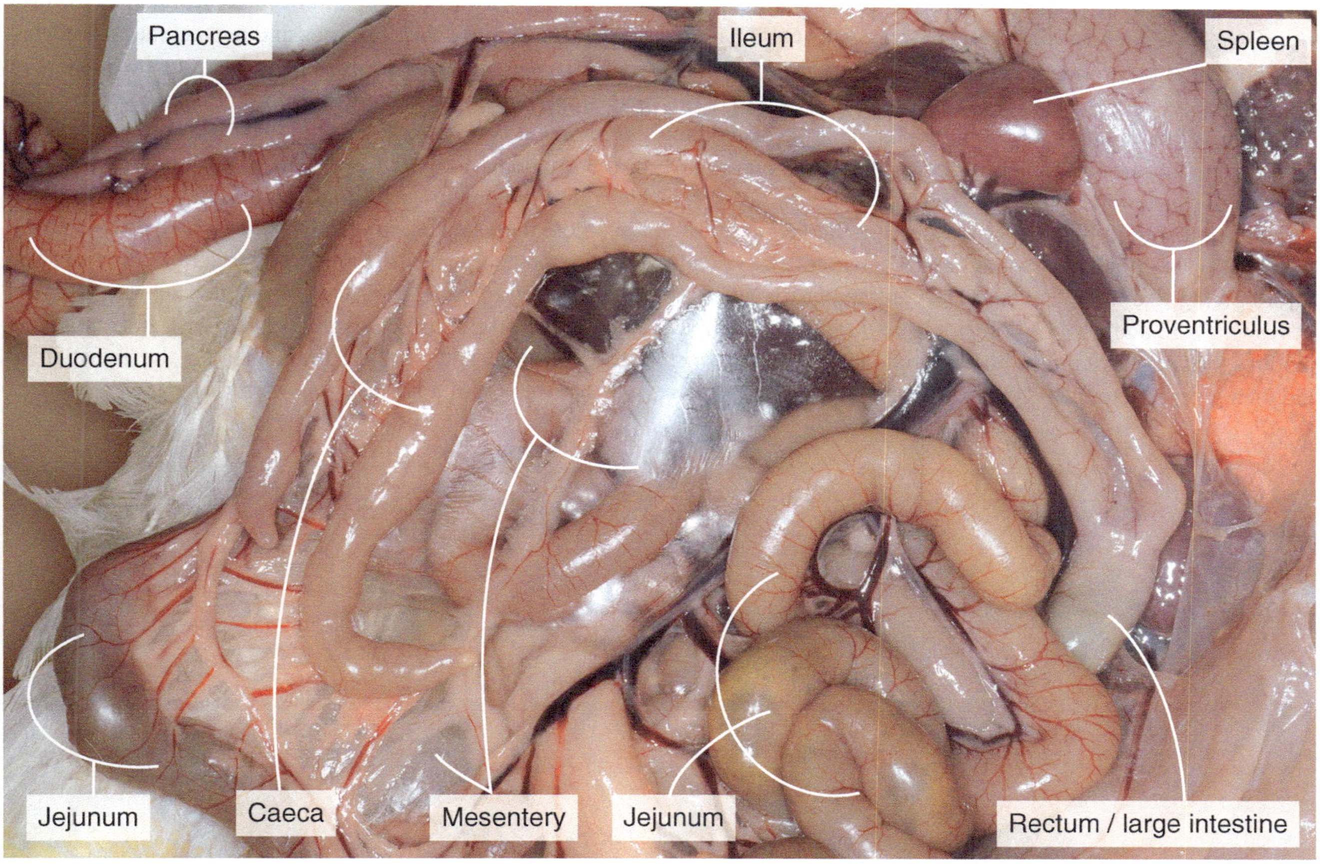

Fig. 11.40 The long paired colic caeca branching at the border of the small and the large intestine

The large intestine and *rectum* are relatively short. Find the cloaca at the end of the rectum (Fig. 11.41).

Open the *cloaca* and note the three divisions: a large ventral portion or *coprodeum* into which the rectum opens, separated by a fold from the *urodeum* into which the gonadal ducts and ureters open and a small dorsal compartment or *proctodeum* a chamber which opens to the anus. On the dorsal wall of the proctodeum near the rim is the thick-walled glandular pouch or *bursa of Fabricius* which is large in young birds but small or absent in adults. This organ is the site of haematopoiesis and necessary for the development of the immune system (Fig. 11.41).

The *aorta* should be traced by reflecting the heart anteriorly. It supplies branches to the oesophagus and body wall before giving rise to the coeliac artery in the peritoneal cavity (Figs. 11.32 and 11.34). Follow the *coeliac artery*. It first supplies the proventriculus with an excellent arterial network and then gives rise to the *left gastric artery* supplying the left side of the gizzard. The main trunk of the coeliac courses over the right side of the spleen and gives rise to a large *splenic artery*. The next artery given off the coeliac is the hepatic trunk which supplies *hepatic arteries* to the liver and a short *cystic artery* to the gall bladder. The coeliac artery after giving rises to the hepatic trunk and branches to form one or more right gastric arteries and a pancreaticoduodenal artery. The *right gastric arteries* pass dorsal to the duodenum and supply the right side of the gizzard. The *pancreaticoduodenal artery* courses between the two lobes of the pancreas in the duodenal loop and gives off many twigs to the pancreas and duodenum. At about the middle of the duodenal loop, note the recurrent *intestinal artery* which crosses to the intestine and colic caeca. Reflect the viscera to the right and note the *anterior mesenteric artery* which leaves the aorta close to the origin of the coeliac (Fig. 11.38). It supplies the small intestine and sends a branch along the intestine to anastomose with the *posterior mesenteric artery*.

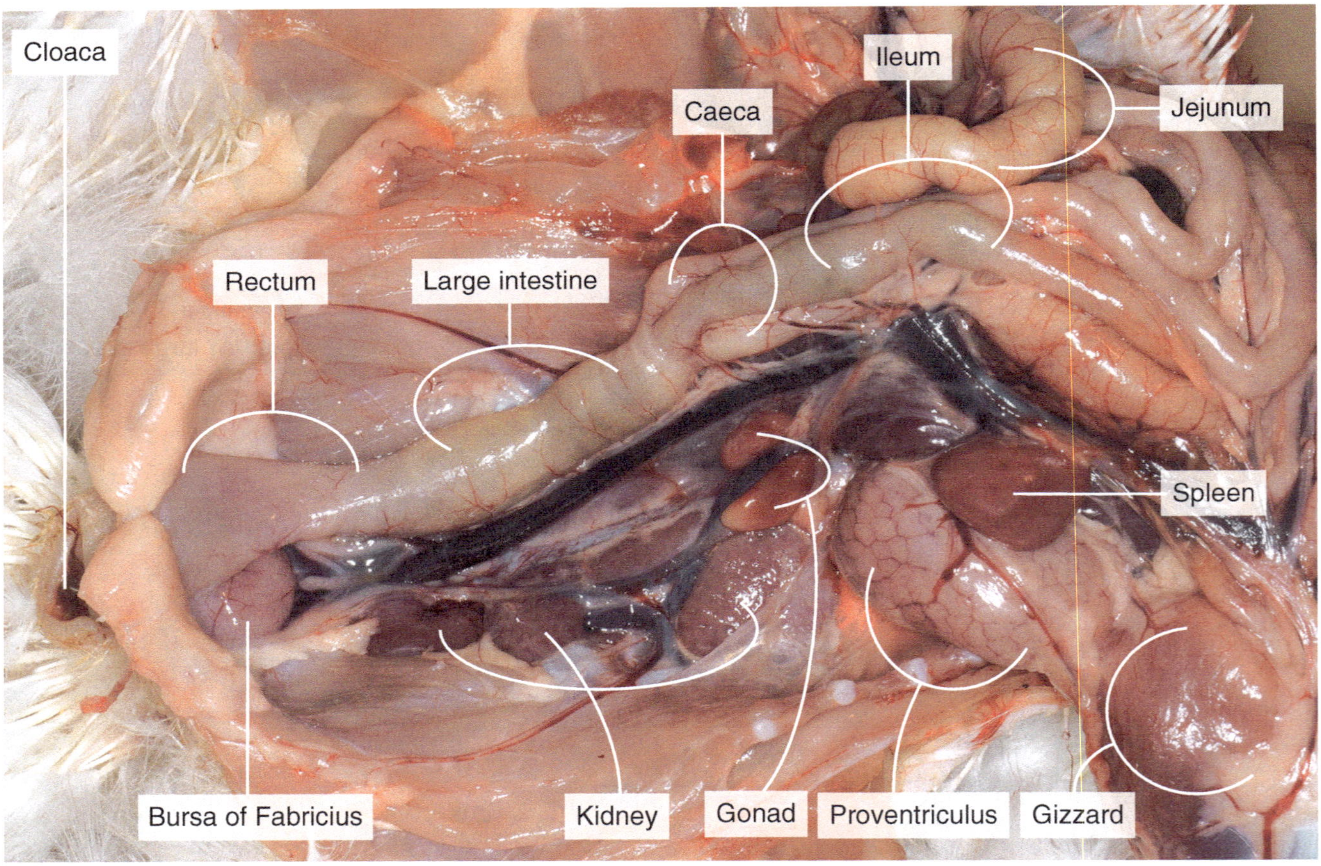

Fig. 11.41 Posterior part of the gut

Reflect but do not remove the viscera to expose the urinary and genital organs. Find the kidneys and the blood vessels supplying the posterior body part behind the parietal wall of the peritoneum. The *kidney* of the chicken is of the metanephric type, each is divided into three lobes and each is provided with a branch of the ureter (Figs. 11.41 and 11.42).

The glomeruli are small and the tubules are exceedingly numerous. The kidneys occupy depressions beneath the synsacrum seen previously. From the medial border of each kidney, the straight *ureter* (metanephric duct) exits between the anterior and middle lobes extending posteriorly to the cloaca. In the female, the ureter opens into the cloaca dorsal to the oviduct, while in the male it opens medial to the ductus deferens on a small papilla. A urinary bladder is lacking. The *renal veins* give rise to the *posterior vena cava* (Fig. 11.42).

The *dorsal aorta* after giving rise to the cranial mesenteric artery passes between the kidneys. A *renal artery* is given off to each cranial lobe of the kidney. On the right side, it supplies the cranial and middle lobes, while on the left in addition to supplying both lobes of the kidney it also supplies the ovary and oviduct. Between all lobes of the kidney, the aorta gives rise to short paired *lumbar arteries*. In the region of the middle lobe of the kidney, the aorta gives rise to the *external iliac arteries*, each of which passes dorsal to the kidney and forks into anterior and posterior branches. The anterior branch supplies the medial and cranial thigh muscles. The posterior branch supplies the abdominal wall, pelvic fat, and enters the limb as the *femoral artery*. The femoral artery anastomoses with the sciatic artery in the region of the knee joint. Between the middle and caudal lobes of the kidneys, the large *sciatic artery* is seen (Fig. 11.42). It is the main supply of the hindlimb. The posterior mesenteric artery leaves the aorta after the sciatic arteries are given off. One branch accompanies the *coccygeo-mesenteric vein* to the intestine, while the other supplies the distal end of the rectum (Fig. 11.42). The terminal end of the aorta trifurcates to form paired *hypogastric arteries* and a middle *sacral artery*. Each hypogastric artery forks to supply the dorsolateral wall of the cloaca and the pelvic musculature lateral to the pubis.

In a juvenile chick, the reproductive system is immature and the sexes are almost indistinguishable (Figs. 11.41 and 11.42). We provide a description of the adult genital system for your information.

Female: Most birds, as well as the chicken, retain only the left ovary and oviduct. The reproductive tract of the female consists of an irregularly shaped left *ovary* containing eggs of various sizes, situated beneath the left kidney, attached by a short mesovarium. Posterior to the ovary, the coiled left *oviduct* supported by the mesotubarium has at its cranial end a flattened funnel-shaped opening, the *ostium*. Demonstrate this opening. The oviduct may be divided into a cranial

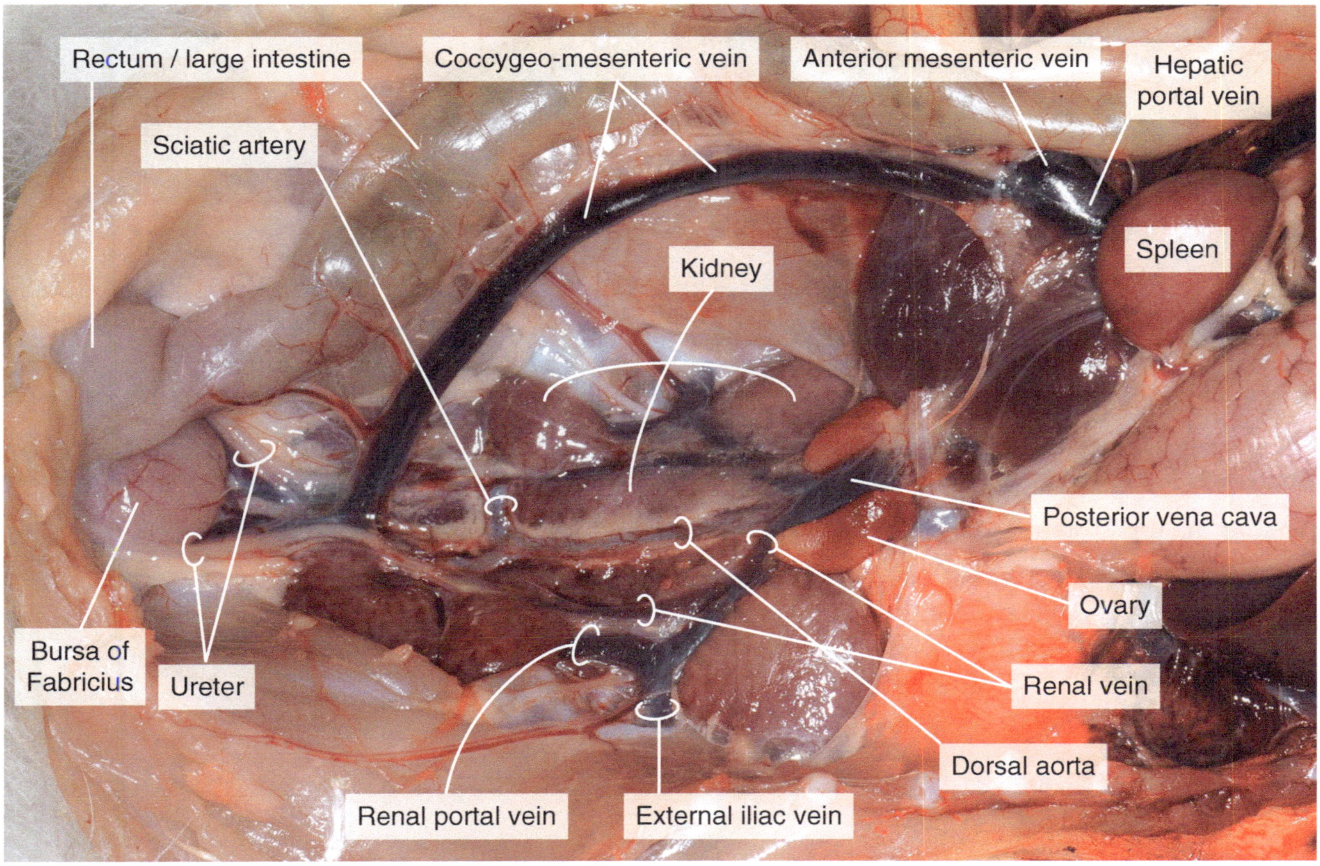

Fig. 11.42 The kidney with posterior part of the circulation

glandular region (albumen secretion), followed by the isthmus (secretes shell membranes and fluid albumen) and then the uterus or shell gland (more fluid albumen and a calcareous shell). The latter opens into the cloaca on the left side.

Male: The male reproductive tract consists of paired *testes*, *epididymides* and *deferent ducts*. Accessory glands are lacking. The testes are oval bodies of about equal size which lie at the anterior end of the kidneys. The seminiferous tubules convey the spermatozoa to the epididymis (not visible grossly) which is continuous with the deferent duct (mesonephric duct). This duct arises on the medial side of the testis as a slender tortuous tube which courses posteriorly parallel to the ureter. Trace both ducts to the cloaca where they open on small papillae in the lateral wall.

Remove the heart and cut long enough stumps for the blood vessels to remain identifiable. The heart is relatively large and four chambered. The apex is directed posterioventrally and is formed largely by the left ventricle (Fig. 11.43).

Externally, the atria are separated by a band of fat from the ventricles. Note the right anterior vena cava entering the lateral anterior portion of the right atrium. The left anterior vena cava enters the right atrium on the dorsal surface of the base of the heart along with the unpaired posterior vena cava.

Transect the apex in one third of the heart's length to see the extent of each ventricle (Fig. 11.44).

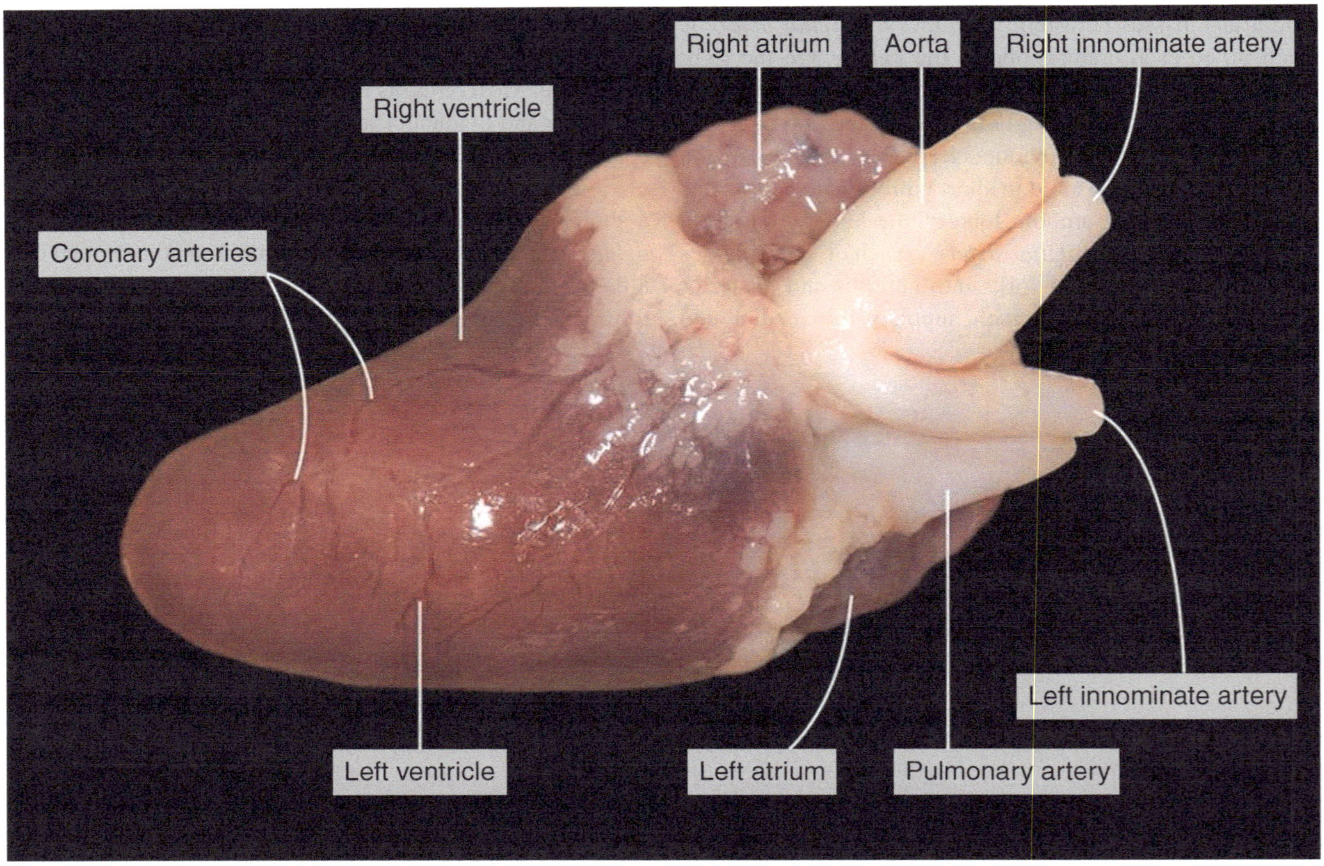

Fig. 11.43 The isolated chicken's heart with the large arteries

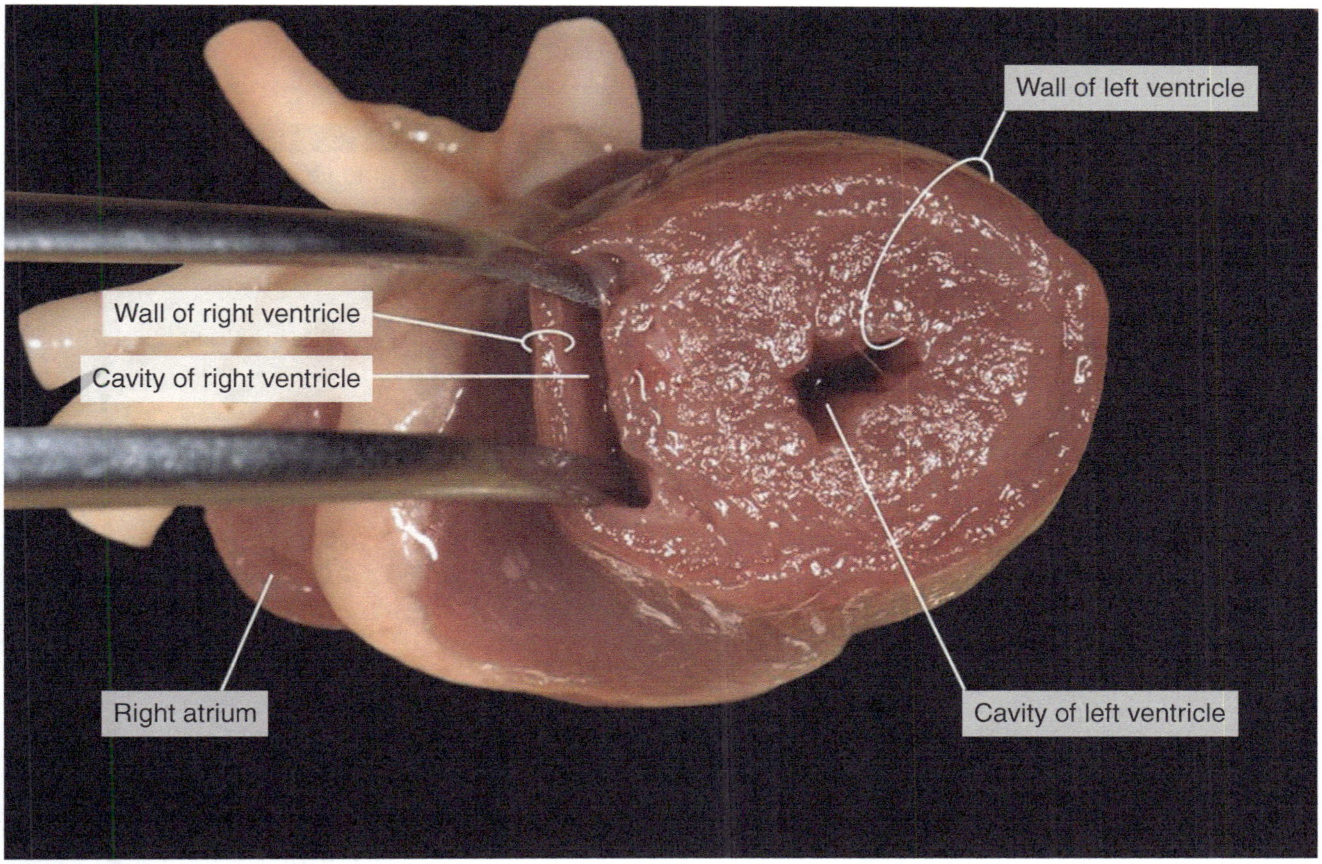

Fig. 11.44 The walls of the two ventricles of the chicken's heart are highly asymmetric

Insert the tip of scissors into the cavity of the right then into the left ventricle and cut towards the corresponding atria. Wash the remnants of blood clots away (Figs. 11.45 and 11.46).

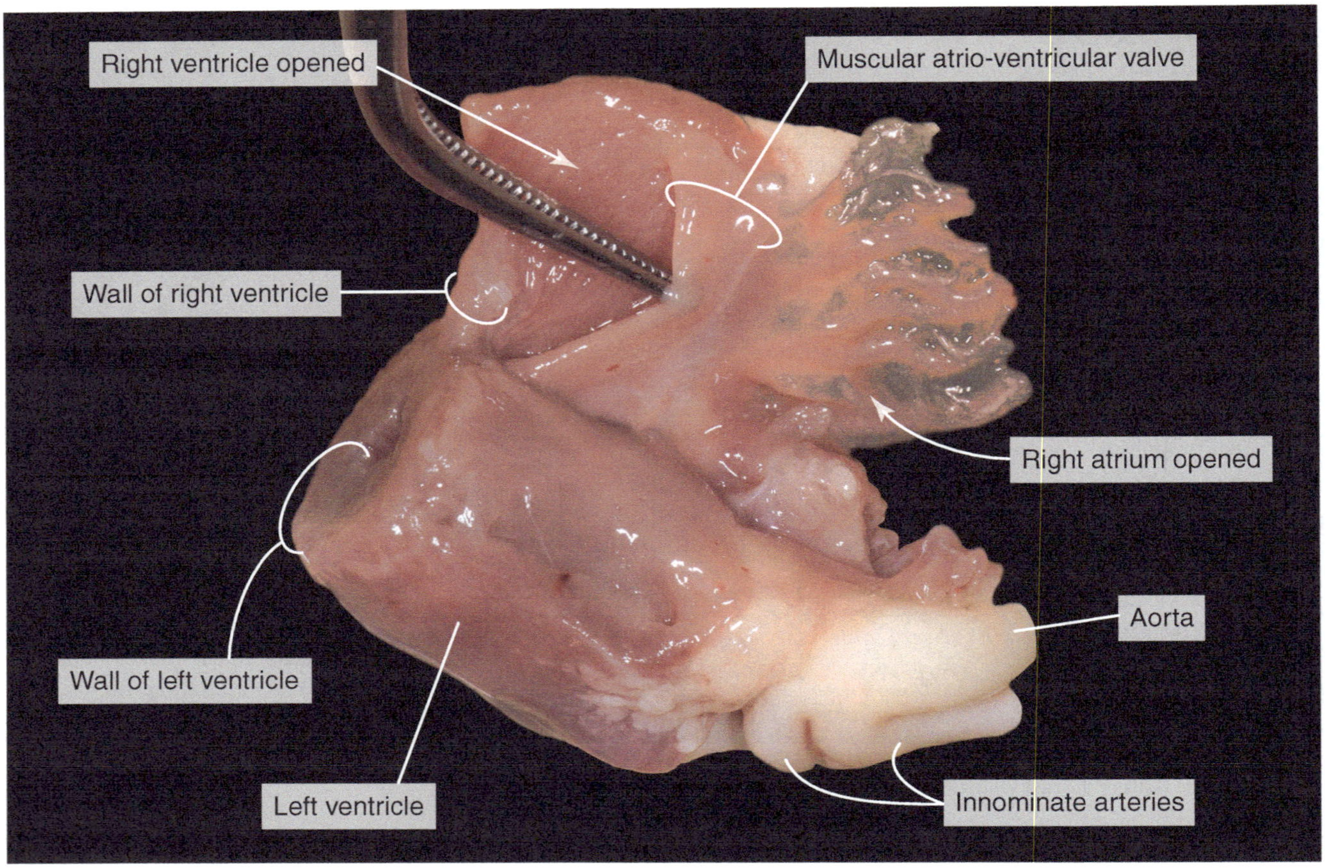

Fig. 11.45 The muscular valve in the atrio-ventricular opening of the right half of the heart

Compare the atrio-ventricular valves of the two halves of the heart: a *muscular valve* can be found in the *right side* (Fig. 11.45), while there is a *tricuspid valve* in the *left side* (Fig. 11.46).

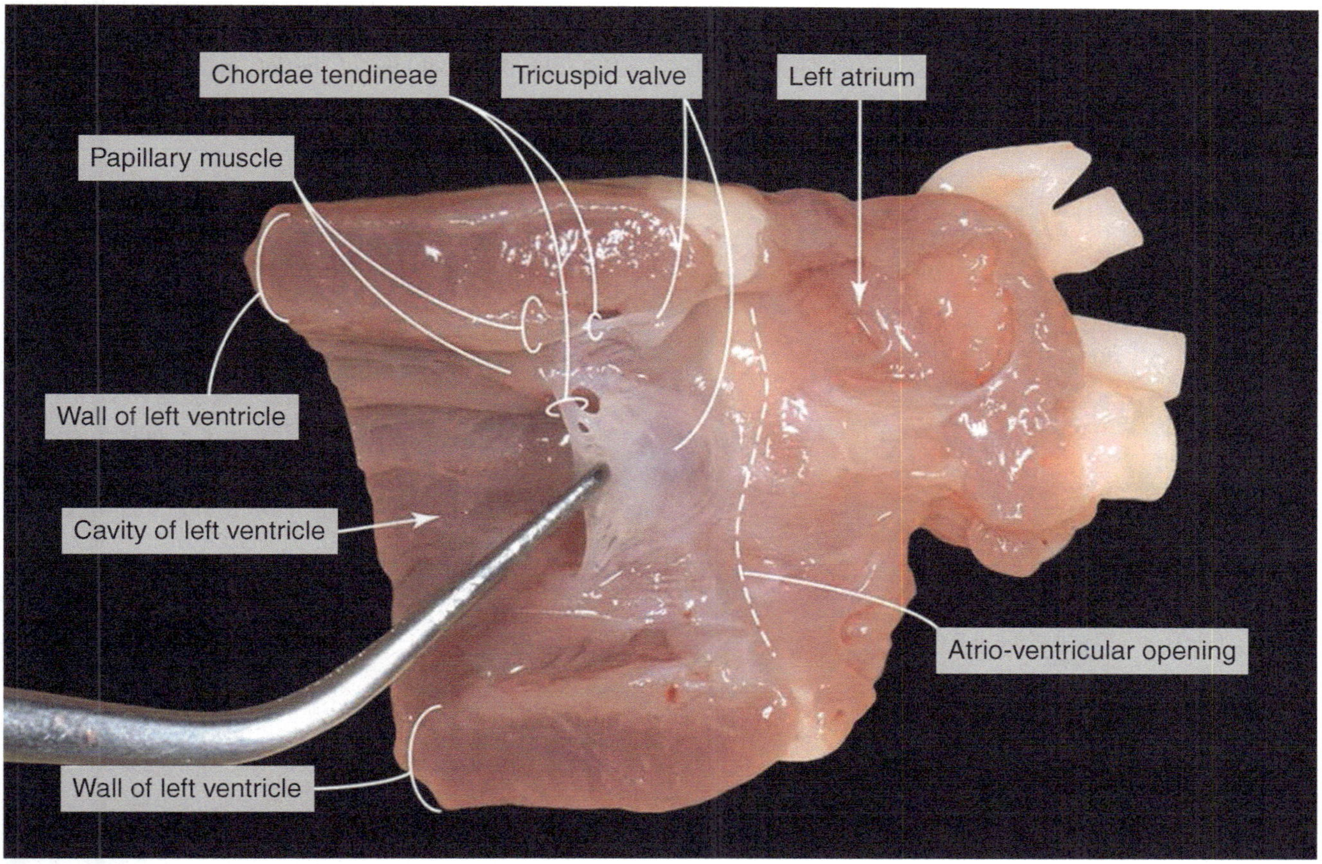

Fig. 11.46 The tricuspid valve in the atrio-ventricular opening of the left half of the heart

The free margin of the tricuspid valve is attached to the *papillary muscles* of the heart wall by *chordae tendineae*, which prevent the valve to flip over into the atrium at contraction of the left ventricle. The wall of the right ventricle is weaker as the resistance of the pulmonary circulation is smaller. Although the volume of the two ventricles is exactly the same, the cavity of the right ventricle is narrower which accounts for the simple muscular right atrio-ventricular valve (Figs. 11.44 and 11.45). Slit open the stump of the aorta back to the ventricle. Observe the three *semilunar valves* in the exit of the aorta and just behind them the orifice of the *coronary arteries* supplying the heart itself (Fig. 11.47).

Note the two openings of the left and right innominate arteries further away.

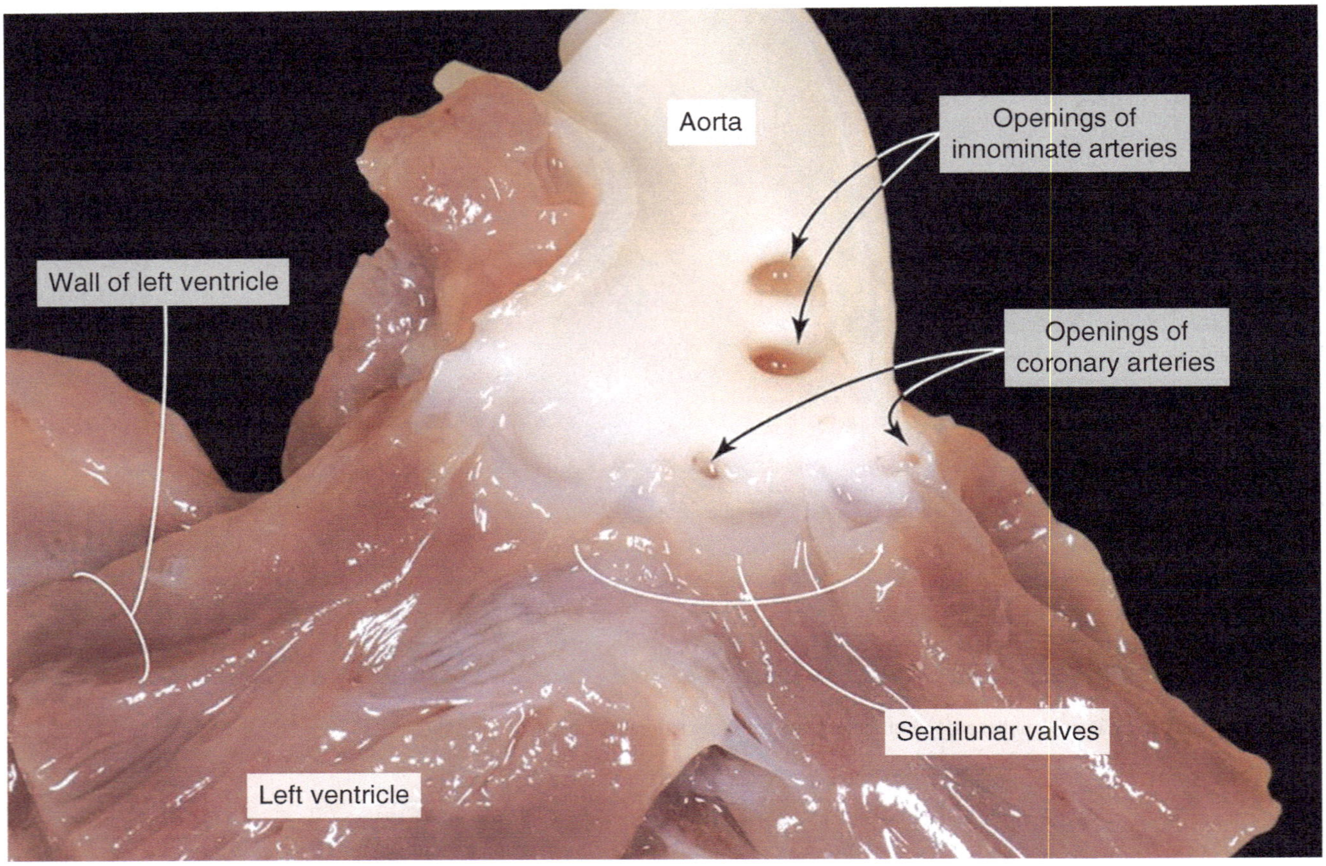

Fig. 11.47 The three semilunar valves in the exit of the aorta

Return to the gizzard; observe the membranaceus tendon on its wall (Fig. 11.48).

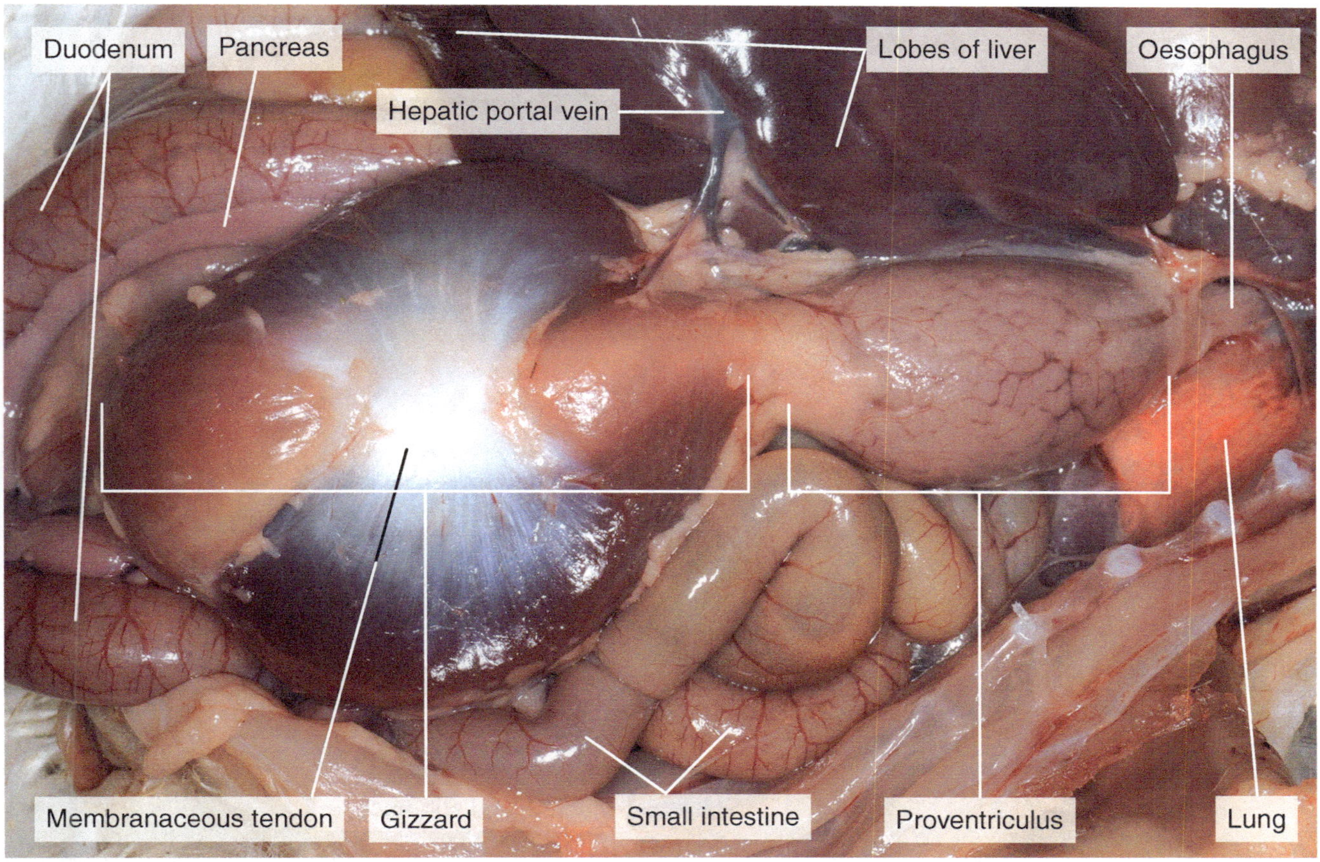

Fig. 11.48 The position of the two-parted stomach of the chicken among the neighbouring organs

Remove both parts of the stomach together and slit the wall of them on the midventral line and wash out the contents (Fig. 11.49).

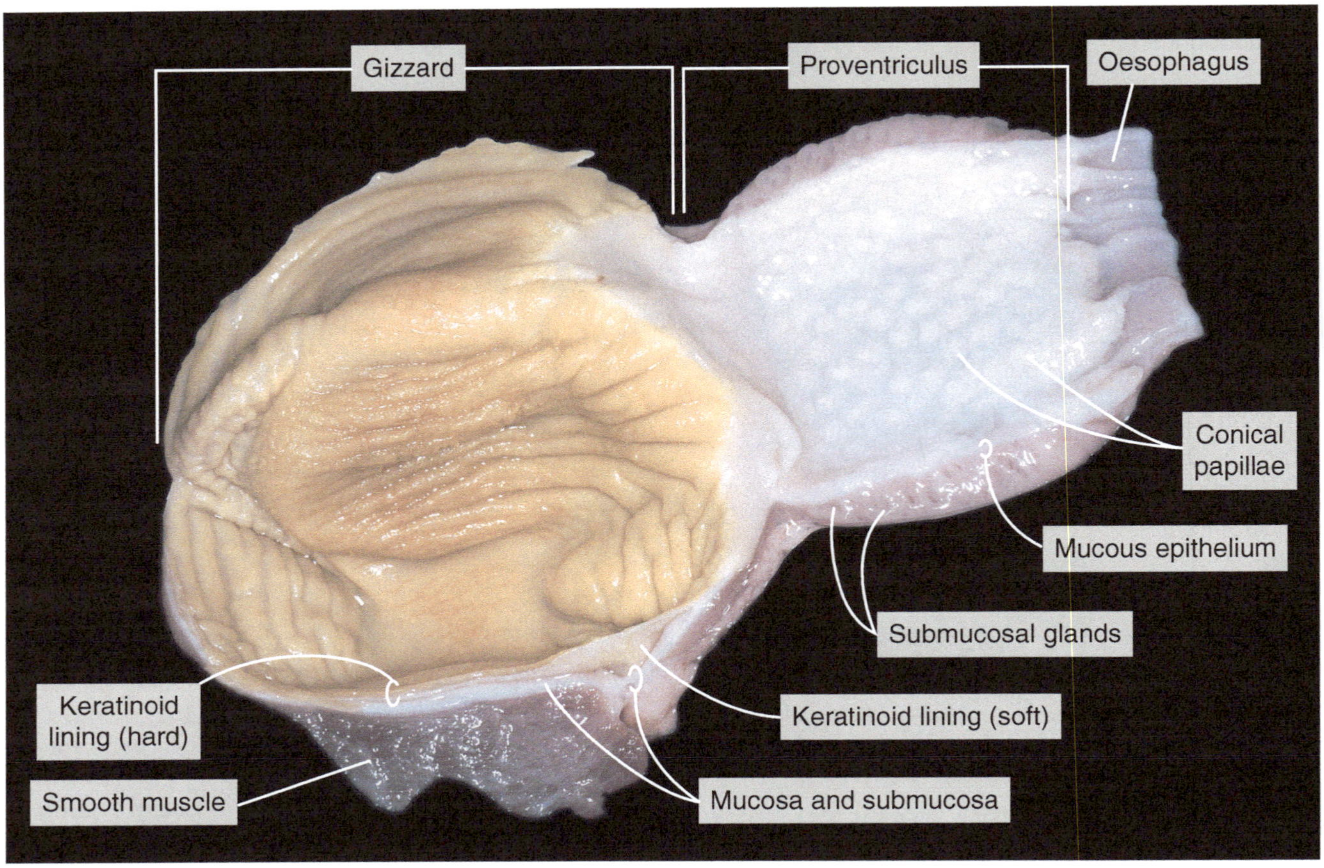

Fig. 11.49 The two-parted stomach of the chicken slit open

Note the glandular wall, soft, slimy surface with conical papillae of the proventriculus (Fig. 11.49). Observe the thick muscular wall with keratinoid lining of the gizzard. Peel off the yellow horny lining of the gizzard demonstrating that it has a soft surface as well. Make an about 3–4 mm thick cross-section slice of the wall of the proventriculus with a sharp scalpel or razor blade, cover with water and examine under a dissecting microscope (Fig. 11.50).

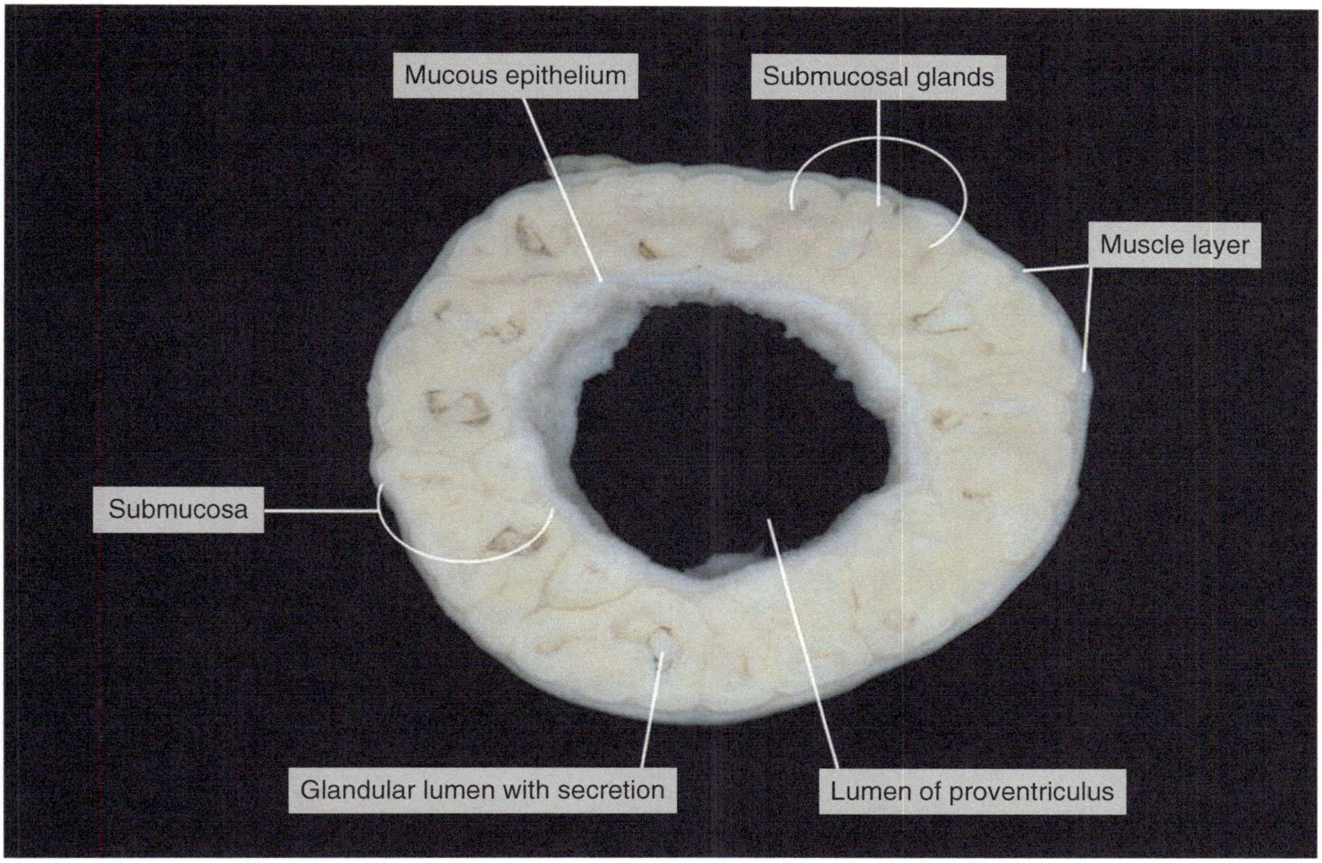

Fig. 11.50 Cross-section slice of the wall of the proventriculus of the chicken (formalin-fixed specimen)

The *stomach* of the chicken is subdivided into two parts, the proventriculus (glandular stomach) and the ventriculus (gizzard). The first part produces enzymes digesting the food, whereas the second serves as a grinder in the absence of teeth.

Proventriculus has a thick mucosa and submucosa occupied by two types of glands (Fig. 11.51).

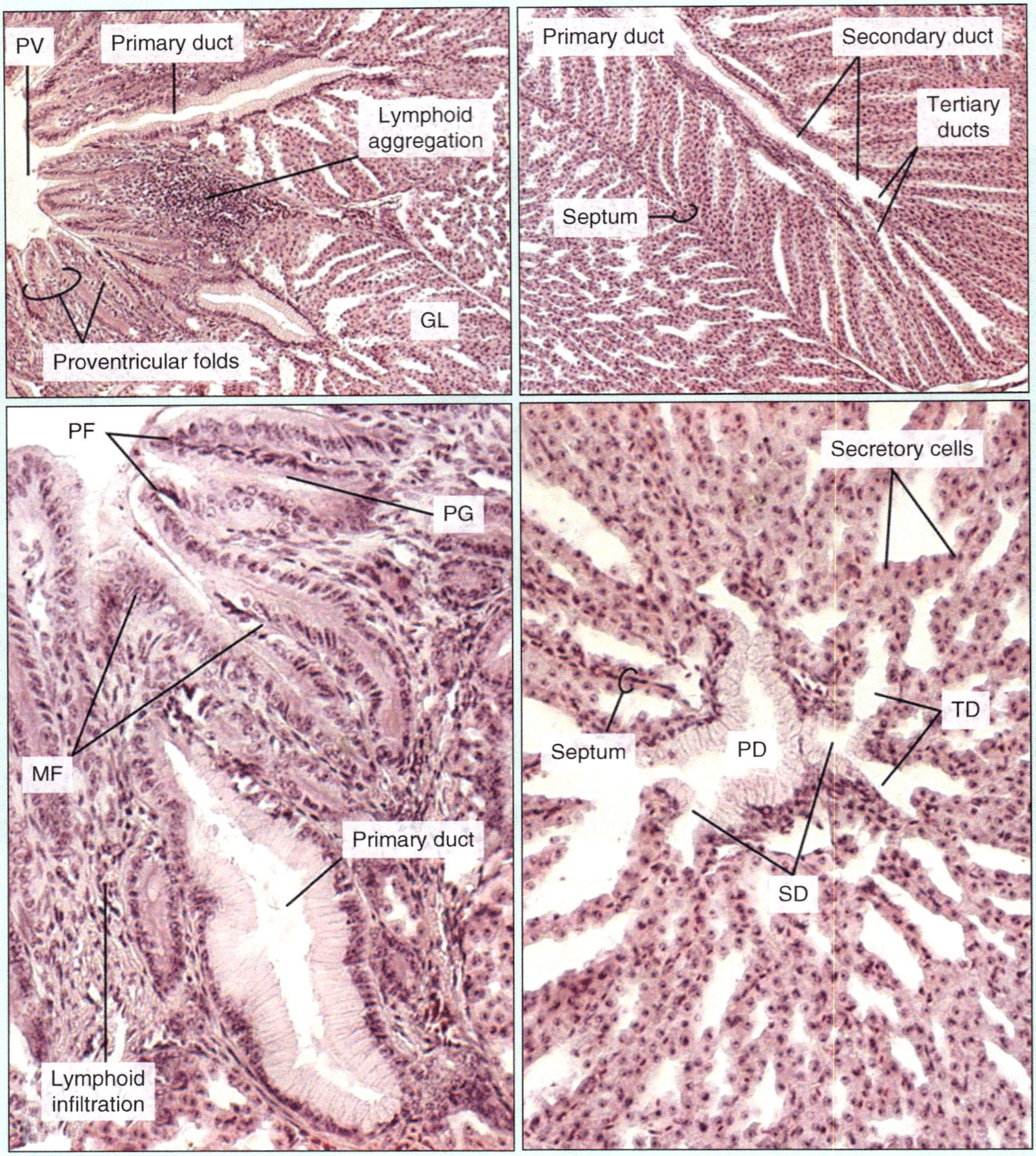

Fig. 11.51 Histological sections of the proventriculus of birds (HE). Left side top and bottom: Mucosal folds of proventriculus; *GL* submucosal glands, *MF* mucosal folds, *PF* proventricular folds, *PG* tubular proprial glands, *PV* proventricular cavity. Right side top and bottom: The duct system of the compound gland; *PD* primary ducts, *SD* secondary ducts, *TD* tertiary ducts

Mucosa forms folds (MF), which are covered by a simple columnar epithelium (Fig. 11.51, top and bottom left). Folds contain several proprial glands (PG) with tubular acini. Cells on the surface and in the proprial glands synthesise and secrete protective mucous onto the surface. Propria supporting epithelial lining contains scattered lymphocytes and lymphoid aggregations (Fig. 11.51, top and bottom left). Submucosal layer is occupied by big compound glands being separated by connective tissue septa (Fig. 11.51, top and bottom right). Cuboidal secretory cells of them synthesise pepsinogen and secrete hydrochloric acid for adjusting the appropriate pH for activating pepsinogen. These cells show acidophil (eosinophil) staining. Tubular acini open into tertiary ducts (TD), which unite and form secondary ducts (SD). The longest primary ducts (PD) collect the secondary ducts belonging to one submucosal compound gland. Primary ducts open on the surface between the mucosal folds. Ductal system has columnar secretory cells with light, basophilic staining. They secrete mucous material. Submucosa forms separating septa among compound glands. Septa contain smooth muscle cells helping the emptying of the secretion. Muscular layer is thin and covered by serosa on the outer surface of the organ.

Make an about 3–4 mm thick cross-section slice of the wall of the gizzard with a sharp scalpel or razor blade, cover with water and examine under a dissecting microscope (Fig. 11.52).

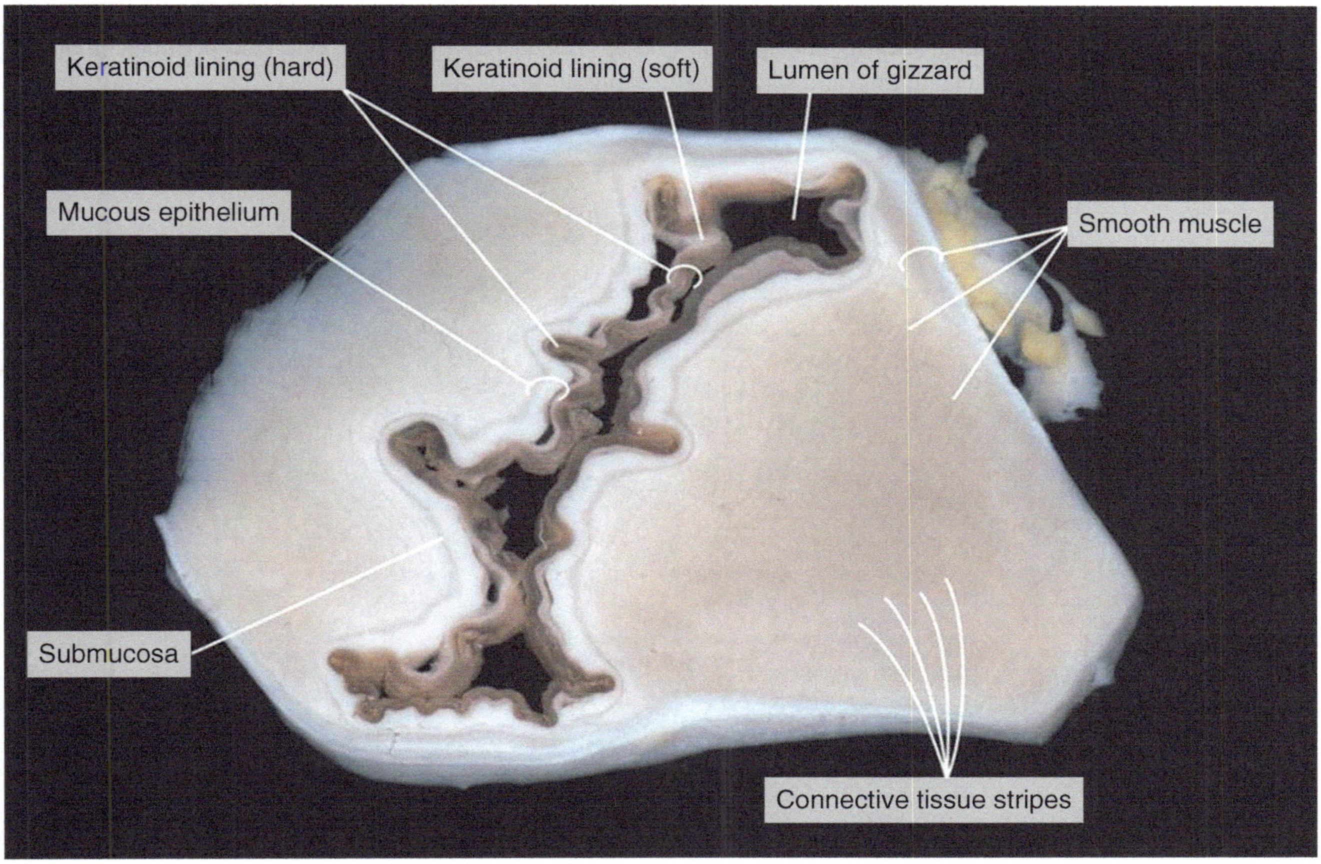

Fig. 11.52 Cross-section slice of the wall of the gizzard of the chicken (formalin-fixed specimen)

The histological composition shows different structure in the *gizzard* (Fig. 11.53).

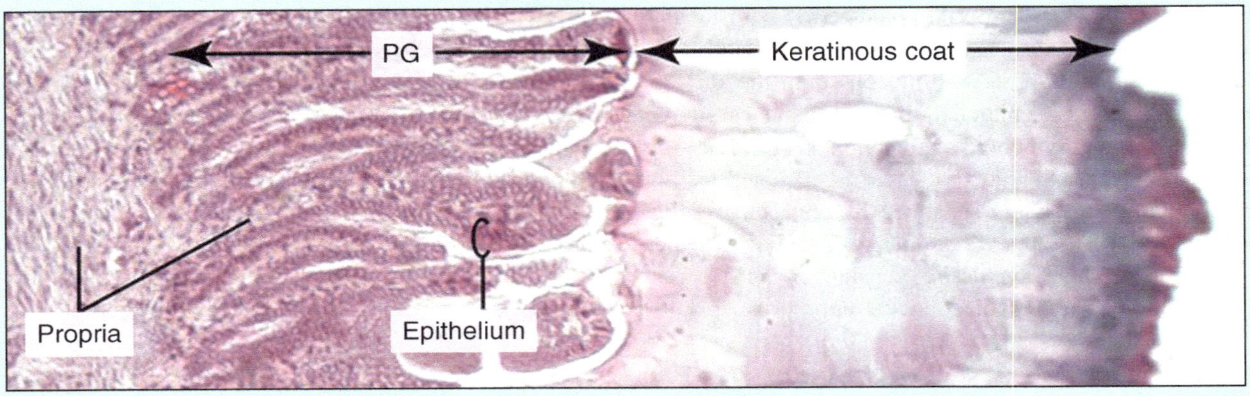

Fig. 11.53 Histological section of the gizzard. *PG* tubular proprial glands

The mucosa of the second part of the stomach is similar: it also forms folds, which are populated by simple tubular glands (PG), embedded in the propria. These glands produce a thick keratinous lining covering the inner surface. It protects the mucosa and gives a hard surface for grinding the softened food. Whereas the submucosa is a thin layer and does not contain any glands, the muscular layers dominate the section (Fig. 11.52). The outer surface is covered by serosa (visceral peritoneum, not shown).

Make an about 3–4 mm thick cross-section slice of the wall of the duodenum together with the pancreas with a sharp scalpel or razor blade, cover with water and examine under a dissecting microscope (Fig. 11.54).

Note the intestinal villi which enlarge the absorptive surface of the duodenum.

Exposure of brain and extraction from the skull on a fresh specimen is time consuming and quite complicated. This part of the dissection should only be attempted by those who have considerable experience in dissection and a good manual dexterity. Clear the occipital region from muscles and periosteum. Insert the sharp tip of the scissors into the foramen magnum and remove the roof of the skull. Advance the scissors only a little in one go and cut short distances. It is important to save the integrity of the dura mater otherwise the brain would be injured. Use the large anatomical forceps to remove pieces of the spongy bone of the skull.

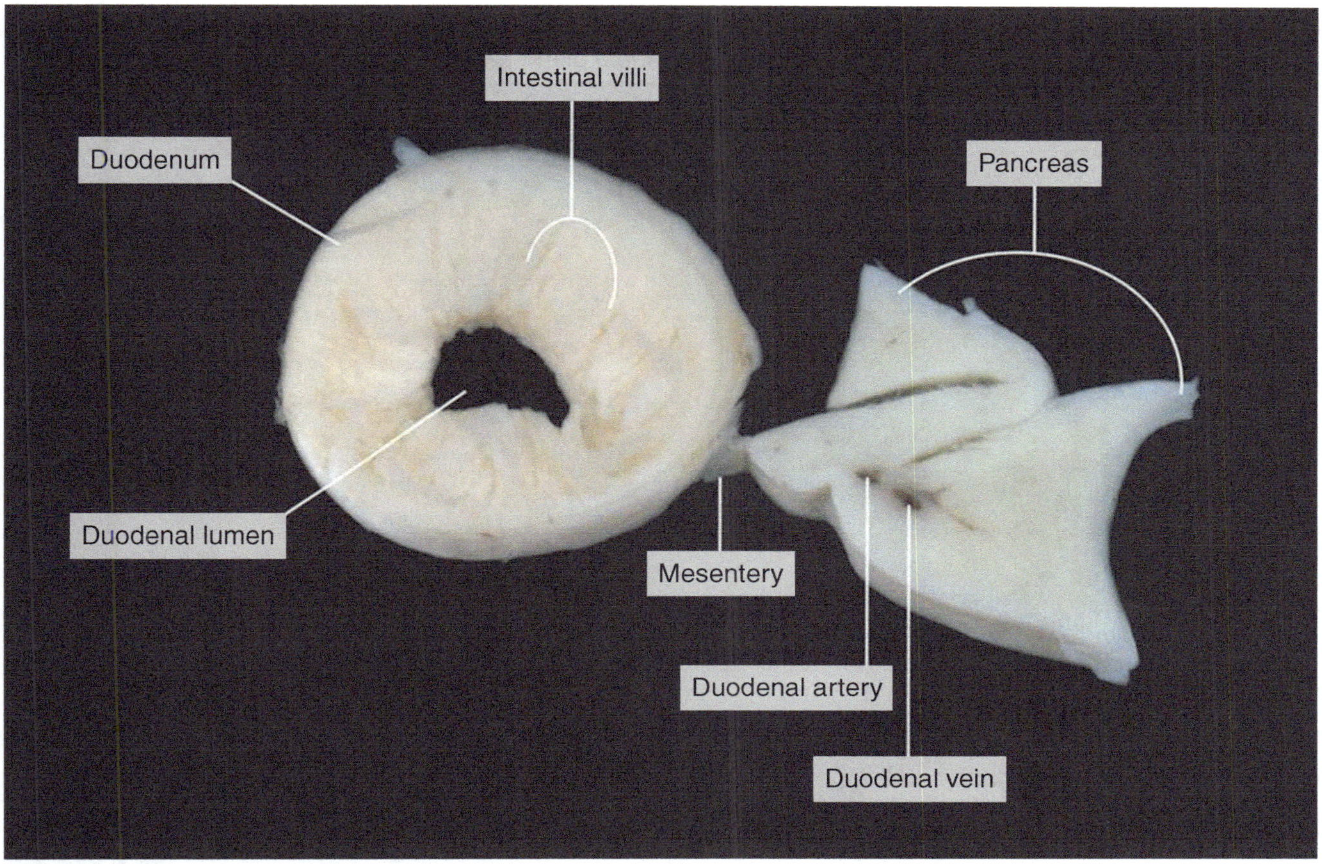

Fig. 11.54 Cross-section slice of the wall of the duodenum and the pancreas of the chicken (formalin-fixed specimen)

The *brain* is relatively large in birds. The cerebral hemispheres, optic lobes and cerebellum are well developed, while the olfactory regions such as the olfactory bulb and the hippocampus are small. This may be because smell is not important in most birds. The best developed senses are visual and auditory. After removal of the skull roof, extract the brain and cover it with water. Identify the *olfactory bulbs*, *cerebral hemispheres*, *pineal gland*, *optic lobes* (midbrain), *cerebellum* and *medulla oblongata* on the dorsal surface of the brain (Fig. 11.55).

Observe a pair of small hillocks, called "Wulst" on the dorsal side of the cerebrum surrounded by two little grooves (vallecula). The Wulst, also known as the sagittal eminence, is a structure in the brain of birds linked to sensory and visual perception. Find the *pineal gland* which regulates daily body rhythms, most notably the day/night cycle. In birds, it is close to the skin and needs no interaction with the eye to register day/night cycles; it is the master clock (this is where the notion of the "third eye" comes from).

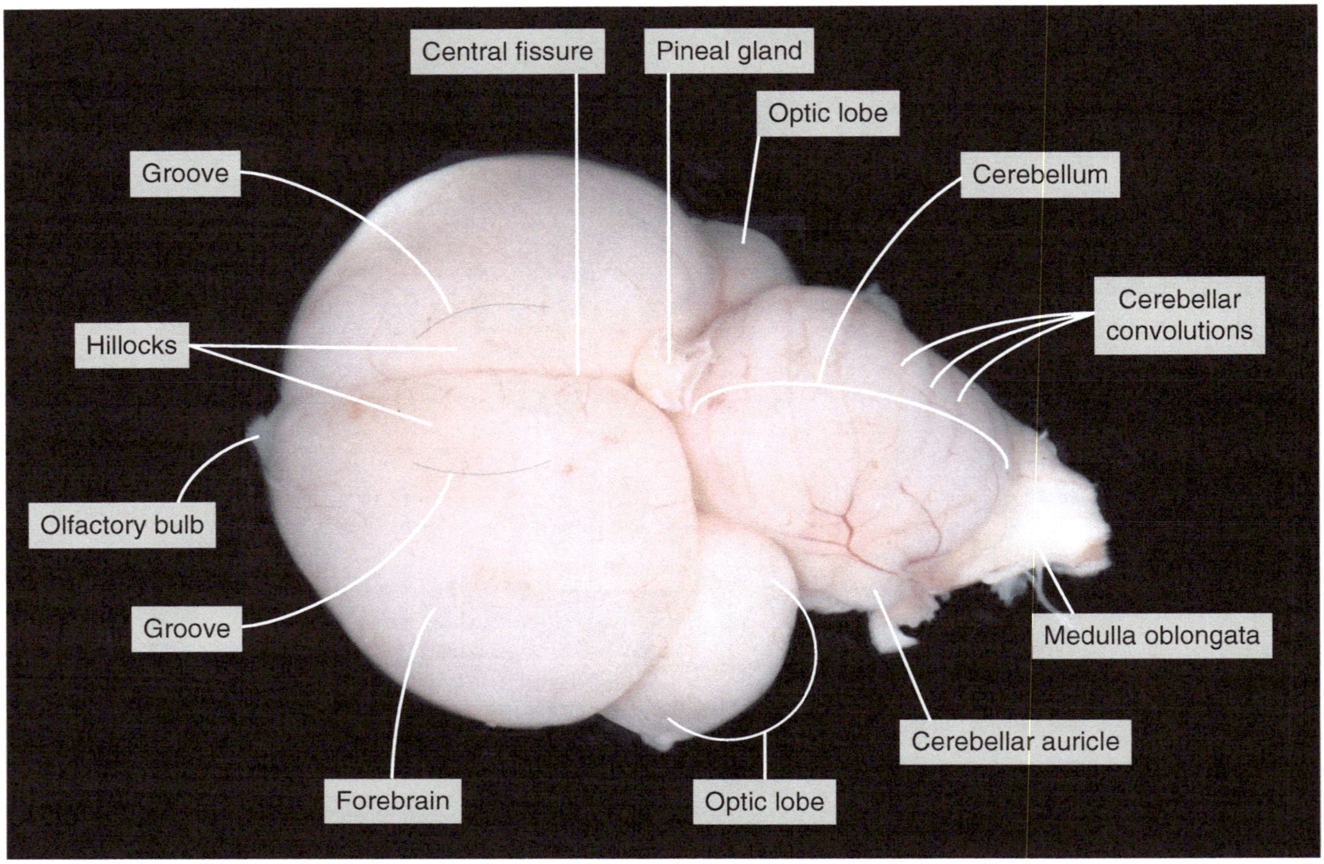

Fig. 11.55 Dorsal view of the brain of the chicken. *Lines* indicate the position of the grooves (valleculae) next to the hillocks (Wulst)

Examine the ventral surface of the brain as well. There are 12 cranial nerves as in mammals (Fig. 11.56).

Locate the studs of the olfactory nerves (cranial nerve I), optic nerves (II), oculomotor nerves (III), *trochlear nerves* (IV), trigeminal nerves (V), abducens nerves (VI), facial nerves (VII), *auditory nerves* (VIII), glossopharyngeal nerves (IX), vagus nerves (X), accessory nerves (XI) and *hypoglossal nerves* (XII). The three nerves in italics are very inconspicuous. If anything is left of the brain at this point, make a mid-sagittal section to observe the ventricles and the large size of the corpus striatum which forms the floor of the lateral ventricle.

The *eye* of a bird is a tight fit for the orbit. The eye of a bird is larger compared to the size of the animal than for any other group of animals, although much of it is concealed in its skull. Its large size is not at once apparent since only a relatively small part of the cornea shows. The eye of birds is not spherical, unlike the mammalian eye. This flatter shape enables more of its visual field to be in focus. Birds with eyes on the sides of their heads, like the chicken, have a wide visual field, useful for detecting predators, while those with eyes on the front of their heads, such as owls, have binocular vision and can estimate distances when hunting. True binocular vision is lacking in birds, because there is a complete decussation of the optic tracts. (In mammals with partial decussation of the optic tracts, we find conjugate eye movements.)

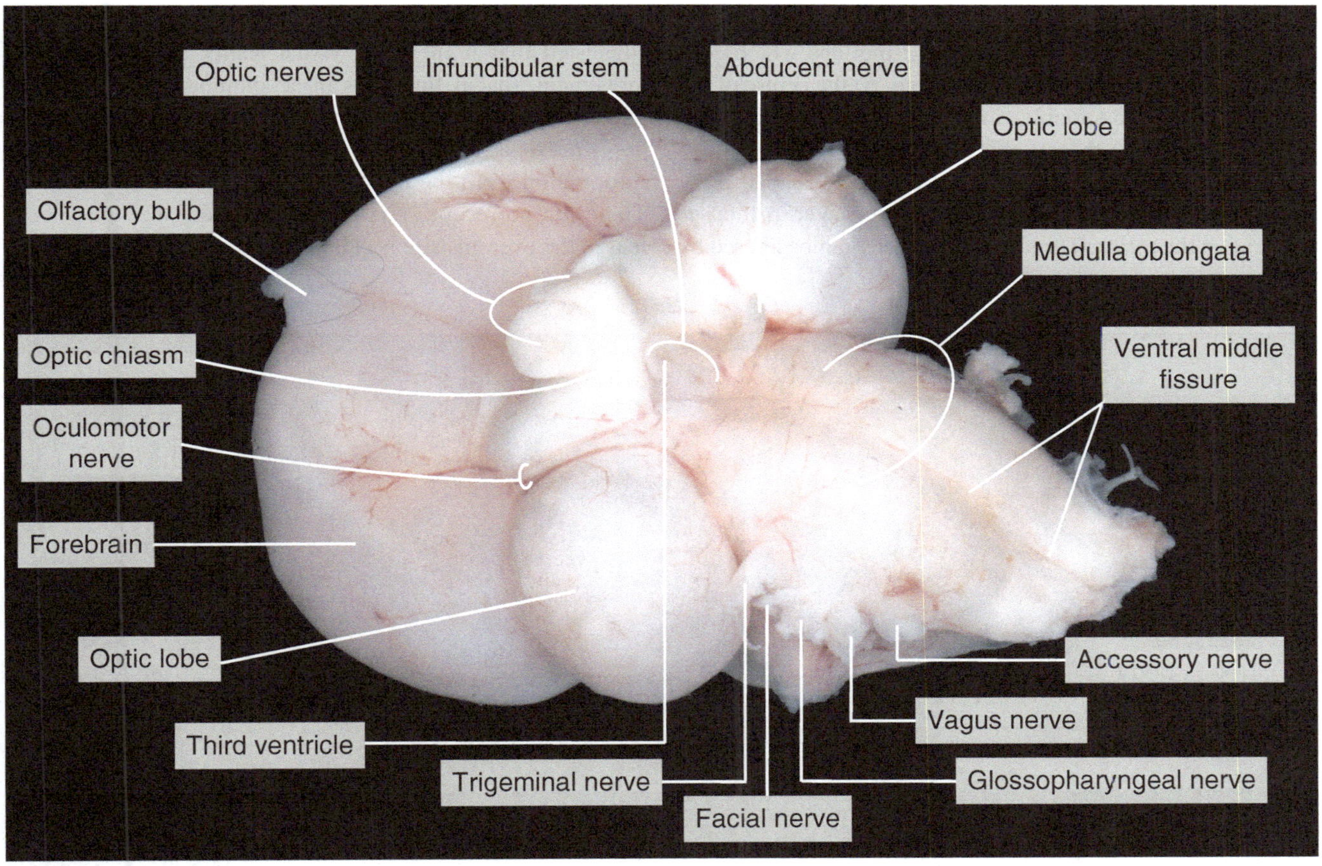

Fig. 11.56 Ventral view of the brain of the chicken. *Lines* show the approximate boundary of the olfactory lobes

For the dissection of the eye, cut the connective tissue around the eye with a pair of small-angled scissors and partly remove the bone from the dorsal side of the orbit. Cut the eye muscles and the optic nerve and remove the eye from the orbit. Cut the eyeball with a sharp scalpel or razor blade in a sagittal plane running through the centre of the pupil and the stud of optic nerve. Cover the section with water and examine under a dissecting microscope. Identify the layers of the eye and note the special characteristics of the bird's eye (Fig. 11.57).

The outer layer of the eye consists of the transparent *cornea* at the front, and the two layers of *sclera*, a tough white collagen fibre layer which surrounds the rest of the eye and supports and protects the eye as a whole. The sclera contains another layer, a hyaline cartilage cup behind the *sclerotic ring* (Figs. 11.2 and 11.57). The sclerotic ring, a circle of bony plates, holds the eye rigid. The *iris* is a coloured muscularly operated diaphragm in front of the lens which controls the amount of light entering the eye. At the centre of the iris is the *pupil*, the variable circular opening through which the light passes into the eye (Figs. 11.12 and 11.57).

The *lens* is a transparent convex or "lens"-shaped body with a harder outer layer and a softer inner layer. It focuses the light on the *retina*. The shape of the lens can be altered by *ciliary process* which is directly attached to lens capsule by means of the *zonular fibres*. In addition to these muscles, some birds also have a second set, Crampton's muscles, that can change the shape of the cornea, thus giving birds a greater range of accommodation than is possible for mammals. The eye contains behind the lens the *vitreous body*, a clear jelly-like substance.

Towards the centre of the retina is the *fovea* which has a greater density of receptors and is the area of greatest visual acuity, sharpest, clearest detection of objects. In 54 % of birds, including birds of prey, kingfishers, hummingbirds and swallows, there is second fovea for enhanced sideways viewing. A characteristic structure of the bird's eye is the *pecten* or fundus oculi (Fig. 11.57). This is a comb-like vascular fold projecting into the vitreous body from the retina. The function of this structure is still unknown. It has been suggested that it increases visual acuity and detection of movement, that it functions in accommodation or perhaps that it is a source of nourishment for the vitreous body. The *optic nerve* is a bundle of nerve fibres which carry light information from the eye to the relevant parts of the brain. There are no receptors in the retina where the optic nerve leaves the eye; this site is called the *blind spot*.

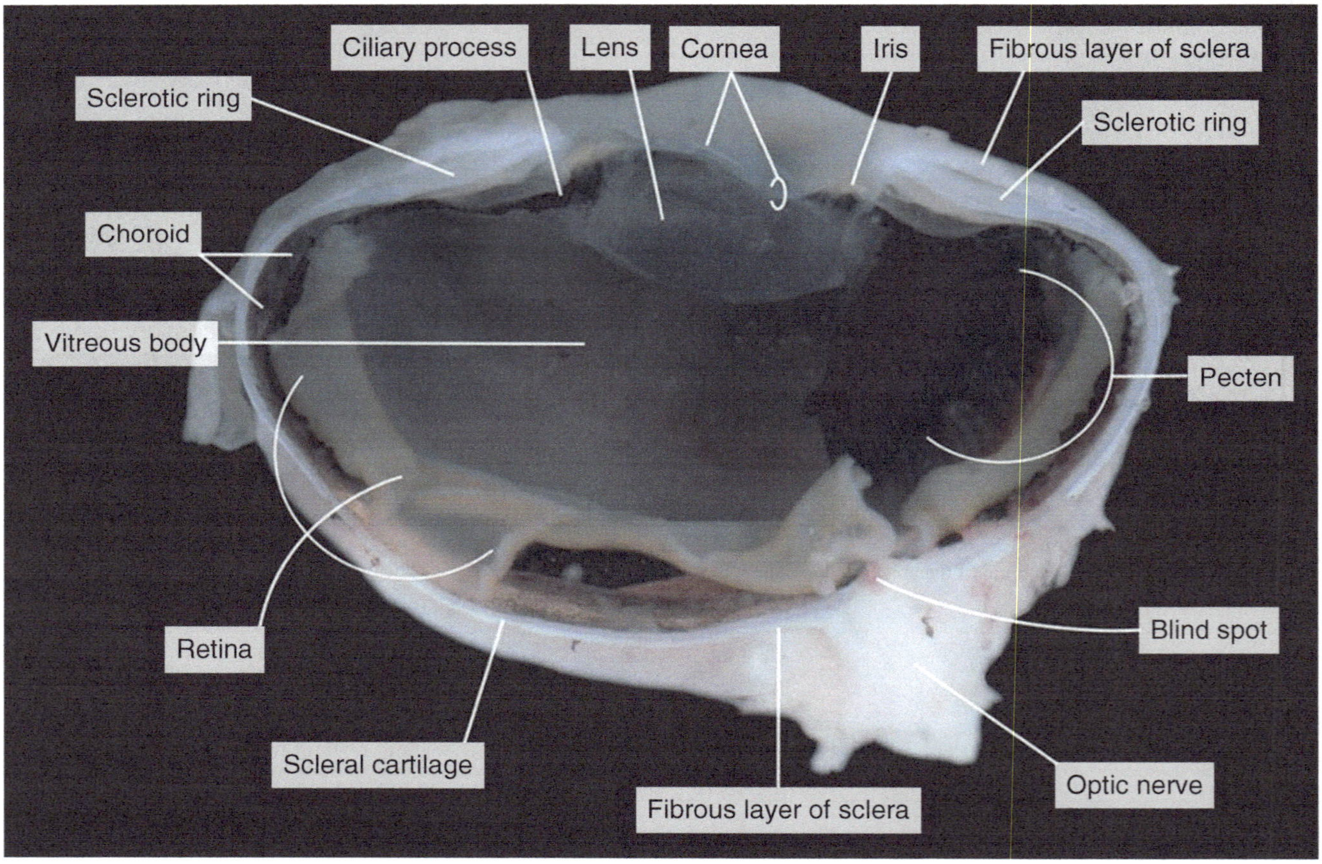

Fig. 11.57 The special bird's characteristics on the sagittal section of the eye: sclerotic ring, cartilaginous layer of the sclera and the pecten. Note retina got loose during preparation

Dissection of the Rat (*Rattus norvegicus*) 12

- *Availability:* Rats are most frequently used as type specimen for mammalian dissection because they are readily available and they possess the typical mammalian body plan. Most of what you learn on the rat is applicable to the anatomy of other mammals, such as humans. Rats can be obtained from pet shops, biological supply companies or pharmaceutical firms.
- *Anaesthesia:* Dissection is best performed on a freshly killed specimen. Killing experimental animals is one of the most unpleasant tasks in science, and it is imperative to do it as humanely as possible. The inhalation method is a widely used technique, which is the least painful and least stressful for the animal. Put the rat in a suitable size airtight container, fill it with carbon dioxide (CO_2) gas from a cylinder and wait until the animal becomes unconscious and suffocates. The complete procedure takes about 10 min. The use of ether is also acceptable but not recommended because it is a respiratory irritant that is considered stressful to animals and it poses an explosive hazard also.

P. Lőw et al., *Atlas of Animal Anatomy and Histology*, DOI 10.1007/978-3-319-25172-1_12

Skeleton All vertebrate skeletons can be divided into two major parts: the *axial skeleton* and the *appendicular skeleton*. The rat has a fairly generalised mammalian skeleton (Fig. 12.1).

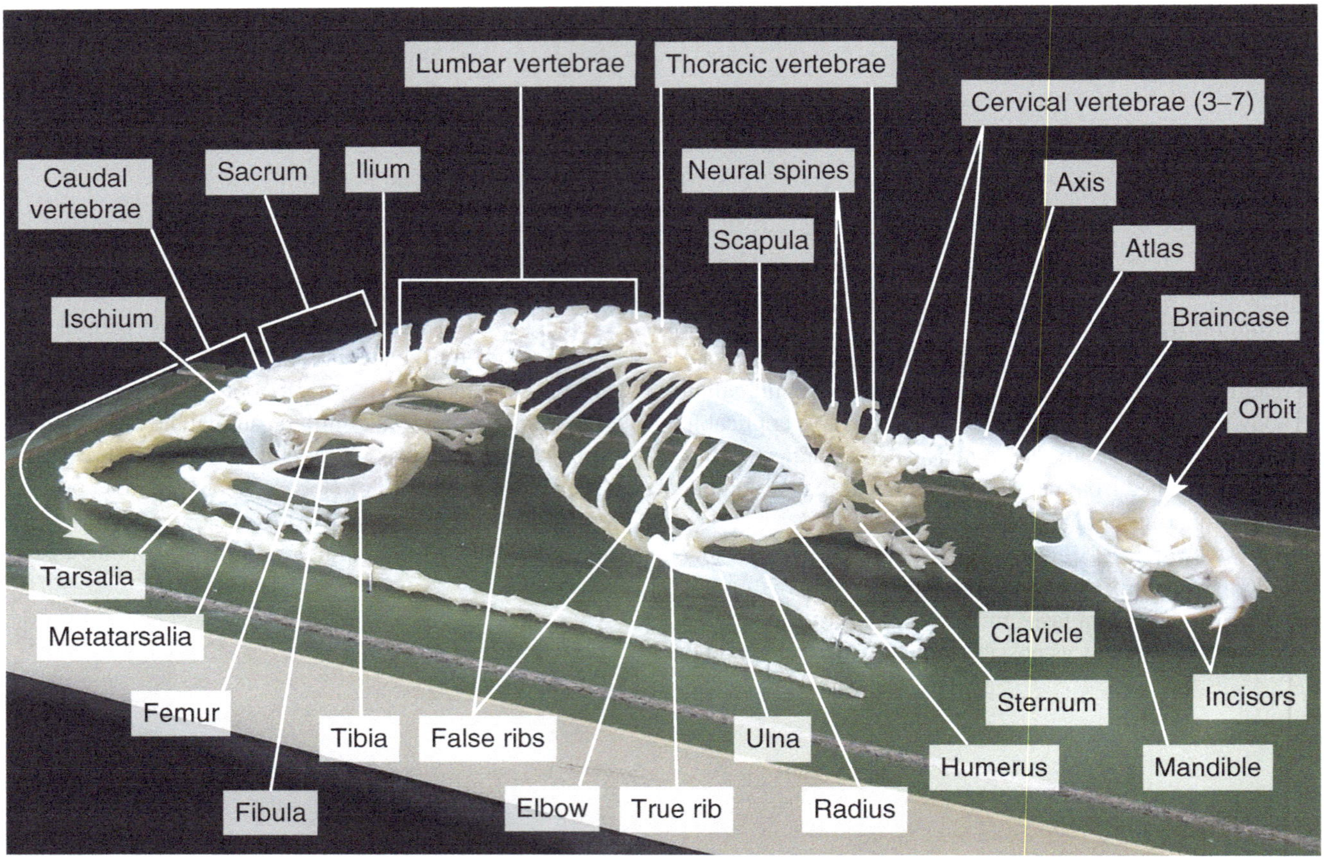

Fig. 12.1 Mounted skeleton of the rat

The axial skeleton consists of the skull, the vertebral column, the sternum and the ribs. The *skull* has two major regions. The *viscerocranium* (facial region) houses the nose and eyes and provides support for the jaw. The more posterior *neurocranium* (cranial region) of the skull houses the brain and ear. At the back of the skull is the foramen magnum through which the spinal cord passes into the vertebral column. On either side of the foramen magnum are the occipital condyles. The *secondary palate*, the bony shelf separating the mouth and the nasal cavity, is only well developed in mammals. Air travels to the trachea through the nostrils which open far back in the mouth. This allows the animal to chew continually without interrupting its breathing. Two dentary bones fuse anteriorly to form the lower jaw or *mandible* which provides support for two types of teeth: *incisors* for gnawing and *molars* for grinding (Fig. 12.57).

The *vertebral column* is clearly divisible into five regions. The *cervical region* (neck) consists of seven vertebrae. The first two neck vertebrae, the atlas and axis, are modified to permit rotation of the head. The *atlas* is a bony ring with wide, wing-like transverse processes. The base of these processes form two joints with the two occipital condyles of the skull, which permit up and down movement of the head. The *axis* has a process (dens axis) formed by the fusion of the bodies of the axis and atlas vertebrae. This process fits into the atlas ring, like a peg into a socket, which allows lateral rotation of the head.

The trunk vertebrae can be divided into two regions: the *thoracic region* (12–13 vertebrae) bears *ribs*, of which the anterior ones (true ribs) connect ventrally directly with the *sternum*, the middle ones (false ribs) are connected through a long cartilage to the sternum, while the three posterior ones (floating ribs) are not coupled to the sternum (Fig. 12.1). The *lumbar region* (seven vertebrae) is ribless. The four vertebrae of the *sacral region* are fused together as *sacrum* to provide additional support for the hindlimbs. *Caudal vertebrae* forming the tail are variable in number. The *intervertebral discs* are inserted between adjacent vertebrae as shock absorbers. The centra of the vertebrae have flat terminal faces, as is common in mammals; they are termed *acoelous*.

The head and neck musculature is attached to the long neural spines of the thoracic vertebrae (Fig. 12.1). These point backwards in the anterior region, unlike the short forward-pointing neural spines of the lumbar vertebrae. These two regions also differ in the size of their vertebral body. Those of the lumbar vertebrae are much larger and carry robust transverse processes. However, the lumbar and posterior thoracic vertebrae resemble one another in being able to move in the vertical plane only. This bounding movement is used to increase the length of the stride.

In a land mammal like the rat, the vertebral column, together with its associated muscles and ligaments, acts as a beam to support the total weight of the body off the ground, but between the legs, the body will tend to sag. This leads to compression forces on the dorsal side of the vertebral column and tension forces on the ventral side. The compression forces are resisted by the centra, while the tension forces are counteracted by ligaments and muscles running between the vertebrae and the ribs. The vertebral column is arched gently upwards, which helps effective weight bearing.

The appendicular skeleton consists of the *pectoral* (shoulder) *girdle*, the *pelvic* (hip) *girdle* and the *limbs*. The pectoral and pelvic girdles of the rat show that the legs are turned in under the body to support it.

In rats, the pectoral girdle consists of the *scapula* and *clavicle* (Fig. 12.1). In the pectoral girdle, the muscles are responsible for pulling the forelimb forwards. For that the scapula is wide. The forelimb consists of the long bones, the *humerus*, the *radius* and the *ulna*, and the bones of the wrist (carpalia) and hand (metacarpalia and digits).

Each side of the pelvic girdle is composed of three bones that have fused together to form the hip socket (acetabulum): the *ilium*, the *ischium* and the *pubis* (Fig. 12.1). The elongate ilium extends cranially, the ischium is caudal and the pubis extends ventrally from the ischium. The two pelvic girdle bones attach to the sacrum and meet ventrally along the pubic symphysis. The hindlimb is divided the same way as the forelimb; the long bones are called the *femur* (the small *patella* (kneecap)), the *tibia* and the *fibula* and the bones of the ankle (tarsalia) and foot (metatarsalia and digits) (Fig. 12.1).

First, study the external appearance of the rat. The rat's body consists of the following regions: the head, trunk, appendages and tail (Figs. 12.2, 12.3, 12.4 and 12.5).

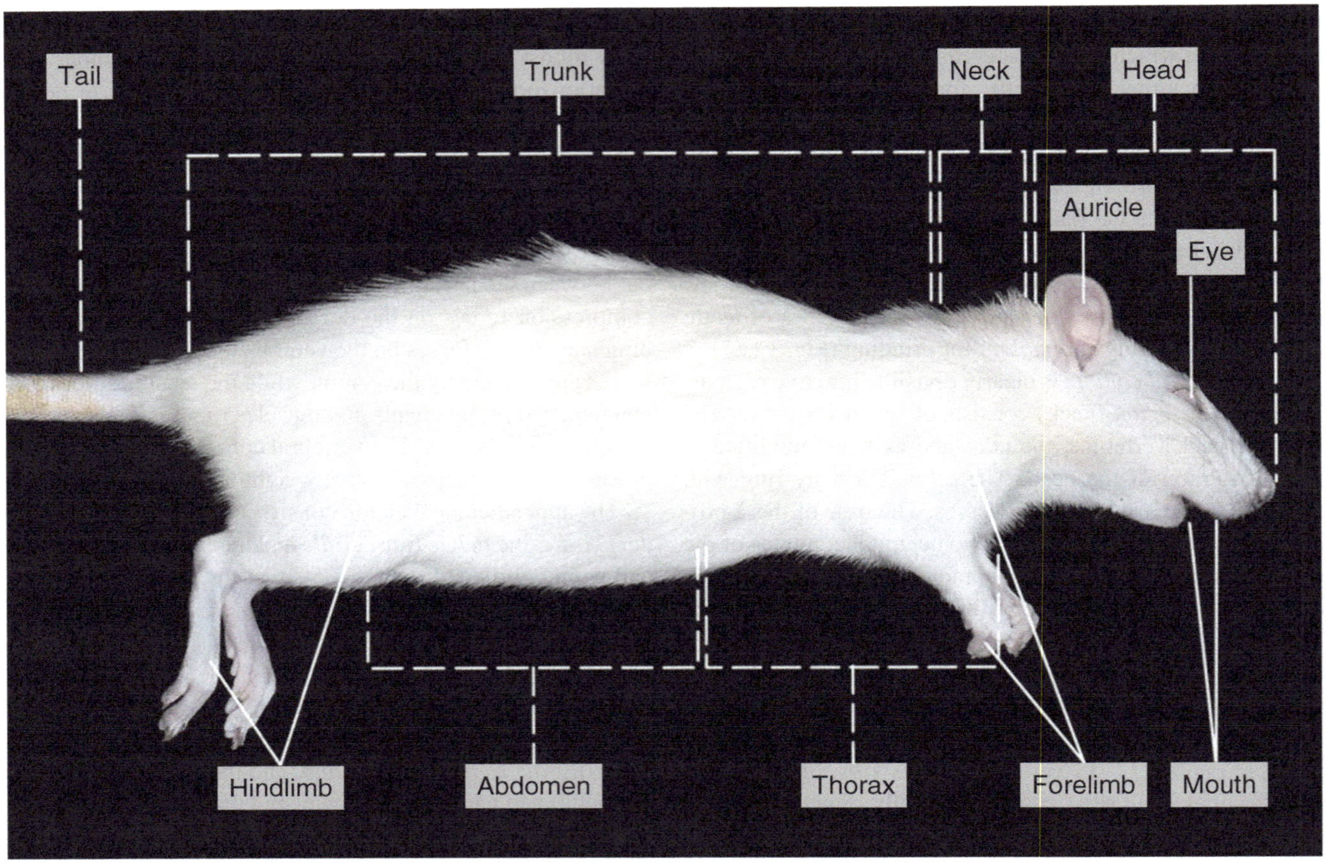

Fig. 12.2 Lateral view of a female rat

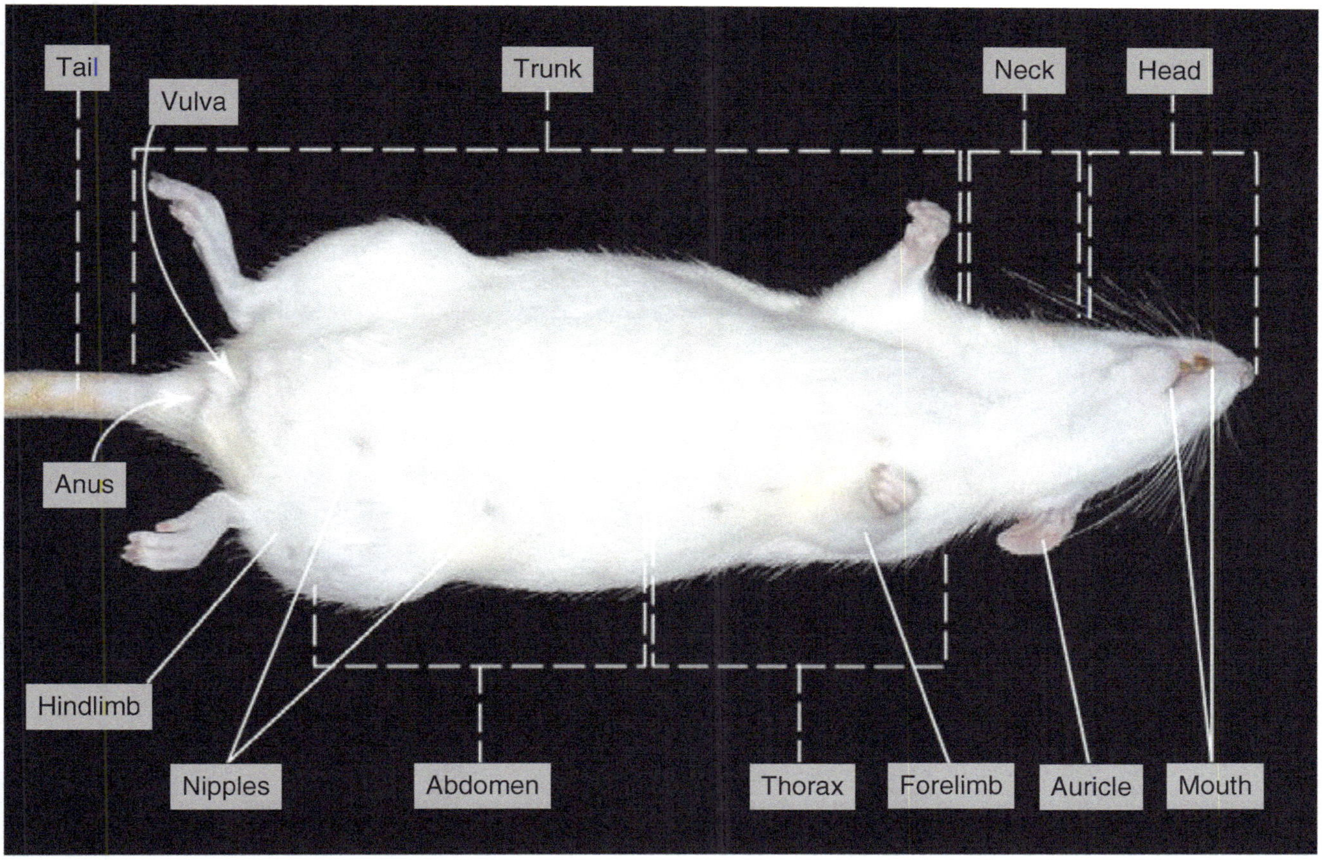

Fig. 12.3 Ventral view of a female rat

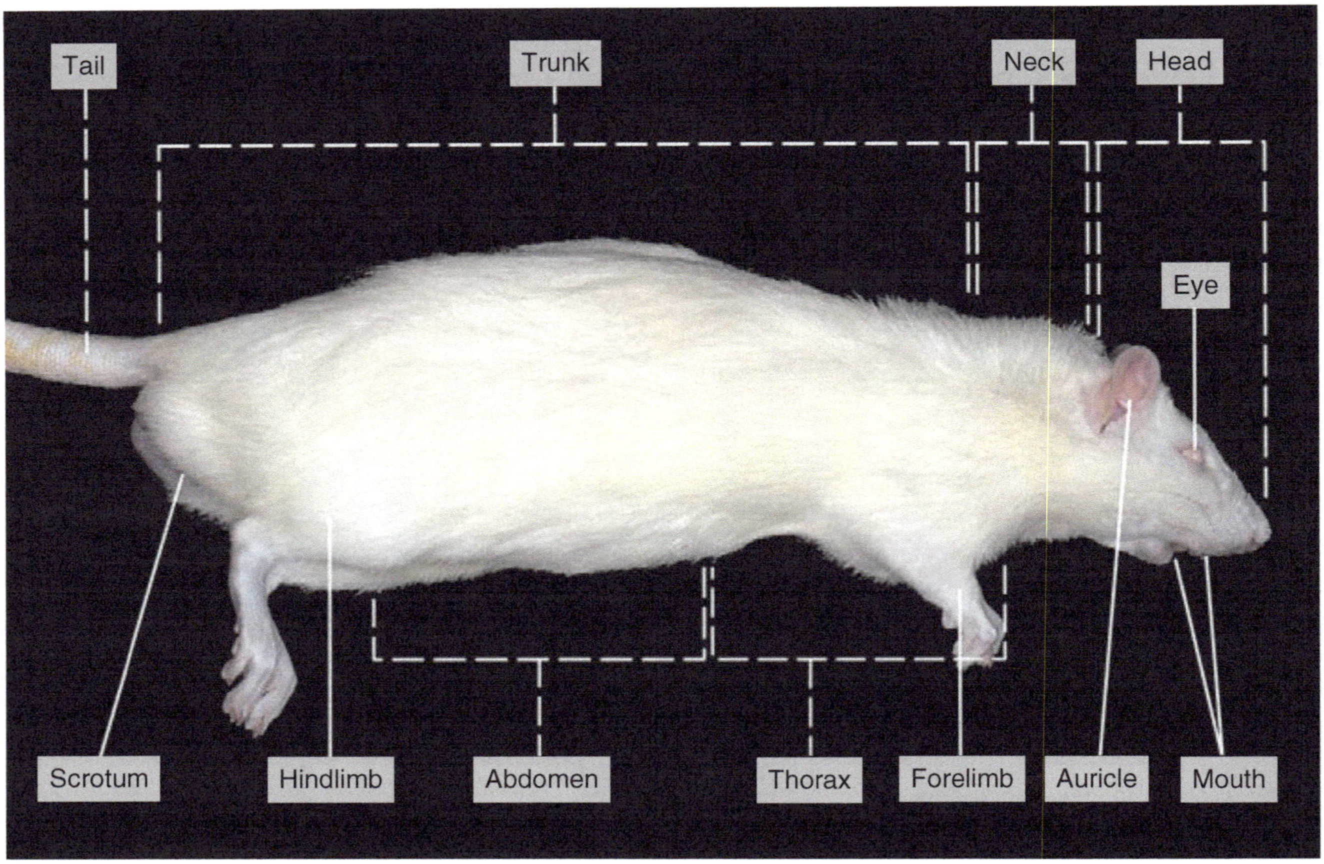

Fig. 12.4 Lateral view of a male rat

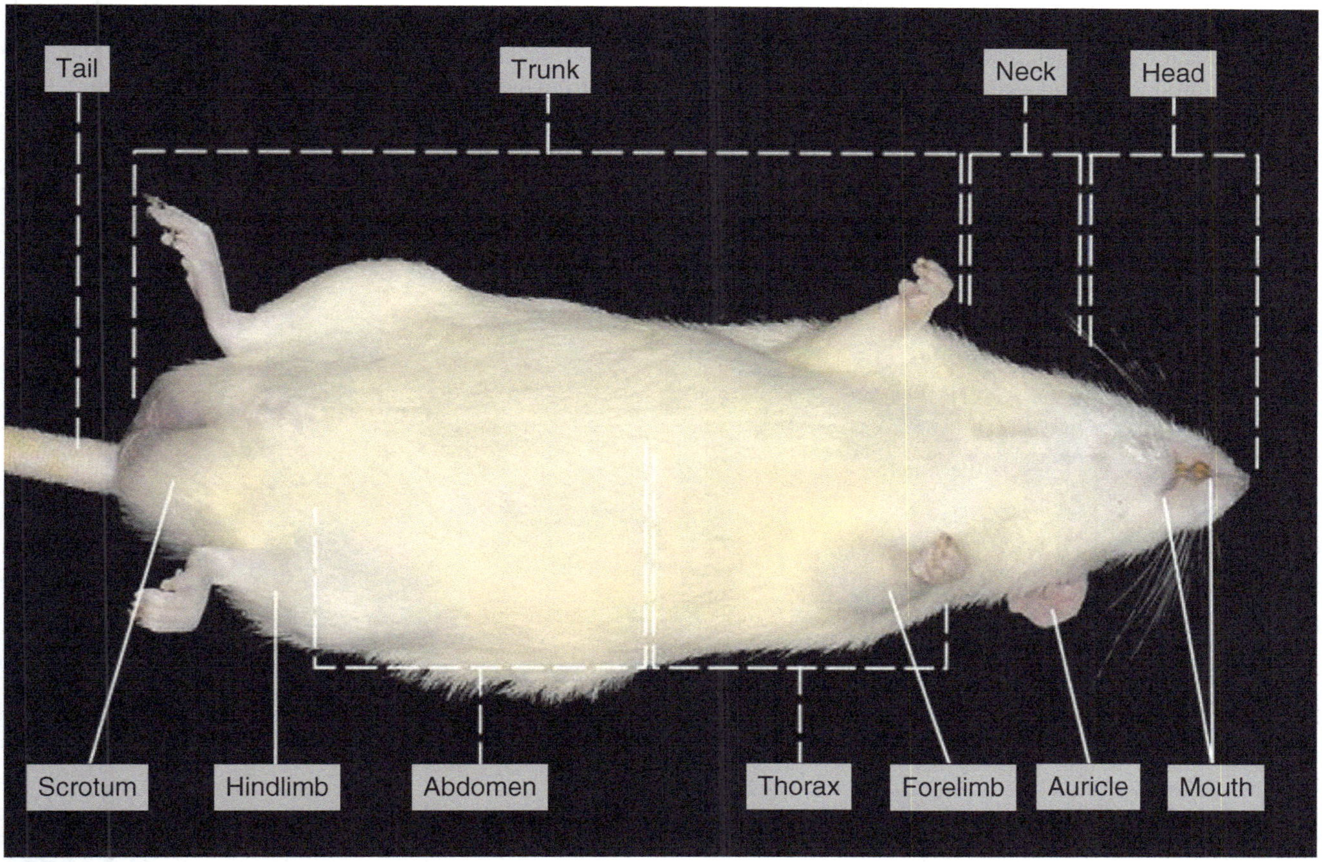

Fig. 12.5 Ventral view of a male rat

The external views of the female and male rat illustrate several general mammalian features: the covering of fur, the separate reproductive and excretory openings, the slender tail, the mammary glands (in the female) and the scrotal sac containing the testes (in the male). The sex of the rat can be determined by looking for the presence (or absence) of nipples (teats), which are only found on female rats (Fig. 12.6), and external testes in the scrotal sac, found on males (Fig. 12.7). The anogenital distance (AGD), the distance from the anus to the genitalia, is shorter in females than in males (Figs. 12.6 and 12.7).

The mammary glands are arranged in two lines on either side of the ventral surface extending from the neck to the lower abdomen (Figs. 12.3 and 12.6).

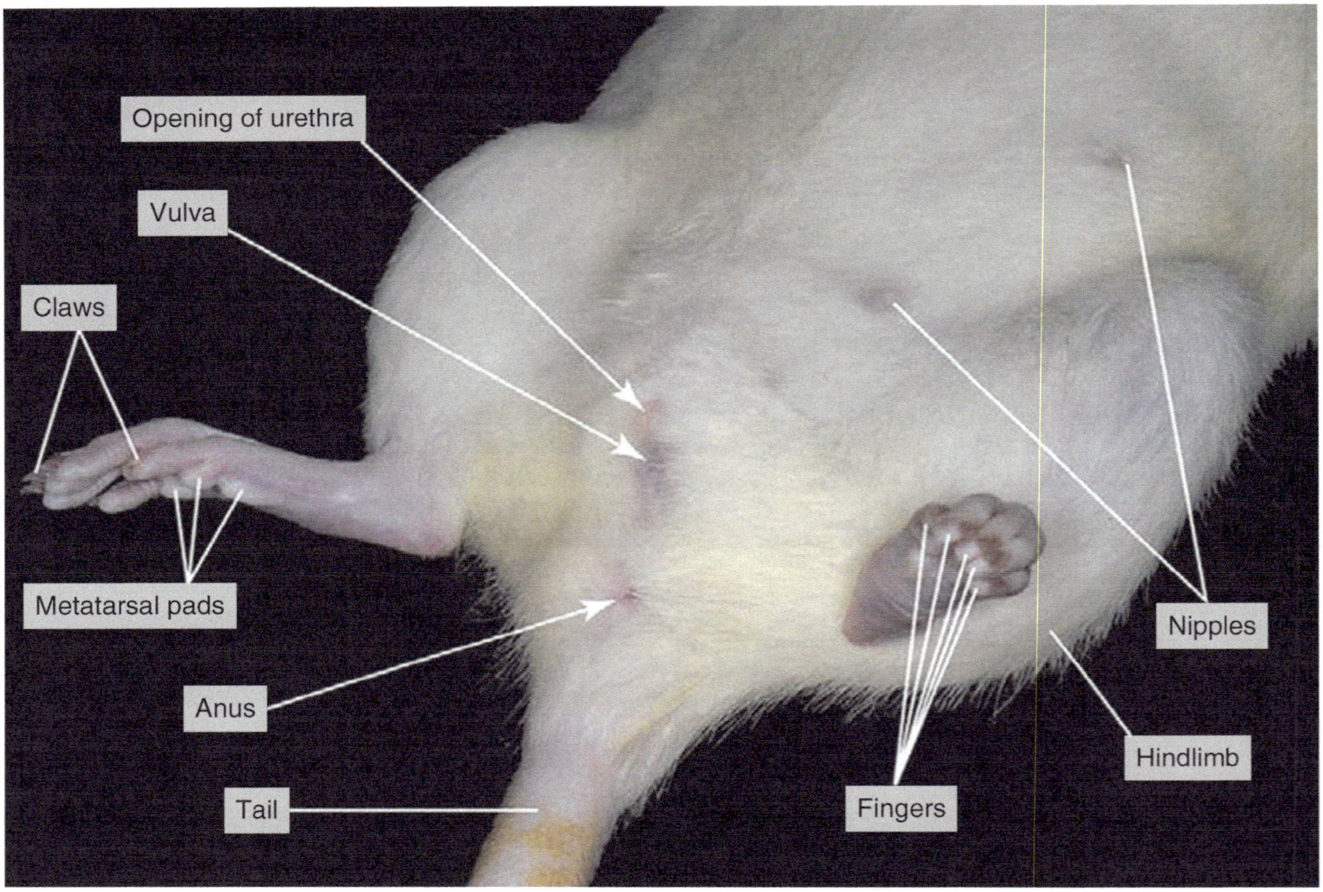

Fig. 12.6 Ventral view of the posterior part of a female rat

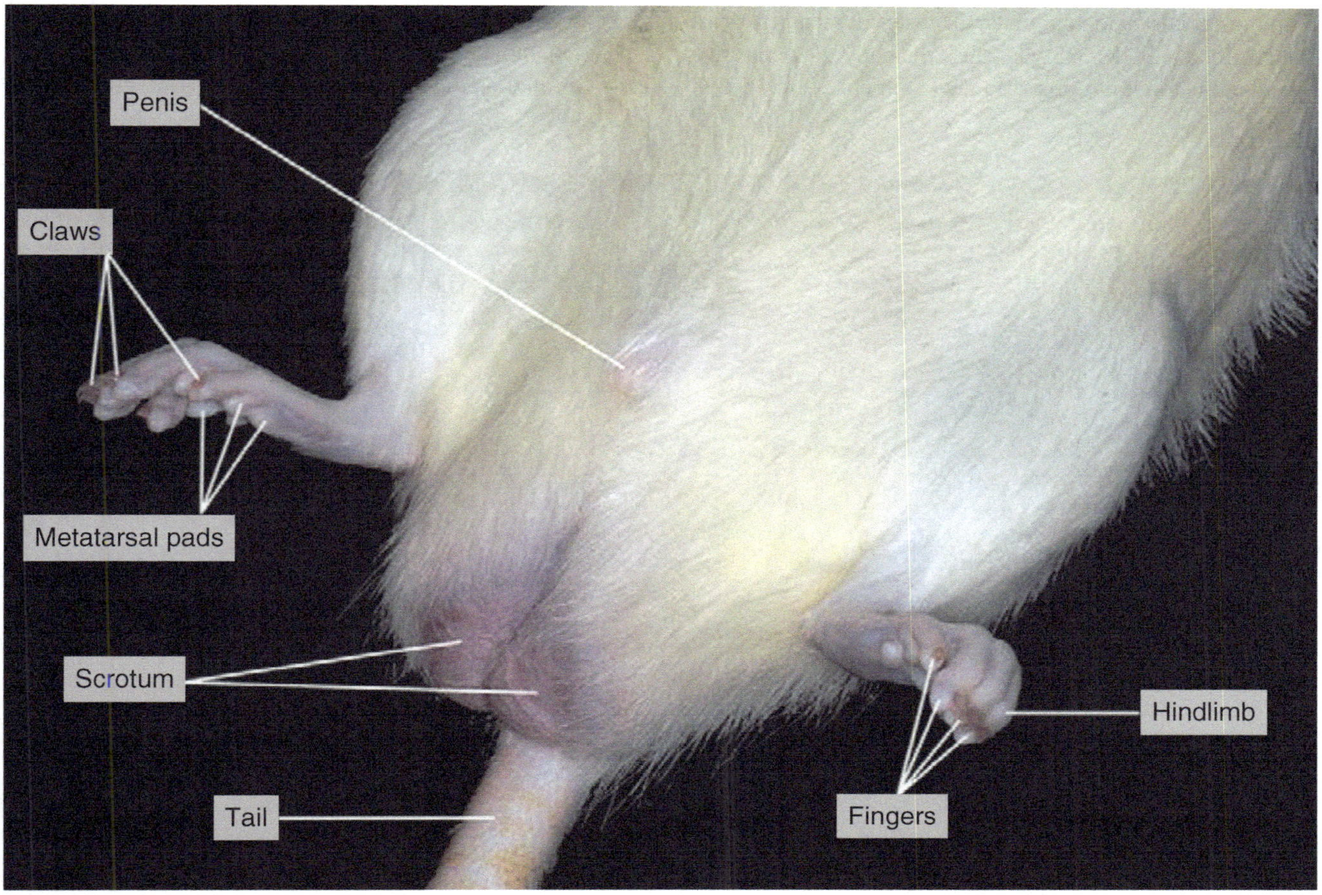

Fig. 12.7 Ventral view of the posterior part of a male rat

In most adult male mammals, the testes descend to a position outside the body cavity because the body temperature of a mammal is too high for sperm development.

Note the hairy coat that covers the rat; this is a characteristic of mammals in general. Observe the *whiskers* (sensory hairs) located on the rat's face, called vibrissae (Fig. 12.8).

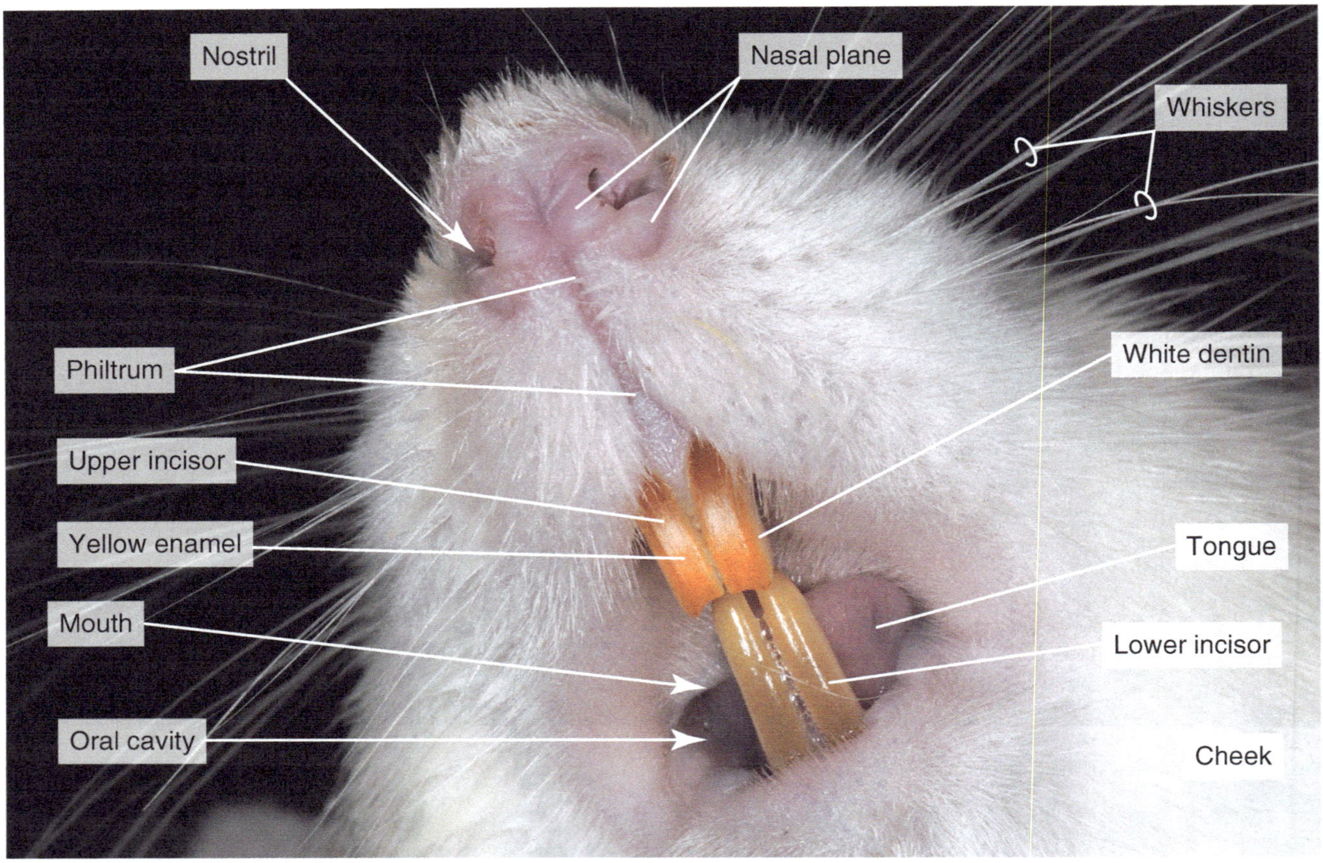

Fig. 12.8 Ventral view of the head of a rat

Rat whiskers are tactile organs very useful to a nocturnal animal; they are very sensitive to touch and may even detect changes in air pressure. While rats scurry for food, often in total darkness, these bristles enable them to maintain contact with walls and other solid objects that will guide their search. The *mouth* has a large cleft in the upper lip (philtrum), which exposes large front incisors (Fig. 12.8). Rats are gnawing mammals, and these incisors will continue to grow for as long as the rat lives. Note the *eyes* with the large *pupil*. The eyes are pink as they contain no melanin. The *nictitating membrane* is found at the inside corner of the eye. This membrane can be drawn across the eye for protection. This membrane allows the rat to blink without closing its eyes. The *eyelids* are similar to those found in humans. The ears are composed of the external part, called the *auricle* (pinna), and the *auditory meatus*, the ear canal. Locate the *anus*, which is ventral to the base of the tail (Fig. 12.6).

The mammalian *skin* differs from that of birds and reptiles by the presence of *hair* and sweat glands and mammary (modified sweat) glands unique to mammals and by the absence of epidermal scales (except the tail, Fig. 12.9).

Examine the *tail*; it has three short hairs under every little *scale*. However, some other rodents, like gerbils, have long hair on their tails and no scales.

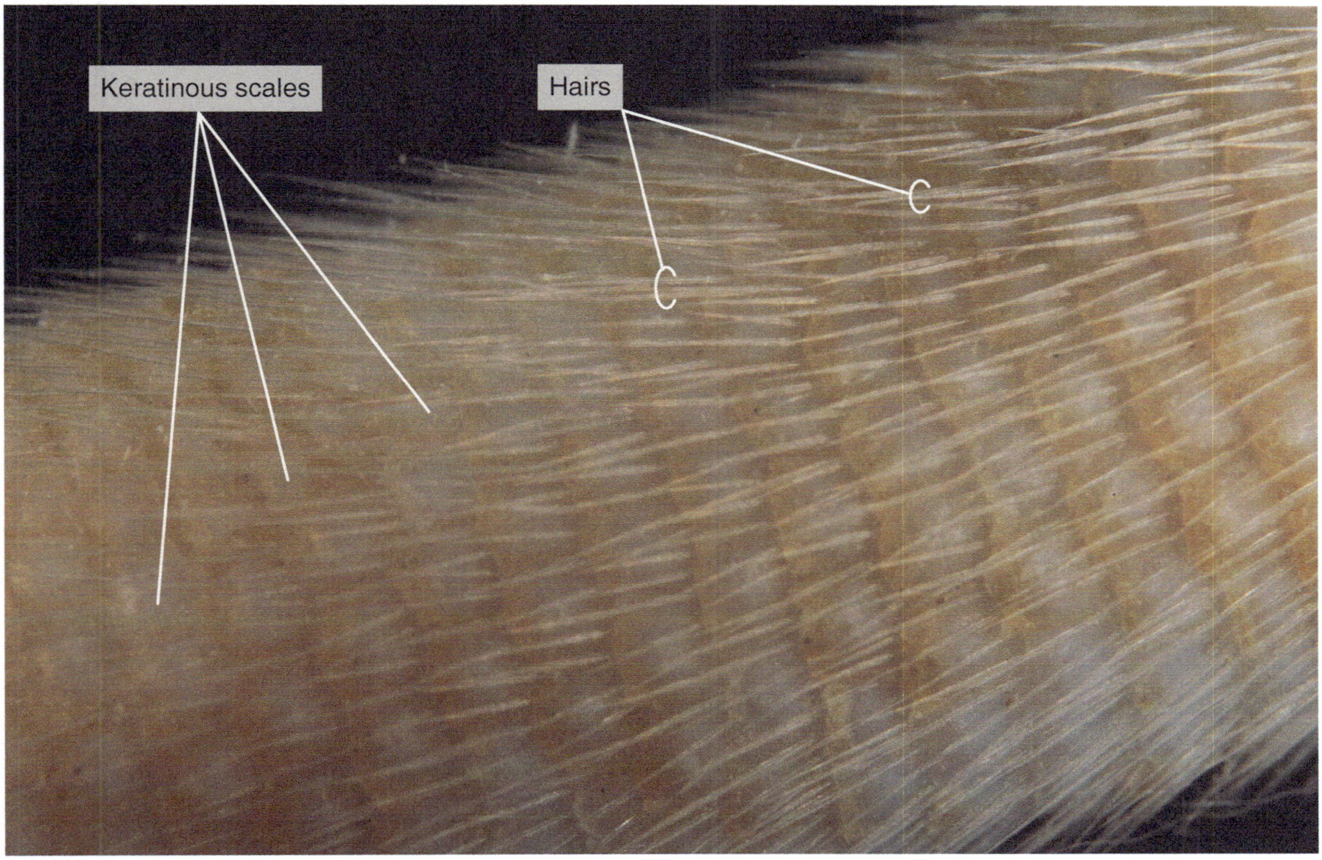

Fig. 12.9 Enlarged view of the tail of the rat with scales and hairs

Although *skin* in different parts of the body is composed fundamentally of similar structure, there are many local variations in parameters such as thickness, mechanical strength, softness, flexibility, degree of keratinisation (cornification), sizes and numbers of hairs, frequency and types of glands, pigmentation, vascularisation and innervation. Two major classes of skin are thin, hairy (hirsute) skin (Fig. 12.10), which covers the greater part of the body, and thick, hairless (glabrous) skin (Fig. 12.12), which forms the surfaces of the soles of the feet. The epidermal appendages (pilosebaceous units, sweat glands and claws) are formed developmentally by ingrowths of the general epidermis; thus, the latter is referred to as the interfollicular epidermis. This one is thin, keratinised, stratified squamous epithelium, which lies on a dermis formed by dense connective tissue containing bundled collagen fibres. The pilosebaceous units form groups (Fig. 12.10, right).

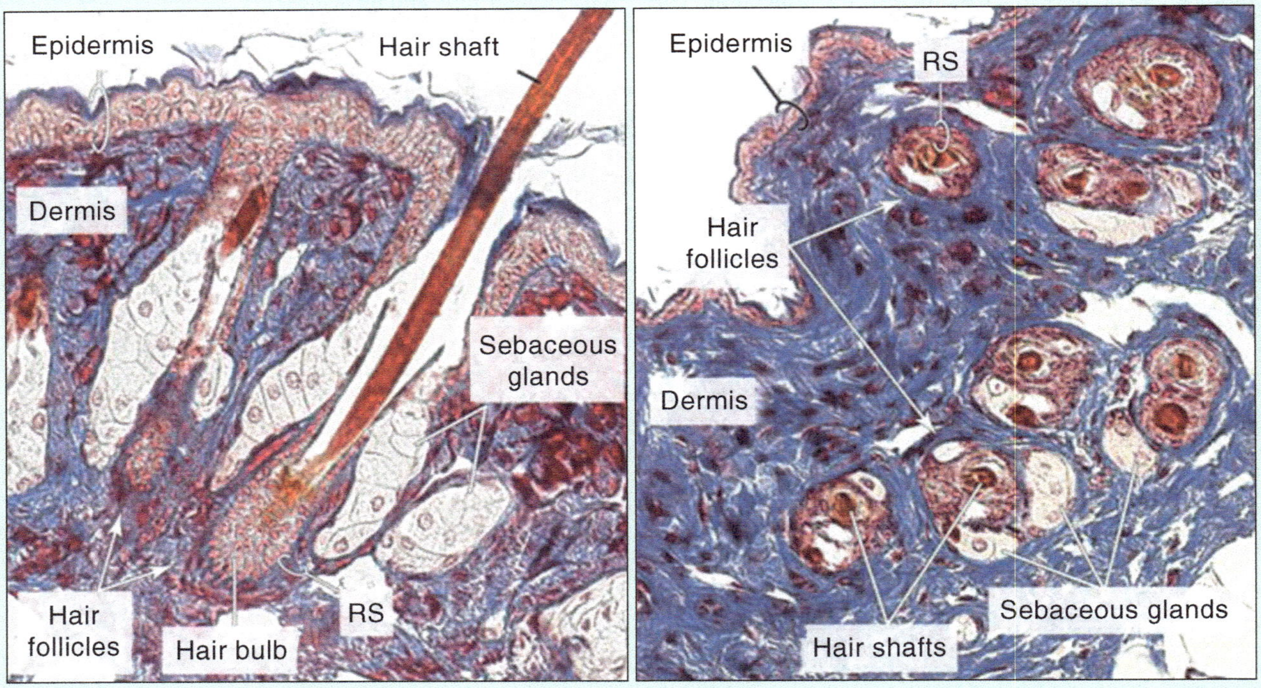

Fig. 12.10 Histological sections of hairy skin with pilosebaceous units (Azan) (*left*, longitudinal section; *right*, cross section). *RS* root sheath

Every unit consists of a *hair* (shaft and bulb), follicle (which is formed by root sheath), sebaceous gland and arrector pili muscle (Fig. 12.11, bottom). The lowest part of hair follicles is the hair bulb, which contains cells producing the hair shaft. The hair follicle is ensheathed by layers of the root sheath. Sebaceous glands open into the follicle – their cells are light, because the microtechnical procedure eliminates lipids from the secretory product, sebum.

The *hair follicle* is surrounded by a strong dermal capsule made of connective tissue (Fig. 12.11, top left). There are two sheaths composed of epithelial cells that outline the follicle: the outer root sheath (ORS) and the inner root sheath (IRS). The latter is continuous with the germinal matrix, which produces the hair shaft. Melanocytes are not seen among the matrix cells, indicating that the rat was albino. *Sebaceous glands* are formed by sebocytes. They release a greasy secretory product, sebum, into the canal of follicle by a holocrine mechanism (Fig. 12.11, top left). Here the entire gland cells detach and disintegrate, and this cell debris forms the sebum.

The *whiskers* (vibrissae) are special hairs for touch perception (Fig. 12.11, bottom). The rodent's primary somatosensory organ consists of arrays of *sinus hairs* on each side of the snout. The sensory function is provided by a cavernous sinus system located in the mesenchyme sheet between the outer root sheet and the dermal capsule of the follicle. The vessels filled with blood push the nerve endings to the follicle and amplify the tiniest movement of the shaft. The organ is innervated redundantly.

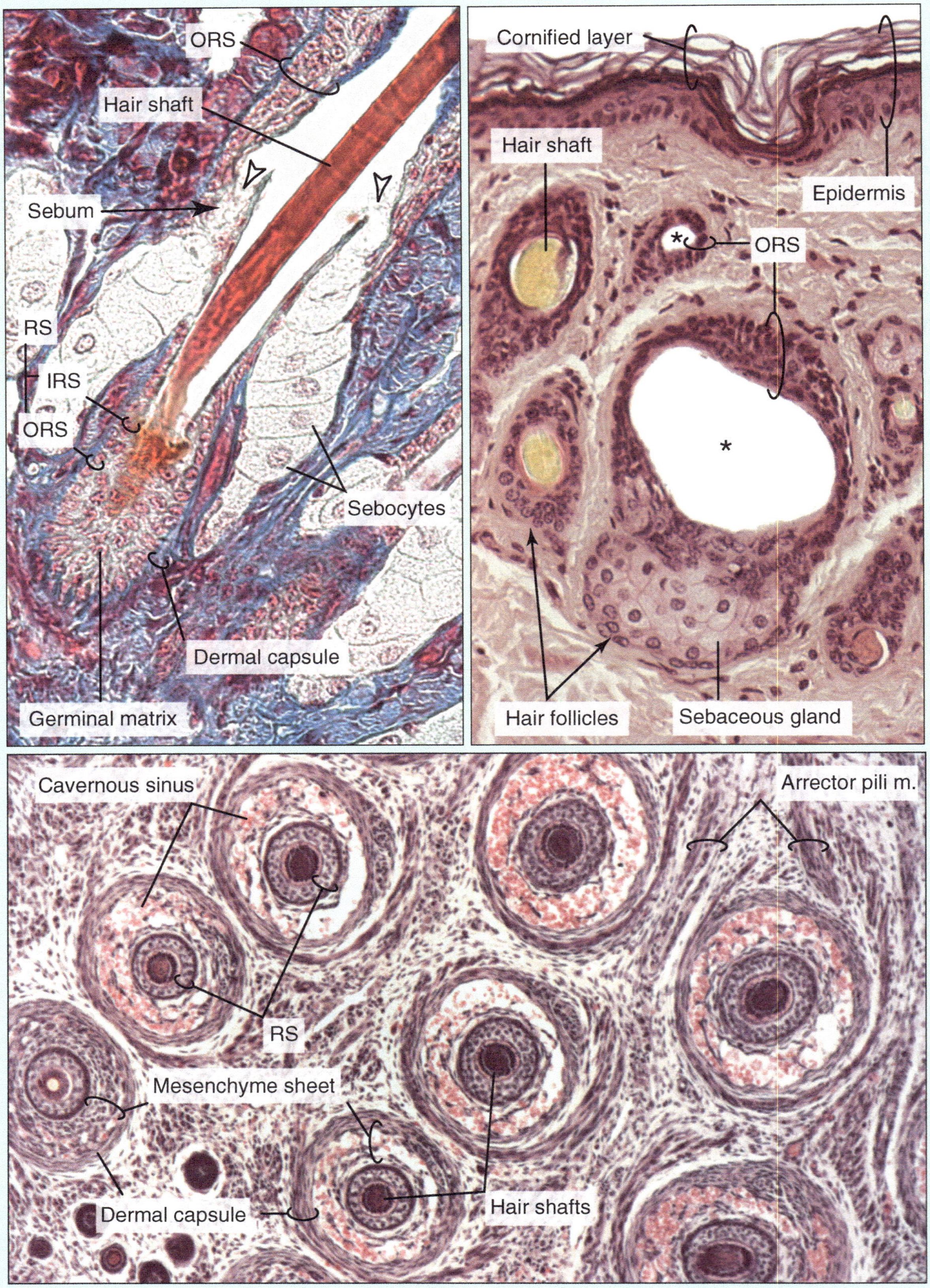

Fig. 12.11 Histological sections of hair follicles of pilosebaceous units and whiskers (*top left* Azan, *top right* and *bottom* HE), *white arrows* (*top left*) cell debris, *asterisks* (*top right*) hair shafts fallen out during the sectioning, *IRS* inner root sheath, *ORS* outer root sheath, *RS* root sheath

The thick epidermis of the hairless (glabrous) skin is supported by several dermal papillae which build up a dermal connective tissue. Layers forming the keratinised squamous epithelium are distinguishable (Fig. 12.12). Its layers from deep to superficial are as follows: basal layer, prickle or spiny cell layer, granular layer, clear layer and cornified layer. The *basal layer* (BL) is composed of columnar, mitotically active cells; they organise columnar proliferative units. Their daughter cells gradually move upwards. The *prickle cell layer* (PCL) consists of layers of closely packed keratinocytes that are connected with each other by numerous cell surface projections, which are linked by desmosomes. This characteristic spiny appearance is an artefact generated by specimen preparation for light microscopy (Fig. 12.12, right). The cytoplasms of keratinocytes cumulate keratin filament bundles. In the next layer, the *granular layer* (GL) the main organelles degenerate, and keratohyalin granules appear as dense spots in the cytoplasm. The *clear layer* (CL) stains more strongly than the cornified layer with acidic dyes (eosin); it is the most refractive optically. The *cornified layer* (CoL) is the uppermost layer. It consists of closely packed, flattened squama. This layer is the final product of epidermal differentiation (cornification). The *dermis* forms several dermal papillae (DP) protruding into the epidermis, enlarging the connecting surface between the two layers.

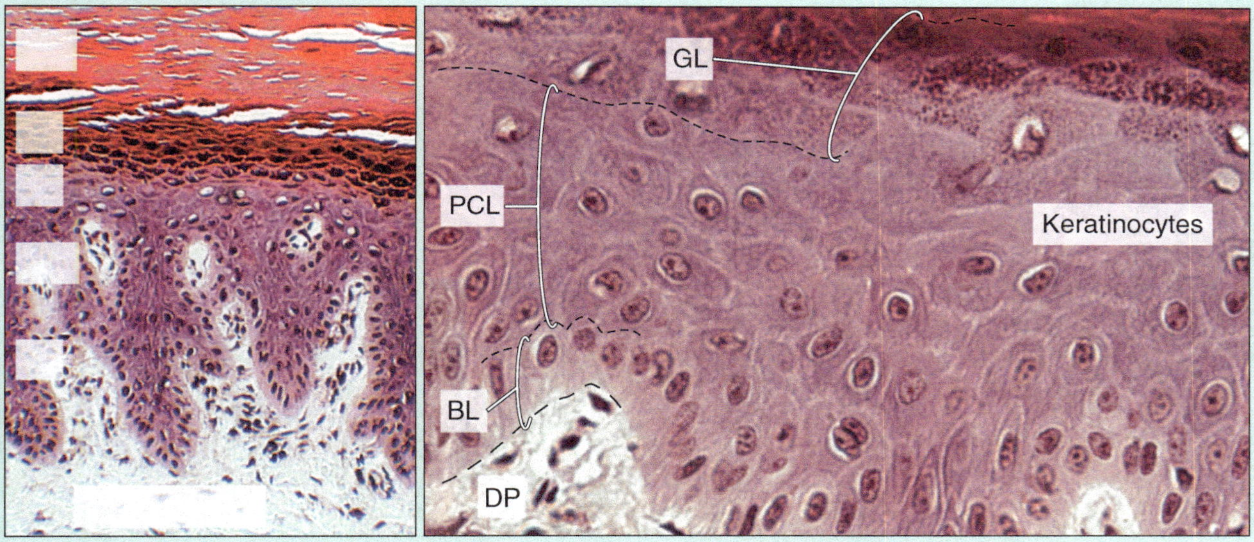

Fig. 12.12 The epidermis of the thick, hairless skin (HE). *BL* basal layer, *CL* clear layer, *CoL* cornified layer, *DP* dermal papilla, *GL* granular layer, *PCL* prickle cell layer

Lay the rat ventral side uppermost on a dissecting tray and attach it in a spread-eagle position by means of strong dissecting pins through the fore- and hindlimbs. Place each pin at an angle to the strain put on it. Lift the skin in the mid-ventral line and make a small incision with a pair of scissors. Slit the skin at the mid-ventral line. Follow the cut patterns illustrated on Figs. 12.13 and 12.14, avoiding the genital area.

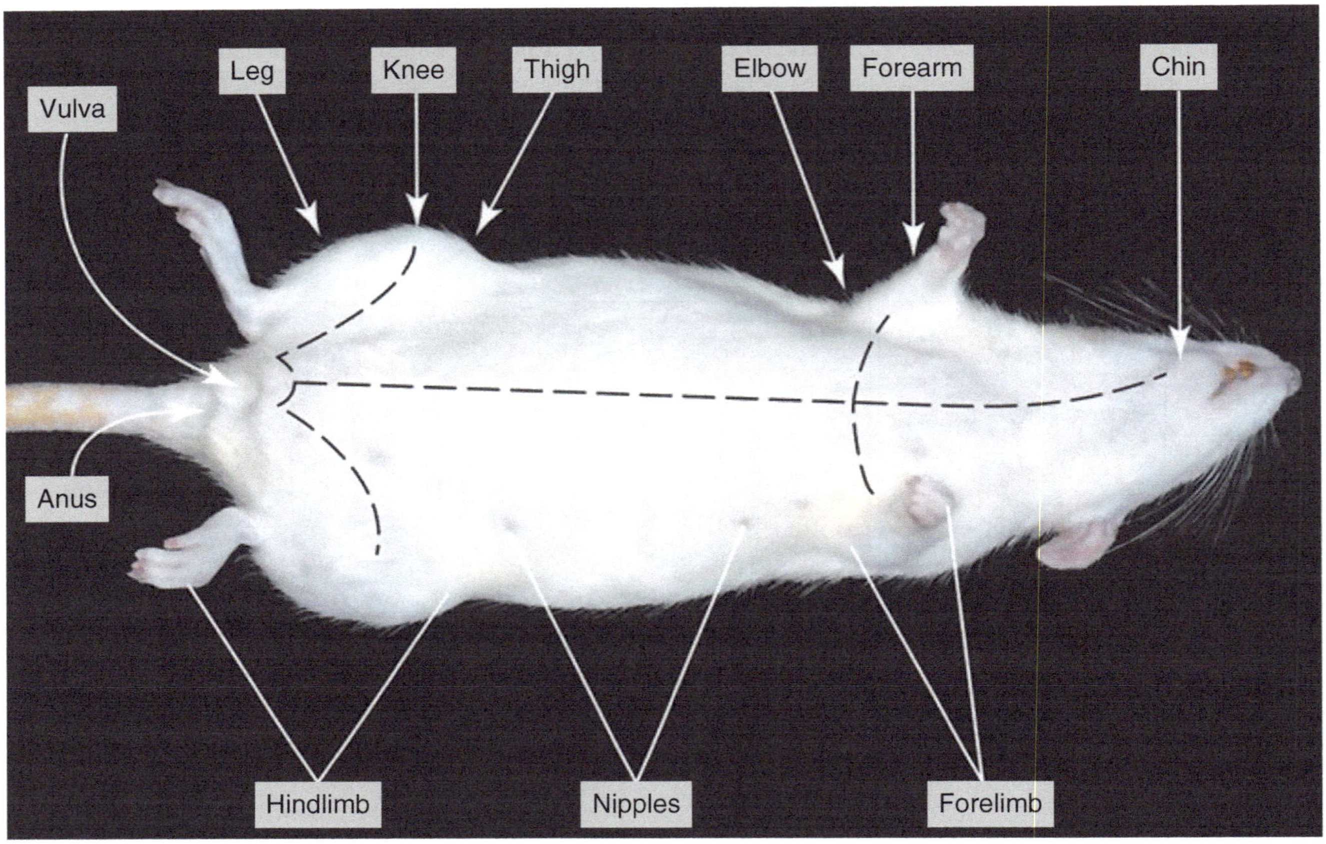

Fig. 12.13 Ventral view of a female rat with a cut pattern for the opening the skin

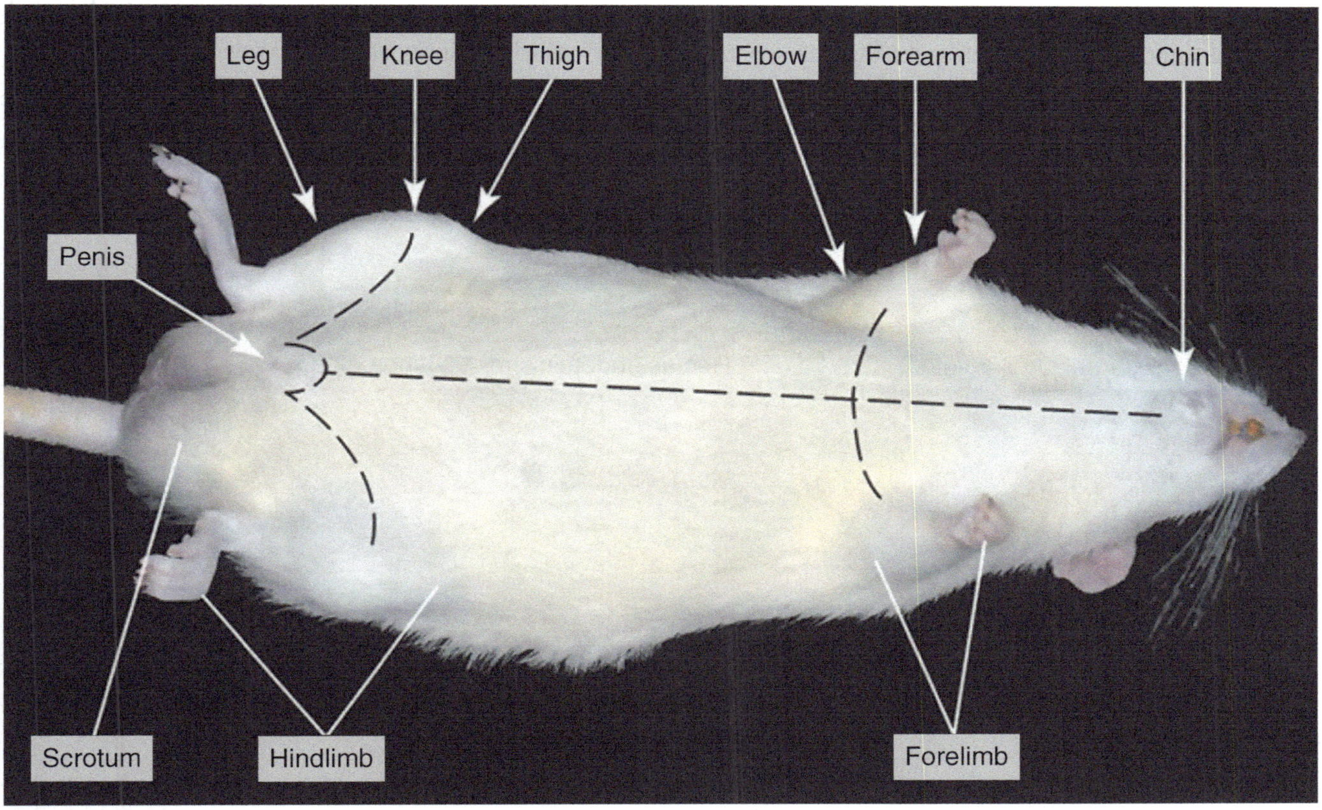

Fig. 12.14 Ventral view of a male rat with a cut pattern for the opening the skin

As you proceed, keep the scissors as horizontal as possible to avoid cutting the body wall under the skin. Continue the initial cut anterior on the abdomen to the level of the lower lip (Fig. 12.15). As you go, reach between the skin and the muscles with your fingers or a blunt probe and separate the skin from the muscles. Be careful not to tear the nerves and muscles in the axillary region. At some points, you may need to use a scalpel to cut muscles or ligaments running to the skin. In the majority of the following figures demonstrating the phases of the dissection, the *anterior* (head) region of the rat is towards the *upper* or *right* margins. In those exceptional cases where the animal is in a different position and the anterior (head) region faces to the left, this is indicated in the legend of the figure (Fig. 12.15).

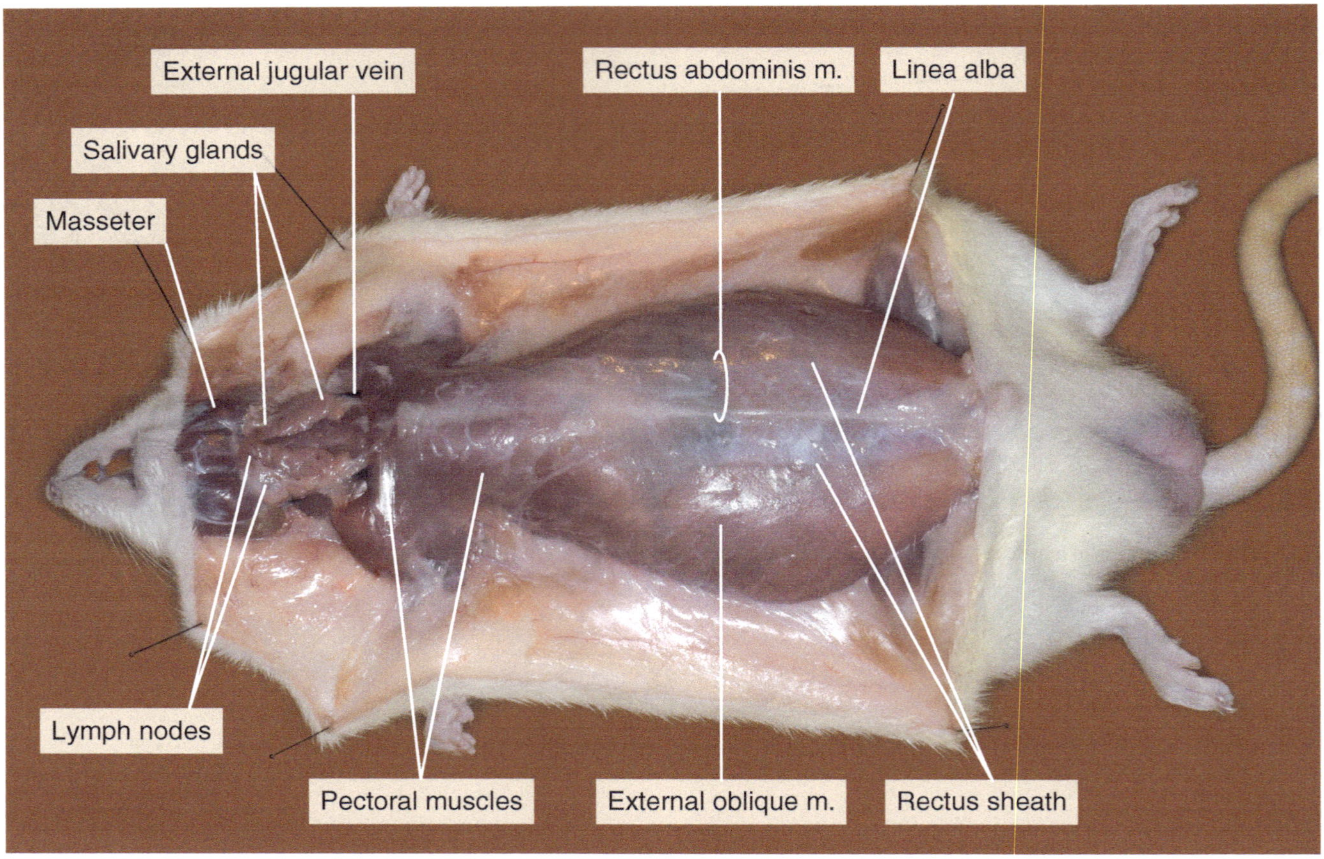

Fig. 12.15 Partly skinned rat with exposed muscular body wall (anterior end is to the left)

There is always much capillary bleeding when dissecting a freshly killed rat. The blood should be kept mopped up with a cloth so that structures retain their own shade of colour and do not become uniformly reddened and thus more difficult to identify. Peel the skin back to expose first the neck region (Fig. 12.16).

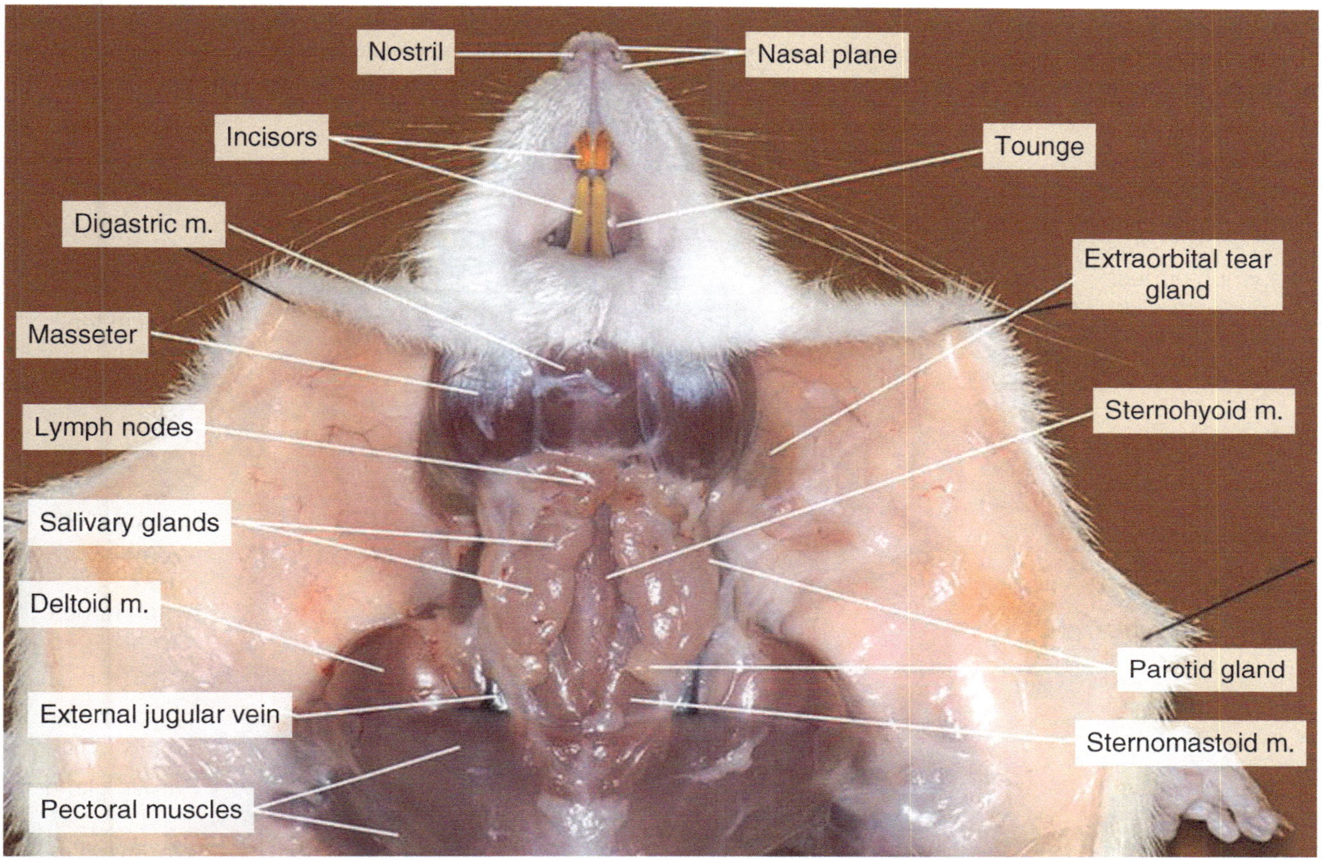

Fig. 12.16 The exposed neck region of the rat

Locate the *salivary glands*, which are on the sides of the neck, between muscles and the skin. They are soft organs that secrete saliva, which contains amylase (an enzyme that breaks down starch). There are three salivary glands: the sublingual, submandibular and parotid glands (Fig. 12.17).

The *parotid gland* is the easiest to find; it lies just beneath the ear and extends to the neck (Fig. 12.17). The parotid gland is the major salivary gland. It empties saliva through the parotid duct into the oral cavity. Try to find the others also. The *submandibular gland* is a salivary gland that secretes thick mucus. The *sublingual gland* is a small salivary gland that empties into the oral cavity behind the lower incisors. Find the *lymph nodes*, which lie anterior to the salivary glands. Lymph nodes are darker and circular and are pressed against the jaw muscles. They are part of the immune system.

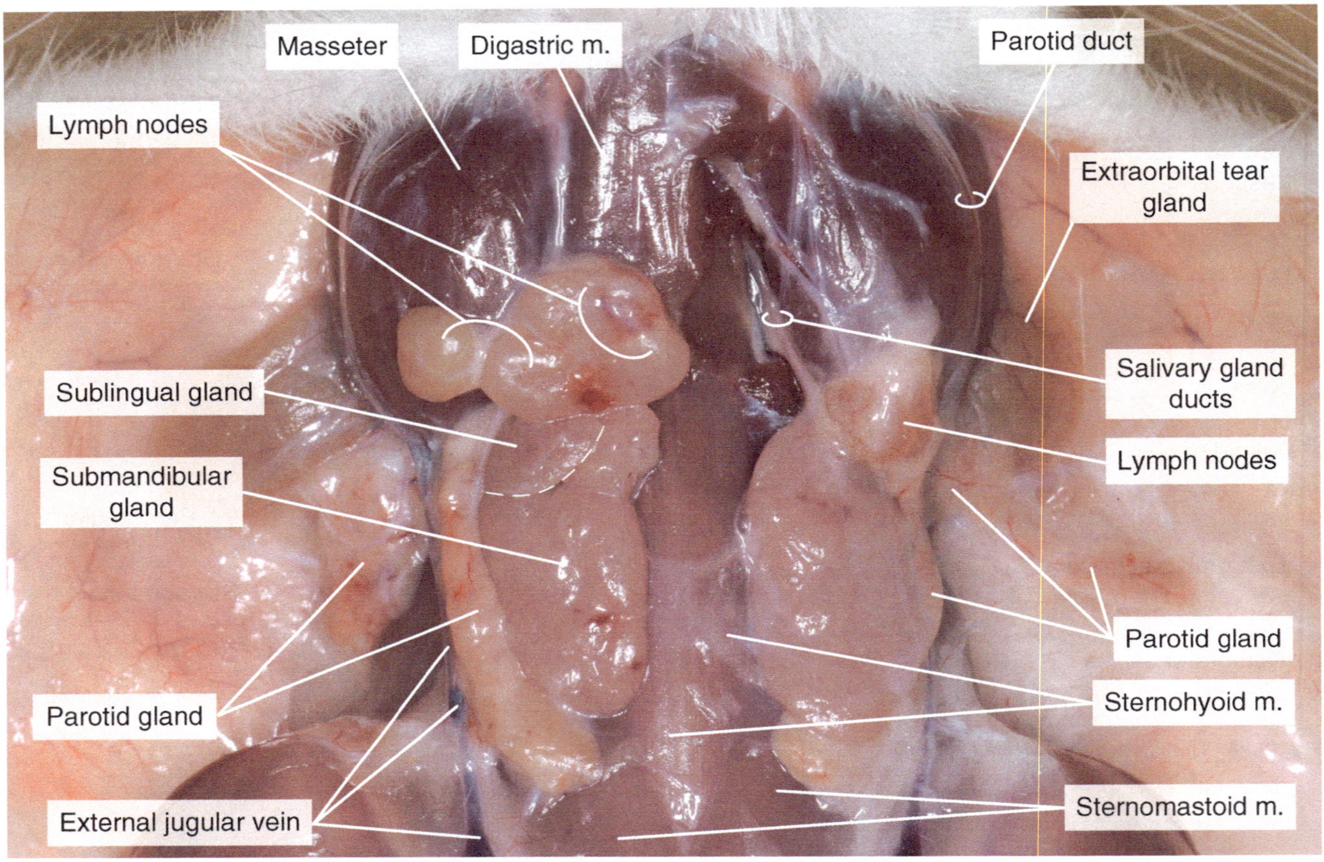

Fig. 12.17 The three salivary glands: the sublingual, submandibular and parotid glands and lymph nodes on the neck. *Dashed line* indicates the boundary between the sublingual and the submandibular glands. Note muscles are partly removed to reveal the medial salivary gland ducts

Salivary glands are compound, tubuloacinar exocrine glands whose ducts open into the oral cavity. The lingual glands are embedded into the connective tissue of the tongue and surrounded by striated muscle fibres (Fig. 12.18, left). They have serous and mucous merocrine acini and ducts. The first type belongs to the circumvallate papilla (glands of von Ebner, Fig. 12.58); the latter ones open onto the surface of the tongue. The cells of the serous acini show darker (basophilic) staining than the mucous cells – the latter appear almost empty, because the routine microscopic procedure dissolves the vast majority of secreted materials.

The *submandibular gland* is almost entirely serous (Fig. 12.18, right). The acinar cells have round nuclei; their cytoplasm is filled with fine PAS-positive secretory granules. The submandibular gland has a special, convoluted duct connecting the acini to the collecting duct system. It is lined with epithelium secreting serous material containing eosinophilic and PAS-positive granules (Fig. 12.18, bottom left, right side and bottom right). Granular duct (GD) shows sexual dimorphism, because its cells are testosterone dependent. The ducts are longer, and the lining epithelium is taller in males than in females.

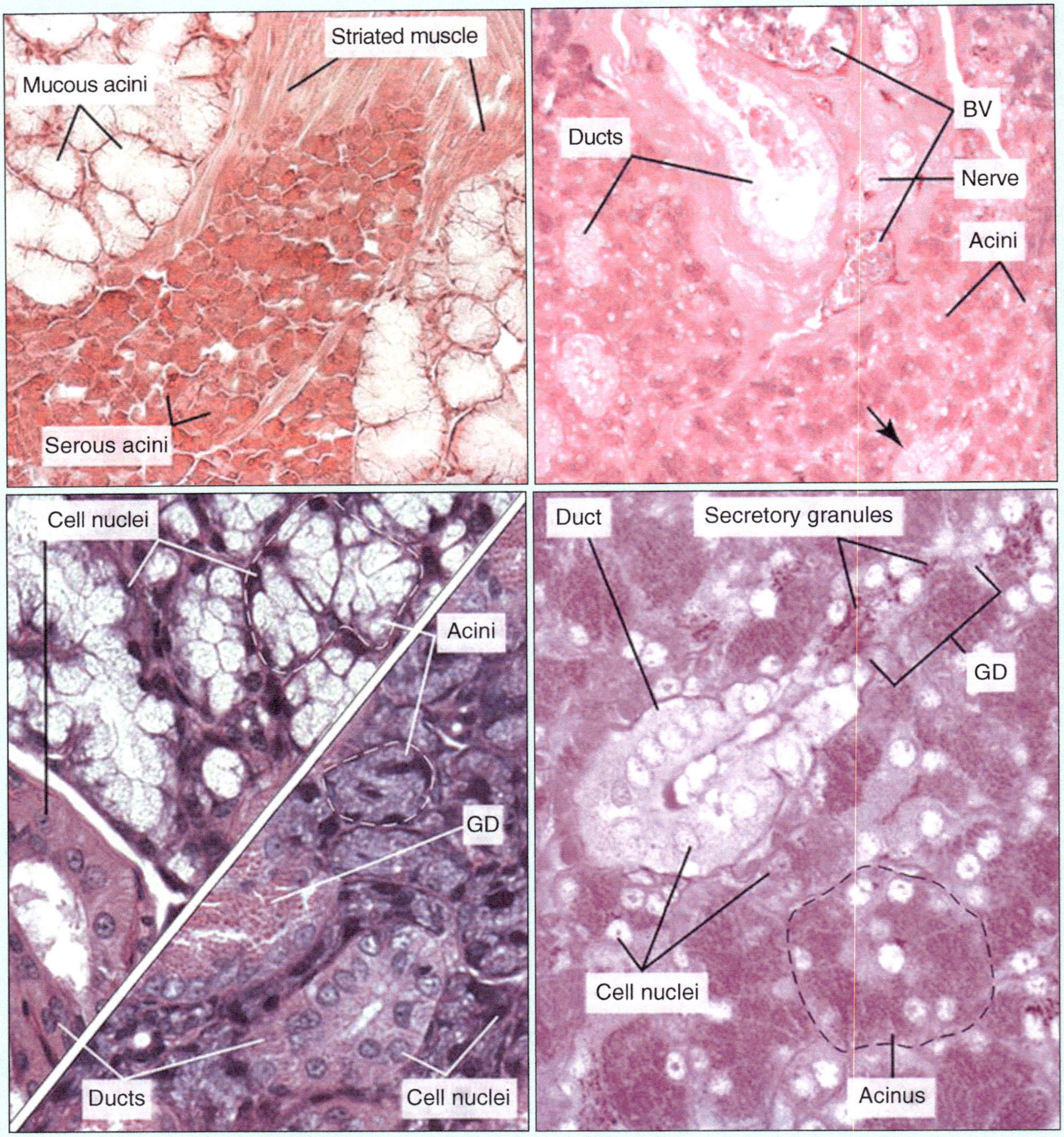

Fig. 12.18 Salivary glands: lingual glands (*top left*, HE) and submandibular gland (*top* and *bottom right*, PAS). The *bottom left* panel is halved for the comparison of the histological composition of the lingual gland (*top left*, HE) and submandibular gland (*bottom right*, HE). *Arrow* beginning of a duct, *BV* blood vessels, *GD* granular duct

Fold out the salivary glands and lymph nodes and tease away the muscles of the neck according to the dashed lines on Fig. 12.19 to reveal the trachea (Figs. 12.20 and 12.21).

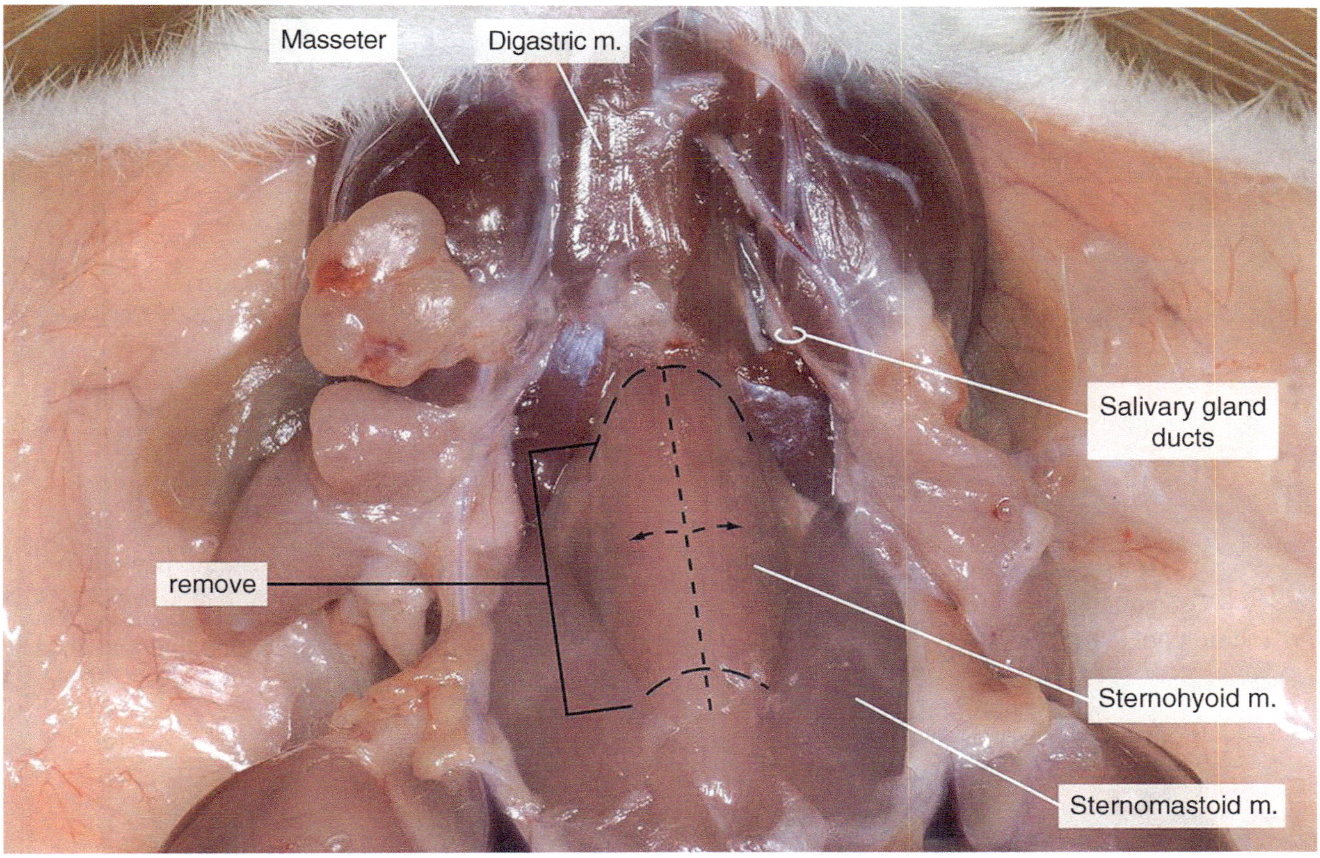

Fig. 12.19 The muscles of the neck, *dashed lines* show the way of removal of the sternohyoid muscles

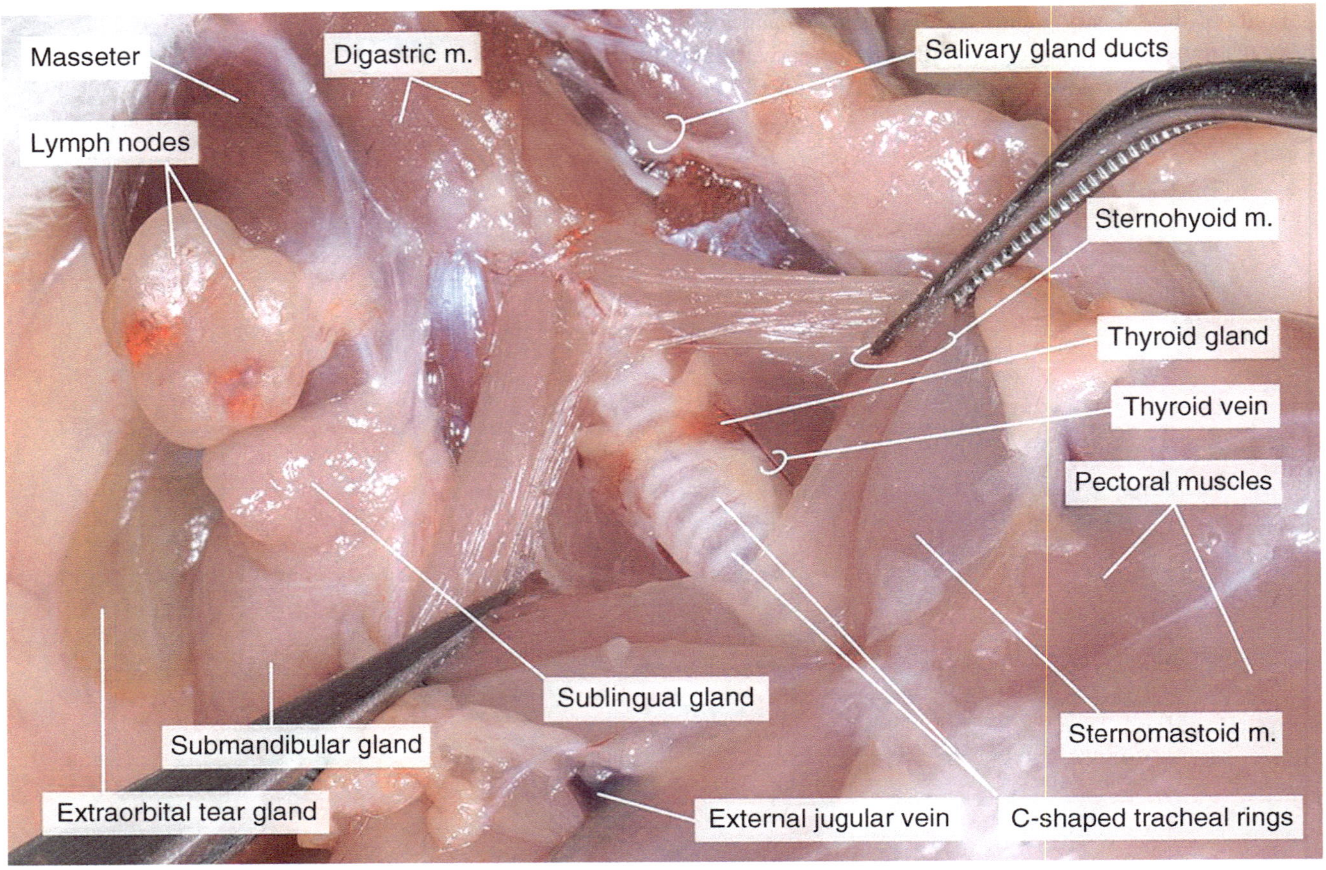

Fig. 12.20 Division of the sternohyoid muscles

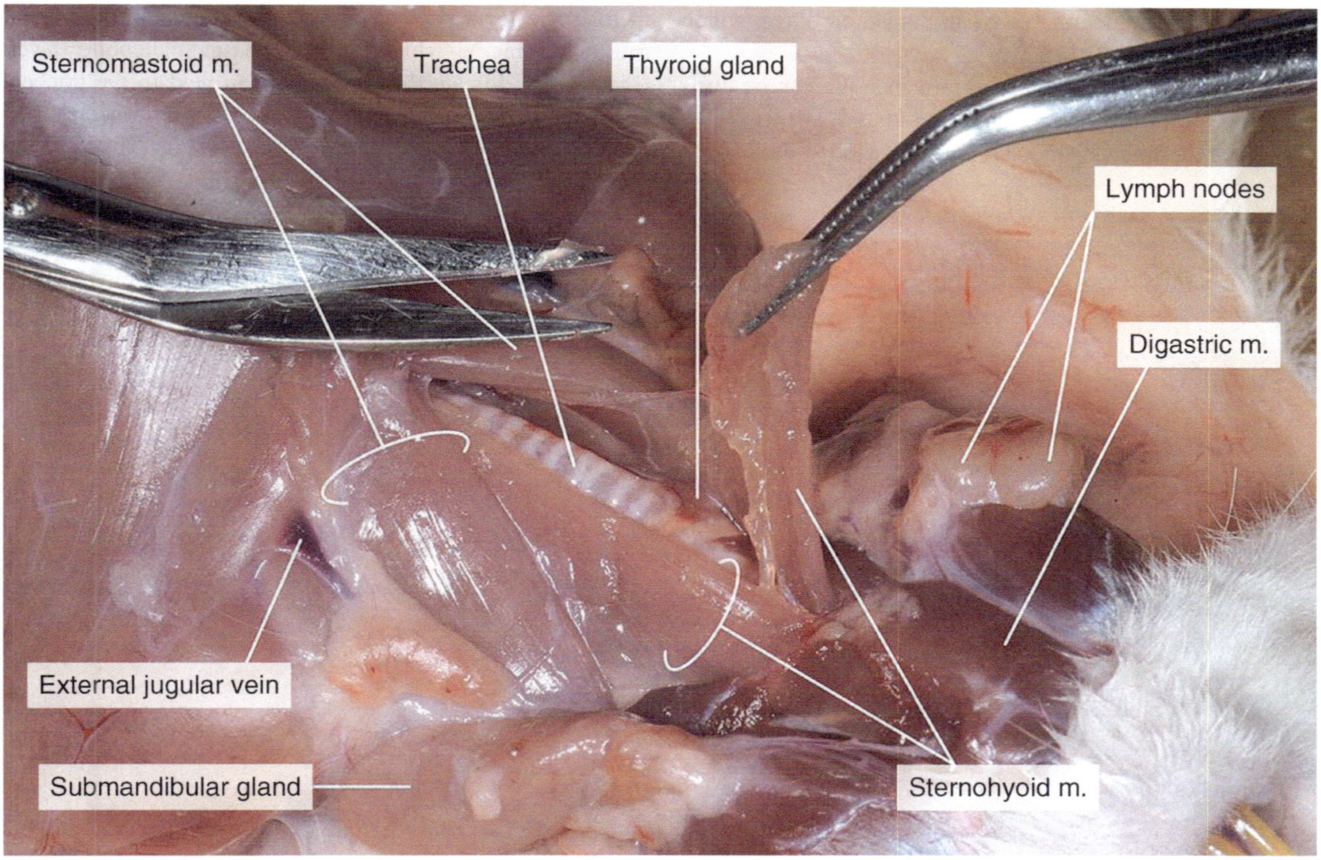

Fig. 12.21 Isolation of the sternohyoid muscles

The *trachea* is identifiable by its ringed cartilage, which provides support (Fig. 12.22).

The oesophagus lies behind the trachea but can be difficult to locate in this area. Locate the *larynx*, which is just anterior to the trachea (Fig. 12.22). Posterior to the larynx, there is the *thyroid gland* producing hormones (thyroxine, triiodothyronine), which set the basic metabolic level (Fig. 12.22). Locate the *carotid vagina* next to the trachea containing three structures: the *common carotid artery*, the *vagus nerve* and the *internal jugular vein* (Fig. 12.22).

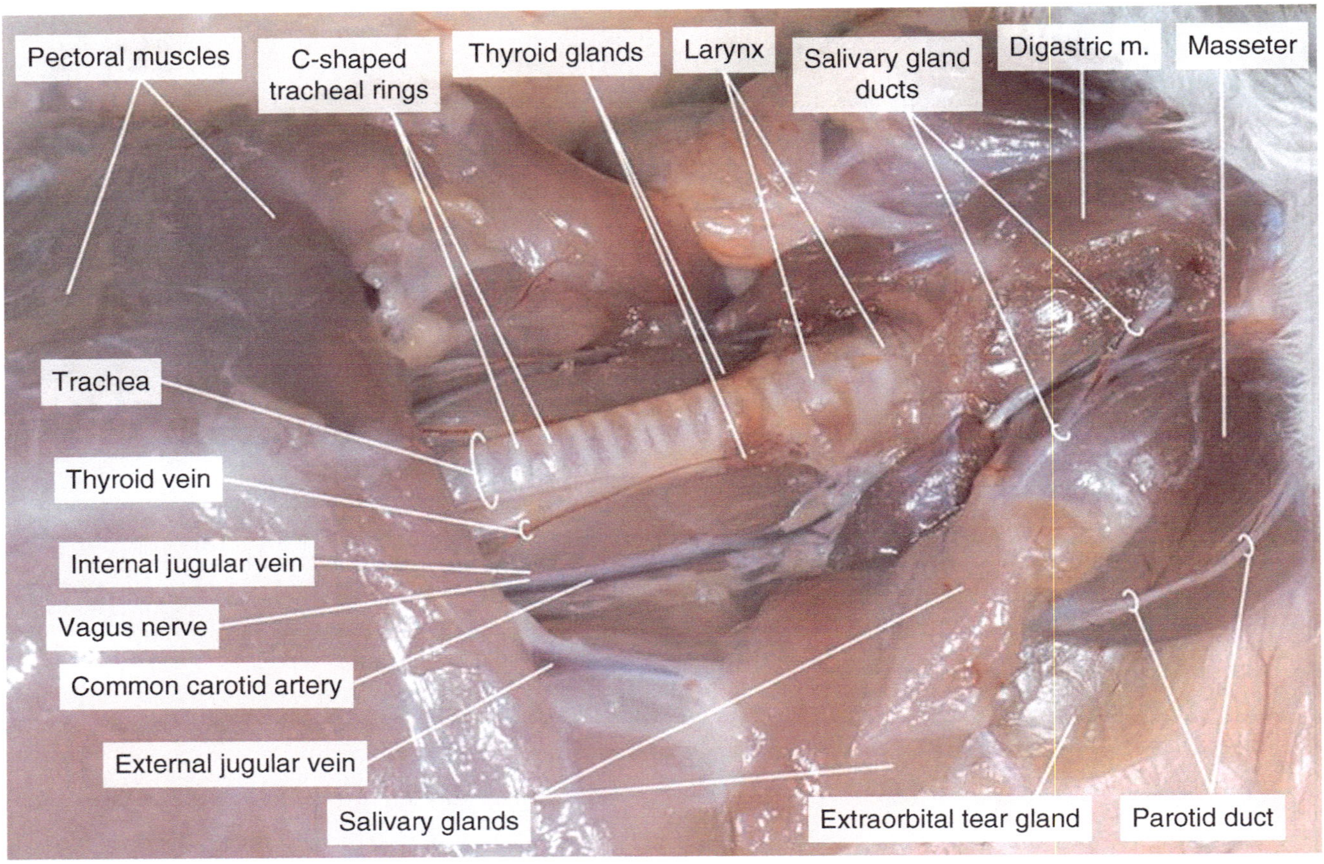

Fig. 12.22 The exposed trachea, larynx, thyroid glands and the carotid vagina lying next to them

Explore the axillary region by removal of pectoral muscles and find *external jugular vein*, the *brachial plexus* with the *brachial vein*. Take an extreme care for the thin-walled *subclavian vein* to prevent bleeding (Fig. 12.23).

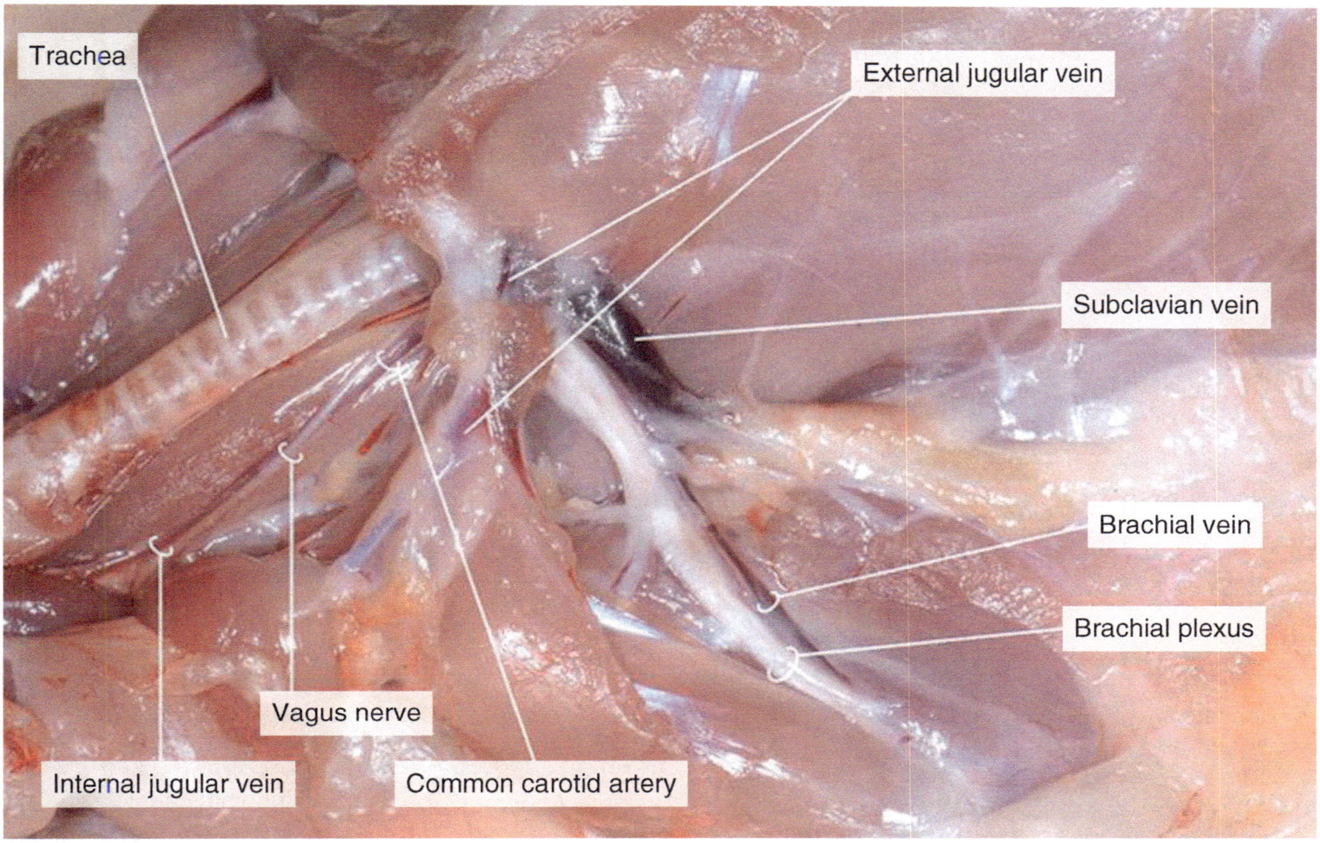

Fig. 12.23 The right axillary region of a rat (anterior end is to the left)

Now you will dissect first the thoracic cavity and the organs in there and then the abdominal cavity and its organs. Follow on the cut pattern on Fig. 12.24 and keep the order of incisions. Cut number 1 has been made already and the neck region was explored.

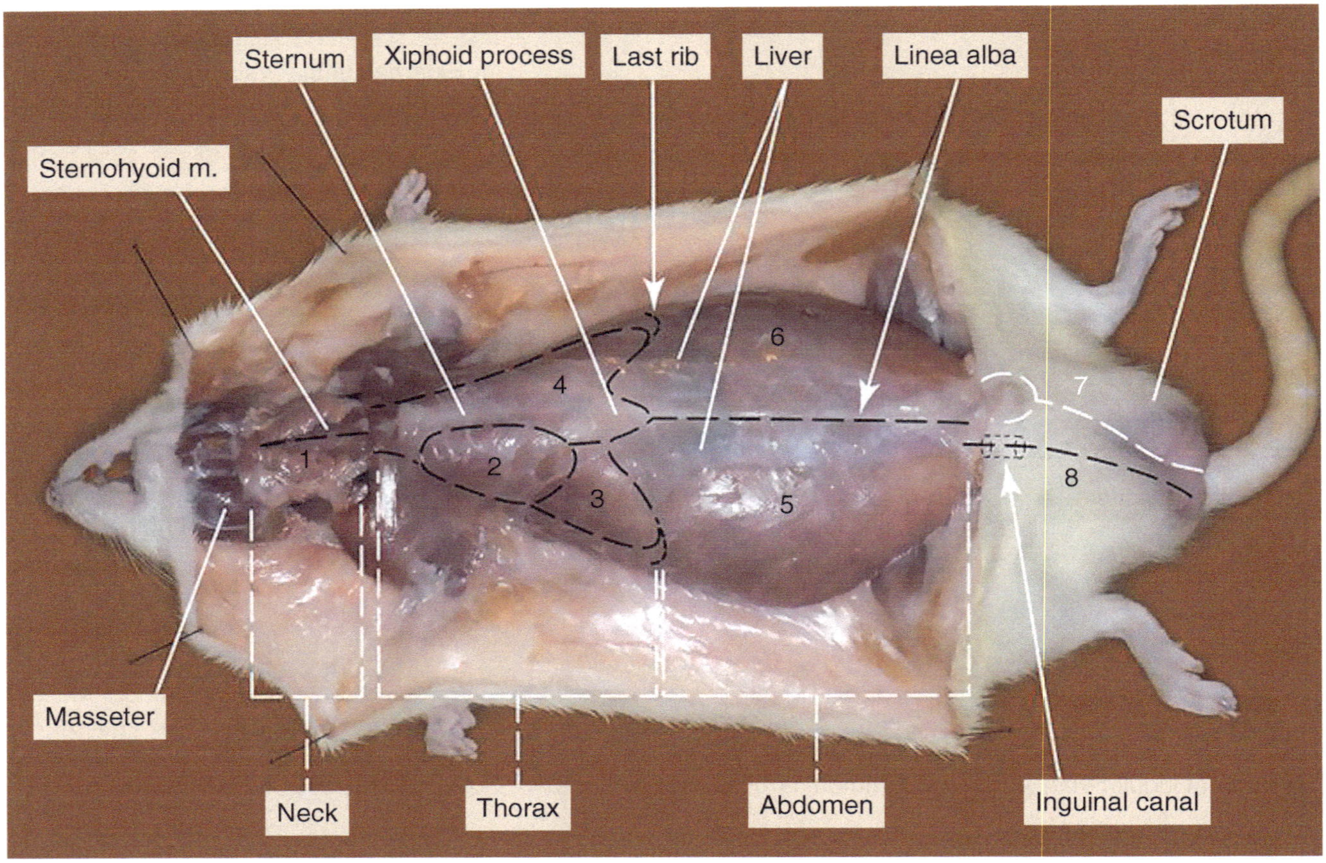

Fig. 12.24 The ventral view of the muscular body wall of a rat with the cut pattern used. Numbers indicate the order of the described cuts. Some organs show through the transparent body wall as orientation points (anterior end is to the left)

Next cut through the ribs and make a window on the thoracic wall. Be careful not to cut too deeply and do not damage the underlying structures. Once you have opened a window on the thoracic cavity, locate the *lungs*, the *mediastinum* and the *heart* and the *thymus gland* in it, and the *diaphragm* (Fig. 12.25).

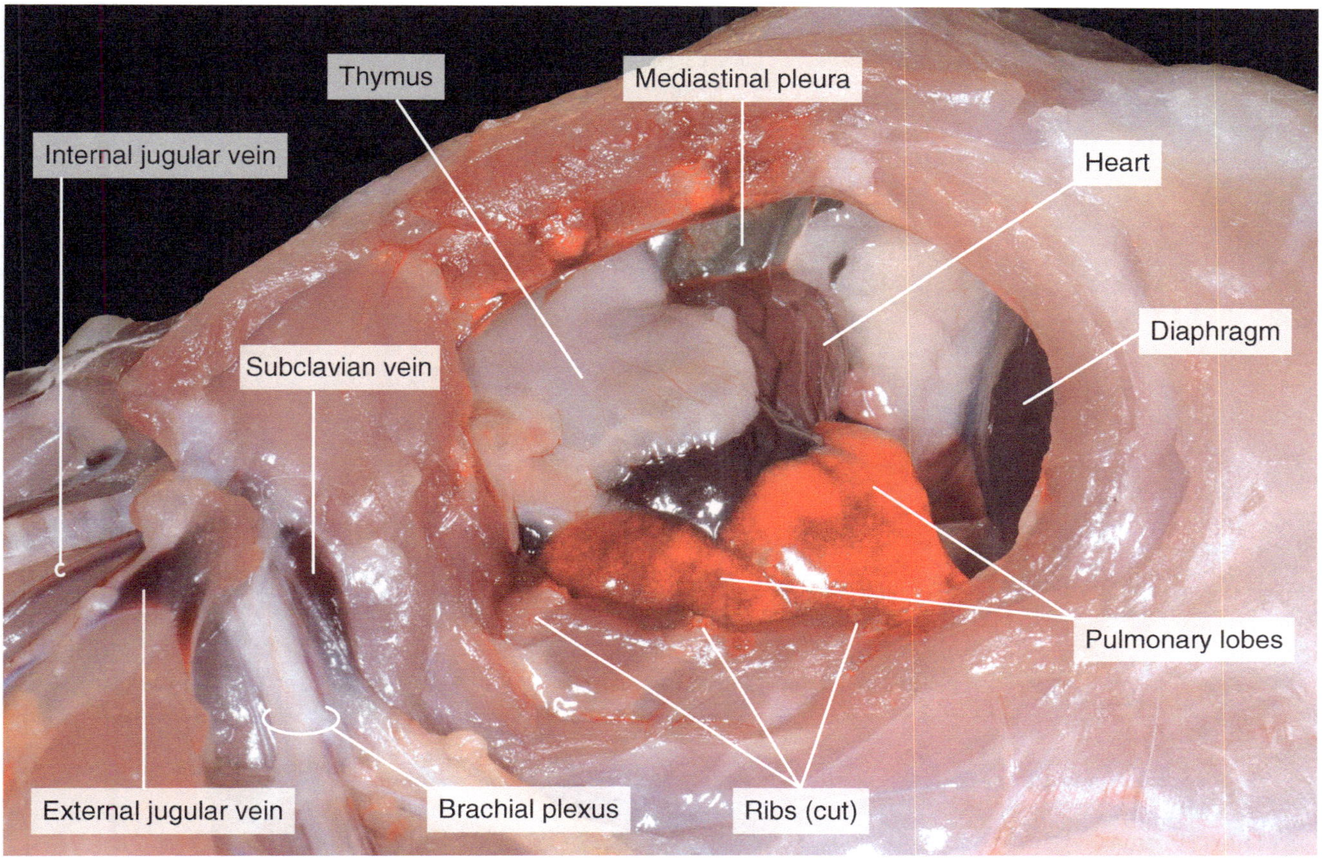

Fig. 12.25 A cut window on the thoracic wall (cutline 2 in Fig. 12.24), through which the organs of the thoracic cavity are visible in their original position (anterior end is to the left)

Complete the incision towards the other side of the thoracic cage and lift the sternum. Note the mediastinal pleura and the mediastinum in a posterior view (Fig. 12.26). The mediastinum is the central compartment of the thoracic cavity surrounded by sheets of the parietal pleura. The mediastinum contains the heart and its vessels, the oesophagus, trachea, phrenic and cardiac nerves, the thymus and lymph nodes.

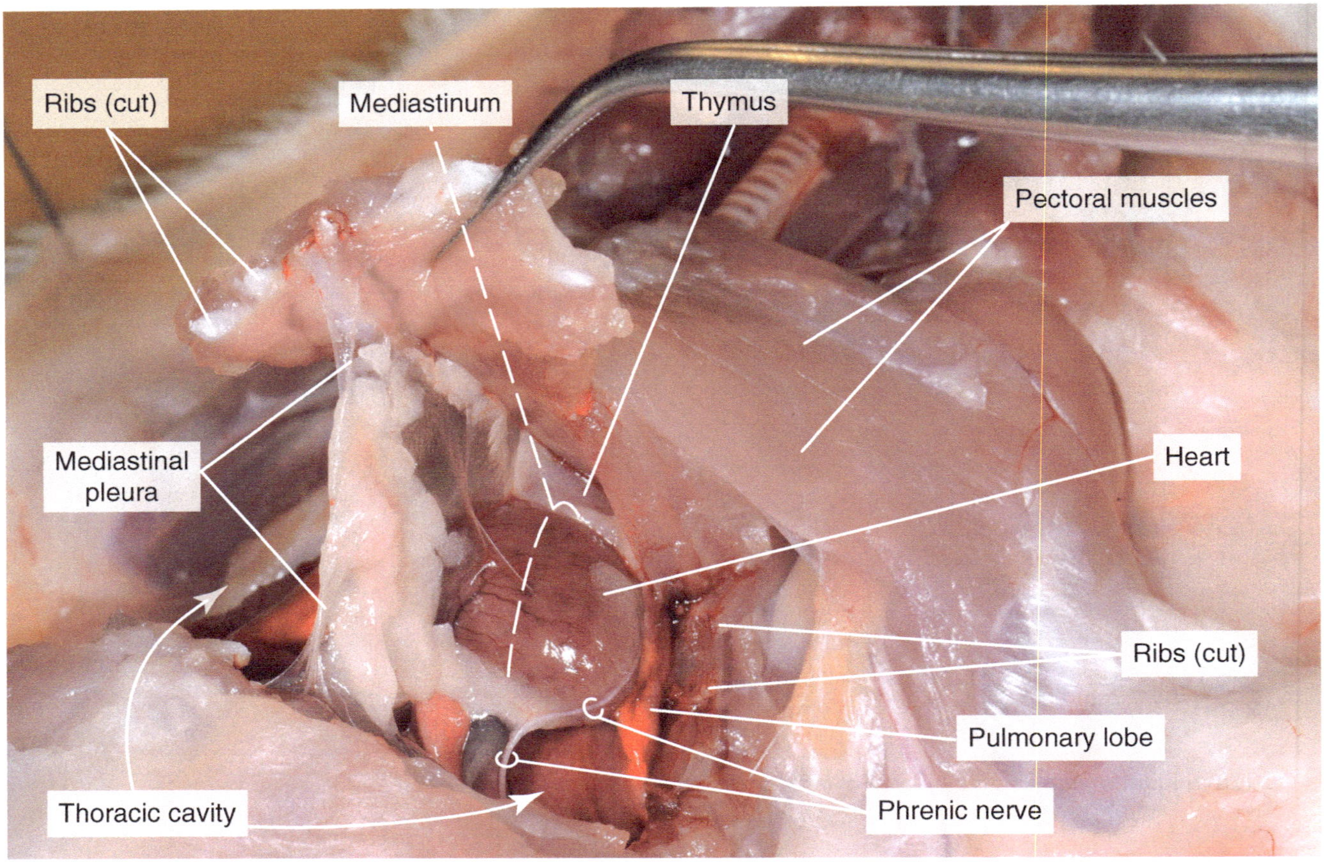

Fig. 12.26 The mediastinal pleura and the mediastinum in a posterior view. (Cutlines 3 and 4 in Fig. 12.24)

Finally, remove the sternum, ribs and pectoral muscles. At the anterior end of the sternum, cut between the two clavicles and be extremely careful to avoid damage of the external jugular vein (Fig. 12.27).

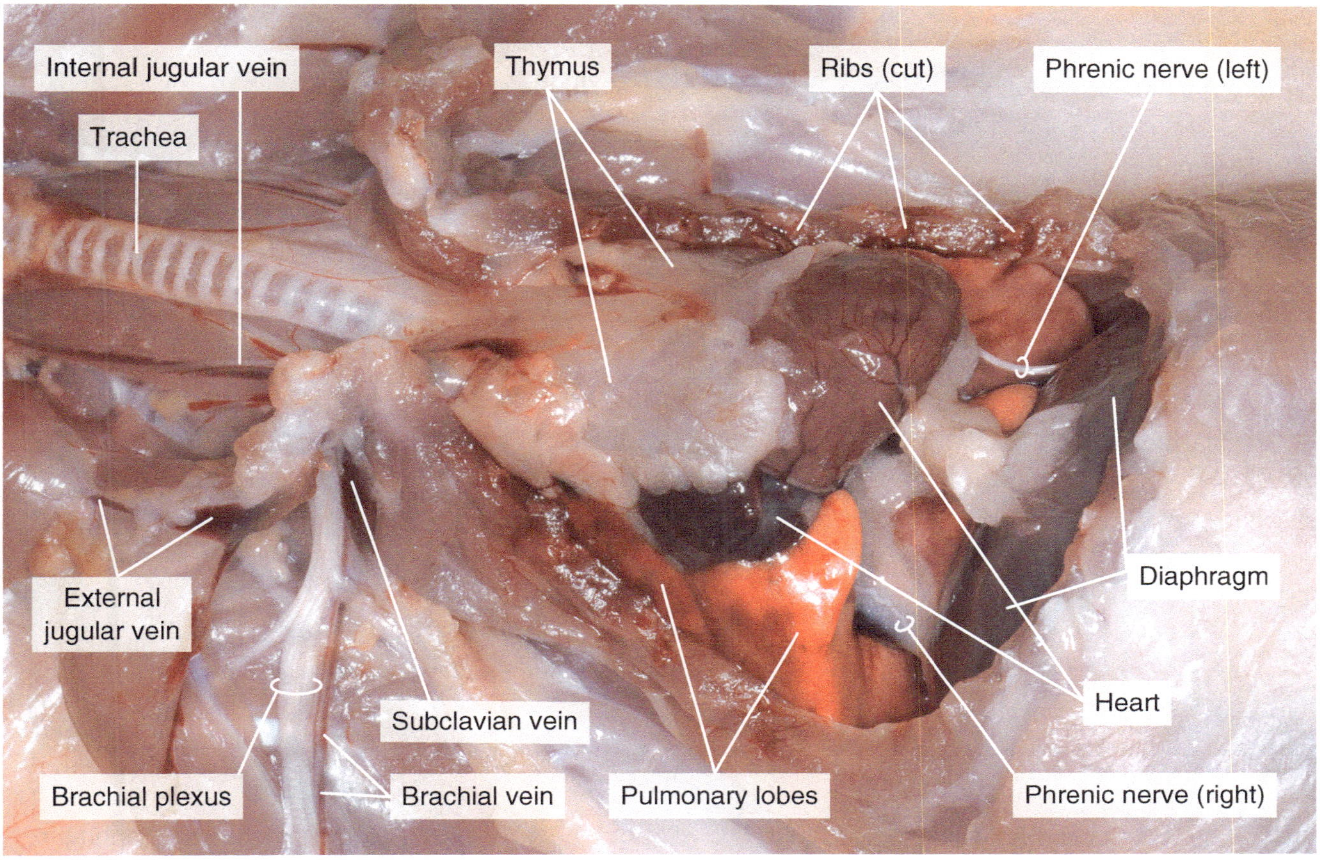

Fig. 12.27 Organs in the thoracic cavity after the removal of the sternum and the ribs (anterior end is to the left)

The *thymus gland* is visible at the upper part of the heart (Fig. 12.27). In some juvenile rats, there may be a thymus gland covering the anterior half of the heart, but in older rats, this gland disappears. Remove the thymus gland very carefully together with the underlying *pericardium*, a tough fibrous sack surrounding the heart (Fig. 12.28).

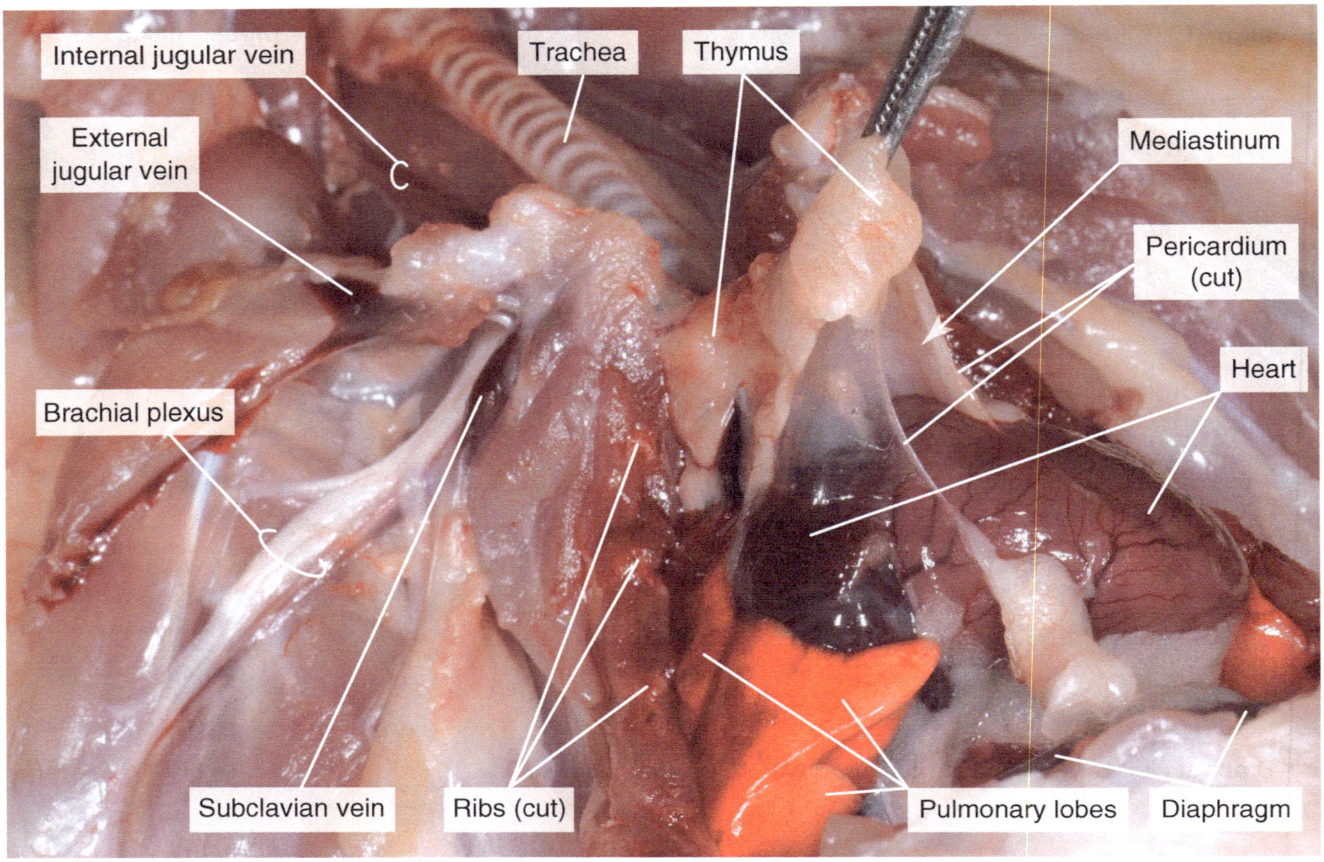

Fig. 12.28 Careful removal of the thymus gland with the pericardium of the heart

Examine the *heart* in the thoracic cavity (right and left in this description refer to the rat's right and left). Find the *right* and *left atria*. Their portions can be seen as dark flaps on top of the heart (Fig. 12.28). The *ventricles* (right and left) cannot be differentiated without opening the heart (see later). The large vessels seen leaving the heart on the anterior surface are from left to right: the *pulmonary trunk*, which soon forks into two large *pulmonary arteries* and the *aorta*, which immediately turns to the left, called *aortic arch* (Figs. 12.29 and 12.30).

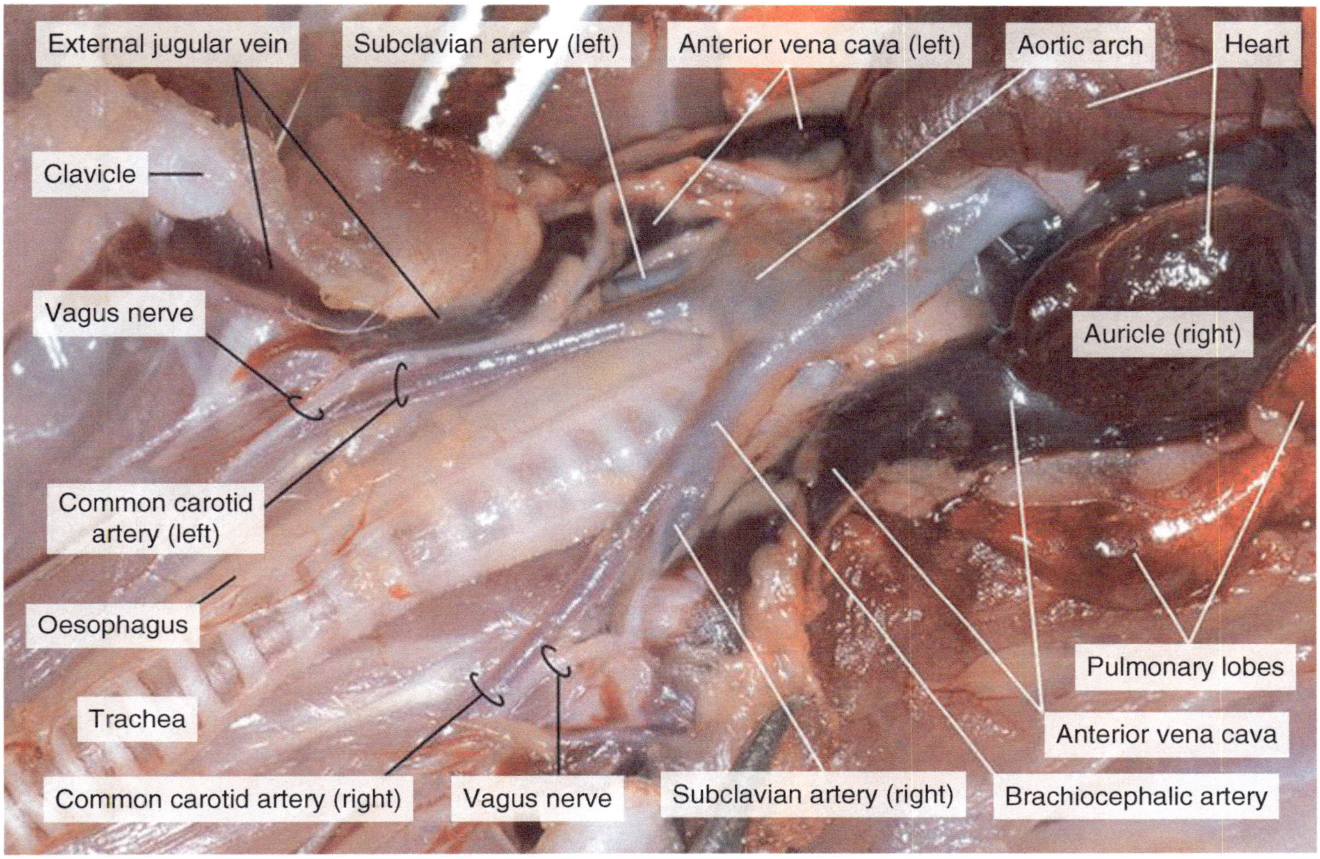

Fig. 12.29 The aortic arch and its asymmetric branches: the left common carotid artery, the left subclavian artery, and the brachiocephalic artery (anterior end is to the left)

The main arteries arise from the aortic arch are asymmetric: There are two arteries on the left, the left *common carotid artery* and the left *subclavian artery*, and there is only one on the right the *brachiocephalic artery*. Trace the brachiocephalic artery; it gives rise to an anterior *common carotid artery* and a lateral *subclavian artery* (Fig. 12.29). The *posterior* and *anterior venae cavae* return blood from the posterior and anterior parts of the body to the right atrium (Fig. 12.30).

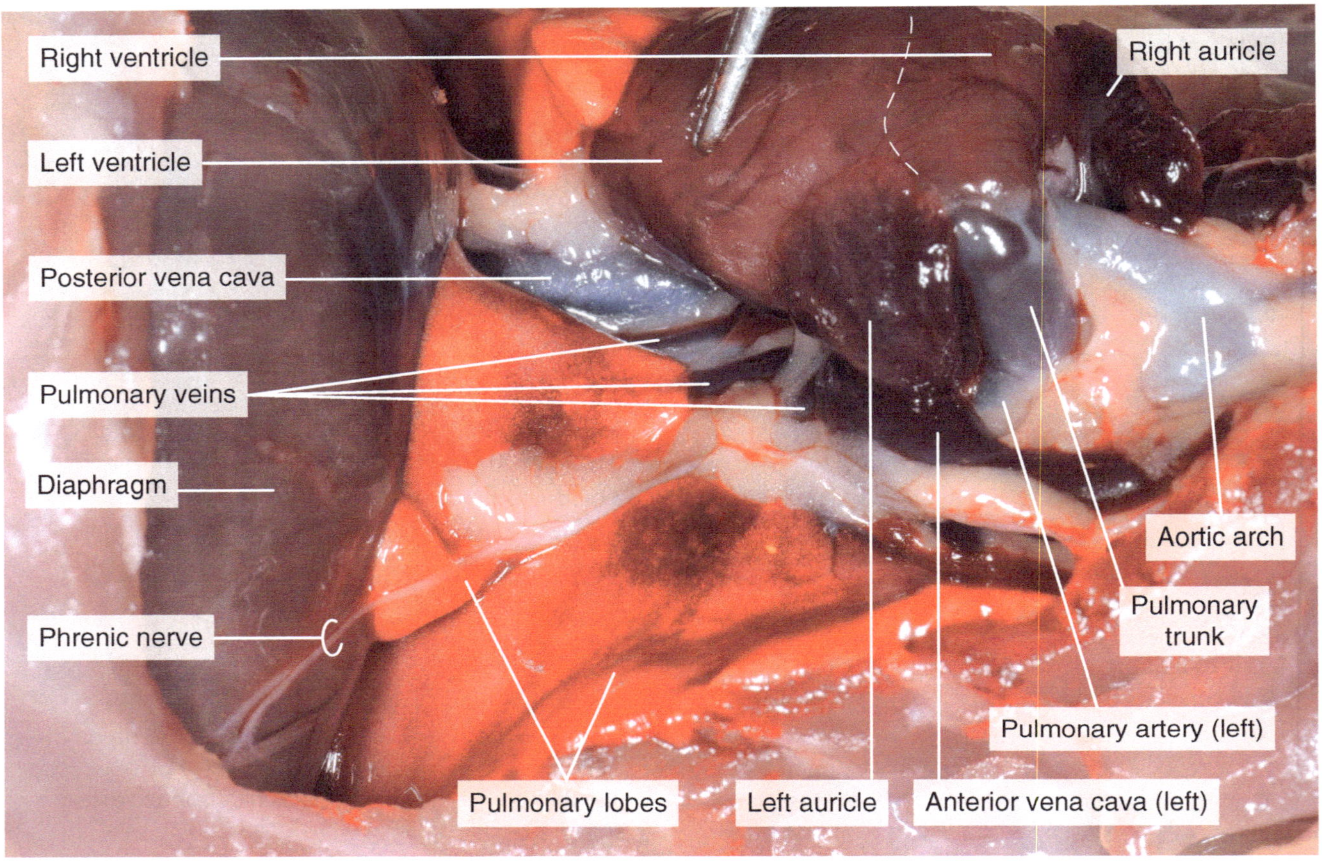

Fig. 12.30 The ventral side of the lifted heart with the large blood vessels. *Dashed line* shows the approximate border of the ventricles. Note the phrenic nerve

Be able to identify any of the blood vessels illustrated on the figures above and below that you can find without cutting any internal organs out of your specimen. The *diaphragm* is a thin sheet of muscle separating the abdomen from the thorax. This muscle is only found in mammals. It is innervated by the phrenic nerve (Figs. 12.30 and 12.31). When contracted, it draws air into the lungs.

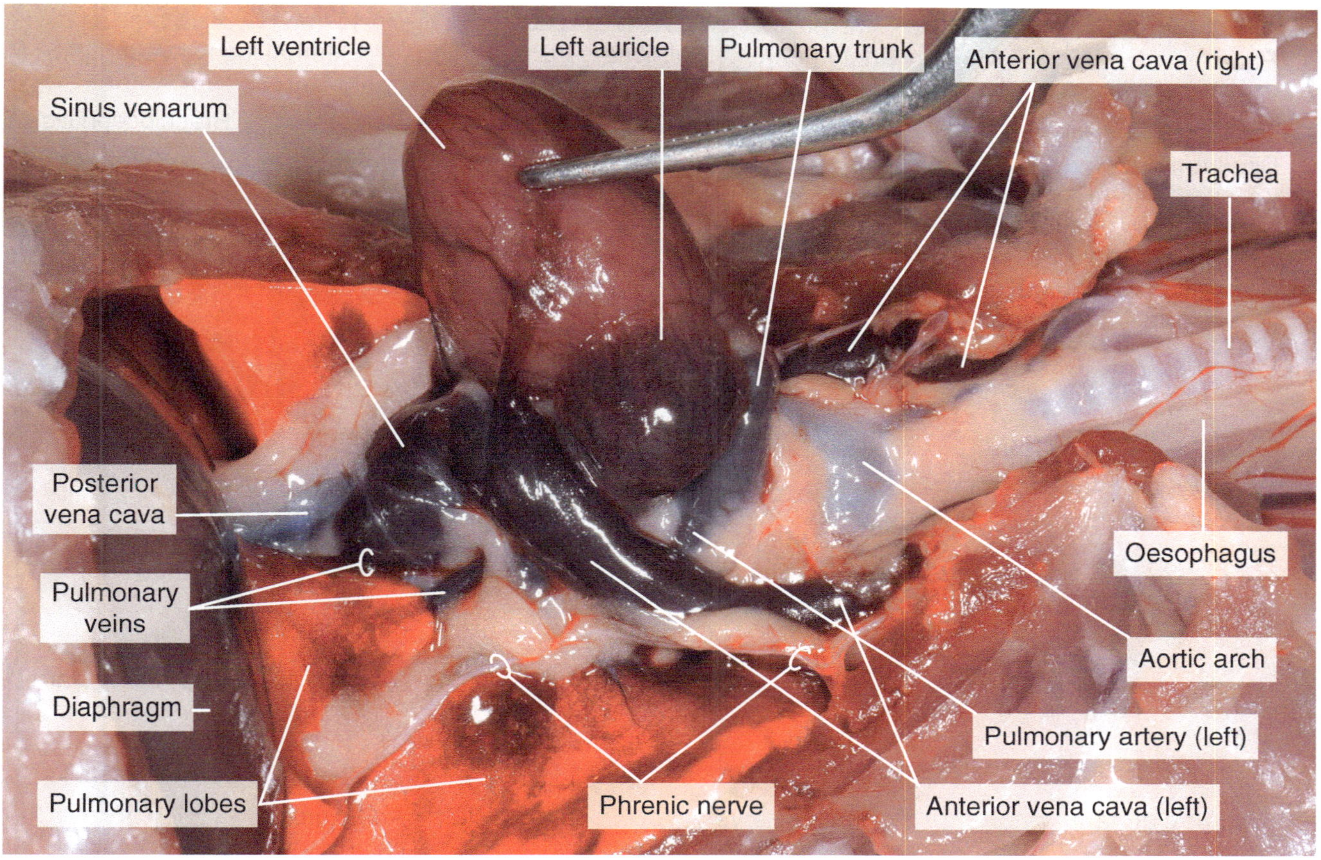

Fig. 12.31 The ventral side of the lifted heart with the large blood vessels

From the right atrium, the unoxygenated blood moves to the right ventricle, from which it is pumped to the lungs via the *pulmonary trunk*. The pulmonary trunk splits into two *pulmonary arteries*, one leading to each lung (Fig. 12.31). Oxygenated blood returns from the lungs through the *pulmonary veins*, which are difficult to see, to the left atrium. From there it moves to the left ventricle, and then it is pumped to the body via the *thoracic aorta* (descending aorta) (Fig. 12.32).

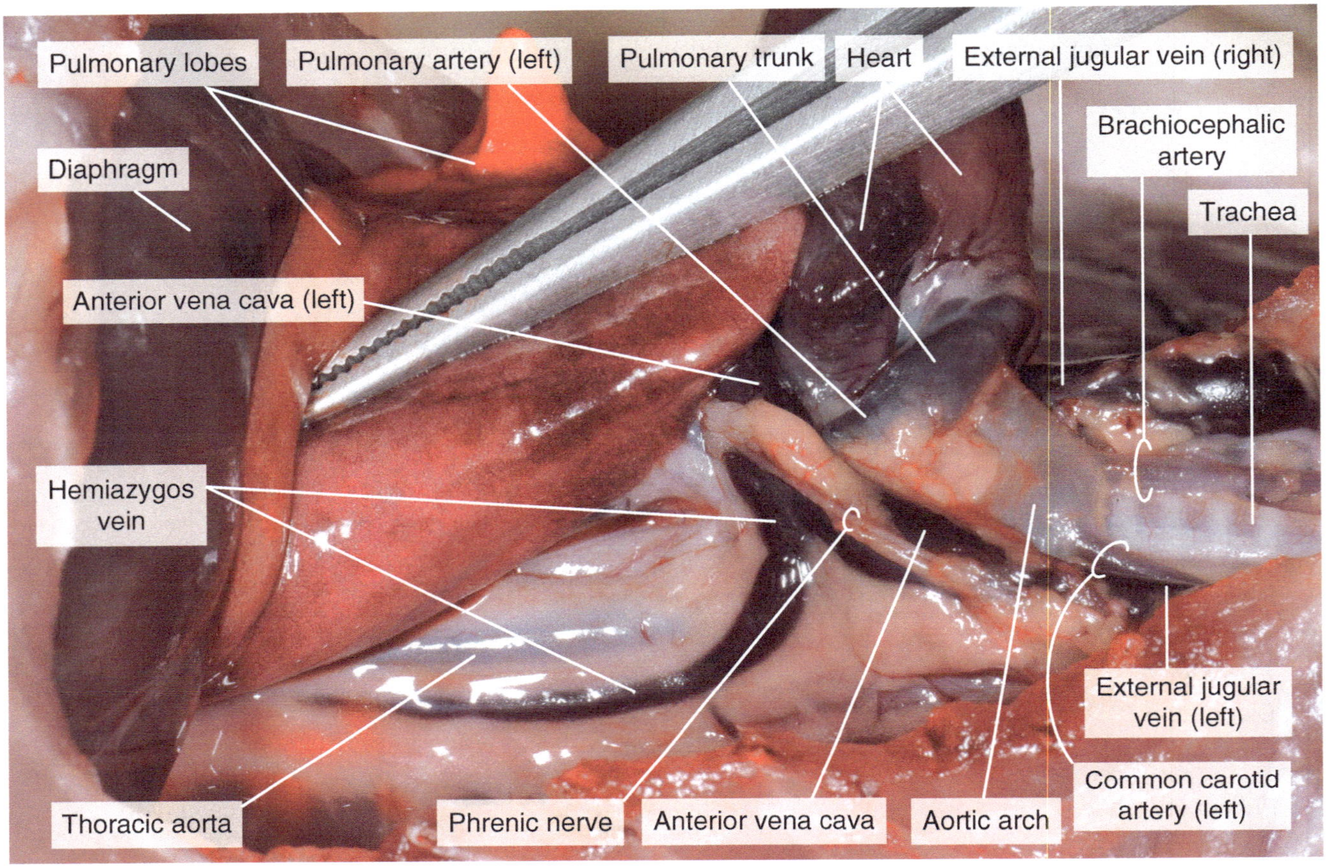

Fig. 12.32 The thoracic aorta and the hemiazygos vein on the left side of the dorsal body wall under the lifted lung lobe

It lies together with the hemi*azygos vein* on the left dorsal body wall behind the parietal pleura. These are impaired structures; they appear only on the left side. Lift the lobes of the lung on the right side, and note the *sympathetic trunk*, which is part of the autonomous nervous system (Fig. 12.33). The sympathetic trunk is paired, but it is difficult to observe on the other side.

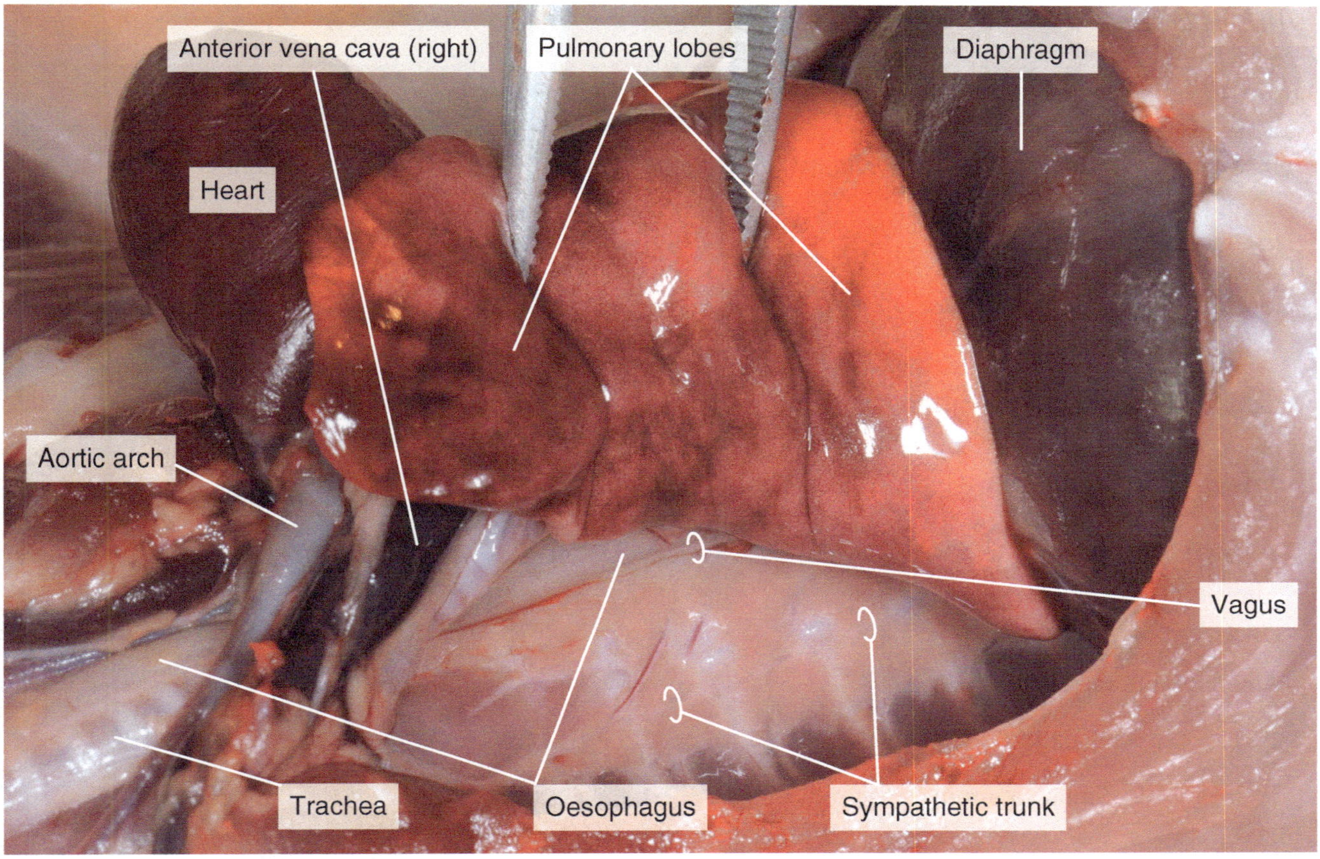

Fig. 12.33 The thoracic part of the sympathetic trunk on the right side of the vertebral column (anterior end is to the left)

Trace the *trachea* as it divides into a left and right *bronchus*, which go into the lungs. Gently move the contents of the thorax around until they can be found. Like the trachea, bronchi have conspicuous rings (Fig. 12.33). The *lungs* are pink, spongy organs that lie on either side of the heart and should take up most of the thoracic cavity (Fig. 12.34).

The lungs lie closer to the back of the rat; you will need to push the ribs to the side to find them. The right lung on the rat has four lobes and the left lung has three. Lungs can be inflated with the help of a 10 ml syringe and a small plastic tube introduced across the glottis or through a small cut on the trachea. Each lung is covered by a thin layer of tissue called the *visceral pleura*. The inside of the thoracic wall, diaphragm and mediastinum is covered by a similar layer called the *parietal pleura*.

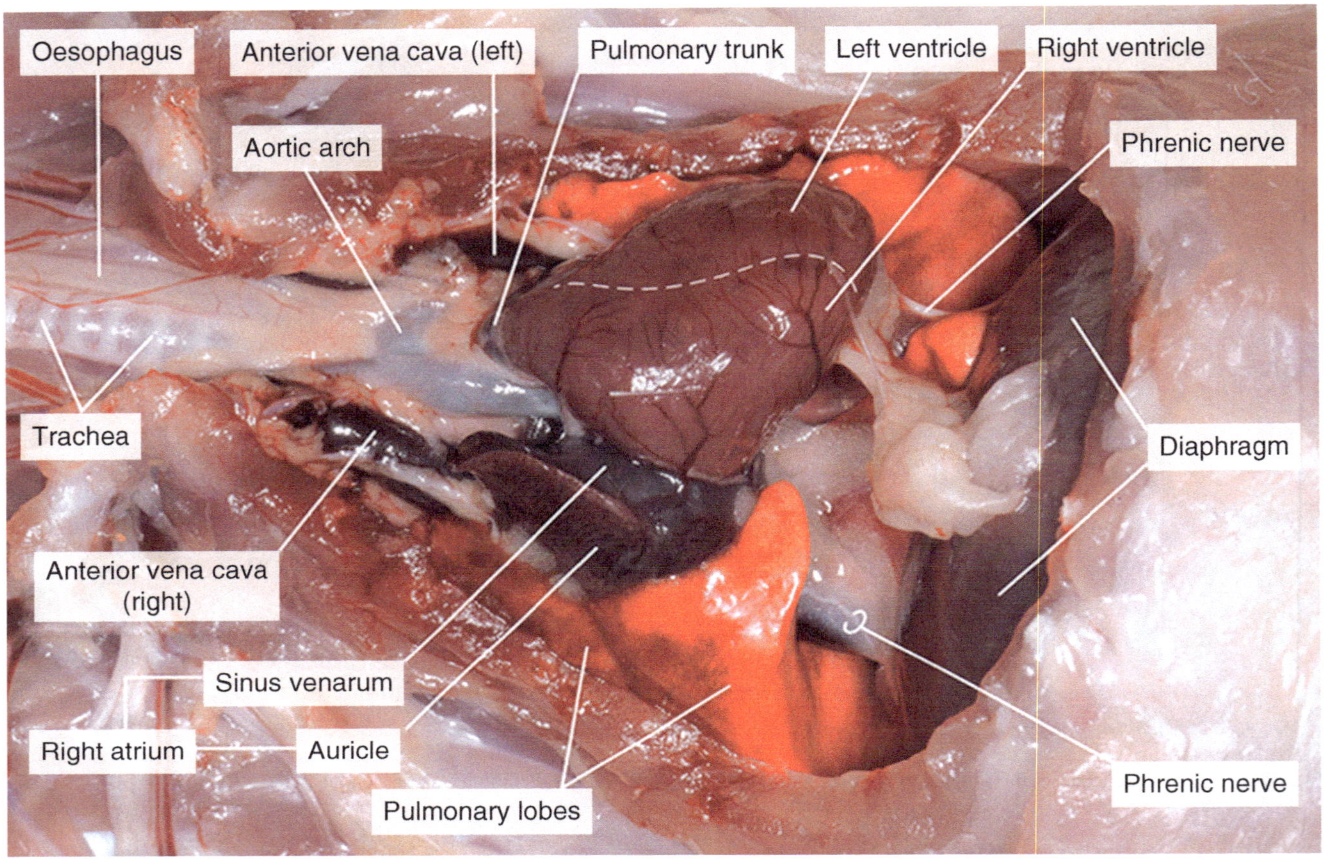

Fig. 12.34 The heart, the large blood vessels and the lung in the thoracic cavity. *Dashed line* shows the approximate border of the ventricles (anterior end is to the left)

The *primary bronchi* entering the mammalian lungs branch into smaller *bronchi* and then *bronchioli*. The smallest bronchioli lead into alveolar ducts where alveoli open (Fig. 12.35). *Alveoli* are thin-walled pouches, which provide the respiratory surface for gas exchange. Their wall is formed by alveolar epithelium and a very thin interalveolar septa containing capillary network. There are rings made of smooth muscle cells at the opening of alveoli, where the septa terminate and broaden (Fig. 12.35, left, dotted circles). The alveolar wall contains two types of epithelial cell (pneumocytes or pulmonocytes). *Type I pneumocytes* are flattened, squamous epithelial cells – they extend thin cytoplasmic part from a thicker perinuclear region and facilitate gaseous diffusion between the lumen of the alveolus and its capillaries (Fig. 12.35, right). Pneumocytes and capillary endothelial cells unite their basal membrane forming a minimal barrier to gaseous exchange. Interalveolar septa with elastic fibre network contain capillaries. *Type II pneumocytes* are rounded cells, which protrude into the alveolar surface (Fig. 12.35, right). They form couples frequently at the angles between alveolar profiles. Numerous characteristic secretory vesicles can be seen in their cytoplasm, which contain precursors of the surfactant layer. The role of the surfactant layer is to reduce the surface tension and keep the small alveoli open against capillary action. Type II pneumocytes are stem cells: they are able to divide. These cells are responsible for the renewal of the damaged respiratory surface by producing new pneumocytes. *Alveolar macrophages* are present within the alveolar lumen – they clear the respiratory spaces of inhaled particles (Fig. 12.35, right). They derive from circulating monocyte precursors.

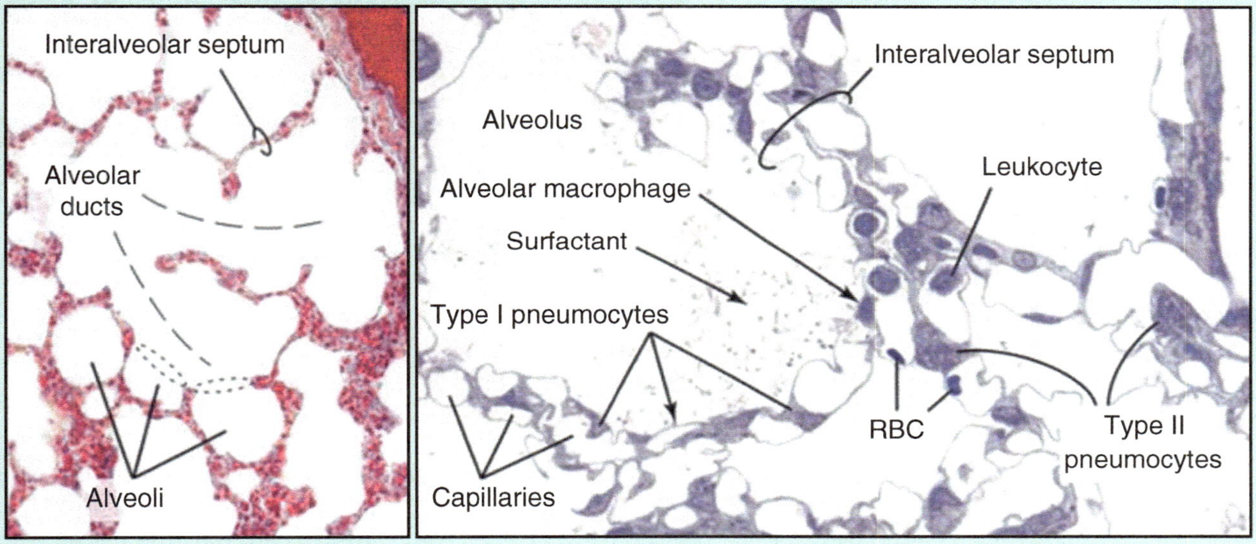

Fig. 12.35 The histology of alveoli (*left*, Azan; *right*, semithin section after perfusion fixation, toluidine blue staining). *Dashed lines* alveolar ducts, *dotted lines* entrances of the alveoli, *RBC* red blood cells

To see the organs of the *abdominal cavity*, begin cutting the muscular abdominal body wall along the linea alba just anterior to the genital opening and continue anteriorly (Fig. 12.24). There find the diaphragm and cut it away from the body wall (Fig. 12.36).

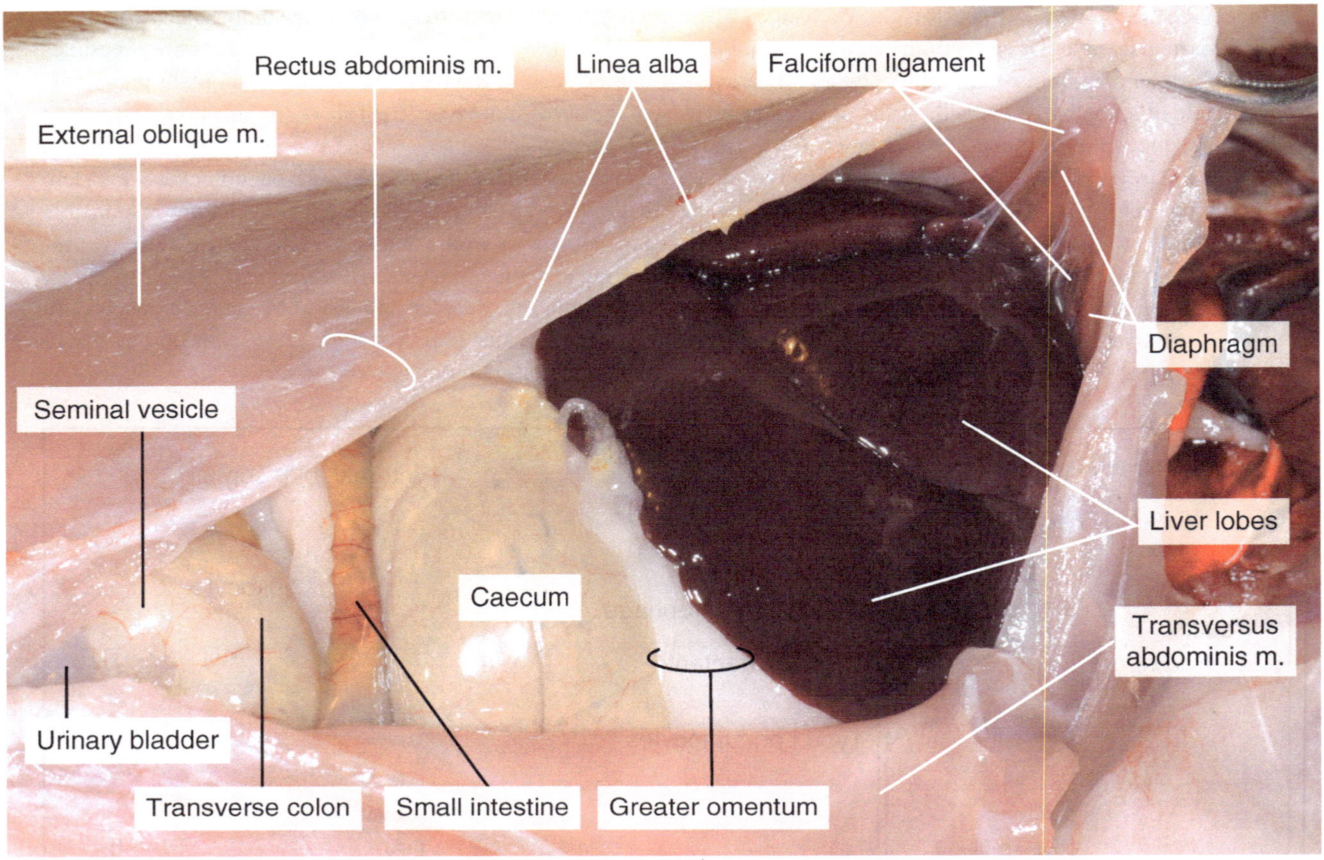

Fig. 12.36 Organs in the abdominal cavity seen through the cut rectus abdominis muscle. Note the falciform ligament of the liver and the diaphragm

At the posterior end of the abdominal cavity, cut laterally through the external oblique muscle until you can pin back the ventral side muscles and expose the organs underneath. If necessary, make another pair of lateral cuts towards the anterior end, but be sure not to cut through any organs. Locate the liver, stomach, spleen, small intestine and large intestine (Fig. 12.37).

In the abdominal cavity, pick up the *oesophagus* as it comes through the diaphragm and note where it enters the stomach. The oesophagus transports food from the mouth to the stomach. It is distinguished from the trachea by its lack of cartilage rings (Fig. 12.37).

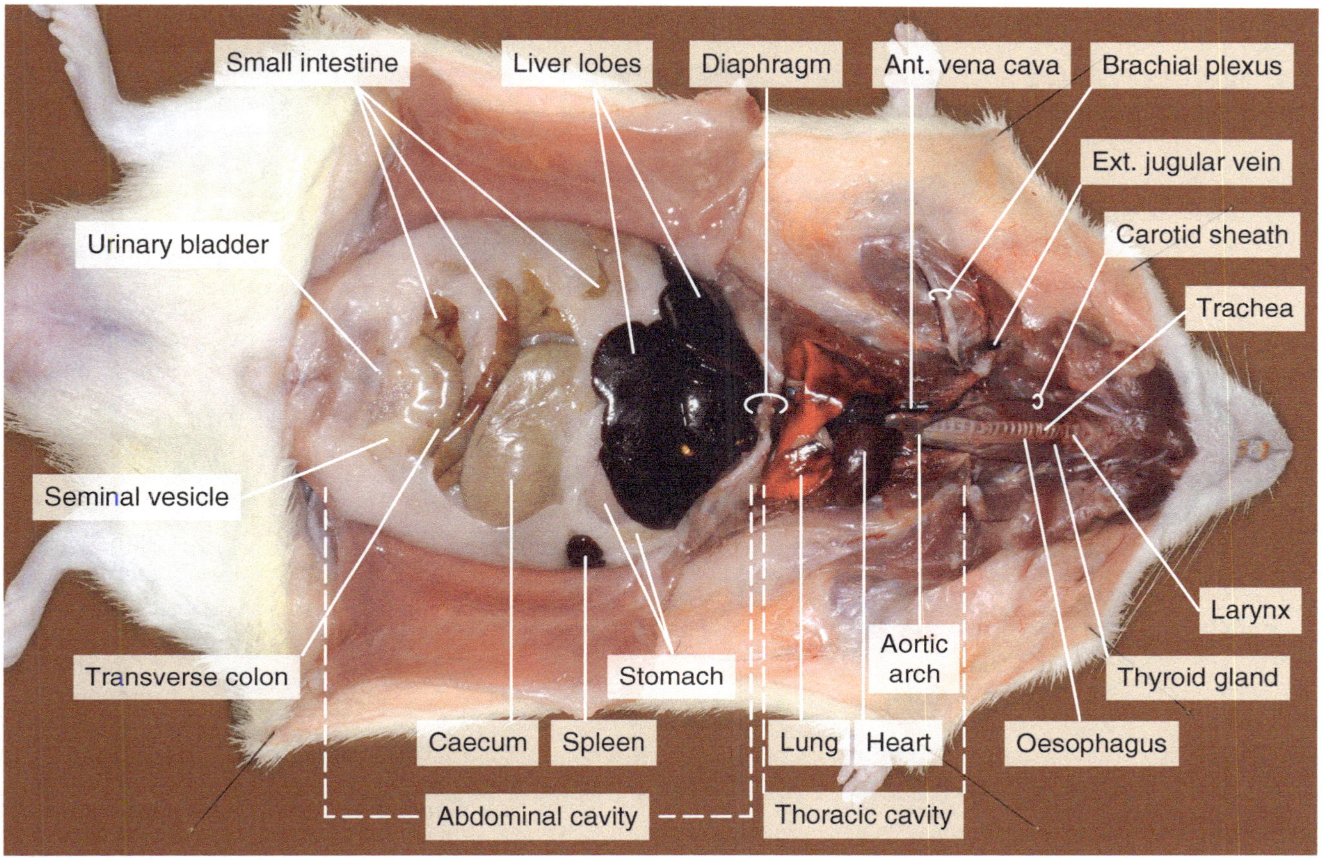

Fig. 12.37 A dissected male rat after the opening of the thoracic and abdominal cavity (organs are in their original position)

Locate the *stomach* on the right side just under the liver. The rat has a compound stomach, which has a single cavity, although it consists of two compartments: forestomach (derived from the oesophagus) and glandular stomach (containing gastric glands). The functions of the stomach include food storage (forestomach), physical breakdown of food and the digestion of proteins (glandular stomach) (Fig. 12.38).

The opening between the oesophagus and the stomach is called the *cardiac sphincter*. The outer margin of the curved stomach is called the greater curvature; the inner margin is called the lesser curvature (Fig. 12.38). At the top of the stomach, you can see the oesophagus where it pierces the diaphragm and joins the stomach. At the posterior end of the stomach, there is a muscular valve called the *pyloric sphincter*; this leads into the small intestine.

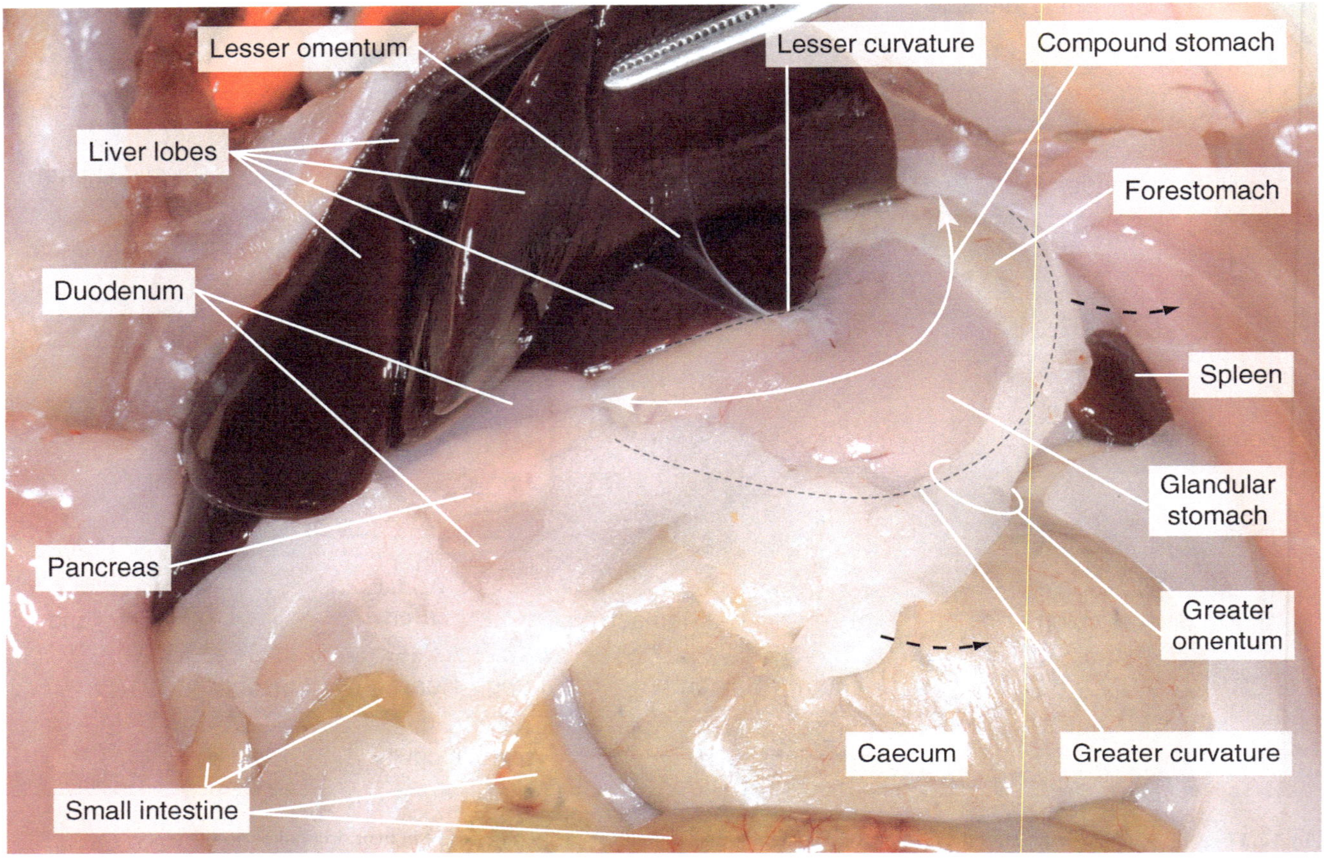

Fig. 12.38 The compound stomach under the liver and neighbouring organs. *Dashed line* margins of the compound stomach partly hidden by other organs, *dashed arrows* direction of stretching the greater omentum for examination (anterior end is to the left)

The *stomach* of the rat is divided into two portions: a non-glandular and a glandular part (forestomach and glandular stomach, respectively) (Fig. 12.39). The mucosal layer of the non-glandular portion is identical with the same layer in the oesophagus: its surface is covered by keratinised, stratified squamous epithelium (SQE) (Fig. 12.39, top left).

The other part (glandular stomach) contains glandular epithelium, which is responsible for the digestion (Fig. 12.39, *top*). Its epithelium forms long, tubular gastric glands (acini) that extend deep into the propria as far as the muscular layer of the mucosa (SML) (Fig. 12.39, *bottom*). Simple columnar epithelium (CE) covers the entire luminal surface including the *gastric pits* (GP); this epithelium is composed of *surface mucous cells* (SMC) (Fig. 12.39, *top right*). These and the *neck mucous cells* (NMC) (Fig. 12.39, *top right*) are typical secretory cells: they synthesise a thick protective layer over the gastric lining. The *zymogenic cells* (chief cells) are the source of the digestive enzymes: their cytoplasm contains several secretory zymogenic granules (Fig. 12.39, *top right* and *bottom*). The *parietal cells* (oxyntic cells) are the source of gastric acid: they are large, oval cells, with centrally placed nuclei (Fig. 12.39, *top right* and *bottom*). Their very fine cytoplasmic pattern is emphasised by the PAS reaction: the staining indicates deep, plasmamembrane invaginations for enlarging the secretory surface.

The *submucosal layer* is formed by loose connective tissue (LCT) and contains large blood vessels (BV), nerves and ganglionated plexus of the organ. The muscular layers (ML) are well developed. The outer surface of the stomach is covered by the visceral peritoneum (serosa) (Fig. 12.39, *bottom, arrowhead*).

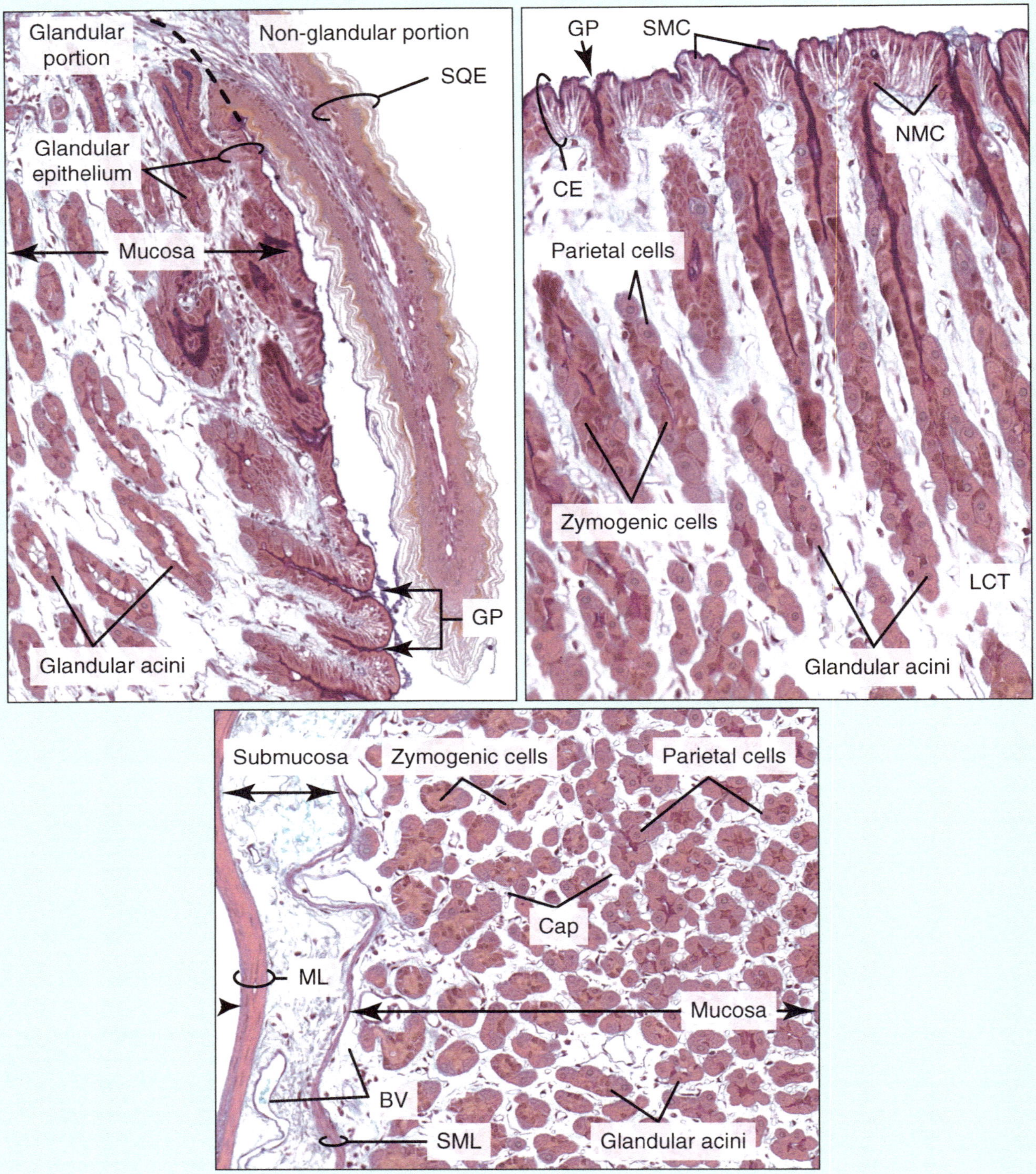

Fig. 12.39 The histological sections of the stomach (PAS). *Black arrowhead* (*bottom*) serosa, *dashed line* (*top left*) boundary between the forestomach and the glandular stomach, *BV* blood vessels, *Cap* capillaries, *CE* columnar epithelium, *GP* gastric pits, *LCT* loose connective tissue, *ML* muscular layers, *NMC* neck mucous cells, *SMC* surface mucous cells, *SML* smooth muscle layer, *SQE* squamous epithelium

The *small intestine* is a slender coiled tube that receives partially digested food from the stomach (via the pyloric sphincter). The term "small" refers to its diameter, not its length. It consists of three sections: *duodenum*, *jejunum* and *ileum*. The duodenum is recognisable as the first stretch of the intestine leading from the stomach; it is mostly straight. In the mesentery that stretches along the stomach, spleen and small intestine, there is a diffuse, bumpy, glandular organ called the *pancreas* (Fig. 12.40).

The pancreas produces digestive enzymes that are sent to the intestine via small ducts (the *pancreatic duct*). The pancreas also secretes insulin, which is important in the regulation of glucose uptake of the cells.

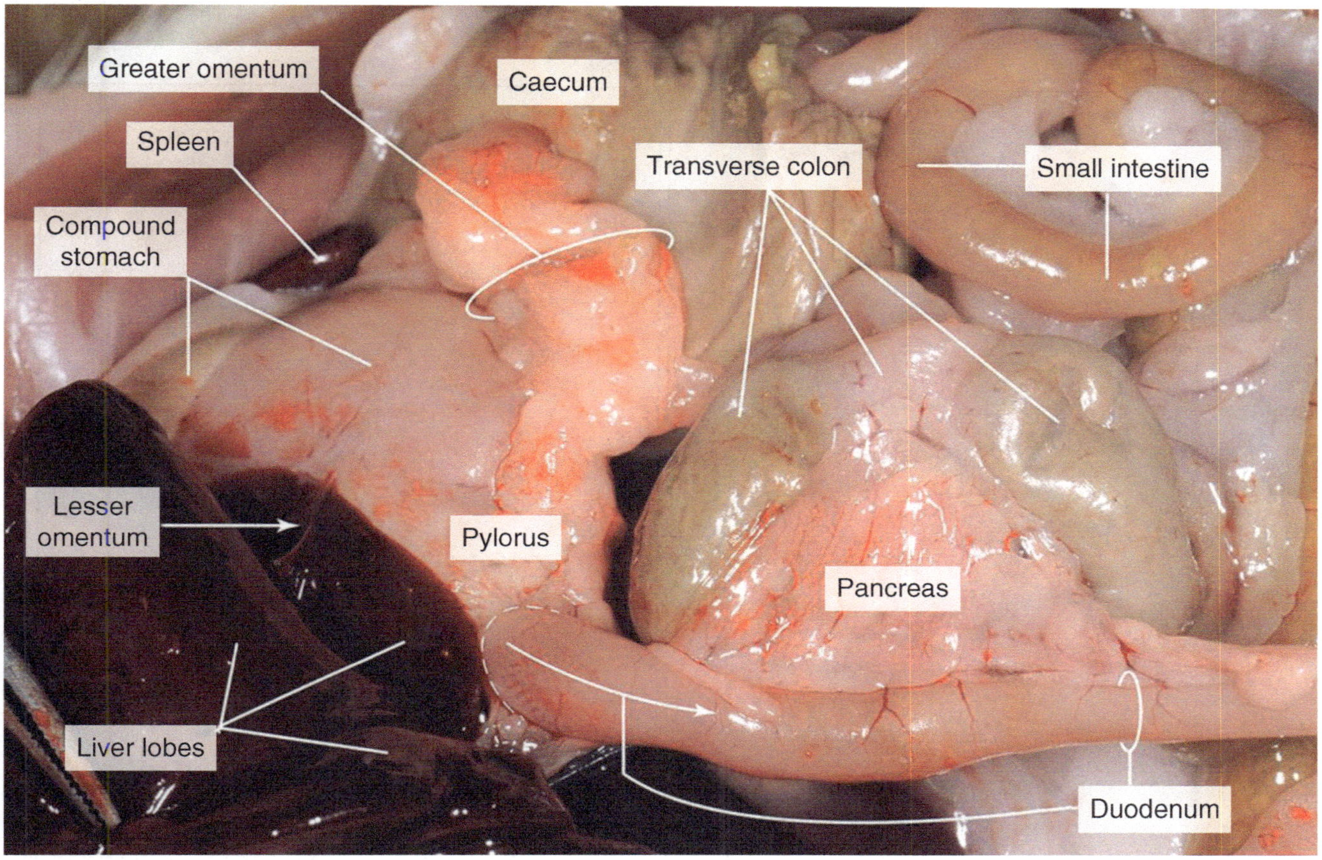

Fig. 12.40 The pancreas in the mesentery that stretches along the stomach and duodenum. *Dashed line* indicates the border between the stomach and duodenum; *arrow* shows the direction of the peristaltic movement (anterior end is to the left)

The *pancreas* is composed of two different types of glandular tissue. The main part is exocrine, in which pancreatic islets of endocrine cells are embedded. The exocrine pancreas is a branched, acinar gland (Fig. 12.41, left). The glandular tissue is lobulated by connective septa. The exocrine cells have pyramidal shape, basal, round nuclei and apical secretory granules. The acinar lumen is invisible. Centro-acinar cells of acini form the beginning of the duct system (distinct small ducts are infrequent). Distally, in the wider ducts, they replaced by cuboidal or columnar epithelium.

Using routine light microscopic technique, the acinar cells stain darker than duct system and endocrine cells. The latter form pale Langerhans islets, which are spherical or ellipsoid clusters of cells embedded in the exocrine tissue (Fig. 12.41, right). Fenestrated capillaries net the islets and deliver the secreted hormones to the target organs.

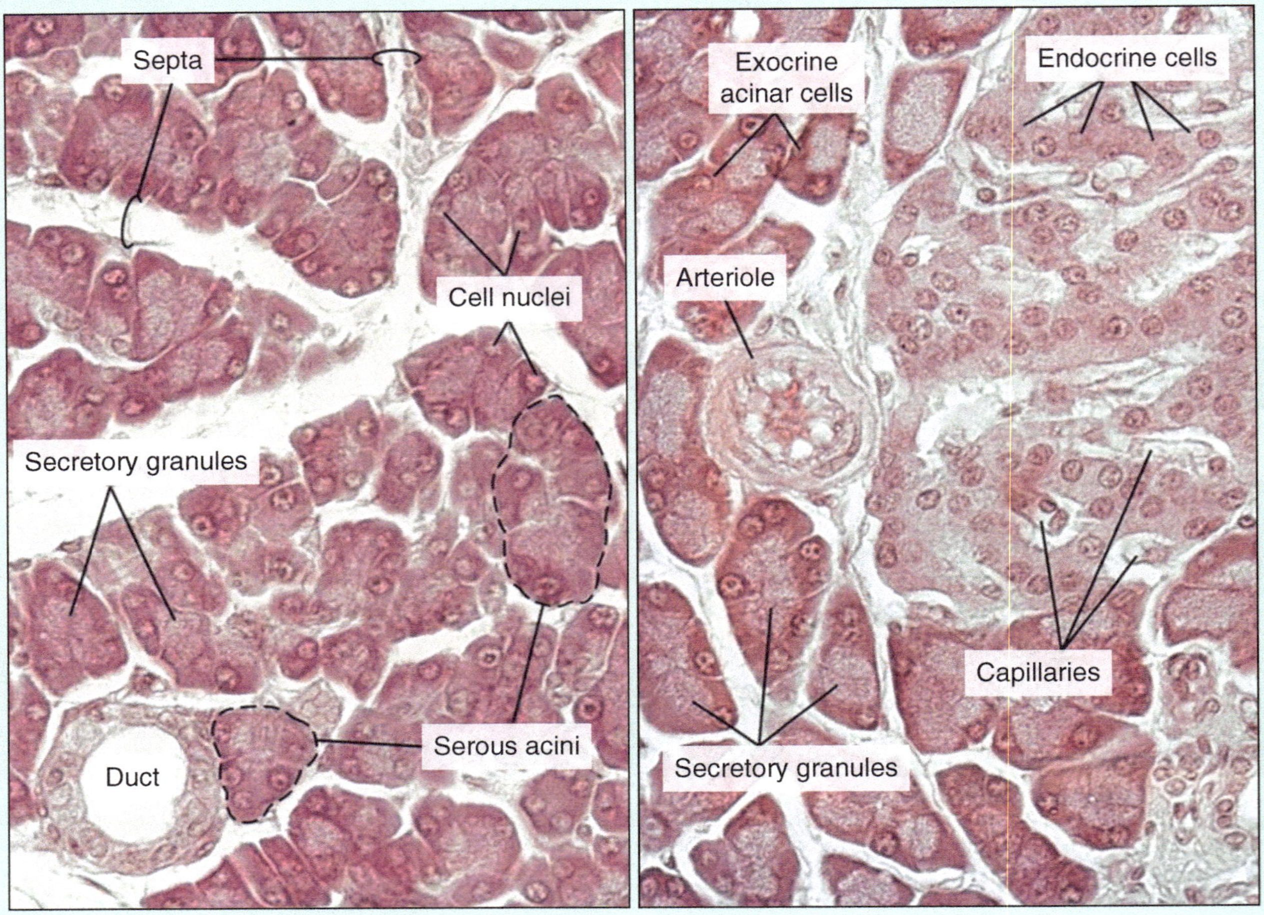

Fig. 12.41 Histological sections of the pancreas: exocrine (*left*) and endocrine (*right*) parts (HE)

Over the stomach and quite conspicuous is the *liver* (Figs. 12.38 and 12.40). The liver is a dark coloured organ suspended just under the diaphragm. It has four lobes. The liver has many functions, one of which is to produce bile, a dark green to yellowish brown fluid that aids the digestion of lipids in the small intestine. The liver also stores glycogen and transforms wastes into less harmful substances. Rats do not have a gall bladder, which is used for concentrating and storing bile in other animals. Therefore, the bile is released through a duct directly into the small intestine, where it acts.

The *liver* parenchyma consists of a complex system of angular *hepatic lobules*, which are formed by epithelial cells, called hepatocytes. *Hepatocytes* are arranged in plates or cords, as seen in two-dimensional sections (Fig. 12.42). Between the plates there are *venous sinusoids*, which anastomose with each other via gaps in the hepatocyte plates. The lobule is surrounded by connective tissue septa, in which branches of the *hepatic artery* (HA) run, *hepatic portal vein* (HPV) and *bile duct* (BD) system are grouped as *portal triads*. From the artery and portal vein, the blood enters the venous sinusoids. The movement of blood prominently slows down to give necessary time for hepatocytes to carry out the major metabolic activities of this organ. Additional cell types possessing storage, phagocytic and mechanically supportive functions are uncertainly identifiable. The blood leaves the lobules through the *central veins*, which are the proximal branches of the *hepatic vein*. Bile is secreted by the hepatocytes. A network of tiny *bile canaliculi* with which every cell is in contact gathers the hepatic secretion. Bile canaliculi are strictly separated from the blood space and join into the *bile ducts* running in the septa among the hepatic lobules in the portal triads. They unite into one or more *hepatic ducts* draining the liver.

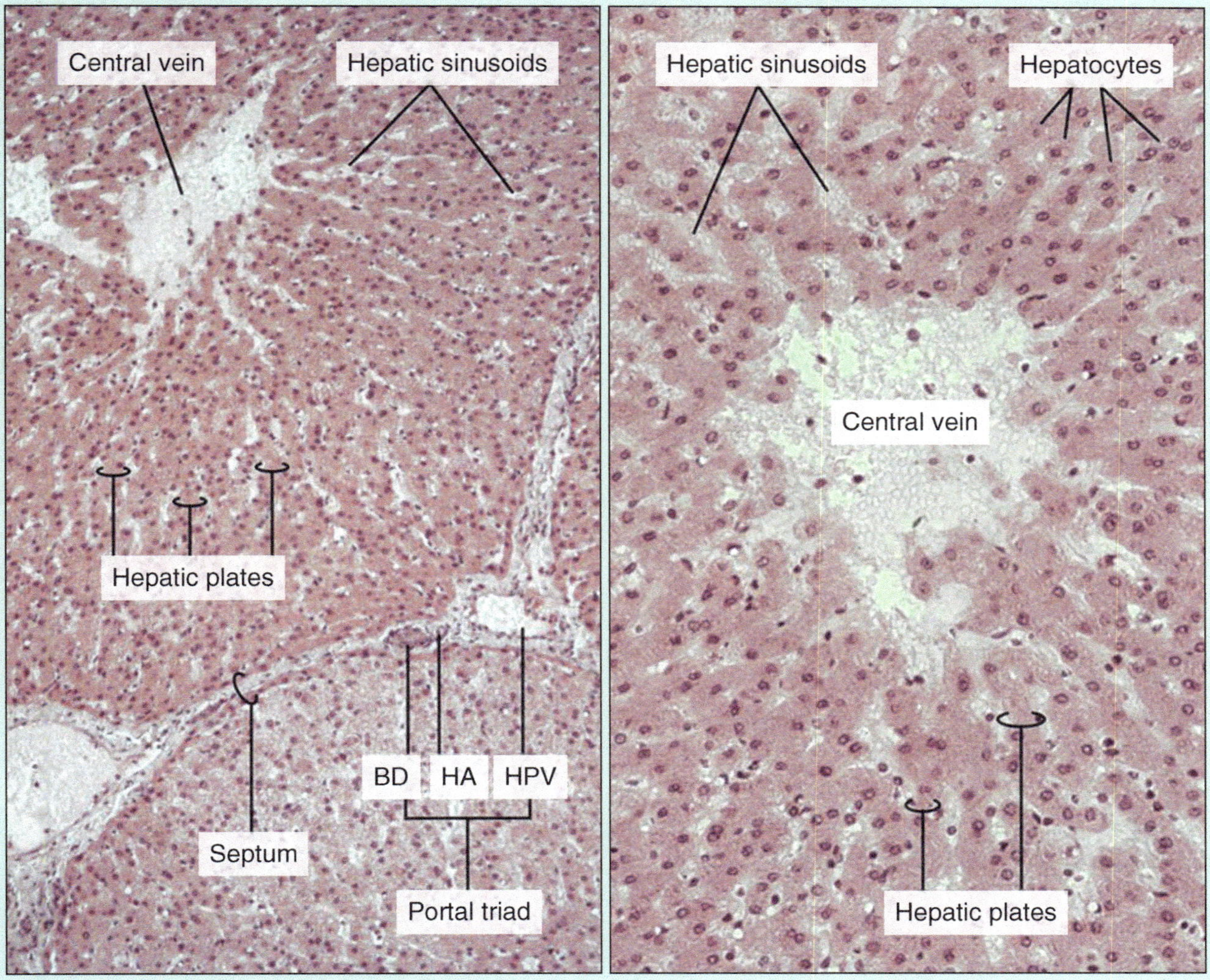

Fig. 12.42 Histological sections showing the main elements of the hepatic lobules (HE). *BD* bile duct, *HA* hepatic artery, *HPV* hepatic portal vein

The *coelom* is the body cavity, which is compressed by internal organs to a narrow, virtual space filled by a serous fluid. The cavity is lined by a membrane called the *peritoneum*. The intestines and other organs are attached to the dorsal body wall by double sheets of peritoneal membranes. The support for the intestine is the *mesentery*. Be careful not to tear the mesentery as you examine the intestine (Fig. 12.43).

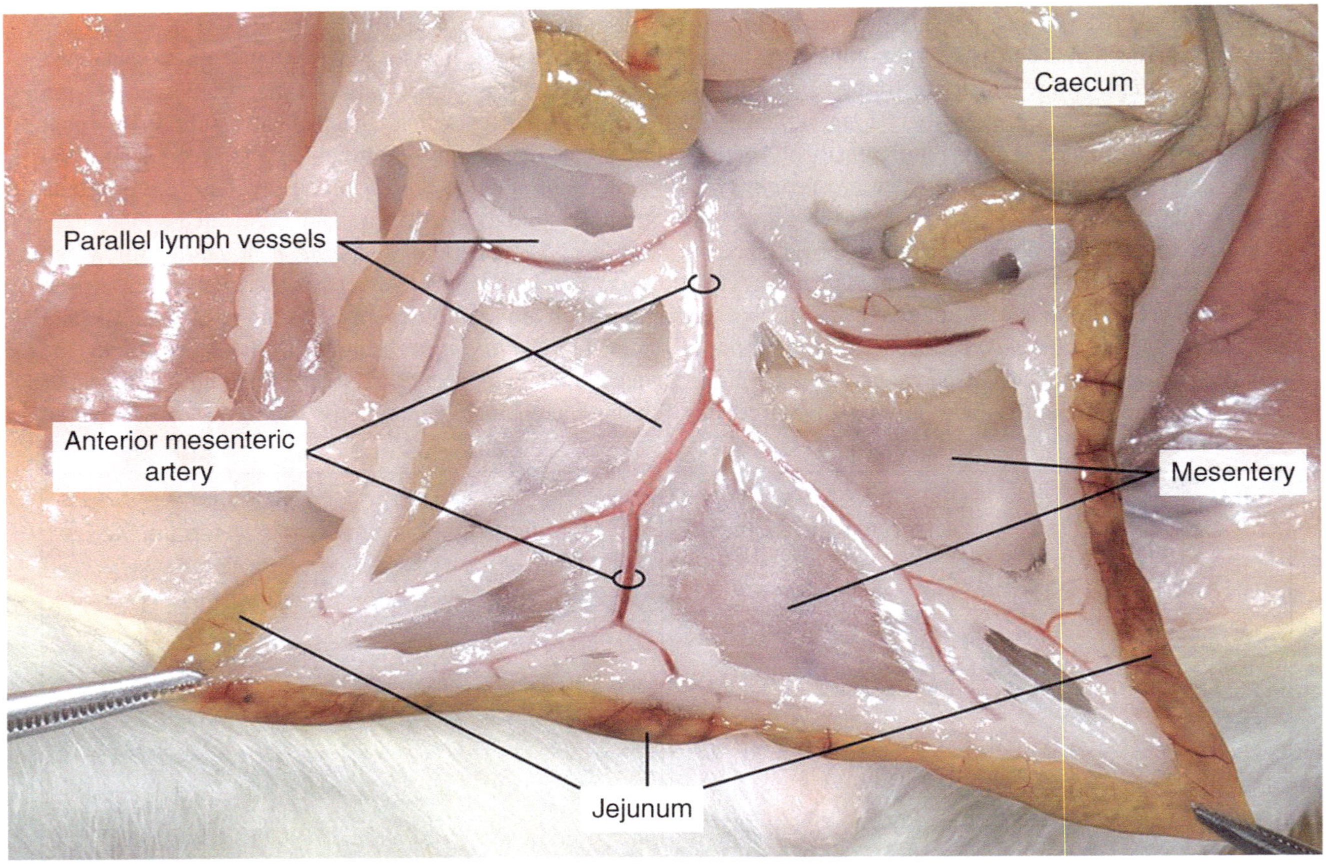

Fig. 12.43 A region of the small intestine (jejunum) with its supporting mesentery (peritoneal fold)

There is a special part of the coelom behind the liver cut away from the rest, called bursa omentalis. There are two attachment sites where the wall of this bursa connects with the peritoneum proper: the *lesser omentum* between the liver and the stomach and the *greater omentum* at the margin of the greater curvature of the stomach (Figs. 12.38 and 12.44). Grab the free edge of greater omentum with your forceps and pull it to observe in full (Fig. 12.38, *dashed arrows*).

Now follow the small intestine all the way along until it becomes the large intestine. The jejunum and ileum are both curly parts of the intestine, with the ileum being the last section before the small intestine becomes the large intestine (Fig. 12.44).

The *spleen* is the flattened, reddish organ lying just to the left and posterior to the stomach (Figs. 12.37 and 12.38). The spleen is about the same colour as the liver and is attached to the greater curvature of the stomach. It is curved and tongue-like and is associated with the circulatory system and functions in the destruction of blood cells and blood storage. A person can live without a spleen, but they are more likely to get sick as it helps the immune system function.

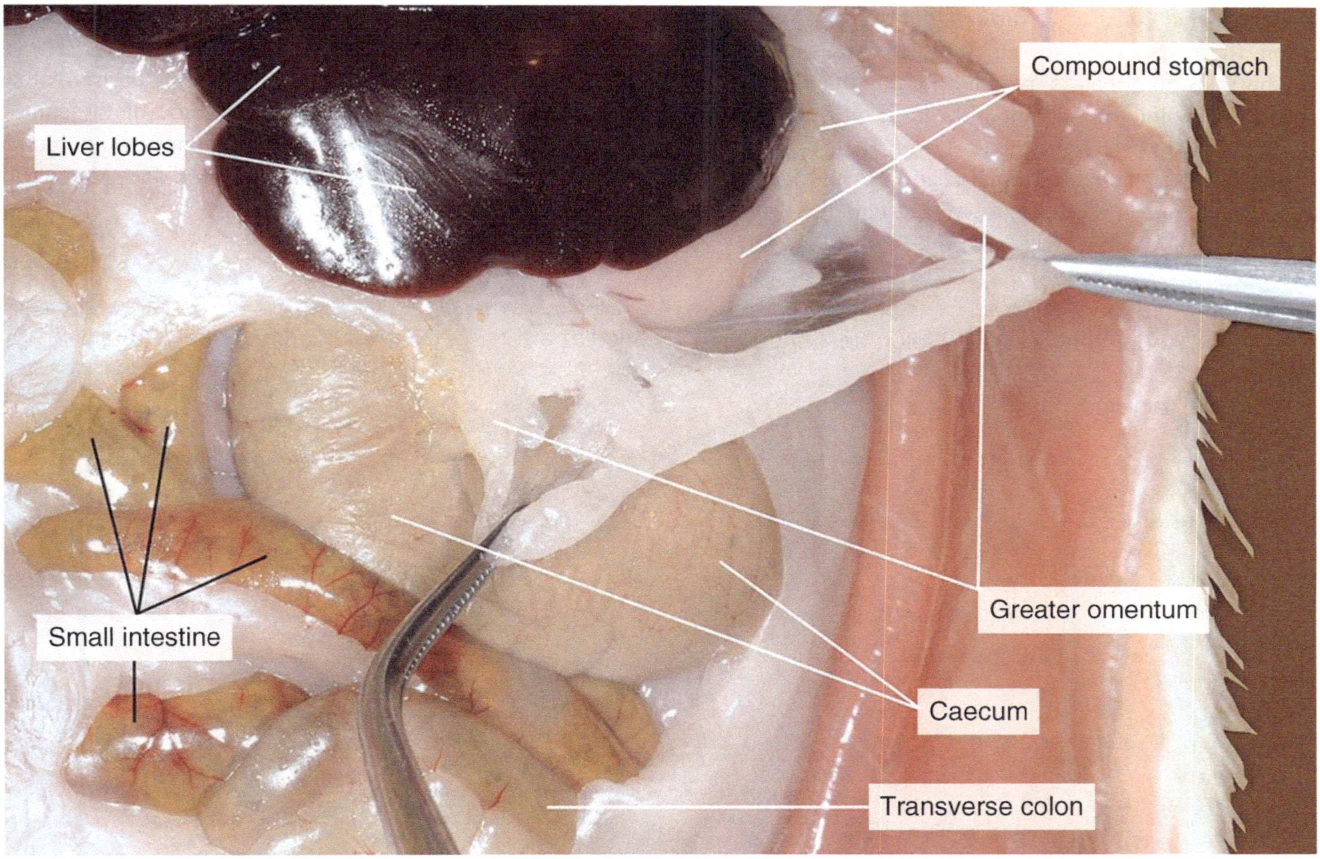

Fig. 12.44 A special fold of the peritoneum and bursa omentalis greater omentum at the greater curvature of the stomach

Locate *large intestine*, which is the large greenish tube that extends from the small intestine and leads to the anus. This is where the final stages of digestion and water absorption occur, and it contains a variety of bacteria to aid in digestion. It has a *caecum* at its beginning, which is a large flattened sac in the lower third of the abdominal cavity (Figs. 12.43 and 12.44). The caecum is a dead-end pouch and is similar to the appendix in humans. The caecum is a place where ingested cellulose is diverted from the main track and is digested by microbial fermentation. Rats, rabbits and hares produce a special faeces formed from the caecum product. They then ingest these faeces again, to digest it a second time. This behaviour is called coprophagy. The large intestine consists of further four sections: ascending colon (here the food travels upward), transverse colon (a short section that is parallel to the diaphragm) and descending colon (the section of the large intestine that travels back down towards the rectum) (Fig. 12.45). The *rectum* is the short, expanded, terminal section of the large intestine, which opens to the exterior through the *anus*. The rectum temporarily stores faeces before they are expelled from the body.

The excretory and reproductive systems of vertebrates are closely integrated and are usually studied together as the *urogenital system*. However, they do have different functions: the excretory system removes wastes and the reproductive system produces gametes (sperm and eggs). The female reproductive system also provides an environment for the developing embryo and regulates hormones related to sexual development.

The primary organs of the excretory system are the *kidneys* (Fig. 12.45).

If you move the intestines to one side, you will see one of the two kidneys embedded in the dorsal body wall and covered by the peritoneum (retroperitoneal position). These organs are large bean-shaped structures located towards the back of the abdominal cavity on either side of the spine. *Renal arteries* and *veins* supply the kidneys with blood (Fig. 12.45). Blood is filtered through the kidneys approximately once in every 45 min.

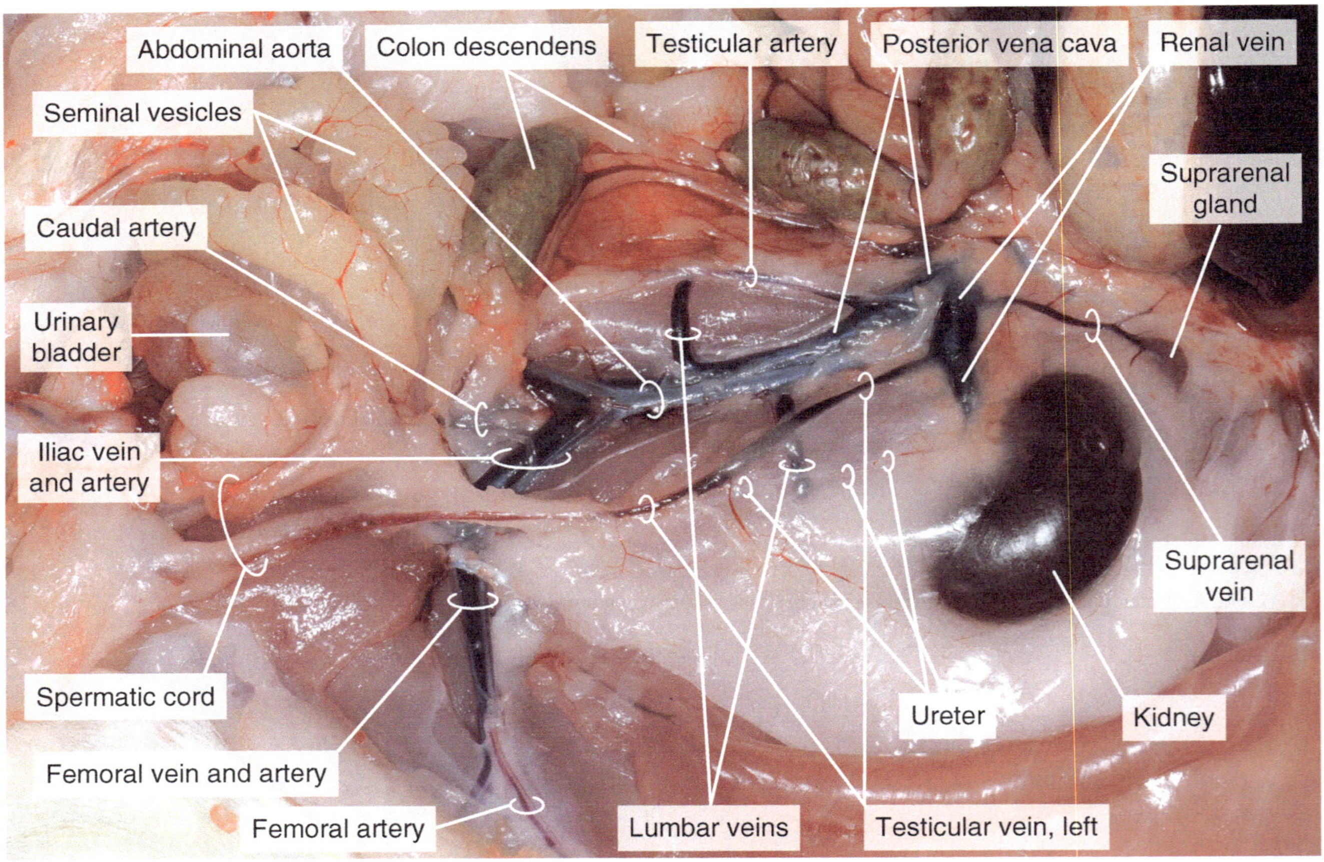

Fig. 12.45 The kidney, the suprarenal gland and the posterior circulation of a male rat

Mammalian *kidney* is composed of two main parts, the *cortex* and the *medulla*. Furthermore, the medulla is subdivided into three territories by their pattern (composition): the outer stripe of the outer medulla (OsOM), the inner stripe of the outer medulla (IsOM) and the inner medulla. The inner medulla forms the renal papilla, which is surrounded by the renal pelvis (Fig. 12.46, top left).

Each area of the kidney contains defined segments of the *nephron*, the functional unit of the kidney, and portions of the collecting duct system. The upper portions of the renal pelvis have specialised folds called fornices. The pelvis is lined by a special pseudostratified epithelium, referred to as urothelium (Fig. 12.46, *top left*). Hilus is formed by the invagination of the renal parenchyme where the ureter leaves the organ and vessels and nerves enter here. There is a narrow space between the renal parenchyme and the pelvis called renal sinus, which is often occupied by white adipose tissue. Original (developmental) architecture of the kidney is lobulated, which is maintained in its circulatory system. The renal artery and the vein give radial branches running from medulla towards the cortex in the septa separating the (original) lobes – these are the greatest vessels in the kidney. These vessels finish at the boundary of the medulla and cortex, where they give arched vessels. Smaller vessels enter the cortex from these arched vessels (they form, e.g. the glomeruli).

Renal corpuscle is the filtering apparatus of the kidney. It is composed of a glomerulus and a Bowman's capsule formed by parietal and visceral epithelia and mesangial cells (Fig. 12.46, *top right*). Vascular pole can be found at the opening of the Bowman's capsule, where the afferent and efferent arteries enter or leave the glomerulus (Fig. 12.46, *top right*, *black arrowhead*). The renal tubule, opening into Bowman's capsule at the urinary pole, has the next segments: proximal tubule with convoluted and straight parts, loop of Henle (with thin descending limb, U-turn and thin ascending limb), distal tubule with straight and convoluted parts and connecting tubule. The latter joins a collecting duct, which finally leads into wider papillary ducts opening on the tip of a renal papilla. Morphology of the epithelial lining of the renal tubule varies in these parts. The thinnest composed of squamous cells is found in the loop of Henle. Taller epithelium with cuboidal cells may show deep basolateral folds, and short apical microvilli can be seen in proximal and distal segments. The distal tubule wall shows a focal thickening formed by cuboidal cells with very regular round nuclei, the macula densa (Fig. 12.46, *top right*).

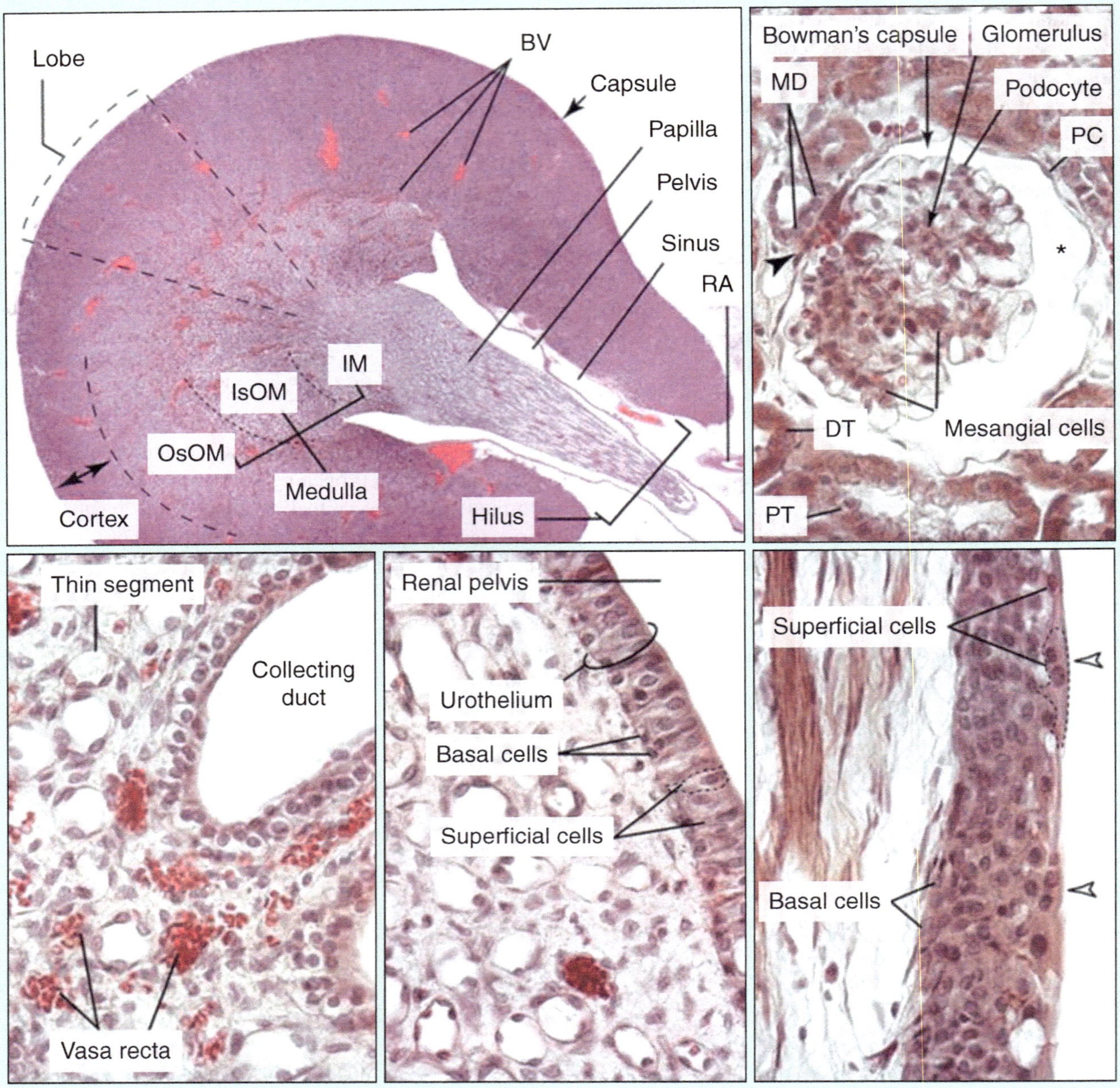

Fig. 12.46 Histological sections of the kidney (HE). *Asterisks* cavity of Bowman's capsule, *black arrowheads* vascular pole of renal corpuscle, *white arrowheads* superficial dome-shaped cells with two nuclei, *BV* blood vessel, *DT* distal tubule section, *IM* inner medulla, *IsOM* inner stripe of the outer medulla, *MD* macula densa, *OsOM* outer stripe of outer medulla, *PC* parietal epithelial cells, *PT* proximal tubule section, *RA* renal artery

The cortex contains glomeruli, Bowman's capsules, convoluting segments of the nephric tubule and connecting and collecting ducts. On the sections, duct profiles are featured by the orientation pattern in the stripes of renal parenchyma. The outer stripe of the outer medulla is formed by the straight segments, whereas the inner stripe of the outer medulla additionally contains thin ascending parts of the loop of Henle. The inner medulla is composed of thin segments and U-turn of nephric ducts. These ducts run parallel each other and the collecting tubules – orderliness of the pattern grows towards the renal papilla.

The *urothelium* is a pseudostratified or transitional epithelium (Fig. 12.46, *bottom right*). Its superficial cells form a barrier between the urine and the renal parenchyme with their special apical plasmamembrane. Superficial cells originate from basal cells hiding on the basal membrane. Two appearance of the urothelium can be seen on the opposing walls of the pelvis. The urothelium covering the surface of the renal papilla is thinner because it is stretched, whereas it is thicker on the opposite side of the pelvis. The superficial cells are the dome-shaped or *umbrella cells* stretching out their apical plasmamembrane to form a continuous barrier (Fig. 12.46, *bottom right, dotted lines*). They may frequently contain two cell nuclei (Fig. 12.46, *bottom right, white arrowheads*).

From the kidneys, the *ureters* lead to the *urinary bladder* (Figs. 12.47 and 12.51). The ureters are small white tubes and are often difficult to find. The urinary bladder empties to the outside through the *urethra*. Locate the urethral opening on both female and male rats. Optionally remove one of the kidneys and cut it lengthwise with a sharp scalpel or razor blade. Notice the very fine veins and arteries within.

The small yellowish glands embedded in the fat atop the kidneys are the *adrenal glands* (suprarenal glands), which secrete adrenaline into the blood during stress (Fig. 12.45).

The *abdominal aorta* after giving rise to the *coeliac artery* and *anterior mesenteric artery* pass between the kidneys. The *renal artery* is given off to each of the kidneys but not in the same level. Then the abdominal aorta gives rise to short paired the *genital arteries* and posteriorly the *lumbar arteries*. At the end, the abdominal aorta forks into the *iliac arteries*, each of which enters the hindlimb as the *femoral artery* (Fig. 12.45).

To expose the genital organs, your initial cut must be lengthened posteriorly as far as the anus, passing on either side the urinary and genital apertures in females or backwards around the penis and between the two halves of the scrotal sac in males (Fig. 12.24, cutlines 7, 8).

Female reproductive organs The reproductive system of female rats is relatively simple. You should examine both female and male systems. In the female, the prominent *uterine horns* extend towards the kidneys (Fig. 12.47).

This duplex uterus is common in some animals and will accommodate multiple embryos (a litter). In contrast, a simple uterus, like the kind found in humans, has a single chamber for the development of a single embryo.

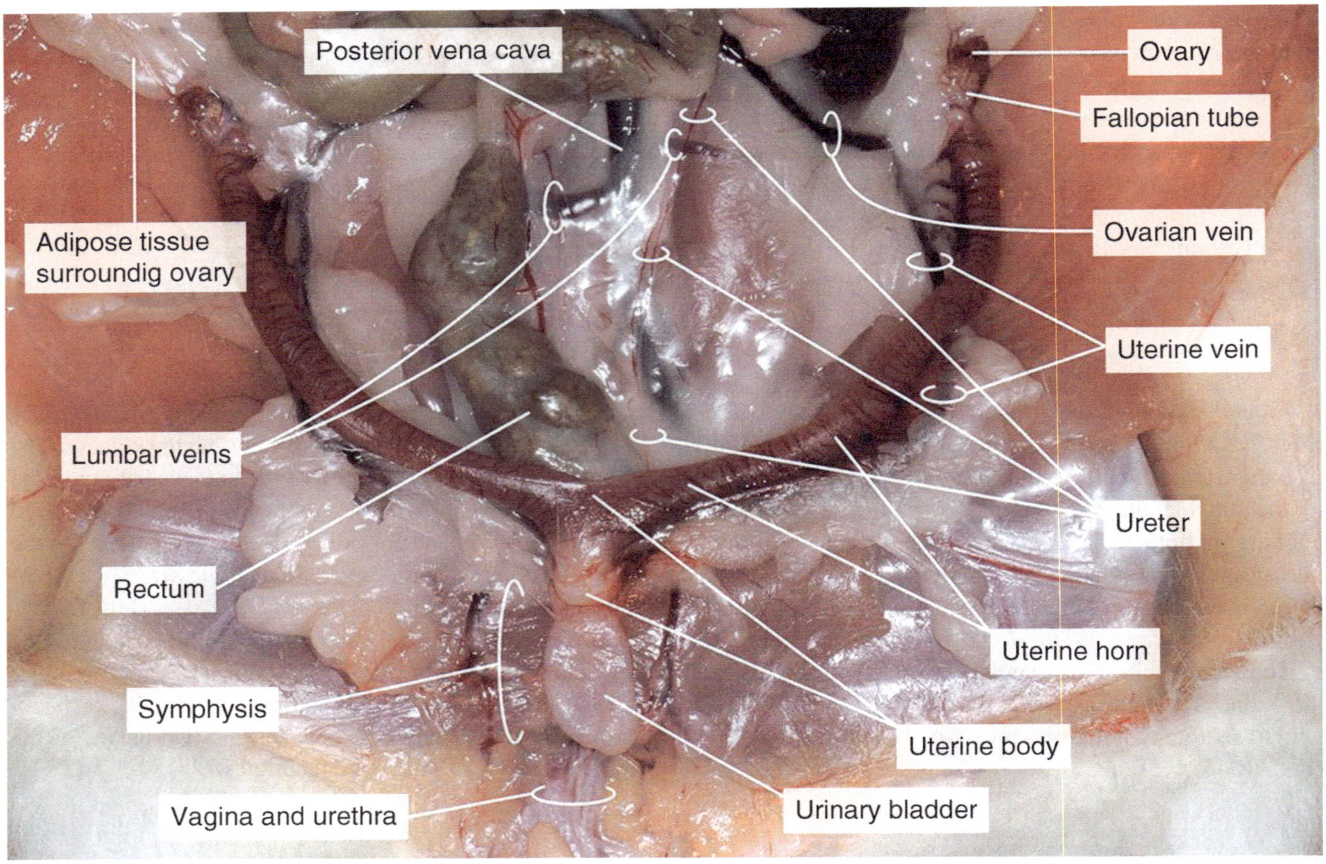

Fig. 12.47 The reproductive system of a female rat

Histological section of a horn of the *uterus* shows a typical pattern (Fig. 12.48). They have inner endometrium, middle myometrium and outer perimetrium (Fig. 12.48, left). *Endometrium* is identical with the mucosa. Its epithelium is a single-layered columnar epithelium, which is continuous with endometrial (uterine) glands running perpendicular to the luminal surface. The stroma consists of a highly cellular connective tissue (spinocellular tissue, like in case of the ovary), which contains blood vessels and scattered lymphocytes. Endometrial glands are formed by columnar cells, which secrete glycoproteins and glycogen (Fig. 12.48, right). The endometrium undergoes dramatic changes during the menstrual cycle, it builds up gradually and its upper layer undergoes necrosis if there is no fertilisation. The *myometrium* is composed of smooth muscle bundles mingled with loose connective tissue. There is a blood vessel plexus between its inner and outer layers. The vessels embedded in muscle layers enter the suspensory ligaments of the organ (Fig. 12.48, left). The p*erimetrium* is identical with the serosa (peritoneum) with a thin connective tissue (propria) and serous epithelium.

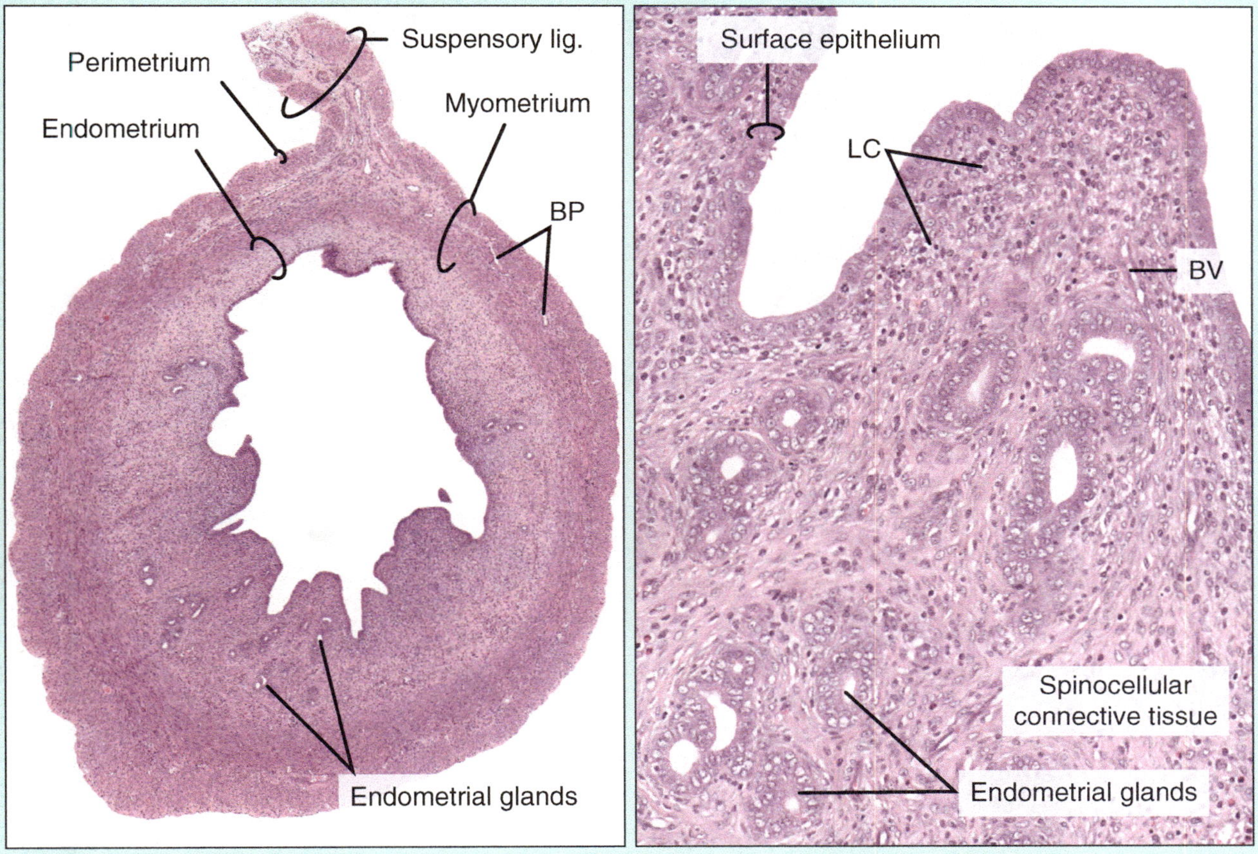

Fig. 12.48 Histological composition of a horn of the uterus (HE). *BP* blood vessel plexus, *BV* blood vessel, *LC* lymphocytes, *lig* ligament

Where the horns join dorsal to the urethra is the *vagina*, which opens to the exterior through the *vaginal opening* (vulva) (Fig. 12.6). On the exterior, the anus is just ventral to the tail, the vaginal opening is ventral to the anus and the urethral opening is ventral to the vagina. At the anterior end of each uterine horn, there is a short, convoluted *fallopian tube* or oviduct, which opens into a transparent, membranous pouch (periovarial sac) around the small, round *ovary* (Fig. 12.49).

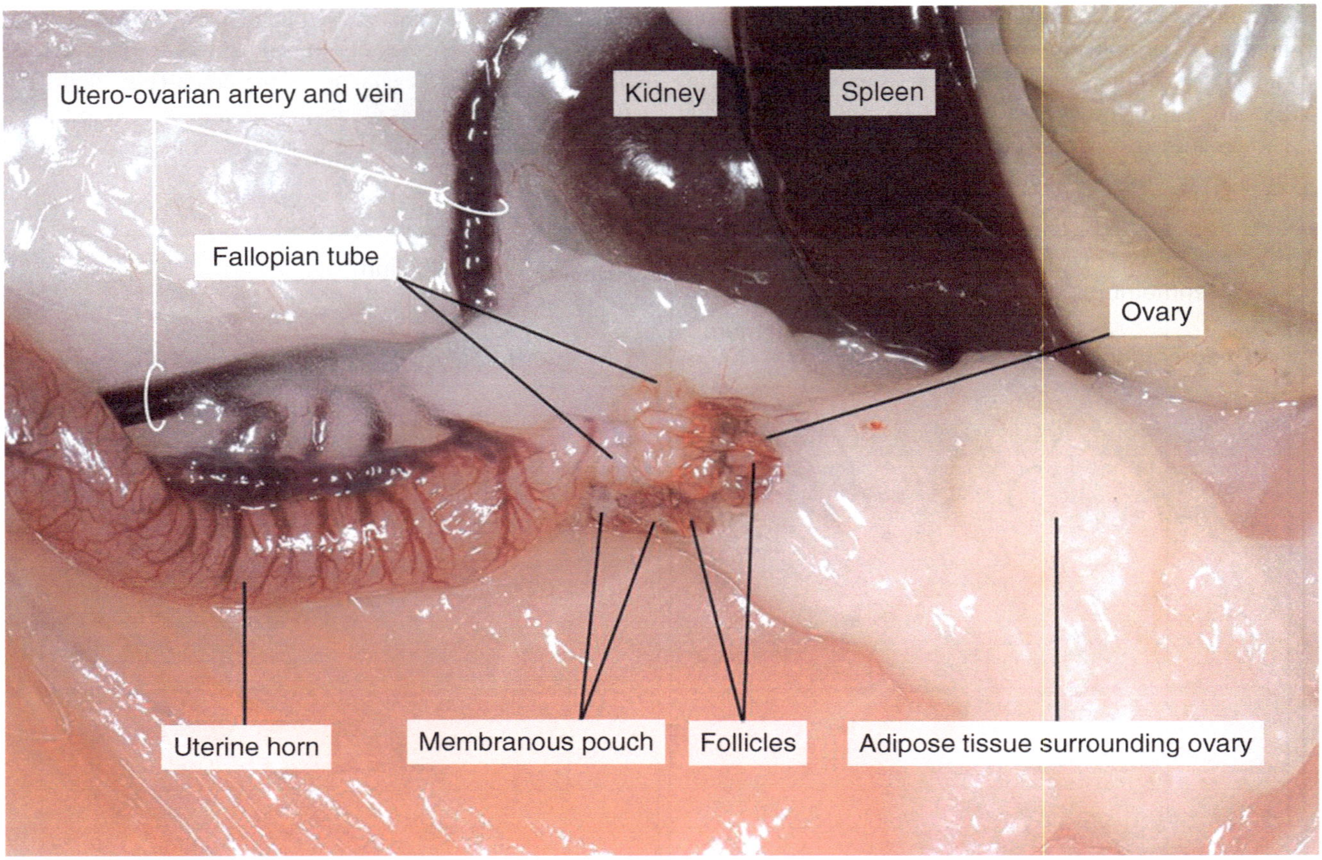

Fig. 12.49 The left ovary, fallopian tube (oviduct) and distal part of the uterine horn in a female rat. The membranous pouch is the periovarial sac

The *ovary* of rodents is surrounded by a thin *periovarial sac* (PS), which separates the organ in the abdominal cavity (Fig. 12.50, *top left*, *bottom left*). The surface of the organ is covered by a single layer of cuboidal epithelium, named as *germinal epithelium* (GE, this is a misnomer, because it is not the source of germ cells). Just beneath this epithelium, there is a collagenous sheet, which is homologous with the tunica albuginea of testis. The inner area is divisible into two parts: the outer *cortex*, which contains developing follicles and corpora lutea (CL), and the inner *medulla*, which receives blood vessels and nerves from the hilus. Stroma is formed by special connective tissue (spinocellular connective tissue) made of fusiform fibroblast-like cells arranged in characteristic whorls (the extracellular matrix almost absent) (Fig. 12.50).

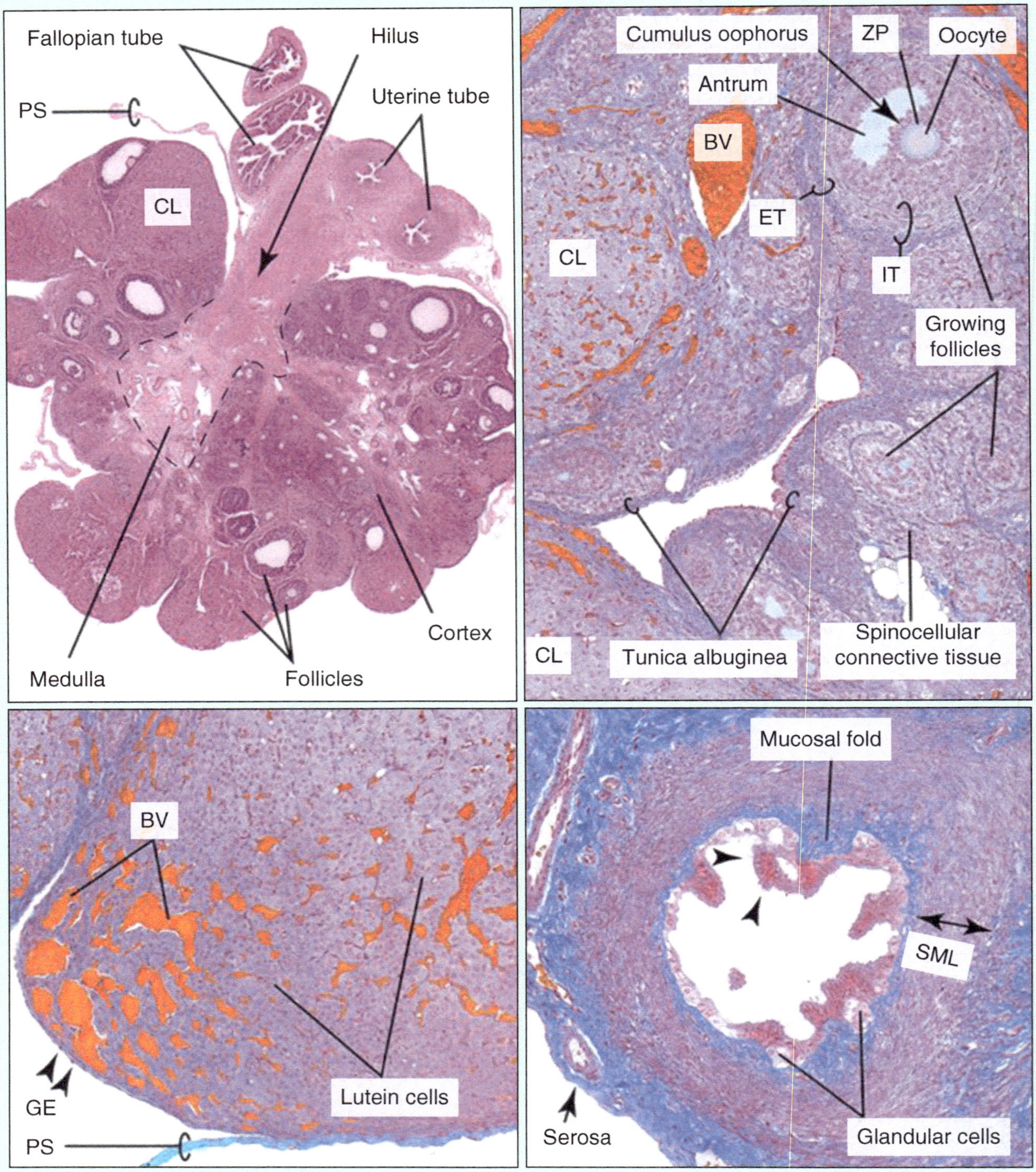

Fig. 12.50 Histological sections of the female reproductive organs, ovary (*top* and *bottom left*) and oviduct (*bottom right*) (*top left*, HE; others, Azan). *Black arrowheads* (*bottom left*) cell nuclei of the germinal cells, *black arrowheads* (*bottom right*) ciliated cells, *BV* blood vessels, *CL* corpus luteum, *ET* external theca, *GE* germinal epithelium, *IT* internal theca, *PS* periovarial sac, *SML* smooth muscle layer, *ZP* zona pellucida

Follicles undergo the stages of folliculogenesis, during which follicular epithelium become multilayered and oocyte enter the meiotic division. *Primordial follicles* are the smallest ones, and they are located in the deeper layer of the cortex. They have unilayered follicular wall composed of flattened granulosa cells. *Primary follicles* are bigger, because *granulosa cells* become cuboidal and start to proliferate. Stromal cells surrounding the follicle begin to differentiate into spindle-shaped cells, which constitute the *theca folliculi*. At the same time, the oocyte increases and secretes a thick layer of proteoglycan-rich material, the *zona pellucida* (ZP). It is clearly seen between the oocyte and the surrounding granulosa cells (Fig. 12.50, *top right*). In the *secondary follicle*, the granulosa cells continue their proliferation resulting to a multilayered follicular wall. Some cavities began to form in it – later these cavities unite and form the *antrum*, filled with antral fluid containing growth factors and hormones secreted by the granulosa layer. There is a thickening in the granular layer, referred to as *cumulus oophorus*, which contains the *oocyte*. The follicular wall becomes multilayered, and the theca differentiates into inner and outer layers. The *internal theca* (IT) cells synthesise androgens, what granulosa cells transform into oestrogens. The *tertiary follicles* are the greatest ones: their antrum is the largest. They move into the superficial layer of the cortex and cause a protrusion on the surface. The tissues between the matured tertiary follicle (*Graafian follicle*) and the germinative epithelium of the ovary are eroded until the follicle ruptures. The ovulating oocyte is surrounded by its zona pellucida, and granulosa cells form corona radiata. After ovulation, the walls of the empty follicle collapse and fold. Its empty cavity becomes invaded by transformed granulosa and internal theca cells (*lutein cells*), capillaries and connective tissue to form *corpus luteum* (CL). Lutein cells contain carotenoid yellowish pigment in their cytoplasm. They synthesise progesterone and oestradiol. In the absence of fertilisation, the corpus luteum atrophies.

The coiled *oviduct* opens in the cavity of the periovarial sac. Its lining epithelium composed of columnar, ciliated epithelial and unicellular gland cells. It has well-developed *smooth muscle layers* (SML) to help spermatozoa locomotion upwards and the oocyte (or fertilised oocyte) moving downwards into the uterus.

Oviducts are extremely tiny and may be difficult to find without a dissecting microscope. Often the ovary is hard to find as well. Try feeling in the fat at the end of the uterus for a small, hard, round body. Some female rats may be pregnant and have small embryos in the uterine horns. If yours is pregnant, cut longitudinally along the horns and observe the embryos.

Male reproductive organs The male reproductive system is more complicated than the female's. Locate the male's *scrotum* and cut longitudinally through just the skin to locate the *testes* (singular: testis) (Fig. 12.51).

Separate the skin from the testes and continue the cut up to the abdominal cavity. On the surface of the testis is a coiled tube called the *epididymis*, which collects and stores sperm cells (Fig. 12.51).

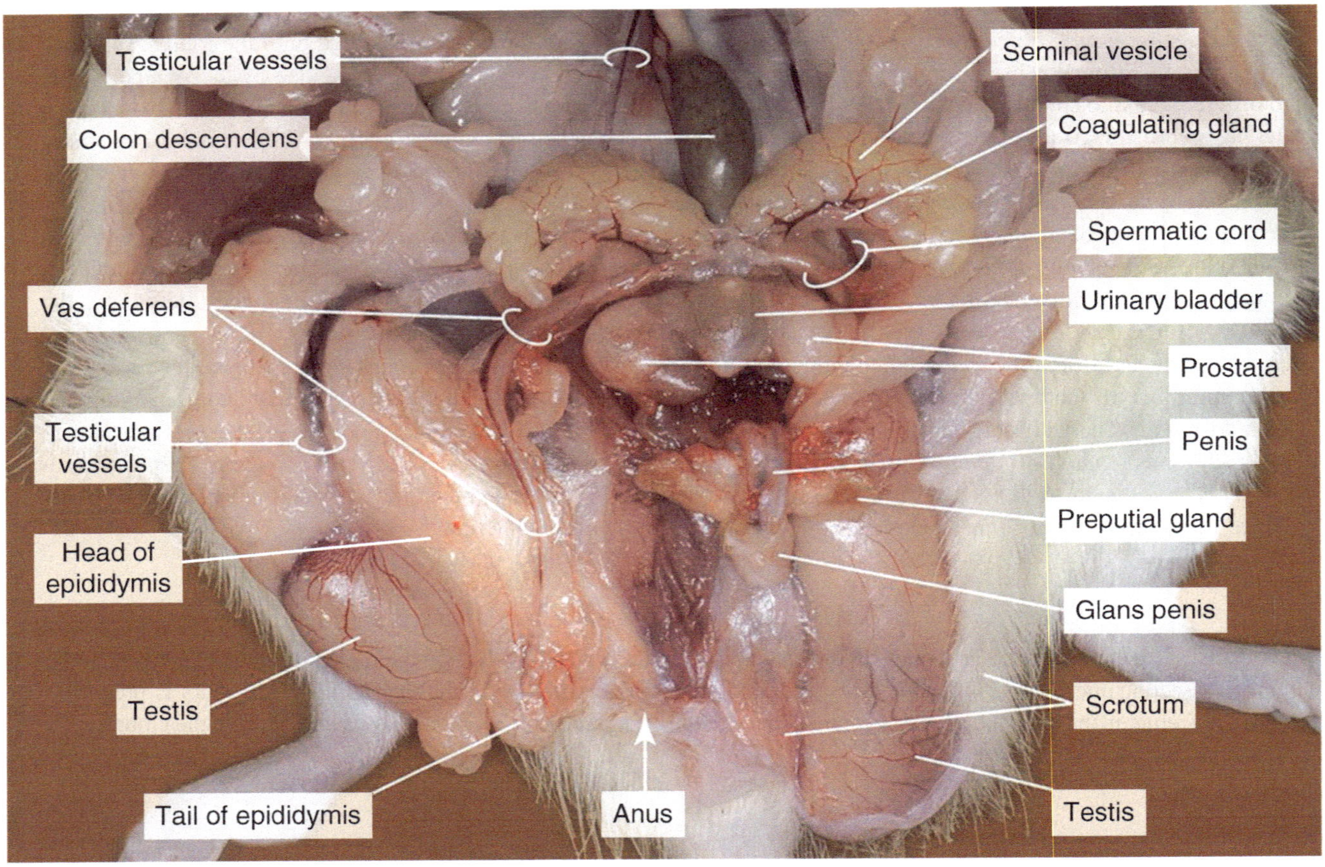

Fig. 12.51 The reproductive system of a male rat

The *testis* has a tight connective tissue capsule named *tunica albuginea*. It is covered by serous epithelium of the visceral layer of the tunica vaginalis (which is the lower end of the peritoneal protrusion, whose formation precedes the descent of the foetal testis from the abdomen to the scrotum). Beneath the tunica albuginea is a thin layer of connective tissue containing the superficial blood vessels. Vessels and genital ducts enter or leave the testis at the mediastinum. Septa (referred to as interstitial tissue) from the mediastinum extend internally to partition the testis into lobules. Each lobule contains convoluted seminiferous tubules, in which spermiogenesis occurs. Pattern of the neighbouring tubules shows slight differences (their segments that are seen in micrographs showing different territories of the same section were in different stages of developmental sequence of spermatogenesis).

The original epithelium (in foetus) lining the seminiferous tubules is formed by the supporting cell type, called *Sertoli cells* (S). They have large, light, euchromatic, pear-shaped nuclei. Adjacent Sertoli cells join laterally to each other by tight junctions, which create a diffusion barrier between the extratubular and luminal compartments. This is the blood-testis barrier. Their lateral and apical plasma membrane forms complex recesses, which surround developmental sperm cells (Fig. 12.52).

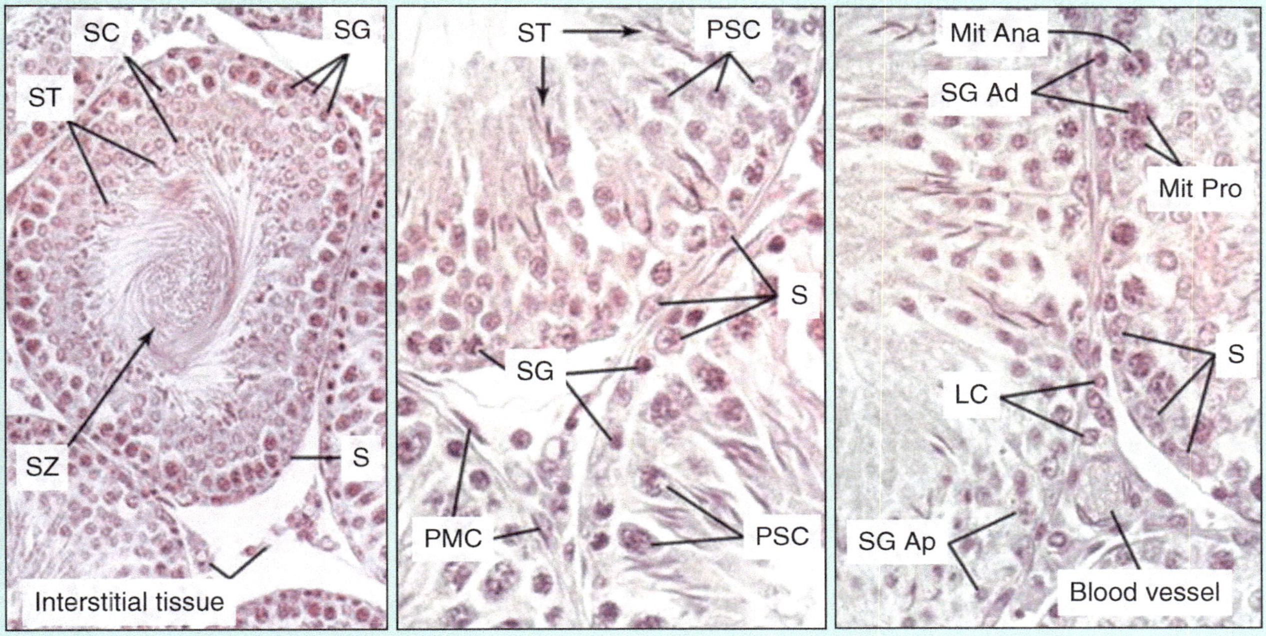

Fig. 12.52 Histological section of the seminiferous tubules in the testis (HE). *LC* Leydig cells, *Mit Ana* mitotic anaphase, *Mit Pro* mitotic prophase, *PMC* peritubular myoid cells, *PSC* primary spermatocytes, *S* Sertoli cells, *SC* spermatocytes, *SG* spermatogonia, *SG Ad* dark spermatogonia, *SG Ap* pale spermatogonia, *ST* spermatids, *SZ* spermatozoa

Spermatogonia (SG), as stem cells, are situated on the basal membrane (Fig. 12.52, *middle*). The three basic types of spermatogonia are the dark type A cells (SG Ad) with dark, condensed nucleus; the pale type A cells (SG Ap) with pale, euchromatic nucleus; and the type B cells. Dark spermatogonia divide mitotically to maintain the stem cell population. Some divisions generate pale cells, which also divide mitotically but remain linked in clusters by fine cytoplasmic bridges. These are the precursors of type B cells. The latter leave the basal membrane and enter meiotic prophase as primary spermatocytes (Fig. 12.52, *middle*). *Primary spermatocytes* (PSC) are large cells with large round nuclei, in which the nuclear chromatin is condensed into dark, threadlike, coiled chromatids. These cells become secondary spermatocytes during the first phase of meiotic division. Only a few secondary spermatocytes can be seen in sections because they rapidly undergo the second meiotic (equatorial) division to produce *spermatids* (ST). The series of these divisions, referred to as spermatocytogenesis, is responsible for the multiplication of the cell number. Spermatids do not divide but gradually mature into *spermatozoa* (SZ) by a series of nuclear and cytoplasmic changes known as spermiohistogenesis. All of these maturational changes take place, while the spermatids remain closely associated with Sertoli cells (S) (Fig. 12.52). Spermatids become elongated, flagellated cells. Finally, their excess cytoplasm is detached as residual body that is phagocytosed by Sertoli cells. During this process, spermatids loose their interconnected cytoplasmic bridges and separate from each other as morphologically mature spermatozoa, which are released into the tubule. They form characteristic whirl-like pattern in the centre of the lumen (Fig. 12.52, left). The abovementioned sperm cell generation cycle occurs at any levels of the seminiferous tubule. Stages in the cycle are characterised by the presence of different combinations of cells.

Seminiferous tubules are surrounded by *peritubular myoid cells* (PMC) and separated by interstitial connective tissue (Fig. 12.52, middle). This matrix contains *Leydig cells* (LC), which are responsible for androgen hormone synthesis (Fig. 12.52, right). The myoid cells are contractile, and their rhythmic activity propels non-motile spermatozoa through the tubule towards the rete testis.

Epididymis is separated from the testis by its own capsule and the tunica albuginea, which are perforated only by the tubules of rete testis. Epididymis has three main parts. Its *head* connects to the rete testis and contains several ducts, which are lining epithelium formed by cuboidal-columnar cells. *Body* of the epididymis is formed by one folded epididymis duct (Fig. 12.53). Its epithelium contains two main cell types. *Principal cells* (PC) or main cells are tall cylindrical cells with basal, oval nuclei. They bear long, nonmotile microvilli termed stereocilia (Fig. 12.53, *right*, *arrowhead*). Stereocilia function to enlarge the surface for resorption of fluid from the testicular secretions. In addition, the principal cells secrete essential glycoproteins for the maturation of spermatozoa and endocytose various components of the seminal fluid. *Basal cells* (BC) lie on the basal membrane, and they are thought to be the precursors of principal cells. The thickness of the epithelium varies from the head towards the tail of the organ. These duct system is enveloped by thin connective tissue sheet and well-developed smooth muscular layer, which propel spermatozoa towards the tail by peristaltic contractions. The third (last) part of the organ is its *tail*, in which the epididymal duct straightens to form the vas deferens.

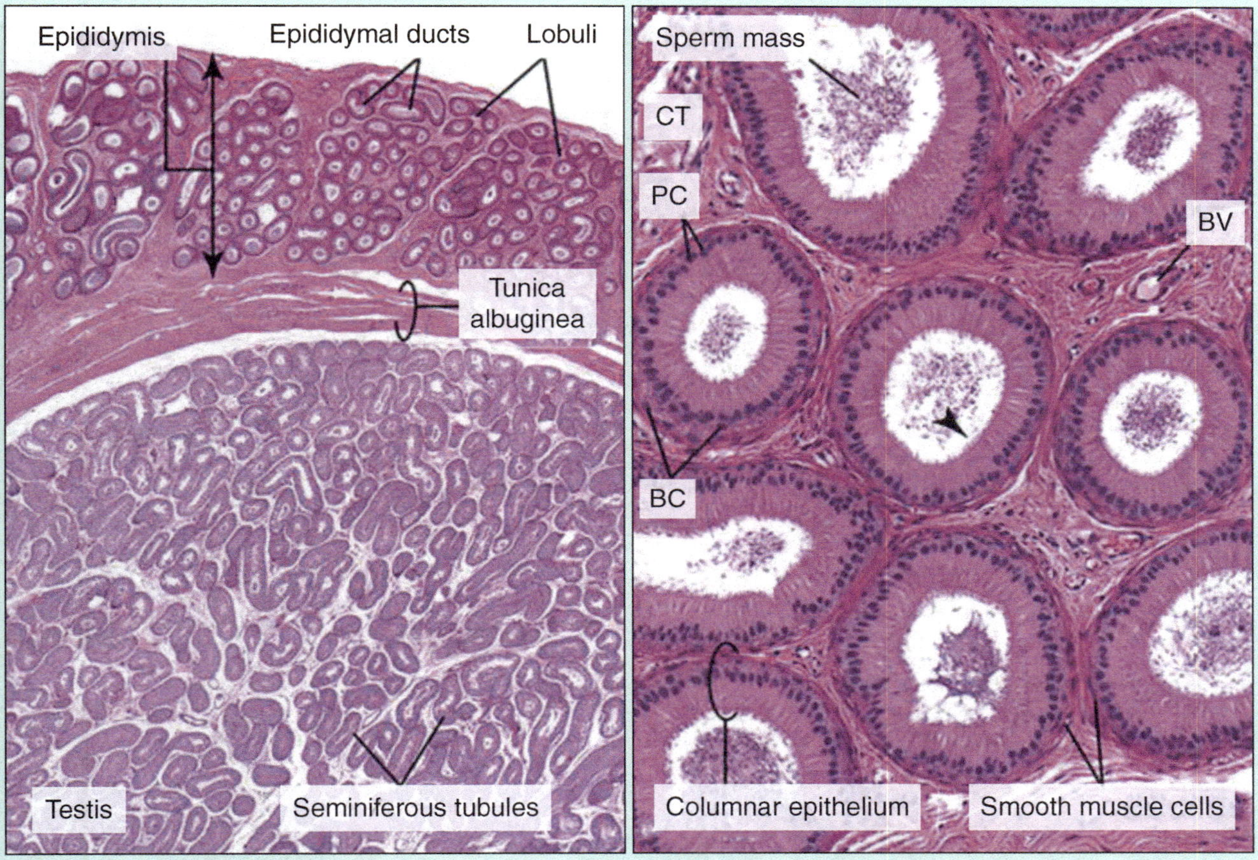

Fig. 12.53 Histological section of the body of epididymis (HE). *Arrowhead* stereocilia, *BC* basal cells, *BV* blood vessel, *CT* connective tissue, *PC* principal cells

Find the *vas deferens* that leads from each testis to the urethra (Fig. 12.51). You will probably have to cut through some muscles in the pelvis area to locate some of these structures. Be careful as you do so. To either side of the bladder are two sets of glands. The smaller, round, gland below the bladder is the *prostate gland,* and it is partially wrapped around the penis. The larger pair of glands are the *seminal vesicles*, which are actually two different glands — the *vesicular glands* and the *coagulating glands* (Fig. 12.51). The seminal vesicles and the prostate gland secrete materials that form the seminal fluid (semen). Follow the urethra as it goes through the *penis*. In males, the bladder and the testes both empty to the outside through the urethra.

After completing the dissection, remove the *heart* from the pericardial sack. You will need to sever the arteries and veins connecting the heart to the circulatory system. Do this slowly and carefully so that you do not cut more than is necessary. Leave as much of the veins and arteries attached to the heart as possible. Identify the aorta, left and right atrium and left and right ventricle. Carefully insert your probe into these openings and work it into the centre of the heart. Transect the apex in one third of the heart's length to see the extent of each ventricle (Fig. 12.54).

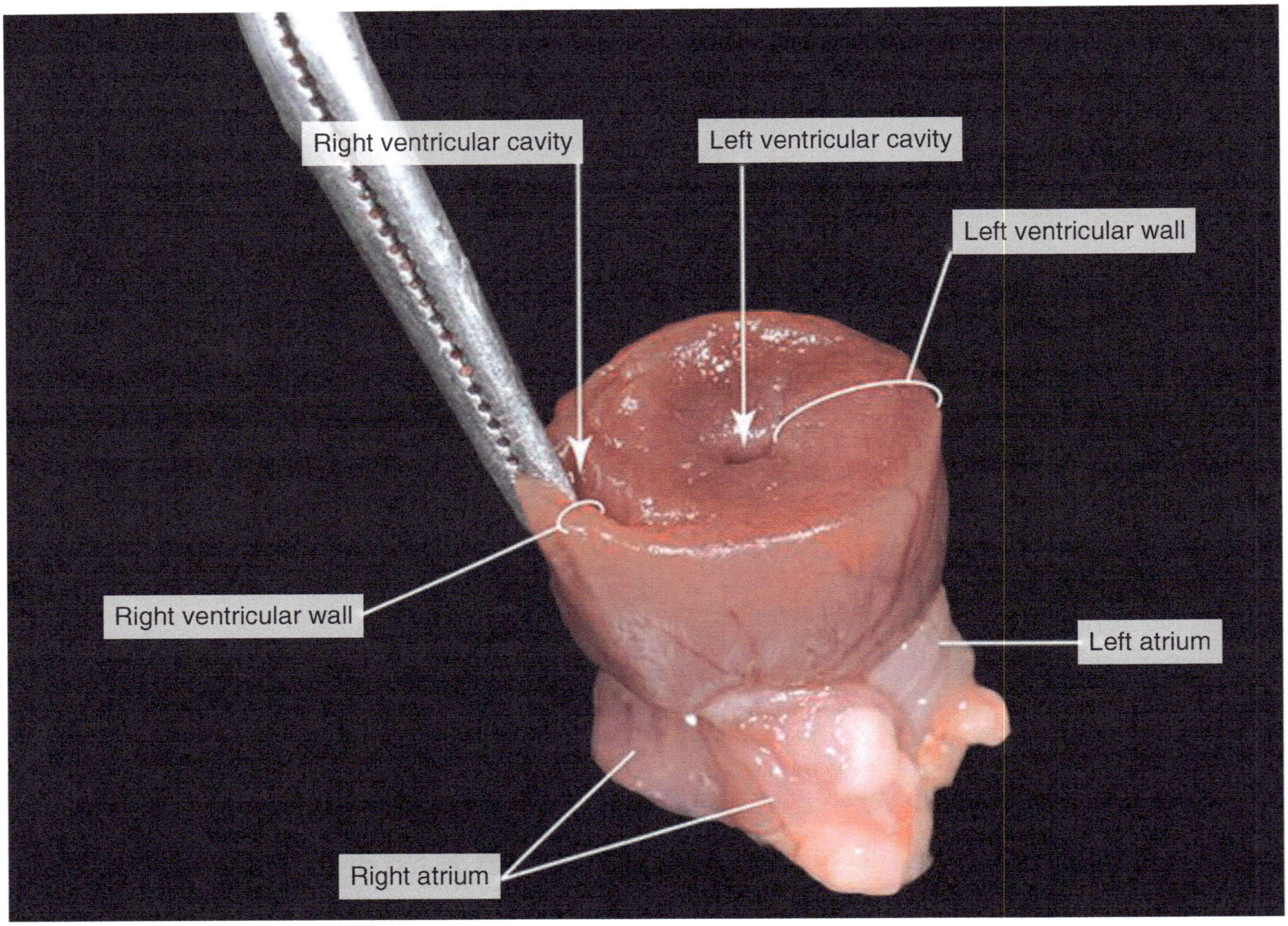

Fig. 12.54 The walls of the two ventricles of the rat's heart are highly asymmetric

Note that the *left ventricle* occupies the larger half of the heart; the *right ventricle* is much smaller as the resistance of the pulmonary circulation is smaller. Make an incision from the left atrium to the left ventricle with your scissors. Wash the remnants of blood clots away. Try to locate the *bicuspid valve*, which open and close the ventricle (Fig. 12.55). A similar set of valves (tricuspid valves) is found between the right atrium and its corresponding ventricle.

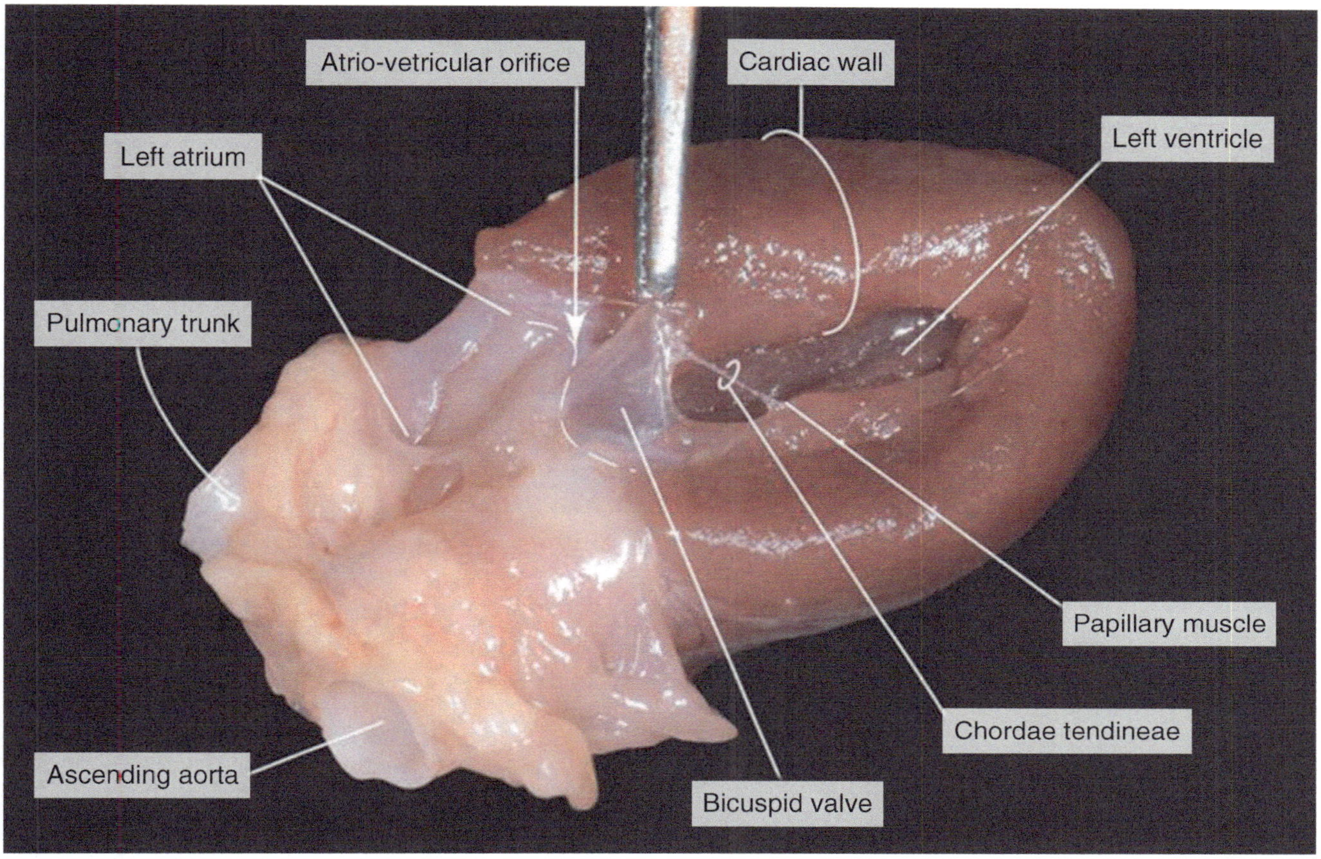

Fig. 12.55 The bicuspid valve in the atrio-vetricular opening of the left half of the rat's heart

When the ventricles contract, blood is forced against bicuspid valves, forcing them to shut and preventing the flow of blood back into the atria. The chordae tendineae prevent the valves from flapping back into the atria, which would permit the backflow of blood. Finally, cut from the stump of the aorta back to the left ventricle with your scissors. Try to locate the three *semilunar valves* in the exit of the aorta, which prevent the blood flowing back into the ventricles (Fig. 12.56). A similar set of valves is found at the junction of the pulmonary artery and the right ventricle.

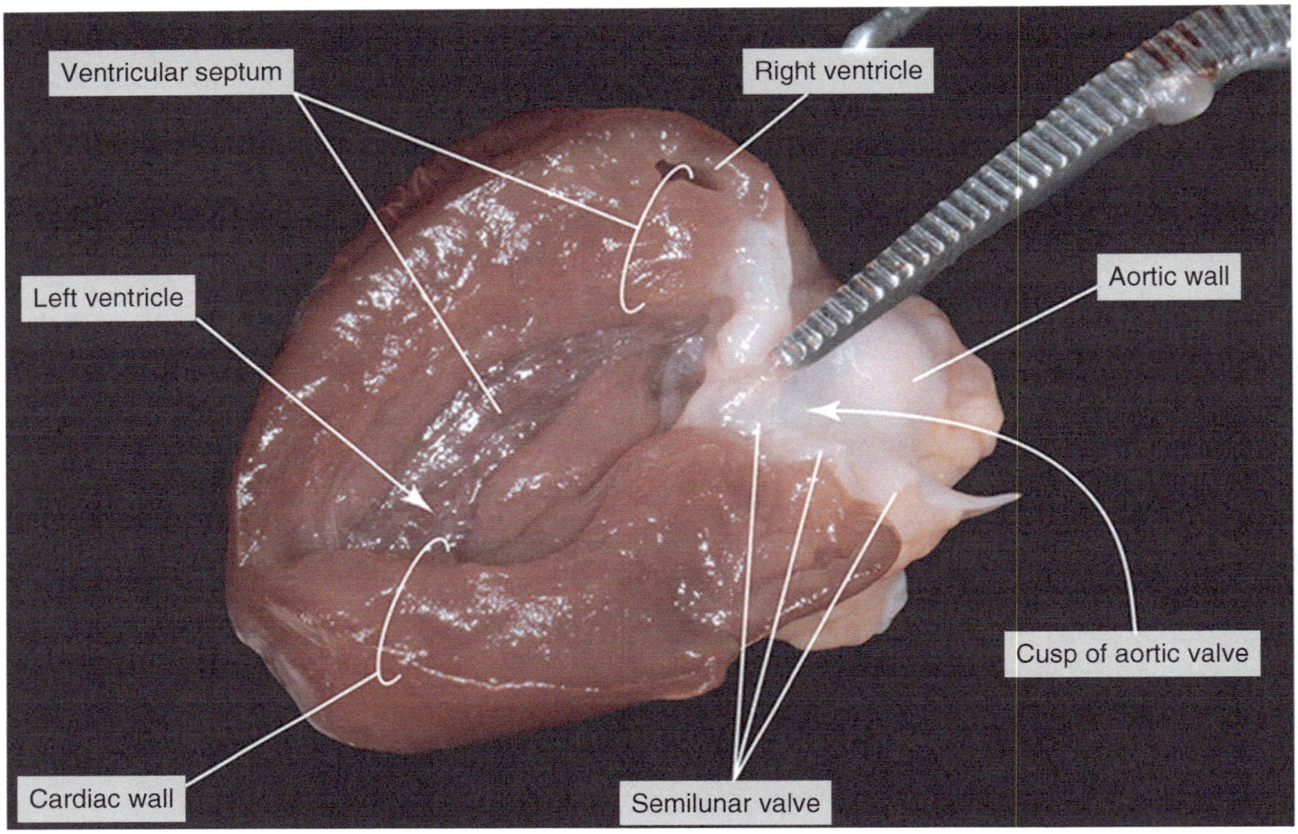

Fig. 12.56 The three semilunar valves in the exit of the aorta

After that you can isolate the tongue-pharynx-larynx-trachea-lung complex. Start by cutting with your scissors from the corners of the mouth back to the ears. Pry the jaws apart and complete the cut so that the mouth can be opened and the pharynx exposed (Fig. 12.57).

Note the *soft palate* and the ridged *hard palate* anterior to it (Fig. 12.57). At the base of the tongue is an opening, the *glottis*, into the *larynx*. Observe the *tongue*; note its three parts. It is covered with filiform and fungiform papillae. The single largest papilla is the circumvallate papilla at the posterior of the root part.

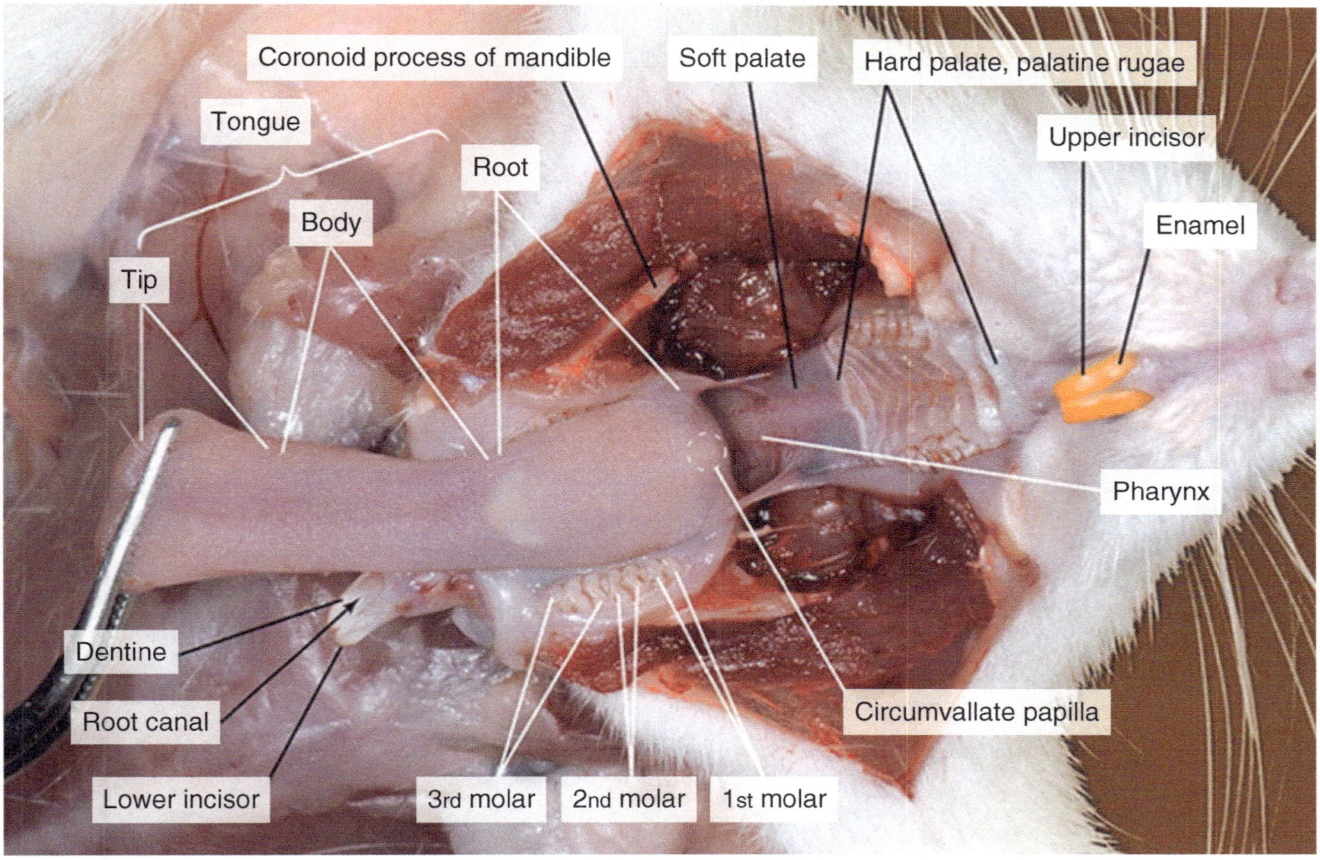

Fig. 12.57 Widely opened mouth of the rat by cutting the muscles and the jaws

Circumvallate papilla is a large cylindrical structure on the dorsal surface of the tongue (Fig. 12.58, left). The papilla is covered with keratinised *stratified squamous epithelium* (SQE). It is surrounded by a circular sulcus formed by epithelial walls. Numerous *taste buds* are scattered in both walls of the sulcus, and small serous glands (glands of von Ebner) open into the base of the sulcus (OSG). Taste buds sense the substances dissolved in the secretion of these serous glands. The taste buds (see *dotted line on bottom right*) are formed by pale *receptor cells* (RC) with heterochromatic nucleus and *supporting cells* (SC) with dense nucleus. The *taste pore* (TP) opening into the sulcus is formed by apical surface of both cell types. The papilla lies on propria of the mucous layer containing nerve branches innervating the epithelium.

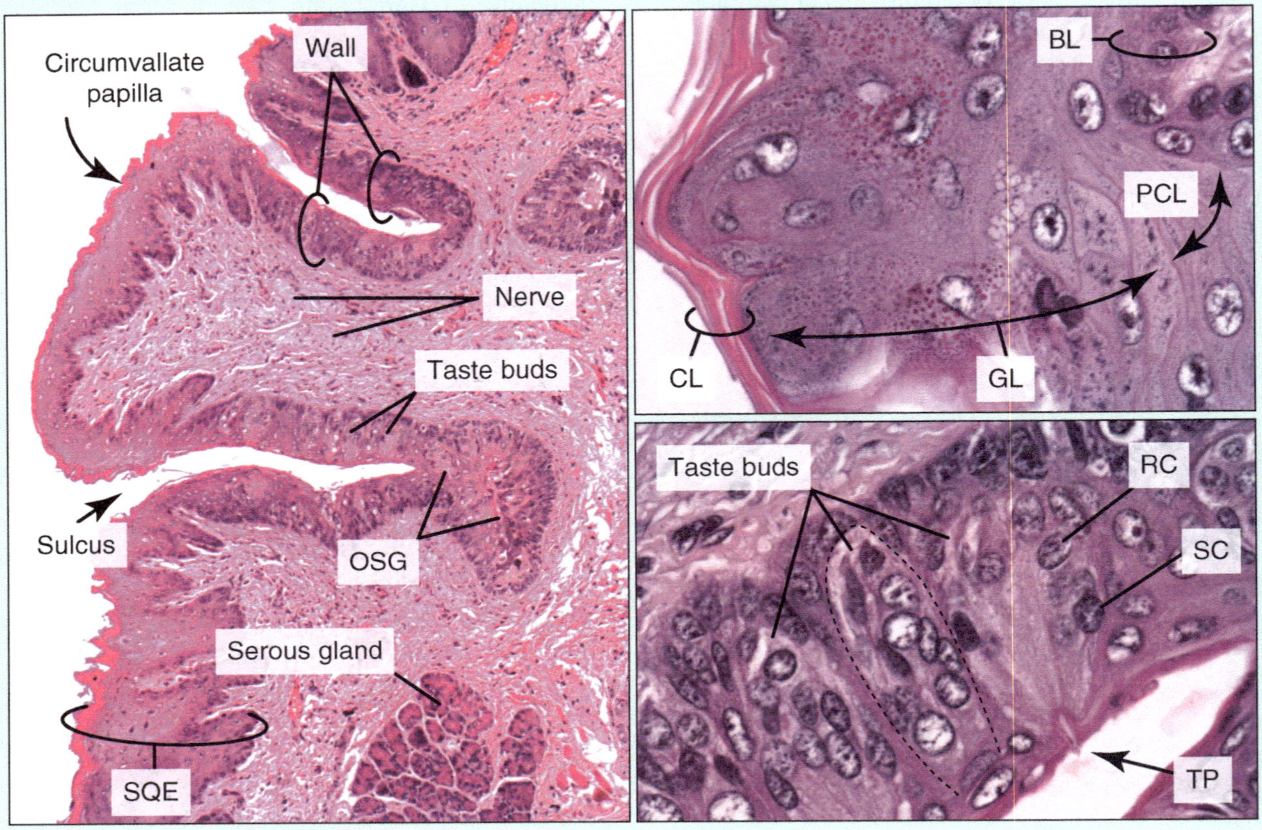

Fig. 12.58 Section through a circumvallate papilla (HE). *BL* basal layer, *CL* cornified layer, *GL* granular layer, *OSG* opening of serous glands, *PCL* prickle cell layer, *RC* receptor cells, *SC* supporting cells, *SQE* keratinised stratified squamous epithelium, *TP* taste pore

Cut out the base of the tongue and the dorsal wall of the pharynx, and isolate the trachea together with the oesophagus and bronchi with the lobes of the lung (Fig. 12.59).

To see the trachea and the *oesophagus* just dorsal to the trachea, you may need to cut through the throat muscles on the ventral side. The *trachea* extends from the larynx to the point of branching just before the lungs.

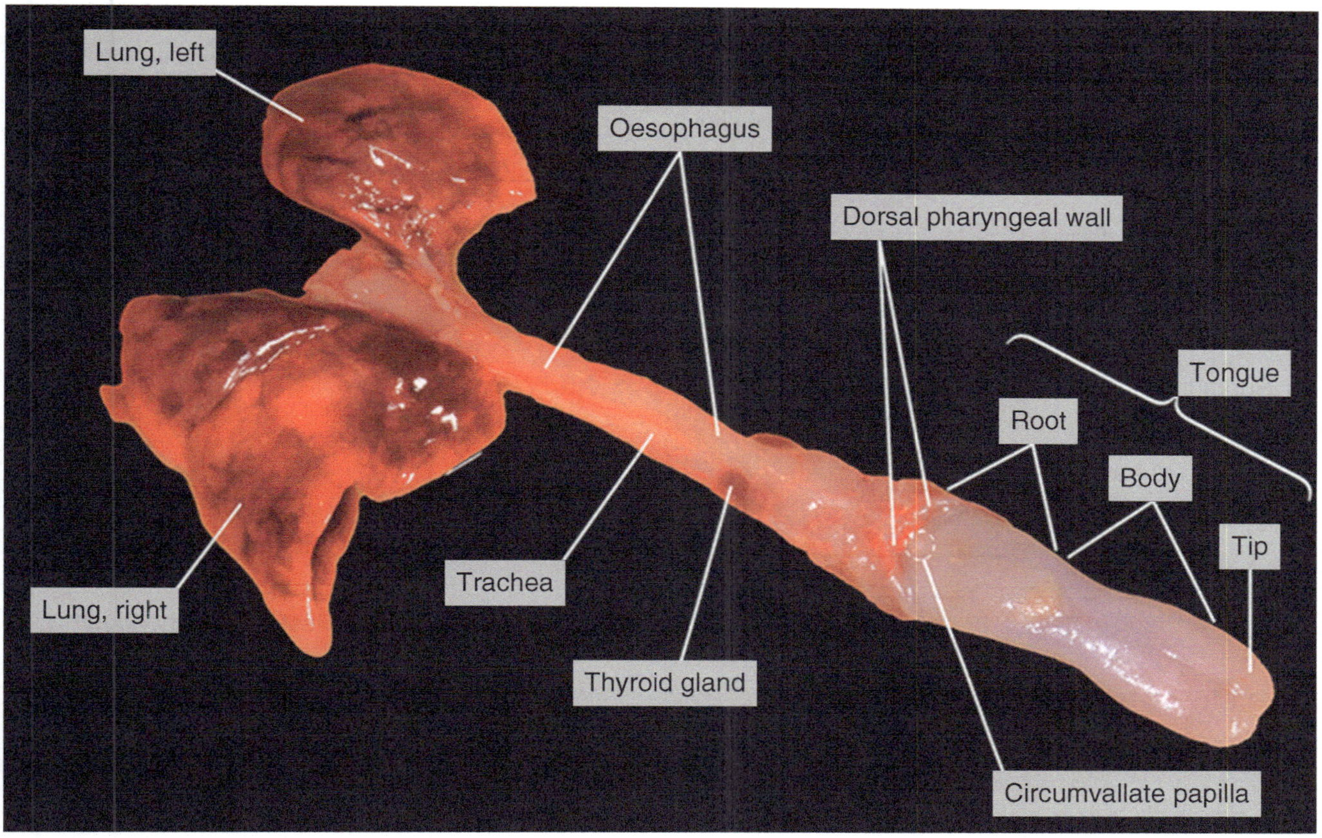

Fig. 12.59 Dorsal view of the isolated tongue-pharynx-larynx-trachea-lung complex

The trachea is the greatest organ of extrapulmonary air-conducting system (Fig. 12.60). Its wall contains a set of incomplete rings (C-shaped) of hyaline cartilages. Mucosal epithelium is pseudostratified and predominantly ciliated (Fig. 12.60, arrowheads) and contains interspersed mucus-secreting goblet cells. Propria of mucosal layer and submucosa are rich in patrolling lymphocytes – some leucocytes can be seen in the mucous layer. They derived from mucosa-associated lymphoid tissue (MALT). Submucosal glands have regional distribution. They may be founded at the opening of the tracheal cartilages. Hyaline cartilage has typical composition showing chondrons, interterritorial matrix and perichondrium with flattened chondroblasts. Opening of C-shaped cartilages are bridged by smooth muscle. The outer layer of the organ is formed by adventitia, which connect the trachea to the neighbouring organs.

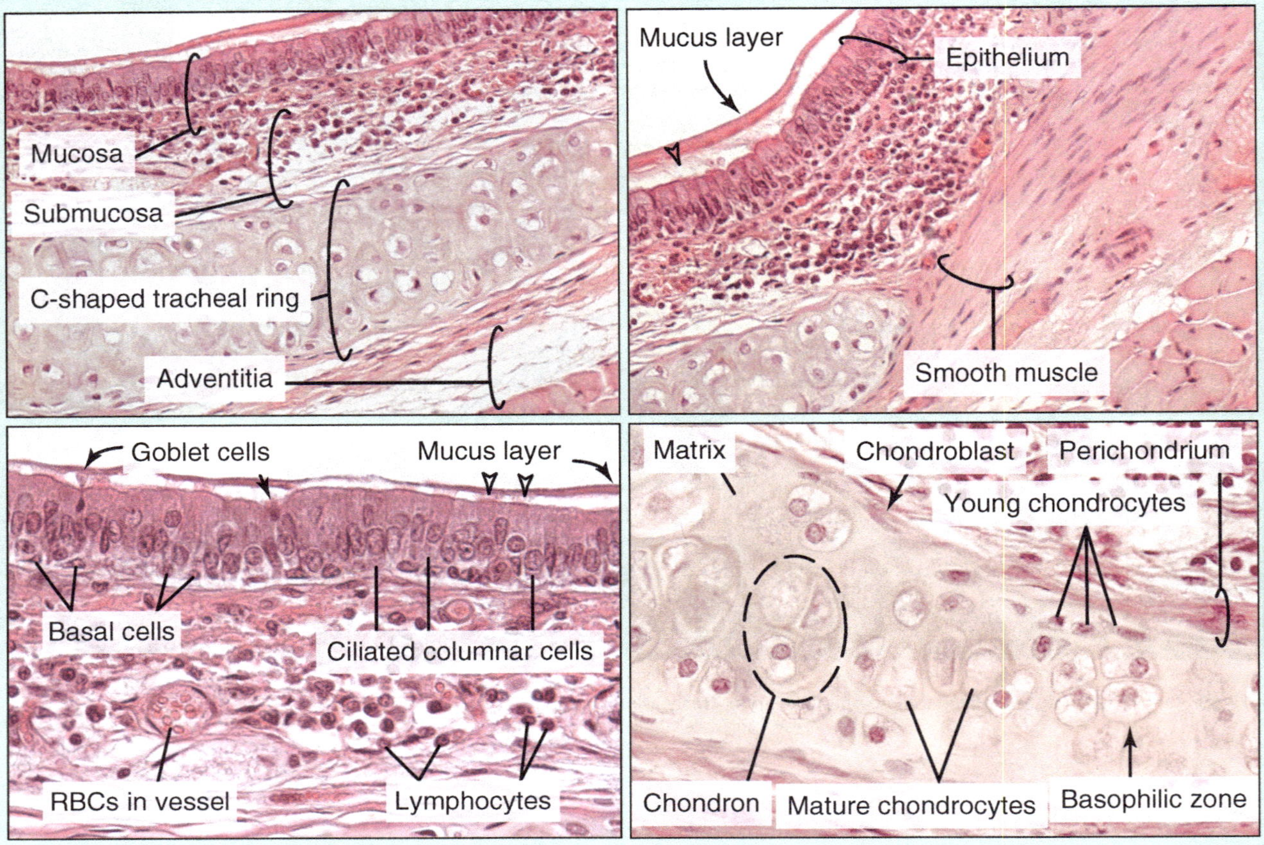

Fig. 12.60 The histological composition of trachea (HE). *Arrowheads* cilia, *RBC* red blood cells

The *epiglottis* is the tiny triangular flap that guards the glottis (Fig. 12.61).

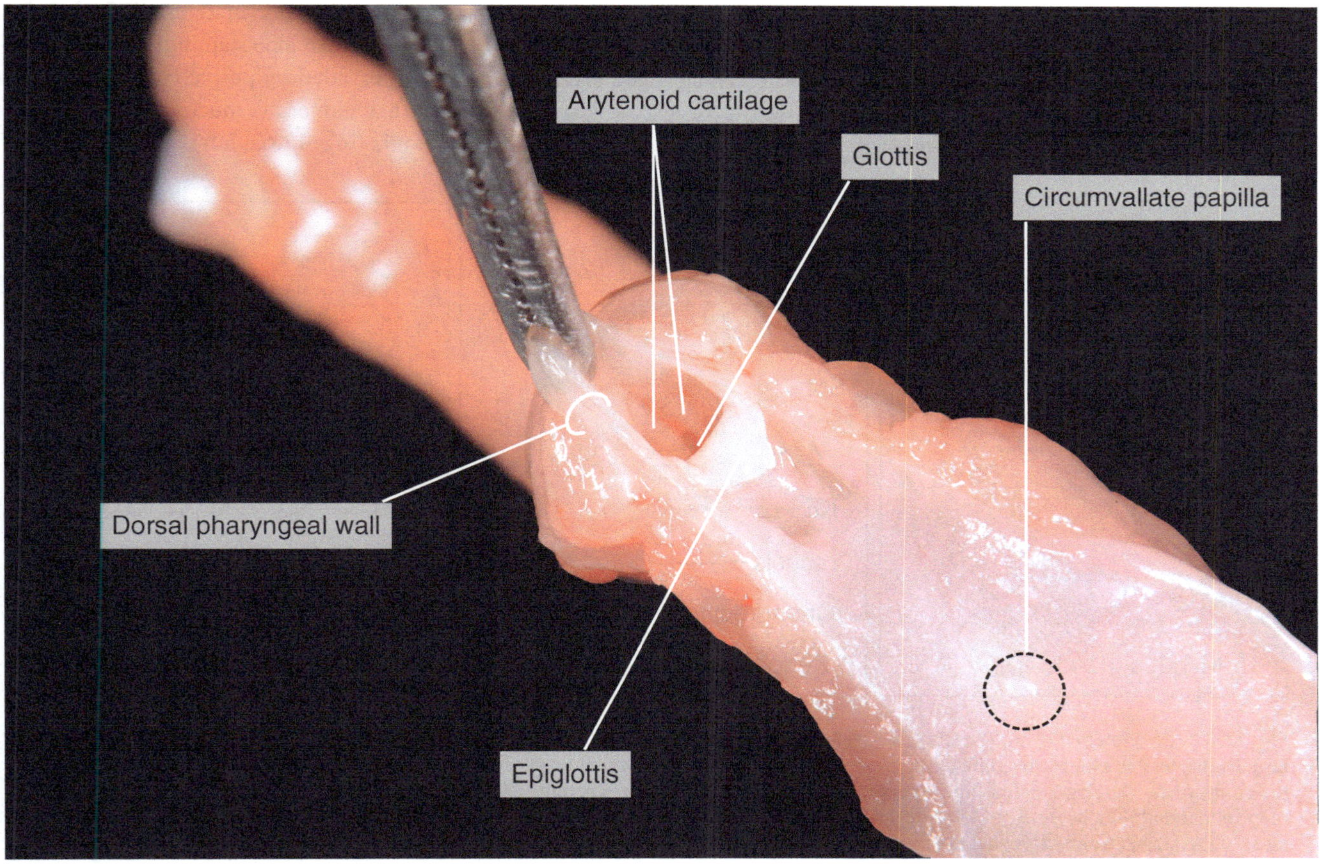

Fig. 12.61 Position of glottis and epiglottis posterior to the root of the tongue

The *epiglottis* is a thin plate of elastic fibrocartilage. Its mucosa (M) has squamous, multilayered epithelium with keratinisation on pharyngeal and without keratinisation on laryngeal surface (Fig. 12.62). Two types of epithelium meet on the apex. Propria is a loose connective tissue with collagen fibre (CF) bundles (pinkish after basic fuchsine staining), capillaries and many immune cells (Fig. 12.62, right, mast cells with dark, blue granules). The submucosa (SM) is featured by fewer cells, but greater blood vessels. Elastic fibres (EF) appear in this layer. Perichondrium shows outer fibrillary and inner cellular layers. Cartilage is formed by chondrocytes containing lipid droplets (LD, showing green colour after alcian blue staining) and glycogen granules. Matrix is pervaded by purple elastic fibres. The laryngeal mucosa has numerous mucous glands, on both sides of the epiglottis (Fig. 12.62, left).

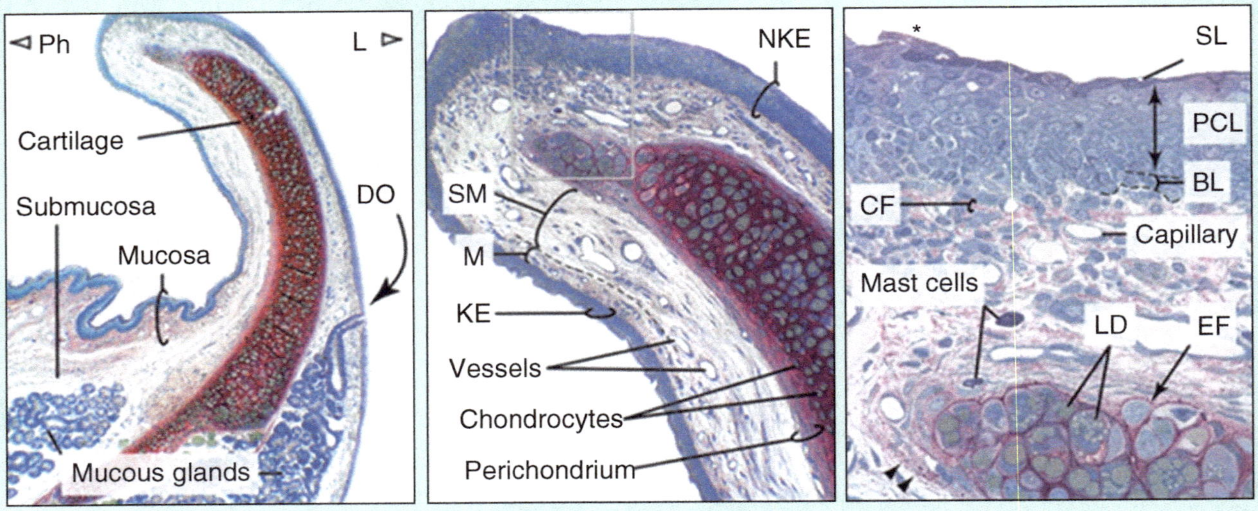

Fig. 12.62 Longitudinal histological section of the epiglottis (semithin section after perfusion fixation, alcian blue-basic fuchsine staining). Right panel is a higher magnification micrograph of the portion of the middle panel outlined by the rectangle. *Asterisk* (*right*) indicates the point, where the non-keratinised epithelium replaces the keratinised epithelium, *black arrowheads* elastic fibre bundles in the vicinity of the perichondrium, *white arrowhead* and *Ph* direction of pharyngeal cavity, *white arrowhead* and *L* direction of laryngeal cavity, *BL* basal layer, CF collagen fibres, *DO* duct opening, *EF* elastic fibres, *KE* keratinised, squamous epithelium, *LD* lipid droplets, *M* mucosa, *NKE* non-keratinised, squamous epithelium, *PCL* prickle cell layer, *SL* superficial layer, *SM* submucosa

Find the *larynx* at the anterior end of the trachea. The larynx is the voice box and allows rats to make squeaking noises (Fig. 12.63).

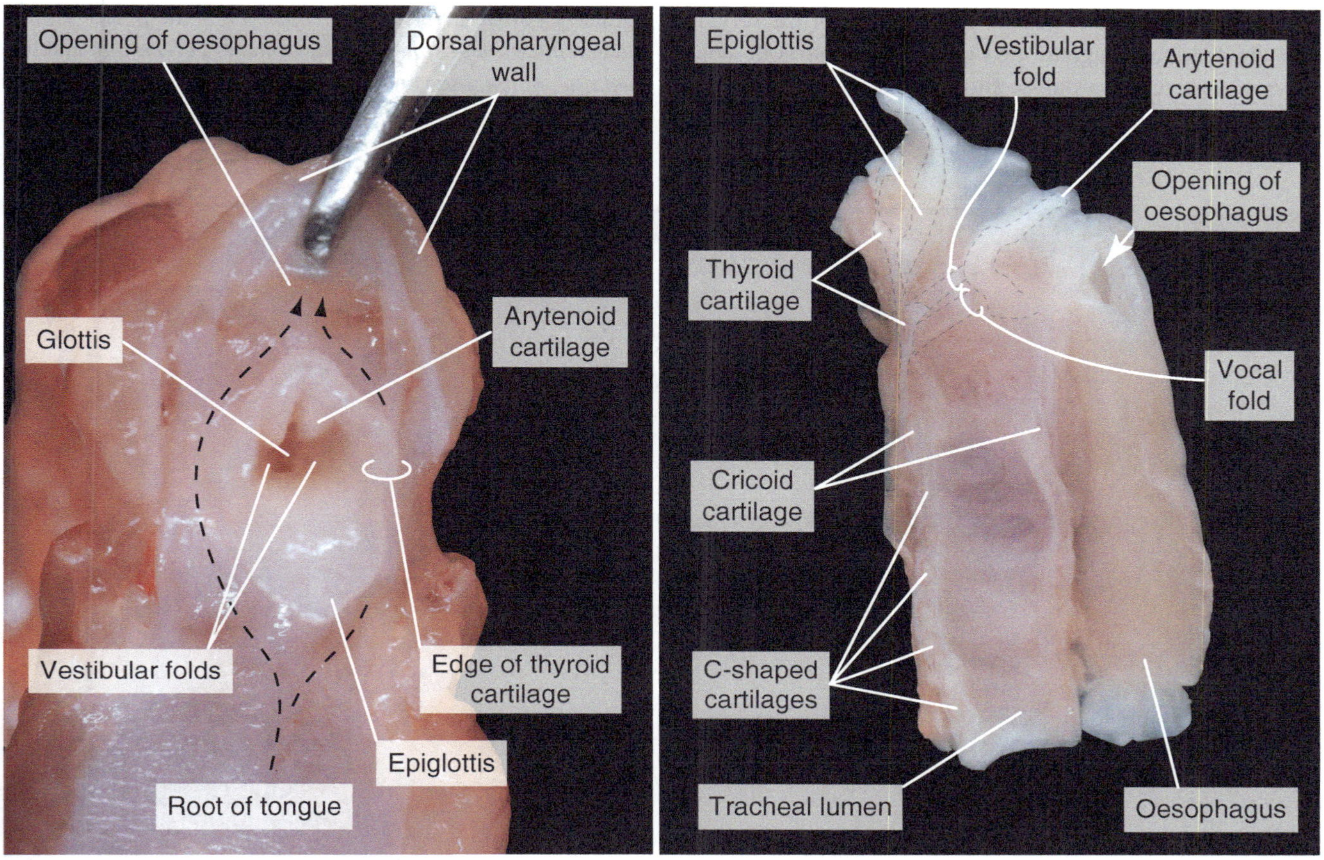

Fig. 12.63 Structure and role of the larynx. *Left* dorsal view of the cartilages of the larynx. *Dashed line* shows the course of the food in the pharynx from the oral cavity to the oesophagus. *Right* median-sagittal longitudinal section of the larynx and trachea. Note the rudimentary vocal folds

As for the last attempt, exposure of the *brain* and its extraction from the skull on a fresh specimen is complicated and requires patience. This part of the dissection should only be endeavoured by those who are skilled and have considerable experience in dissection. Clear the occipital region from muscles and periosteum. Insert the sharp tip of the scissors into the foramen magnum and cut laterally on both sides. Advance the scissors only a little in one go and cut short distances. Grab the middle part with an anatomical forceps and remove the roof of the skull. It is important to save the integrity of the dura mater; otherwise, the brain would be injured.

After removal of the skull roof, extract the brain by cutting the cranial nerves and cover it with water. Identify the *olfactory bulbs*, the *cerebral hemispheres*, the *pineal gland*, the *cerebellar hemispheres*, the *vermis of cerebellum* and the *medulla oblongata* on the dorsal surface of the brain (Fig. 12.64).

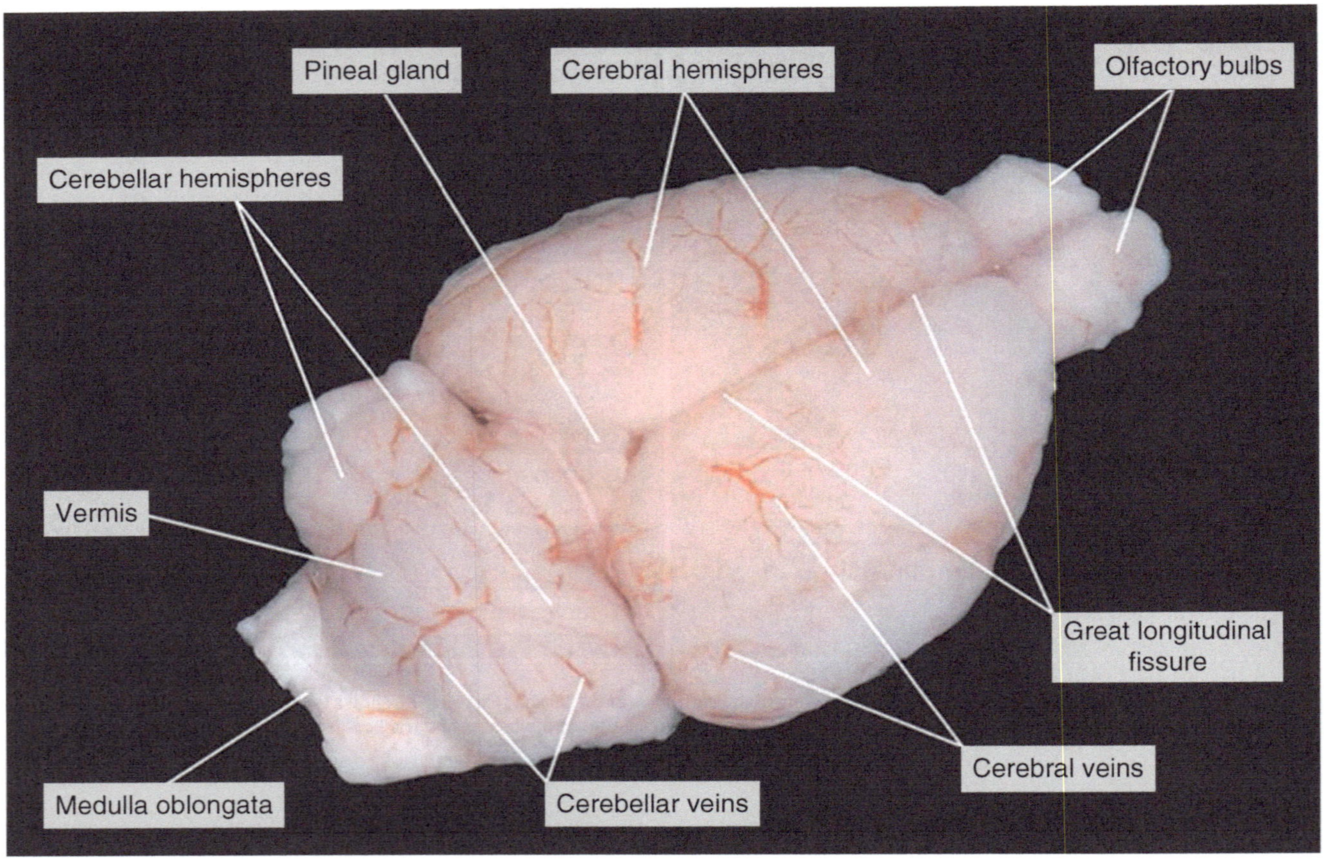

Fig. 12.64 Dorsal view of the brain of the rat

The *cerebral hemispheres* are greatly enlarged by the development of the neocortex. The hemispheres are convoluted to increase the surface area in large mammals but remain simple in the rat. They have taken over the functions of the basal nuclei, some brainstem centres and the tectum and are involved in newer and higher levels of activity, like correlation, association and learning. This increased control of coordination by the cerebrum has resulted in a further elaboration of the thalamus as a relay centre. The *cerebellum* is also large and convoluted and a distinct *pons* connects it with the cerebral hemispheres. The olfactory regions such as the *olfactory bulb* are well developed in rat, reflecting the importance of the sense of smell. Find the *pineal gland* (epiphysis), which regulates the day/night cycle. Examine the ventral surface of the brain as well (Fig. 12.65).

If you worked careful enough, you may find the *pituitary gland* (hypophysis), which has three parts: pars anterior, pars intermedia and pars posterior. Together with the *hypothalamus*, this gland is the coordinator of the hormonal system of the rat. Hypothalamus is highly differentiated; it is concerned with maintaining homeostasis by the regulation of temperature, blood pressure and respiration. There are 12 *cranial nerves* in mammals; however, the studs of them are mostly inconspicuous. Only the olfactory nerve (cranial nerve I) fibres, the optic nerves (II) and the trigeminal nerves (V) can be identified.

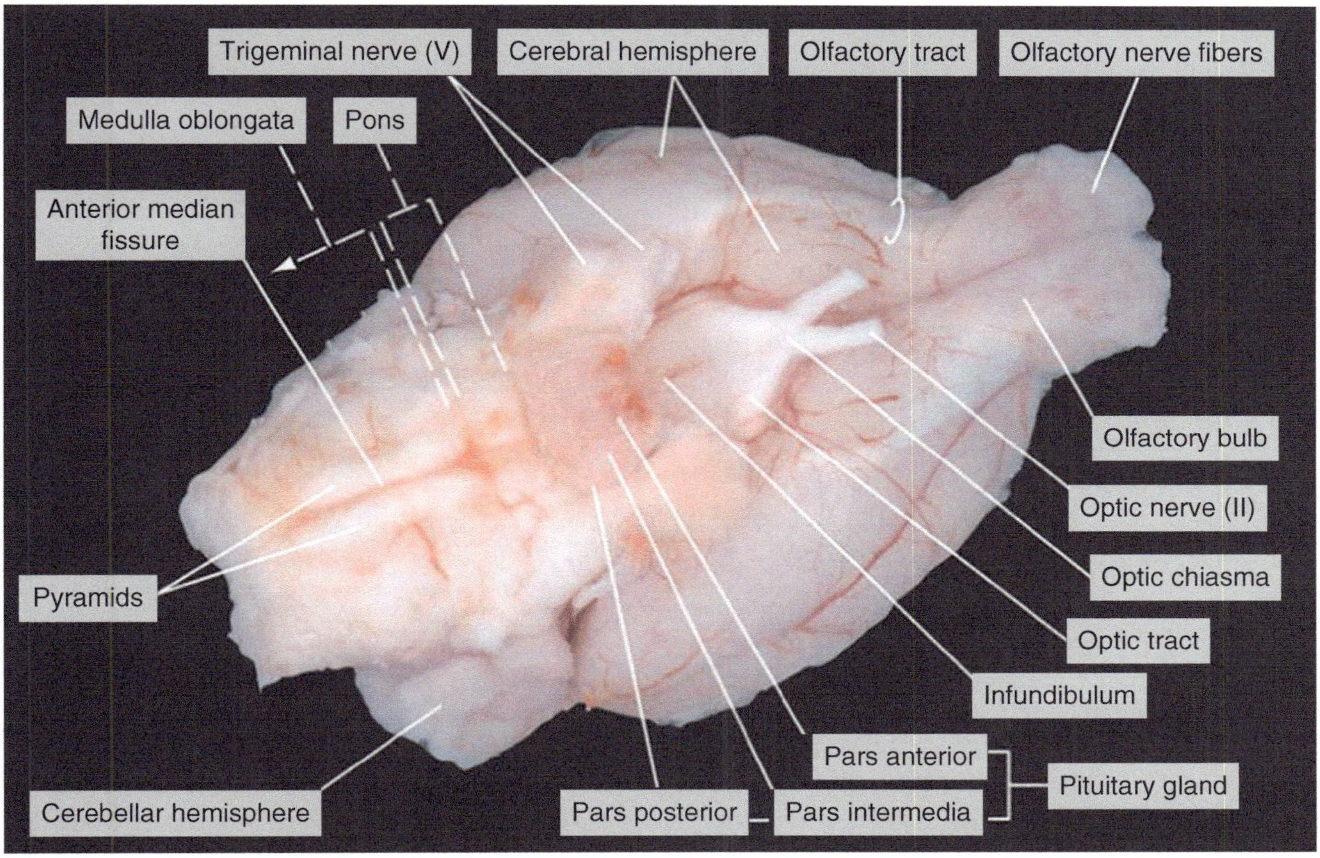

Fig. 12.65 Ventral view of the brain of the rat

Bibliography

Cleveland Jr PH, Roberts LB, Allan SL, I'Anson H. Laboratory studies in integrated principles of zoology. 16th ed. Boston: McGraw-Hill Education; 2013. ISBN 0077508882.

Freeman WH, Brian B. An atlas of invertebrate structure. London: Heinemann Educational; 1971. ISBN 0435603159.

Gillian MK, David RN. Custance: colour atlas of vertebrate anatomy: an integrated text and dissection guide. Oxford: Blackwell Scientific Publications; 1982. ISBN 0950767603.

Rowett HGQ. Guide to dissection. London: John Murray Ltd; 1982. ISBN 0719533163.

Zboray G, Molnár K, Pálfia Z, Kovács Z, Kriska G. Atlas of comparative sectional anatomy of 6 invertebrates and 5 vertebrates. Wien: Springer-Verlag; 2010. ISBN 978-3-211-99762.

P. Lőw et al., *Atlas of Animal Anatomy and Histology*, DOI 10.1007/978-3-319-25172-1

Index

Note: Page numbers followed by f indicate figures. Irregular plural form of terms is after the word preceded by pl.

P. Lőw et al., *Atlas of Animal Anatomy and Histology*, DOI 10.1007/978-3-319-25172-1

H

I

GPSR Compliance

The European Union's (EU) General Product Safety Regulation (GPSR) is a set of rules that requires consumer products to be safe and our obligations to ensure this.

If you have any concerns about our products, you can contact us on ProductSafety@springernature.com

In case Publisher is established outside the EU, the EU authorized representative is:

Springer Nature Customer Service Center GmbH
Europaplatz 3
69115 Heidelberg, Germany

Batch number: 10808278

Printed by Printforce, the Netherlands